T0190347

Lecture Notes in Computer Science 12846

More information about this subseries at http://www.springer.com/series/7407

Alessandro Abate · Andrea Marin (Eds.)

Quantitative Evaluation of Systems

18th International Conference, QEST 2021
Paris, France, August 23–27, 2021
Proceedings

Editors
Alessandro Abate
University of Oxford
Oxford, UK

Andrea Marin
Ca' Foscari University of Venice
Venice, Italy

ISSN 0302-9743 ISSN 1611-3349 (electronic)
Lecture Notes in Computer Science
ISBN 978-3-030-85171-2 ISBN 978-3-030-85172-9 (eBook)
https://doi.org/10.1007/978-3-030-85172-9

LNCS Sublibrary: SL1 – Theoretical Computer Science and General Issues

This Springer imprint is published by the registered company Springer Nature Switzerland AG
The registered company address is: Gewerbestrasse 11, 6330 Cham, Switzerland

Preface

It gives us great pleasure to open these proceedings of the 18th International Conference on Quantitative Evaluation of SysTems. QEST 2021 was hosted within QONFEST 2021, and held virtually during August 23–27, 2021. The event was co-located with CONCUR, FORMATS, FMICS, and other workshops.

The QEST conference series has a long and rich history, as can be seen at https://www.qest.org. Most recently, it was held in Vienna (Austria, in virtual mode), Glasgow (UK), Beijing (China), Berlin (Germany), and Quebec City (Canada). Further information on QEST 2021 can be found on the conference webpage at https://www.qest.org/qest2021/.

The 34 members of the International Program Committee (PC) helped to provide at least three reviews for each of the 47 submitted contributions. Based on the reviews and PC discussions, 23 high-quality papers - two of them as short contributions and two of them as tool papers - were accepted to be presented during the conference. The overall acceptance rate for the conference was thus just below 50%. The contributions were bundled into eight thematic sessions, covering the following topics in verification and evaluation: Probabilistic Model Checking, Learning and Verification, Abstractions and Aggregations, Stochastic Models, Quantitative Models and Metamodels, Queueing Systems, Simulation, and Performance Evaluation. These contributions appear as papers in the ensuing proceedings.

The program chairs plan to edit a special issue of the journal ACM TOMACS, where the authors of selected papers will be invited to contribute significantly revised and extended versions of their manuscripts containing new results.

QEST 2021 did not host a poster session (as is common for the conferences in the series) due to difficulty of interaction and limited time in the virtual format, but Best Paper awards were presented, according to QEST policies and tradition.

A highlight of QEST 2021 was the presence of two high profile invited speakers, amongst those of QONFEST:

– Boudewijn Haverkort from Tilburg University the Netherlands, giving a lecture on the topic of "Performance Evaluation: Model or Problem Driven?".
– François Baccelli from Inria, France, and the University of Texas at Austin, USA, contributing with a seminar on the topic of "Stochastic Geometry based Performance Analysis of Wireless Networks".

Short contributions on the topics of the two keynotes appear in these proceedings.

Another highlight of QEST 2021 was the introduction of an optional Repeatability Artifact Evaluation process for accepted papers, providing to authors feedback on their shared codebase associated to the submitted article. This initiative was much in line with similar ones at cognate verification conferences, and was aimed at increasing the open sharing of reproducible scientific software-generated results. A total of 14 papers participated in the repeatability evaluation (this was obligatory for tool papers), 12 of which were finally found to be repeatable (up to different degrees of completeness). A special badge marks them in the ensuing proceedings. The repeatability evaluation committee was co-chaired by Arnd Hartmanns and David Safranek.

A few words of acknowledgment are due. First and foremost, thanks to the authors for entrusting their best work to QEST 2021. The review process clearly showed that the conference was able to put the bar for acceptance really high, which makes us very proud. Our thanks go to the QEST steering committee and previous conference chairs for their help and feedback on the organization process. We were also particularly pleased with the interest in the repeatability evaluation, and thank the repeatability evaluation committee and chairs (Arnd Hartmanns and David Safranek) for the truly exemplar work, and all the authors who participated in this exercise, which was novel for QEST. Sincere thanks to the local organizing committee (in particular, Benoit Barbot for the administration of QONFEST), to the steering committee of the QEST conference series (in particular its chair, Enrico Vicario), and to Marco Paolieri for the conference website and for the event publicity.

Finally, we wish to thank all the PC members and additional reviewers for their hard work in ensuring the quality of the contributions to QEST 2021, and to all the participants for contributing to this memorable event.

August 2021

Alessandro Abate
Andrea Marin

Organization

Program Committee Co-chairs

Alessandro Abate	University of Oxford, UK
Andrea Marin	Ca' Foscari University of Venice, Italy

Program Committee

Ebru Aydin Gol	Middle East Technical University, Turkey
Luca Bortolussi	University of Trieste, Italy
Davide Bresolin	University of Padua, Italy
Peter Buchholz	TU Dortmund, Germany
Laura Carnevali	University of Florence, Italy
Giuliano Casale	Imperial College London, UK
Pedro R. D'Argenio	Universidad Nacional de Córdoba and CONICET, Argentina
Susanna Donatelli	University of Turin, Italy
Maryam Elahi	Mount Royal University, Canada
Marco Gribaudo	Politecnico of Milan, Italy
Ichiro Hasuo	National Institute of Informatics, Japan
András Horváth	University of Turin, Italy
David N. Jansen	Chinese Academy of Sciences, China
Nils Jansen	Radboud University, The Netherlands
Peter Kemper	William & Mary College, USA
William Knottenbelt	Imperial College London, UK
Jan Kretinsky	Technical University of Munich, Germany
Gethin Norman	University of Glasgow, UK
Meeko Oishi	University of New Mexico, USA
Marco Paolieri	University of Southern California, USA
Dave Parker	University of Birmingham, UK
Tuan Phung-Duc	University of Tsukuba, Japan
Elizabeth Polgreen	University of Edinburgh, UK
Sylvie Putot	LIX, Ecole Polytechnique, France
Anne Remke	University of Muenster, Germany
Sabina Rossi	University Ca' Foscari Venice, Italy
Miklos Telek	Budapest University of Technology and Economics, Hungary
Ufuk Topcu	University of Texas at Austin, USA
Mirco Tribastone	IMT School for Advanced Studies Lucca, Italy
Benny Van Houdt	University of Antwerp, Belgium
Verena Wolf	Saarland University, Germany

Katinka Wolter	FU Berlin, Germany
Lijun Zhang	Chinese Academy of Science, China
Paolo Zuliani	Newcastle University, UK

Repeatability/Artifact Evaluation Committee

Arnd Hartmanns (Chair)	University of Twente, The Netherlands
David Safranek (Chair)	Masaryk University, Czech Republic
Elvio Gilberto Amparore	University of Turin, Italy
James Fox	University of Oxford, UK
Anastasis Georgoulas	University College London, UK
Mirco Giacobbe	University of Oxford, UK
Matej Hajnal	Masaryk University, Czech Republic
Mohammadhosein Hasanbeig	University of Oxford, UK
Pushpak Jagtap	KTH Royal Institute of Technology, Sweden
Sebastian Junges	University of California, Berkeley, USA
Bram Kohlen	University of Twente, The Netherlands
Luca Laurenti	TU Delft, The Netherlands
Riccardo Pinciroli	Gran Sasso Science Institute, Italy
Fedor Shmarov	University of Manchester, UK
Simone Silvetti	Esteco SpA, Italy
Matej Trojak	Masaryk University, Czech Republic

Additional Reviewers

Amparore, Elvio
Azeem, Muqsit
Bacci, Edoardo
Backenköhler, Michael
Badings, Thom
Bharadwaj, Sudarshanan
Budde, Carlos E.
Cairoli, Francesca
Carbone, Ginevra
Cubuktepe, Murat
Degiovanni, Renzo
Delahaye, Benoit
Eisentraut, Julia
Fu, Jie
Gros, Timo
Grossmann, Gerrit
Groß, Dennis
Horvath, Illes
Incerto, Emilio
Junges, Sebastian
Krüger, Thilo
Linard, Alexis
Liu, Depeng
Menzel, Verena
Meszaros, Andras
Mohr, Stefanie
Niehage, Mathis
Ortiz, Kendric
Peng, Guang
Piazza, Carla
Pilch, Carina
Pinchinat, Sophie
Priore, Shawn
Putruele, Luciano

Ramírez-Cruz, Yunior
Randone, Francesca
Roy, Diptarko
Savas, Yagiz
Schmidl, Christoph
Sedwards, Sean
Sivaramakrishnan, Vignesh
Thorpe, Adam
Trubiani, Catia
Vandin, Andrea
Yu, Yue

Stochastic Geometry Based Performance Analysis of Wireless Networks (Abstract of Keynote)

François Baccelli

Inria Paris, France, and UT Austin, USA
francois.baccelli@inria.fr

Extended Abstract

Stochastic Geometry is commonly used for analyzing spectrum sharing in large wireless networks. In this approach, network elements, such as users and base stations, are represented as point processes in the Euclidean plane, and interference fields as spatial shot-noise processes. The analytical machinery of stochastic geometry and basic formulas of information theory can then be combined to predict important spatial statistics of such networks.

The talk will first exemplify this approach by showing how to derive the distribution of the Shannon rates obtained by users in two fundamental models, the Poisson dipole model, which is a mathematical abstraction for a large device to device network, and the Poisson-Voronoi model which is an abstraction for a large cellular network. A few variants of these now classical models will be also discussed, like multi-tier cellular networks, or networks leveraging beam-forming.

The talk will then exemplify how to introduce birth-and-death type dynamics in this stochastic geometry framework. This will be illustrated through recent results on the simplest model in this class. In this model, users arrive according to a Poisson rain process on the Euclidean plane and leave with a stochastic intensity proportional to their instantaneous Shannon rate.

Contents

Abstractions and Aggregations

Stochastic Models

Keynote Speaker

Performance Evaluation: Model-Driven or Problem-Driven?

Boudewijn R. Haverkort(✉)

Tilburg School of Humanities and Digital Sciences, Tilburg University, Tilburg, The Netherlands
b.r.h.m.haverkort@tilburguniversity.edu

Abstract. In this paper I address the divide that has emerged between the field of performance evaluation and the field of computer and communication system design. After looking back briefly on the history of the field, I subsequently reflect on the reasons why the field of performance evaluation has become so isolated. I then continue with a set of eight recommendations, based on my experience in performing projects with industry, that will help in reconnecting, and that will result in a better uptake of the newest techniques and tools in the field of design of computer and communication systems. Following these recommendations will probably push scientists a little out of their comfort zone, however, I feel the potential extra reward of seeing our work truly applied is more than worth it.

Keywords: Computer systems · Communication systems · Digital twin · Performance evaluation · Model-driven design · Modeling and simulation · System design

1 Introduction

Going back as far as the mid 1960's, a large variety of performance evaluation techniques have been proposed and used to address design and dimensioning questions for computer and communication systems. In the 1970's a number of workshops on the topic of computer performance evaluation started to emerge, that over the years evolved into successful conference series that last until today, such as IEEE Mascots [17], ACM Sigmetrics [1] and also QEST [20]. In the same field, *Performance Evaluation* has been a well-acclaimed journal since the beginning of the 1980's. However, when we carefully investigate the situation in the 1970's, we observe that many of the researchers active in the then emerging field of performance evaluation, were actually also active in the field of computer and communication system design. Indeed, very often their key focus was system design, for which they used the state-of-the-art performance evaluation techniques of that time, or even developed these techniques themselves. Classical examples in this context are the work on time-sharing systems in the 1960 by Alan Scherr [9], the work on closed queuing networks, mean-value analysis and

A. Abate and A. Marin (Eds.): QEST 2021, LNCS 12846, pp. 3–11, 2021.
https://doi.org/10.1007/978-3-030-85172-9_1

polling models, that evolved from the work on local area networks and the IBM token ring [5,22], or the work on packet-switching and time-sharing systems [15].

In the 1970's, also the classical textbooks on operating systems still used and devoted considerable time on system performance evaluation, cf. [4,6]. However, later books on operating systems, like Silberschatz and Galvin [23] (widely used in the 1990's) or Tanenbaum and Bos [25], do not bring any modelling aspects anymore, not even on such topics as (disk) scheduling or memory management.

A similar remark can be made for books on communication systems. Classical books of the early 1990's on data and communication networks, like Bertsekas and Gallager [3] and Walrand [27], do still contain a lot of material on performance evaluation (simple queues, Little's law, queuing network, Markov chains, etc.). Also when ATM network became popular, queuing models were being presented as important tools in the design of such systems, cf. [18]. However, in the 1996 edition of Tanenbaum's widely used textbook on computer networks [24], there is only one subsection devoted to performance issues, largely in a qualitative way, with focus on measurement and implementation issues. In the now widely used textbook by Kurose and Ross [16], some quantitative issues (delays, round-trip times, etc.) are addressed, but queuing models and the like are not addressed at all.

A similar remark can be made for computer architecture. Even though the widely used textbook by Hennessy and Patterson [13] has "quantitative" in its title, almost no performance evaluation models are used in it; Sect. 7.8 (only 10 pages) presents the M/M/1 and the M/M/m queue in a fairly isolated way, almost at the end of the book.

We can only conclude that the methods and tools that have been developed in the recent past, let alone the ones we are developing these days, are not being considered important enough by the above mentioned system-oriented scholars, to present them as useful (design) tools in their scholarly books. But, if "our methods and tools" (such as can be found in [12]) are not, or only to very limited extent, being addressed, we should ask ourselves: What has happened, what has gone wrong? Why is it that methods and techniques considered important and valuable by us, do not receive the uptake in scholarly books, let alone in industrial practice? In what follows I will try to shed some light on this issue, and bring some recommendations that might improve the situation.

2 On the Insularity of Quantitative Methods

The observations that I shared in the previous section are, in fact, not new. Some 35 years ago, Domenico Ferrari, a computer science professor at Berkeley and author of a well-known textbook on computer systems performance evaluation [10], already pointed to the "insularity of performance evaluation", in his (maybe forgotten) paper [11]. To a large extent, his considerations still apply. Ferrari puts forwards a number of reasons that might be the cause why performance evaluation has developed too much independently from computer systems and computer network design and engineering. He observes that (i) young computer scientists are often not enough skilled (taught) in performance evaluation, hence, no surprise that such methods are not used; (ii) that the field of computer system design is still very young, yet the need for proper design methods would require a much more reflective attitude towards system design than to go for just quick wins as most engineers would do; (iii) that computer systems are too complex to allow for an all-encompassing mathematics-based design theory, as we have for pure mechanical or electronics systems; and, finally, (iv) that many computer scientists see computer system design more as an art than as science and engineering.

Looking back on these considerations, the only observation that appears to be valid today is the third one. Indeed, computer systems, especially the ones that we design and work with today, with millions of lines of code, distributed over multiple computing nodes, etc., are of a complexity that goes beyond anything else made by humans. And although that level of complexity was less apparent in the mid 1980's, yet, at that time already, the field of performance evaluation appeared to have lost contact with the field of computer system design. And the complexity and sheer size of systems has increased at exponential pace since then. Are we ever going to reconnect? Is there any hope that we find the right models, at the right level of abstraction to appeal to system designers and engineers, yet, at the same time allow for in-depth analysis and design trade-offs?

Ferrari also points to a few advantages of the insularity of performance evaluation. He claims that the field itself, especially the work focusing on advanced queuing network models and algorithms, has developed tremendously fast, actually faster due to the independence from specific problems that needed to be addressed. He, sort of, argues that the researchers, for instance those working on algorithms for product-from queuing networks, could develop their field so quickly because they were not bothered by solving practical design problems. This observation might actually be true, focus is good to attain deep scientific insights, yet, one might wonder what the long-term advantage of such a strategy really is. The question that really matters to me is whether these deep insights and specialized algorithms are of use in the design process of computer and communication systems, the field that initiated the work on that type of models. We did solve the models right (in the sense of 'correctly'), but the true question is, whether we solved the right models?

Ferrari does spend considerable time on discussing the disadvantages of the lost connection between performance evaluation and computer system design. For one, he sees that the field (of performance evaluation) often does not have available the required techniques or tools, or only does make these techniques available at a point in time when system designers already struggle with other challenges. Although things have changed tremendously in this respect, with all the state-of-the-art tooling that is currently available, I do think the conditions under which these tools can be used, that is, the modelling conditions and restrictions, often hamper practical application for non-experts. Secondly, Ferrari observed that some members of the performance evaluation community are more "mathematicians in disguise"; their work has value, but probably less so in the context of performance evaluation. I fully recognize this; in my role as editor for *Performance Evaluation* [19], I have handled many papers that could not be considered performance evaluation papers, but rather papers on queuing theory or Markov chain theory; nice as contribution to journals such as *Queuing Systems* or *Applied Probability*, but, if you ask me, not for *Performance Evaluation*.

In the last section of his paper, Ferrari pleads to focus more strongly on what he calls *applied performance evaluation*, in which the field much more co-operates and co-evolves with the engineering and design disciplines for computer and communication systems. He also stresses the importance of the connection to system measurements (his field of specialty), with which I fully agree. Many computer scientist have lost the connection with experimental work, something to reconsider, as argued firmly by Walter Tichy [26]. But how can all this achieved?

3 Observations and Recommendations

Following the observations and plea of Ferrari, but also based on my experience in applying advanced modelling and analysis techniques, including performance and reliability evaluation techniques and (timeless) formal description techniques, in an industrial context, I would like to make a number of recommendations as well. Please do get me right here; these are experiences that I gathered, based on the many projects that I was involved in with Dutch high-tech systems industries [8], and based on various European and national projects with industry involvement. I have no "formal proof of correctness" for my observations and recommendations, nor are all my observation fully in line with each other. I just like to share what I experienced and hope you can use these observations to your benefit.

1. Do **cooperate with true system designers**, working in true system design departments of companies. Getting involved with them and their projects, and keeping connection with them is probably more difficult than with scientists working in research labs (although that type of lab seems to be disappearing), however, the return-on-investment is very high. You really get to know what bothers them in their work (in a technical sense), what quantitative concerns they have to deal with, in the end helping you in finding real challenging research topics, both from an application point of view, and from a scientific point of view. *Reality is complex enough to provide interesting cases!* In the end, seeing your work actually being used in practice is most rewarding.
2. Do not start with pushing your modelling method or your tool, but **do start with their problem**, hence, the title of this paper. Too often, researchers take the capabilities of their modelling tool as the starting point. Very often, however, the actual problem does not directly fit the tool. Your tool or technique might be of use, but most probably, a different method or tool is more appropriate. Yet, also in that context there are interesting research challenges ahead. Trying to push your tool scares people away. If you only have a hammer, everything looks like a nail.

3. In industrial practice, **professional tools** are being used, with professional support, connecting to company-wide software suites. Companies do not want to make their design processes dependent on software form a university research group; they want an integrated design and engineering approach, connecting analysis tools to compilers, documentation, version management, etc., as well as a 24/7-telephone number for support. Hence, to increase the uptake of a new method, make sure that your method is being applied "under the hood" from within a professional tool-chain that is being used already. This can be as simple as providing an excel-based user-interface.
4. If there is no way to connect easily to an existing tool already in use, **domain-specific languages** (DSL's) have shown to be a low threshold means to provide access to an advanced performance evaluation or model checking tool. The possibility to express a problem in a language that is close to the problem world the designer or engineer is working in, really helps in practical uptake of a new method. Tools based on the Eclipse modeling framework provide ample means to quickly develop and implement DSL's, and connect them to underlying specialist tools, cf. [2].
5. **Keep it simple**, in both input and output. Often, our advanced techniques and tools have many options that are not easy to use, especially not for novice users. Being able to shield such options, helps in making people use the tools and techniques. This also is true for the output of tools. I have noticed that simple "traffic-light style" output or Pareto-curves showing trade-offs work very well in practice, better than detailed tables or elaborate graphs. In the end, if we want our methods and techniques to be used, we should listen to what is needed (call it market-pull), and not try to push technology we think is needed or more accurate. Do not forget that the model input, especially many of the system parameters involved, are often just rough estimates; the result of a quantitative analysis based on such input therefore cannot be the exact values being computed, but rather the trend one can observe, or the comparison between multiple scenarios.
6. I have also experienced that there often is great appreciation for very **simple analytical models** and computations, that allow for fast trade-off analysis. Do use them, learn to appreciate them, even if you know they are crude, or if you know not all conditions to apply them have been met. For many design trade-offs, these are good enough. Allen Scherr, pioneer in the development of time-sharing systems and of the application of simple queuing models in the 1960's, was a great advocate of simple easy-to-use models. In a nice interview [9] he claims: "blind, imitative simulation models are by and large a waste of time and money. To put it in a more diplomatic way, the return on investment isn't nearly as high as it is on a simpler, analytic-type models". We sometimes appear to be so much in love with our models, that we continue to refine, polish and "improve" them; however, it is questionable whether anyone is in need for these improvements, besides ourselves. Smart and simple engineering rules, supported by (our) deep insight, is probably what is more in demand.
7. Often, there is confusion about whether the quantitative results are obtained using discrete-event simulation, numerical analysis or truly analytical. For

one, note that in different fields of science and engineering, the term "simulation" means different things: is it discrete-event simulation, does it refer to the numerical computations for solving differential equations, or does it refer to a relation between state-transition systems? Based on these different backgrounds, a lot of confusion can arise as to how accurate results are. Results from a discrete-event simulation are, in fact, statistics, whereas solving differential equations numerically leads to exact results (within given bounds of accuracy). And even here, theoreticians would not call the latter results exact! Hence, it is important to stress the **notion of uncertainty in the results**, stemming either from parameter uncertainty, from statistical effects, or from simplifying assumptions in the model itself. In this regard, engineers are often interested in the robustness of their solution under slight parameter changes; allowing for easy to do sensitivity analysis is deemed very useful.

8. Finally, for true applicability of new methods and techniques, the **ability to scale** to truly realistic-size problems is of utmost importance. Discrete-event simulation, especially at modern-day computer speeds, does overcome many of the practical problems encountered when applying performance evaluation techniques to real systems. Of course, a discrete-event simulation is and remains to be a statistical experiment, however, accepting that and carefully dealing with that is often easier than trying to deal with modeling limitations of "more advanced" numerical or analytical techniques.

4 Towards Digital Twins

As a short but possibly prolific aside, I would like to briefly touch upon the new developments around the notion of a digital twin, which is currently widely embraced by the high-tech industry. A full formal theory about digital twins does not yet exist; informally, a *digital twin* is a virtual (digital) representation that serves as the real-time digital counterpart of a physical object or process (definition from Wikipedia). A digital twin typically encompasses a computerized model, from which important quantities related to performance, energy-use, reliability and the like can be extracted, using simulations or analytical or numerical techniques. But a digital twin is more than just a static model. The idea is that a digital twin will accompany a real system throughout its lifetime, meaning that it is connected to the real system, so that system parameters can be monitored continuously and fed back into the model, and that the model can be used to investigate adjusted system settings when needed. Often, the digital twin is developed as part of the system design process. When a real system is delivered, an appropriate digital twin is being delivered as well, allowing the customer to exercise "what-if"-questions before changes in system setting are performed. Furthermore, by connecting multiple digital twins, forms of (federated) learning can be applied, that give manufacturers better understanding of true system operations, and allow for better (preventive) maintenance strategies. Although still in a development phase, it appears that the notion of a digital twin allows for many of the recommendations I made above to be followed, thus giving the performance evaluation community ample means to connect more firmly to real system designers and engineers; the digital twin can be seen as a shared object that facilitates communication (between academic researcher, industrial designer or engineer and system applier) and connection (between methods, tools, techniques, measurements). In the recently started Digital Twin program we try to do just that [7].

5 Epilogue

To summarize, I like to stress that the performance evaluation community has done great things over the last 50 years, however, a disconnect from the field of the design of systems continuous to exist, as has been pointed at already 35 years ago. A different way of working, starting closer to what is really needed in system design, could be very beneficial for the performance evaluation community, in order to enable a better uptake of their methods and techniques, their results, in the real world of computer and communication system design. This requires to set the problems faced in system design, by true system designers and engineers, in a more central position, instead of the models being developed; we need to go from model-based to problem-based! In the end, seeing our methods and tools being used in real practice would be much more rewarding than only having more papers on ever-more conferences. The observations and recommendations I shared might be of use to change focus. The new notion of digital twins might be helpful in this process of change as well. And when doing so, I also expect that more students will be interested in the topics we teach in performance evaluation classes; they might be more willing to go through the math, if they see that the methods and techniques being taught are truly used in practice, just as mechanical engineering students are willing to dive into finite-element methods, because they know that these methods are actually being used when a new engine is designed.

I would like to finish with a comment on academic appreciation. One might hear a concern that the more applied work that I advocate here is less suitable for publication in top journals or conferences, hence, does not help individual scientists in their academic career. I firmly disagree with this observation, as well as with the line of thinking that lies underneath it. First of all, it would be very strange to think that real-world problems are too simple to base good scientific work upon. Work that is practically applicable is not necessarily scientifically less appealing! Secondly, in all fields of engineering, the applicability of the methods and tools being developed to real-life problems is considered normal, so why would that be different in the field of computer and communication system performance evaluation? Third, I regard it as a serious task for senior faculty to properly value the applied type of work as described above, also in the context of selection and promotion committees. Scientific quality cannot be evaluated by just counting publications or pointing to an h-index. Instead, it is about investigating what really has been achieved, under which circumstances, and what impact is made on the field, theoretically or practically. Making impact is, at least in the Netherlands, the third formal task of a university, next to education and research, hence, it should be valued properly as well.

Acknowledgements. First of all, I would like to thank all the lecturers and fellow scientists that I have been privileged to work with and to learn from over the last 35 years. It is this cooperation and interaction that makes science move forward! Secondly, I would like to thank the program committee chairs for QEST'2021, Alessandro Abate and Andrea Marin, for inviting me as keynote speaker, as well as for their constructive criticism on an earlier version of this paper.

References

1. ACM Special Interest Group for Computer Performance Evaluation. https://www.sigmetrics.org/, https://dblp.org/db/conf/sigmetrics/index.html. Accessed 5 July 2021
2. van den Berg, F.G.B.: Automated performance evaluation of service-oriented systems. Ph.D. thesis, University of Twente. https://research.utwente.nl/en/publications/automated-performance-evaluation-of-service-oriented-systems (2017). Accessed 5 July 2021
3. Bertsekas, D.P., Gallager, R.G.: Data Networks, 2nd edn. Prentice-Hall, Upper Saddle River (1992)
4. Brinch Hansen, P.: Operating System Principles. Prentice-Hall, Englewood Cliffs (1973)
5. Bux, W.: Modeling token ring networks-a survey. In: Herzog, U., Paterok, M. (eds.) Messung, Modellierung und Bewertung von Rechensystemen. Informatik-Fachberichte, vol. 154, pp. 192-221. Springer, Heidelberg (1987). https://doi.org/10.1007/978-3-642-73016-0_13
6. Coffmann, E.G., Denning, P.J.: Operating Systems Theory. Prentice-Hall, Englewood Cliffs (1973)
7. Digital Twin Research. https://www.digital-twin-research.nl/. Accessed 5 July 2021
8. Embedded Systems Institute: I served as scientific director during 2009–2013. https://esi.nl/. Accessed 5 July 2021
9. Frenkel, A.K.: Allan L. Scherr, big blue's time-sharing pioneer. Commun. ACM **30**(10), 824–828 (1987)
10. Ferrari, D.: Computer Systems Performance Evaluation. Prentice Hall, Upper Saddle River (1978)
11. Ferrari, D.: Considerations on the insularity of performance evaluation. IEEE Transactions on Software Engineering **12**(2), 21–32 (1986)
12. Haverkort, B.R.: Performance Evaluation of Computer Communication Systems: A Model-Based Approach. Wiley, Hoboken (1998)
13. Hennessy, J.L., Patterson, D.A.: Computer Architecture: A Quantitative Approach, 3rd edn. Morgan Kaufman, Burlington (2003)
14. Journal of Applied Probability. Cambridge University Press. https://www.cambridge.org/core/journals/journal-of-applied-probability. Accessed 5 July 2021
15. Kleinrock, L.: Queueing Systems, Volume II: Computer Applications. Prentice-Hall, Upper Saddle River (1987)
16. Kurose, J.F., Ross, K.: Computer Networking: A Top-Down Approach, 7th edn. Pearson, London (2017)
17. IEEE Modeling and Simulation of Computer and Telecommunication Systems. https://dblp.org/db/conf/mascots/index.html. Accessed 5 July 2021
18. Onvural, R.O.: Asynchronous Transfer Mode Networks. Artech House, Norwood (1994)

19. Performance Evaluation. Elsevier. https://www.journals.elsevier.com/performance-evaluation. Accessed 5 July 2021
20. Quantitative Evaluation of Systems. https://www.qest.org/, https://dblp.org/db/conf/qest/index.html. Accessed 5 July 2021
21. Queueing Systems: Theory and Applications. Springer. https://www.springer.com/journal/11134. Accessed 5 July 2021
22. Reiser, M., Lavenberg, S.S.: Mean-value analysis of closed multi-chain queuing networks. Journal of the ACM **27**(2), 313–322 (1980)
23. Silberschatz, A., Galvin, P.B.: Operating System Concepts, 4th edn. Addison-Wesley, Readin (1994)
24. Tanenbaum, A.S.: Computer Networks, 3rd edn. Prentice-Hall, Hoboken (1996)
25. Tanenbaum, A.S., Bos, H.: Modern Operating Systems. Pearson, London (2014)
26. Tichy, W.T.: Should computer scientists experiment more? IEEE Computer **31**(5), 32–40 (1998)
27. Walrand, J.: Communication Networks: A First Course. Aksen Associates Incorporated Publishers, Stamford (1991)

Probabilistic Model Checking

A Modest Approach to Dynamic Heuristic Search in Probabilistic Model Checking

QEST Repeatability and Artifact Evaluation 2021 Accepted

Michaela Klauck[1(✉)] and Holger Hermanns[1,2]

[1] Saarland University, Saarland Informatics Campus, Saarbrücken, Germany
[2] Institute of Intelligent Software, Guangzhou, China
{klauck,hermanns}@cs.uni-saarland.de

Abstract. This paper presents Modysh, a probabilistic model checker which harvests and extends non-exhaustive exploration methods originally developed in the AI planning context. Its core functionality is based on enhancements of the heuristic search methods *labeled real-time dynamic programming* and *find-revise-eliminate-traps* and is capable of handling efficiently maximal and minimal reachability properties, expected reward properties as well as bounded properties on general MDPs. Modysh is integrated in the infrastructure of the Modest Toolset and extends the property types supported by it. We discuss the algorithmic particularities in detail and evaluate the competitiveness of Modysh in comparison to state-of-the-art model checkers in a large case study rooted in the well-established *Quantitative Verification Benchmark Set*. This study demonstrates that Modysh is especially attractive to use on very large benchmark instances which are not solvable by any other tool.

1 Introduction

Markov decision processes (MDPs) are the base model for probabilistic model checking. A variety of probabilistic model checkers are being developed, and are supported by orchestrated initiatives like the QComp competition [13,18] and the quantitative verification benchmark set QVBS [24]. While in probabilistic model checking MDPs often reflect concurrency phenomena, they have a longer tradition in the context of *sequential decision making under uncertainty* [6,27].

Depending on the modelling context, MDPs are usually decorated with rewards or costs. The term reward is traditionally used if the goal is to maximize the earnings. In the dual context of costs, the spendings are usually to be minimized, under the assumption that decisions in the MDP are controllable. Instead, in a setting where the MDP results from concurrent interleavings it can also be natural to ask for the maximal cost lurking or the minimal reward obtainable, since here the decisions need to be assumed as being uncontrollable. Typical properties of interest in this context include (max, min) reach probabilities w.r.t. a set of goal

This work has received support by the ERC Advanced Investigators Grant 695614 POWVER, by the DFG Grant 389792660 as part of TRR 248 CPEC, and by the Key-Area Research and Development Grant 2018B010107004 of Guangdong Province.

A. Abate and A. Marin (Eds.): QEST 2021, LNCS 12846, pp. 15–38, 2021.
https://doi.org/10.1007/978-3-030-85172-9_2

states as well as (max, min) expected rewards (costs) which are accumulated until reaching a goal state. These properties can also include bounds on the number of steps until reaching a goal or enforce a certain reward (cost) amount to be accumulated on the way to the goal. Iterative methods like *value iteration* are the standard solution to calculate results for these property types. In its basic form, value estimates for each state in the state space are updated synchronously based on the values of their successors until convergence is reached [6].

Heuristic search methods [4,8,9,21] try to compute such optimal values based on only a small fraction of the states, sufficient to answer the considered property. These methods exploit state-wise estimates of the optimal value, for only a subset of the state space. The order in which values are updated is made dependent on their current value estimates, in an approach called *asynchronous value iteration*.

This paper presents a probabilistic model checker that harvests modified versions of asynchronous value iteration based on heuristic search. The core components are the *labeled real-time dynamic programming* (LRTDP) [8] and *find-revise-eliminate-traps* (FRET) [31] procedures. LRTDP tries to find the optimal values by continually updating the current best solution of the state value estimates on single exploration paths. Only one state's value is updated at each step. FRET is needed to guarantee convergence of LRTDP to the optimal value in special MDP structures. It eliminates cycles to guide LRTDP to the correct solution. While contributions to this research line are manyfold (see Sect. 5), they are quite fragmented w.r.t. assumptions on property types and model characteristics. We instead take care to support most of the established property types, from reach probabilities to reward expectations (but no long-run averages), also including bounded versions, on general MDP structures efficiently.

As a result, our tool MODYSH considerably enlarges the property types supported by heuristic methods. The new elements and their integration are described in detail in this paper. A large empirical evaluation shows that MODYSH is competitive relative to state-of-the-art model checkers and is able to solve benchmark instances which are too large to be solved by other tools. MODYSH is shipped as an extension component to the MODEST TOOLSET [22] inside which it can be considered as an alternative to MCSTA [16,19,23], which is an explicit-state probabilistic model checker based on traditional value iteration. The toolset is available for Windows, Linux and Mac OS. Integrating MODYSH into it brings the benefit that the same input languages and operating systems are supported, and it opens the MODEST TOOLSET for property types not supported thus far.

Outline. In Sect. 2 we review the theoretical background. Sect. 3 introduces heuristic search approaches and discusses how LRTDP and FRET can be extended and modified such that they are applicable to general MDP structures and properties. Sect. 4 presents a large empirical evaluation demonstrating that MODYSH is competitive, outperforming state-of-the-art model checkers especially on very large state spaces with a parallel structure. We conclude with a short discussion of our achievements.

2 Theoretical Background

Before looking into the details of the heuristic search techniques implemented in MODYSH, we introduce the theoretical background. A probability distribution over a (countably in-)finite set X is a function $\mu : X \to [0,1]$ s.t. $\sum_{x \in X} \mu(x) = 1$. We denote by $\mathcal{D}(X)$ the set of all probability distributions over X.

A *Markov Decision Process (MDP)* is a tuple $\mathcal{M} = \langle \mathcal{S}, \mathcal{A}, \mathcal{P}, \mathcal{R}, s_0, \mathcal{S}_* \rangle$ consisting of a finite set of *states* $\mathcal{S}$, a finite set of *actions* $\mathcal{A}$, the partial *transition probability function* $\mathcal{P} : \mathcal{S} \times \mathcal{A} \to \mathcal{D}(\mathcal{S})$, a reward function $\mathcal{R} : \mathcal{S} \times \mathcal{A} \times \mathcal{S} \to \mathbb{R}_0^+$ assigning a reward (or cost) value to each triple of state, action, state, a single *initial state* $s_0 \in \mathcal{S}$, and a set of absorbing *goal states* $\mathcal{S}_* \subseteq \mathcal{S}$.

An action $a \in \mathcal{A}$ is *applicable* in a state $s \in \mathcal{S}$ if $\mathcal{P}(s, a)$ is defined. In this case we denote by $\mathcal{P}(s, a, t)$ the probability $\mu(t)$ of state t according to $\mathcal{P}(s, a) = \mu$. We denote by $\mathcal{A}(s) \subseteq \mathcal{A}$ the set of all actions that are applicable in s. We restrict to MDPs where for each state s, $\mathcal{A}(s)$ is nonempty, which is no restriction as per the following. A state s is called *terminal* if $|\mathcal{A}(s)| = 1$ and for this $a \in \mathcal{A}(s)$ it holds that $\mathcal{P}(s, a, s) = 1$ and $\mathcal{R}(s, a, s) = 0$. All goal states g are assumed to be terminal, which forces to stay in g forever without accumulating further reward. Terminal states not contained in $\mathcal{S}_*$ are called *dead-ends*.

For a given MDP $\mathcal{M}$, a function $\pi : \mathcal{S} \to \mathcal{A}$ with $\pi(s) \in \mathcal{A}(s)$ for each state s is called a (memoryless) *policy*, used to determine the next action to take for any given state. We later extend this when focussing on specific, bounded properties. The *accumulated reward* over an infinite sequence of states $\zeta = (s_i)_{i \in \mathbb{N}}$, called *path*, induced by a policy π through $\mathcal{M}$ is defined by $\rho(\zeta) = \sum_{i=0}^{\infty} \mathcal{R}(s_i, \pi(s_i), s_{i+1})$. For the finite prefixes τ of such a path, called finite paths, the reward summation constituting $\rho(\tau)$ is truncated accordingly. We let $Paths(\mathcal{M})$ denote the set of all paths through $\mathcal{M}$ rooted in its initial state s_0. Each policy π induces a probability space on the set of infinite paths through $\mathcal{M}$ in the usual way [25] and this in turn induces well-defined probability measures for each of the finite paths τ, and similarly for the accumulated reward measures $\rho(\tau)$. States from which $\mathcal{S}_*$ can not be reached with positive probability regardless of the policy π are called *sink states* and collected in $\mathcal{S}_\perp$. This set can be precomputed by a simple fixpoint computation (checking for each state the LTL property $\Box\neg\texttt{goal}$ where $\texttt{goal}$ identifies all states in $\mathcal{S}_*$) in the underlying graph. This graph G over $\mathcal{S}$ is spanned by the edge set $E = \{(s, t) \mid \exists a \in \mathcal{A} : \mathcal{P}(s, a, t) > 0\}$.

The subgraph G_π induced by policy π is obtained by restricting the edge set of G to $\{(s, t) \mid \mathcal{P}(s, \pi(s), t) > 0\}$. π is *almost-sure* if the probability of reaching $\mathcal{S}_*$ it induces is 1 regardless of the initial state. If instead that probability is guaranteed to be positive, π is called *proper*. A *cycle* is a path in G starting and ending in the same state. A *strongly connected component (SCC)* in G is a subset of states V such that $\forall (s, t) \in V \times V$ a path from s to t exists. A bottom SCC (BSCC) B is a SCC of maximal size from which only states in B are reachable.

Measures of Interest. We denote by P^π the probability measure induced by π and by E^π the expectation of the accumulated reward ρ w.r.t. measurable sets of paths starting in s_0. We define the extremal values $P_{\max}(\Pi) = \sup_\pi P^\pi(\Pi)$ and $P_{\min}(\Pi) = \inf_\pi P^\pi(\Pi)$, as well as $E_{\max}(\Pi) = \sup_\pi E^\pi(\Pi)$ and $E_{\min}(\Pi) =$

$\inf_\pi E^\pi(\Pi)$, for measurable $\Pi \in Paths(\mathcal{M})$ and $\pi \in \Pi$. We consider the following property types, echoing what JANI supports [12,28], with $opt \in \{\max, \min\}$:

- *MaxProb* and *MinProb*: $P_{opt}(S_U\ \mathcal{U}\ \mathcal{S}_*) = P^{opt}(\{\tau \in Paths(\mathcal{M}) \mid \exists s \in \mathcal{S}_* : \tau = (s_i)_{i=0}^{\infty} \wedge s = s_j \wedge \forall k < j : s_k \notin \mathcal{S}_* \wedge s_k \in S_U\})$ is the max/min probability of eventually reaching a goal state and all states visited before being in S_U. $P_{opt}(S_U\ \mathcal{U}\ \mathcal{S}_*)$ will be abbreviated as $P_{opt}(\diamond\ \mathcal{S}_*)$.
- maximal/minimal expected rewards: $E_{opt}(S_U\ \mathcal{U}\ \mathcal{S}_*) = E^{opt}(\{\tau \in Paths(\mathcal{M}) \mid \exists s \in \mathcal{S}_* : \tau = (s_i)_{i=0}^{\infty} \wedge s = s_j \wedge \forall k < j : s_k \notin \mathcal{S}_* \wedge s_k \in S_U\})$ is the maximal or minimal reward expectation of eventually reaching a goal state. Note that reward ∞ is accumulated for non-almost-sure policies.
- step bounded properties: $P_{opt}(S_U\ \mathcal{U}_{[l,u]}\ \mathcal{S}_*)$ is the maximal or minimal probability of reaching a goal state in $[l, u]$ steps defined as $P^{opt}(\Pi_{[l,u]})$ where $\Pi_{[l,u]}$ is the set of paths that reach a goal state in $[l, u]$ steps while only passing through S_U. Similar for step bounded expected reward properties.
- reward bounded properties: If a reward structure is defined, $P_{opt}(S_U\ \mathcal{U}_{[l,u]}\ \mathcal{S}_*)$ is the extremal probability of reaching a goal state with accumulated reward in $[l, u]$ defined as $P^{opt}(\Pi_{[l,u]})$ where $\Pi_{[l,u]}$ is the set of paths having a prefix τ with accumulated reward in $[l, u]$ containing a goal state and only passing through S_U before. Similar for reward bounded expected reward properties. Bounds with open intervals are also supported (for all bounded properties).

3 Dynamic Heuristic Search

Value Iteration. The problems discussed above are in practice often solved using *value iteration.* This is a variant of dynamic programming where a value is assigned to each state by a value function $V : \mathcal{S} \to \mathbb{R}$ which specifies the current approximation of the value of this state. The value function is placed in an iterative procedure updating the states' values depending on the values of their successors. These values are refined until convergence to the least fixpoint. In many situations this fixpoint $V^* = \lim_{n\to\infty} V_n$ corresponds to the optimal value one is looking for, from which the optimal policy can be extracted. Usually, the value function is calculated greedily via the *Bellman function* [5] (similar for maximum): $V_{i+1}(s) = \min_a \sum_{s'\in\mathcal{S}} \mathcal{P}(s, a, s') \cdot (\mathcal{R}(s, a, s') + V_i(s'))$ (1) where a value of 1 is assigned to goal states and 0 to dead-ends. A value function is *admissible* if it is an optimistic estimate of the correct final value. This means, if we try to minimize, the value function V is admissible if always $V(s) \leq V^*(s), \forall s \in \mathcal{S}$. If we instead maximize, a value function with $V(s) \geq V^*(s), \forall s \in \mathcal{S}$ is admissible. A *greedy policy* is always defined w.r.t. a value function V. For each state the greedy policy always picks the action leading to the successor state with the best value according to the value function. This action may not be unique which means there can be multiple greedy policies. A *greedy graph* G_V of graph G with respect to value function V is the superposition of all G_π induced by any greedy policy π w.r.t. V, so it is the combined reachability graph of all greedy policies.

Heuristic Search. The approach we generally pursue is based on the heuristic search algorithm LRTDP [8], a heuristic search dynamic programming optimization of standard value iteration. To find an optimal policy, up to a prespecified accuracy ε, starting in an initial state, it attempts to avoid exploring the entire state space and delivers the requested values for the initial state only, instead of for all states as in standard value iteration. It constantly keeps updating a *current best solution*, a partial value function providing the current state value estimates. In each round only a single state is selected for an update. These updates are obtained by repeatedly sampling *trials*, i. e., executions starting in the initial state, and ending once a state is reached for which an update does not change the value estimate by more than ε. While doing so, the optimal policy is constructed incrementally by extending a partial policy step by step. A partial policy π will be called *closed* for a state $s \in \mathcal{S}$, if $\pi(t)$ is defined for every state $t \notin (\mathcal{S}_\perp \cup \mathcal{S}_*)$ that is reachable (with positive probability) from s by following π.

Traps. A *trap* [30, p. 171 ff.] is a BSCC not containing a goal state. In our approach *traps are defined on the greedy graph* G_V induced on G by value function V. We distinguish *permanent traps* which are also BSCCs of G, i. e., there is no non-greedy policy which would lead out of the trap. In contrast, *transient traps* are SCCs, but not BSCCs of G, so there is a policy leading out of the trap.

Convenience MDPs. The planning literature has identified a number of model classes with convenient properties and initially arbitrary rewards in $\mathbb{R}$. A *Stochastic Shortest Path* ($\mathcal{SSP}$) MDP [6] is an MDP admitting (i) at least one almost-sure policy and (ii) inducing expected accumulated reward ∞ for each not almost-sure policy π. The latter corresponds to G_π containing no reachable cycle on which (in the MDP) the accumulated reward does not increase. Assuming the former, the latter can trivially be enforced by restricting to models with reward function confined to positive values (possibly except at goal states). As an apparent relaxation, Bertsekas [7] later introduced condition (i′) and (ii′) which replace the role of *almost-sure* policies by *proper* policies in (i), respectively (ii), but showed them to be (pairwise) equivalent. In a *Generalized Stochastic Shortest Path* ($\mathcal{GSSP}$) MDP [30] the first condition (i) is kept while the second condition is further relaxed by instead assuming that (ii″) for each policy and state the expected sum of negative rewards is bounded from below. This relaxation in particular supports zero-reward cycles, while it precludes cycles with alternations of positive and negative rewards that cancel out. Condition (ii″) can trivially be enforced by restricting to models with a reward function confined to non-negative values, as we do. Our contribution relinquishes condition (i) and (i′) of $\mathcal{SSP}$ and $\mathcal{GSSP}$, i. e., we do not rely on the existence of almost-sure or proper policies.

Algorithm Overview. We introduce our algorithmic contributions in the sequel one-by-one. All modifications, adaptions and extensions made to the original versions are marked in blue. If existing, the original version of modified lines is stated in comments of the form $\triangleright \ldots$ The base algorithm expects as inputs the state s of the MDP for which to evaluate the property, the result precision ε and uses flags dependent on the property class to be evaluated. `max-rew` is *True* if a maximal expected reward property is evaluated, otherwise it is *False*, analogously for `min-rew`. We do not use explicit flags for indicating (max or min) reachability probabilities because there are no code fragments specific to these property types. We assume that the initial and current value function, V_0 and V_i, are always globally accessible.

In fact, the original algorithmic contributions have been made without a specific focus on reachability probabilities, which as long as zero reward values are supported, can actually be cast into reward accumulations. We here make an explicit distinction between these cases for the purpose of better explanability and for the purpose of more direct and hence faster implementation in MODYSH.

3.1 Reachability Properties

For reachability properties `max-rew` and `min-rew` are set to *False*. We first concentrate on calculating *MinProb*, i. e., $P_{\min}(S_U \, \mathcal{U} \, \mathcal{S}_*)$. We detail our modifications to the original version of the algorithm in order to enable that condition (i′) and thus (i) can be dropped . Afterwards we turn to *MaxProb* and show how FRET-LRTDP can be modified to solve these kind of properties on general MDPs, too. Kolobov et al. [31] already provided a reduction to show that FRET in combination with LRTDP is applicable to general *MaxProb* properties, even if condition (i′) is violated. We will give an alternative proof, based on the proof for *MinProb*, demonstrating that our implementation is also valid for general MDP types as defined above, not only for problems having at least one proper policy.

We denote by $V^\pi : \mathcal{S} \mapsto [0,1]$ the goal-reachability probabilities induced by π. Goal states $\mathcal{S}_*$ have probability value 1 while sinks and other states enforced to be avoided have probability value 0. This corresponds to the fact that if a partial policy π is closed for s, V^π constitutes the least fixpoint of Equation (2).

$$V^\pi(s) = \begin{cases} 1 & \text{if } s \in \mathcal{S}_*, \\ 0 & \text{if } s \in \mathcal{S}_\perp \cup \overline{S_U} \setminus \mathcal{S}_*, \\ \sum_{s' \in \mathcal{S}} \mathcal{P}(s, \pi(s), s') \cdot V^\pi(s') & \text{otherwise.} \end{cases} \tag{2}$$

Minimum Reach Probability. For *MinProb* properties the objective is to find the minimal probability to reach a state in $\mathcal{S}_*$ if initialized in s_0 and while avoiding the complement of S_U. We are ultimately interested in the value

$$V^*(s_0) = \min_{\pi:\pi \text{ closed for } s_0} V^\pi(s_0). \tag{3}$$

An admissible initialization for this case is a valuation of 0, except for goal states which get a value of 1. Using a reward function defined as $\mathcal{R}(s,a,s') = 1$ if $s \notin \mathcal{S}_* \wedge s' \in \mathcal{S}_*$ and 0 otherwise and then applying the Bellman equation (1) of synchronous value iteration will iteratively fill the partial policy bottom up. Spelled out for our case, this amounts to replacing the third line of (2) by

$$\min_{a \in \mathcal{A}(s)} \sum_{s' \in \mathcal{S}} \mathcal{P}(s,a,s') \cdot V^\pi(s') \quad \text{otherwise.} \tag{4}$$

which echoes the greedy nature of the computation. However, giving up synchronicity in favor of a heuristic approach is the key to efficiency. The base algorithm for this case we call GLRTDP, a generalization of LRTDP [8, Alg. 4]. The pseudocode is shown in Alg. 1. The algorithm iteratively selects only a single state for a Bellman update in each round. It continually updates a *current best solution*, a partial function providing the current state value estimates and repeatedly runs *trials* (line 4), sample executions of the MDP, starting from the initial state, and ending once a state is reached for which an update does not change the value by more than ε, i. e., ε-consistency is reached (line 13, lines 17-27 are not relevant here). To determine which successor state to follow after state s, GLRTDP considers an action $a \in \mathcal{A}(s)$ *greedy* w.r.t. the current value function (line 14), i. e., one that minimizes Equation (4) for s (cf. Alg. 2, line 2 and 4) [8, Alg. 2], and then selects a successor state (line 16). Picking the next state randomly from the set of successors of the greedy action (cf. Alg. 2, line 9) instead of taking the probability into account is an optimization leading to better performance as noted in probabilistic FAST DOWNWARD [36]. The entire exploration procedure is systematic, i. e., does not starve relevant states if the heuristic function used is admissible. A state, which has not converged so far, will not stay in the greedy graph forever without its value being revised. Therefore, it is guaranteed to converge to an optimal solution. After each trial, those states are labeled as *solved* whose values and those of their descendants have reached ε-consistency (cf. Alg. 3) [8, Alg. 3]. Trials are terminated at solved states. GLRTDP terminates the value update procedure as soon as the initial state is solved (cf. Alg. 1 line 3, 8, and 31).

Alg. 1 General Labeled Real-Time Dynamic Programming (GLRTDP)

```
1:  proc GLRTDP(s: State; ε: float)
2:   max-rew, min-rew = True, if max., resp.
       min. reward property is calculated
3:   while ¬Solved(s) do
4:    GLRTDP-trial(s, ε)

5:  proc GLRTDP-TRIAL(s: State, ε: float)
6:   visited := Empty-Stack
7:
8:   while ¬Solved(s) do
9:    visited.Push(s)
10:   v_old = V(s)
11:   Update(s)
12:   v_new = V(s)
13:   if Is-cons(v_old, v_new, ε) then break
                   ▷ original condition Is-goal(s)
14:   a := Greedy-action(s)
15:   if a ≠ NULL then
16:    s := Pick-next(a, s)
17:   if max-rew
        && visited.Contains(s) then
18:     if Elim-cycle-max-rew() then
19:      V(init-node) = ∞
20:      Solved(init-node) = True
21:      return
22:   else
23:     if min-rew
         && visited.Contains(s) then
24:      if Elim-cycle-min-rew then
25:       s := Merged-node(s)
26:   else
27:    break
28:
29:  while visited ≠ Empty-Stack do
30:   s := visited.Pop()
31:   if ¬Check-solved(s, ε) then
32:    break
```

Alg. 2 Subroutines of GLRTDP and FRET

```
1:  proc GREEDY-ACTION(s: State)
2:   return argMinMax_{a∈A(s)} QValue(a, s)

3:  proc QVALUE(a: action, s: State)
4:   return
       Σ_{s'} P(s, a, s') · (R(s, a, s') + V(s'))

5:  proc UPDATE(s: State)
6:   a = Greedy-action(s)
7:   V(s) = QValue(a, s)

8:  proc PICK-NEXT(a: action, s: State)
9:   pick s' randomly from all successors with
       P(s, a, s') > 0
        ▷ originally with probability P(s, a, s')
10:  return s'

11: proc IS-CONS(s_old, s_new, ε: float)
12:  if abs(s_old − s_new) ≤ ε ||
        s_old = ∞ && s_new = ∞ then
13:   return True
14:  return False
```

This is possible because a value remains ε-consistent if its descendants' and its own value do not change by more than ε anymore (Alg. 3). This is because $V(s)$ can only change by more than ε if the greedy graph starting in s changes or the value of a descendant changes by more than ε. The graph can only change if the value of a state within the graph changes. Updating states outside the greedy graph will never make them part of it, because by the monotonicity property, updates according to the Bellman function can only make the states less attractive. Thus, a state's value can only change by more than ε if a descendant changes by more than ε but then it can not have been marked as solved before.

This algorithm converges faster than classical value iteration because not all states need to be converged or even updated. The termination criterion is similar to ε convergence in simple value iteration. If a cycle (zero-reward cycle in MDPs with rewards) occurs in a policy, it needs to be handled during the construction of trials in GLRTDP to guarantee convergence to an optimal value function. In the *MinProb* case permanent and transient traps have to be treated as dead-ends because in the worst case it is possible to always take an edge back to a state in the cycle instead of leaving the loop, i. e., $P_{\min}$ of eventually reaching the goal is 0. This is done indirectly by the termination criteria and the check before adding a new state (line 13 Alg. 1 and line 37 in Alg. 3). Because of the initialization with 0, values of trap states will lead to a cut immediately, because they never change their value in an update and stay ε-consistent, i. e., the cycle is not explored further and the algorithm concentrates on other branches.

To sum up, when calculating *MinProb* over an MDP, GLRTDP presented in Alg. 1 with an admissible initialization for this case and `Check-Solved()` as in Alg. 3 can be used. We will explain in the following why the combination of GLRTDP solves *MinProb* properties on general MDP structures correctly by converging to the optimal fixpoint. A formal proof can be found in Appendix A.

All greedy policies inspected by GLRTDP at some point end in a goal state or a dead-end state. This could be a real dead-end, i. e., a sink state with only a self-loop or a trap. Because of the initialization their value is already 0. In addition, we tag these states, do not explore them further and propagate their value back through the graph. Cycling forever is not possible because eventually all such cycles in greedy policies are eliminated. Having this, we can state that at some point no more states are left to explore in GLRTDP because all relevant traps are eliminated or a goal or a sink has been found. Then GLRTDP runs until the state values of the current greedy policy is converged up to ε. Even if the greedy policy is not the same in every iteration, at some point it will stay within a set of states which are part of finitely many policies. The values of these states converged close enough to the optimal ones such that the algorithm concentrates on these policies. The value function used in GLRTDP is initialized admissibly and therefore can only monotonically increase and approach the optimal result (fixpoint) from below. When this point is reached, the whole procedure terminates. This fixpoint has to be the only one and therefore has to be optimal because it has already been shown that the Bellman equation only has one fixpoint [7].

Alg. 3 Check-solved Procedure used in GLRTDP

```
1:  proc CHECK-SOLVED(s: State; ε: float)
2:    rv := True
3:    open := Empty-Stack
4:    closed := Empty-Stack
5:
6:    if ¬Solved(s) then open.Push(s)
7:
8:    while open ≠ Empty-Stack do
9:      s := open.Pop()
10:     closed.Push(s)
11:
12:     if Dead-end(s) || Goal(s) then continue
13:
14:     a := Greedy-action(s)
15:     if max-rew || min-rew then
16:       check-∞-loop = False
17:       for each s' s.t. P(s, a, s') > 0 do
18:         if closed.Contains(s') then
19:           check-∞-loop = True
20:       if max-rew && check-∞-loop then
21:         if Elim-cycle-max-rew() then
22:           V(init-node) = ∞
23:           Solved(init-node) = True
24:           return True
25:       else
26:         if min-rew && check-∞-loop then
27:           if Elim-cycle-min-rew then
28:             return False
29:
30:     v_old = V(s)
31:     Update(s)
32:     v_new = V(s)
33:     if not Is-cons(v_old, v_new, ε) then
34:       rv = False
35:       continue
36:     for each s' s.t P(s, a, s') > 0 do
37:       if ¬Solved(s') &&
            ¬In(s', open ∪ closed) then
38:         open.Push(s')
39:
40:   if rv then
41:     for each s ∈ closed do
42:       Solved(s) := True
43:   else
44:     for s ∈ closed do
45:       Update(s)
46:   return rv
```

Alg. 4 Find, Revise, Eliminate Traps (FRET) (*M* is the graph of the MDP)

```
1: proc FRET(M, s, V_0)
2:   V_i := V_0
3:   V'_i := GLRTDP(s, ε)
                   ▷ originally Find-and-Revise(M, V_i)
4:   (V_{i+1}, elim-trap) :=
     Eliminate-Traps(M, V'_i)
5:   while elim-trap do
6:     V_i := V_{i+1}
7:     V'_i := GLRTDP(s, ε)
                   ▷ originally Find-and-Revise(M, V_i)
8:     (V_{i+1}, elim-trap) :=
       Eliminate-Traps(M, V'_i)
```

Alg. 5 Eliminate-Traps (for *MaxProb*)

```
1:  proc ELIMINATE-TRAPS(M, V)
2:    elim-trap := False
3:    V_next := V
4:    G_V := {S_V, A_V} ← Vs greedy graph
5:    SCC := Tarjan(G_V)
6:    CSet := ∅
7:
8:    for each SComp C = {S_C, A_C} ∈ SCC
      do
9:      if ∄(s_i, s_j) ∈ A_G : (s_i ∈ S_C, s_j ∉ S_C)
        && (∄g ∈ G : g ∈ S_C) then
10:       CSet := CSet ∪ {C}
11:
12:   for each C = {S_C, A_C} ∈ CSet do
13:     if ∄a ∈ A, s ∈ S_C, s' ∉ S_C :
        T(s, a, s') > 0 then
14:       for each s ∈ S_C do
15:         V_next(s) := 0
16:       MergeSCC(C)
17:       elim-trap := True
18:     else
19:       A_e := {a ∈ A | ∃s ∈ S_C; s' ∉ S_C :
          T(s, a, s') > 0}
20:       m := max_{s∈S_C, a∈A_e} Q^V(s, a)
21:       for each s ∈ S_C do
22:         V_next(s) = m
23:       MergeSCC(C)
24:       elim-trap := True
25:   return (V_next, elim-trap)
```

Maximum Reach Probability. For *MaxProb* properties, $P_{\max}(S_U\ \mathcal{U}\ S_*)$, the objective is to find the maximal probability to reach a state in S_* if initialized in s_0 while avoiding $\overline{S_U}$. An admissible initialization is 1 except for states from which only dead-end states can be reached, which get a value of 0. $P_{\max}$ can be calculated by changing the initialization and replacing the occurrences of min by max in equations (3) and (4). In MODYSH, we use a combination of GLRTDP (Alg. 1, `max-rew`=`min-rew`=*False*) and a modified version of FRET (Alg. 4 and 5), adapted from the originals [31, Alg. 1] to calculate *MaxProb*. The combination is needed to guarantee convergence of GLRTDP for *MaxProb* [30,31]. In FRET iterations of GLRTDP followed by a call to `Eliminate-Traps()` to eliminate zero-reward cycles are performed. In the original version, any Find-and-Revise algorithm is foreseen, we fix GLRTDP (Alg. 4, line 3 and 7) in our implementation. The call to `Eliminate-Traps()` (line 4 and 8) is needed if facing zero-reward cycles, because these may induce convergence of GLRTDP-trials to a non-optimal value by always choosing an action that loops on the cycle and thus the goal is never reached (line 14, 16 in Alg. 1). The trap elimination procedure changes the value function computed in the last iteration of GLRTDP and the graph it is working on, thus guaranteeing progress in its next call (Alg. 4, line 8). This is achieved by finding and eliminating *traps* (cf. Alg. 5). States which are part of a trap are merged into a single new state replacing all trap states.

In contrast to *MinProb*, where traps are handled directly during the trial construction, permanent and transient traps have to be handled differently here. All SCCs in the current greedy policy are collected using Tarjan's Algorithm [37] (Alg. 5, line 5) and it has to be checked if these SCCs are traps (line 8). First, permanent traps (line 13) are dead-ends from which the goal can never be reached. Therefore, all states' values in this SCC can be set to 0 (line 15) and the states of the SCC can be merged into one. If the SCC is a transient trap (line 19), it has to be left to reach the goal eventually. From all states in the SCC it is possible to take the exit with the highest probability value to reach the goal (line 20). Therefore, we merge these states and set the resulting state to this value (line 21). In the next GLRTDP trial this will change the greedy policy, i. e., the cycle is eliminated from the greedy graph. The algorithm terminates if the policy of the last GLRTDP run does not contain a trap anymore.

While the original version of FRET [31] considers in each trap elimination step *all* actions that are optimal according to the current value function, our implementation uses an optimization of Tarjan's algorithm (line 5), called FRET-π [36], considering in the subgraph of the state space inspected during trap elimination only those state transitions chosen into the current greedy policy.

To sum up, when calculating *MaxProb* over an MDP we call FRET, Alg. 4, with GLRTDP, Alg. 1, with an admissible initialization for this case. The trap elimination procedure in FRET is instantiated with Alg. 5. A formal proof of the correctness of this approach for general MDPs in the style of the proof for *MinProb* can be found in Appendix B.

3.2 Expected Reward Properties

Expected reward properties $E_{opt}(S_U \; \mathcal{U} \; \mathcal{S}_*)$, ask for the minimal or maximal (referred to by *opt*) expected accumulated reward when reaching a goal state. For the reachability properties considered thus far, we have been able to ignore the reward function of the MDP or more precisely, assumed it to be 0 except for actions leading to goal states. The calculation of E_{opt} proceeds very much in the same way. Iteratively a variation of the Bellman function updates is performed as presented in Eq. 1, where contrary to the P_{opt}-case (2) rewards are gained by taking a transition. The conceptual variation is that goal states initially get a value of 0 and states $s \in \mathcal{S}_\perp \cup \overline{S_U} \setminus \mathcal{S}_*$ a value of ∞.

Reward maximization. For E_{max} `max-rew` is set to *True*. A trivial admissible initialization is 0 for goal states and ∞ for all others. Because initializing non-goal states with ∞, i. e., the largest possible overapproximation, increases the runtime extremely, we approach an admissible initialization for non-dead-end states from below by starting with a smaller *maxValue*, obtained by *exponential search.* Dead-ends directly get a value of ∞. We execute full GLRTDP runs, as long as one of the final state values after termination is larger than the last *maxValue*, because if this happens, the initialization has not been admissible. In each iteration the new *maxValue* is set to the largest state value increased by 1 and multiplied by 2, which leads to the fastest solution we found in our experiments. Cycles again require a special treatment. Before adding the next state to the current trial (line 16, Alg. 1 and line 38, Alg. 3) it has to be checked if this state closes a SCC in the current greedy graph (independent of the reward accumulated in the SCC) (`Elim-cycle-max-rew()`: line 18, Alg. 1 and line 21 et seq., Alg. 3). If this is the case, the maximal expected reward for this property can directly be set to ∞ because in the worst case always this loop could be taken, i. e., the goal would never be reached.
Reward minimization. For $E_{\min}$ `min-rew` is set to *True* and the value function is initialized admissibly with ∞ for dead-ends and with 0 for all other states. Similar to the $E_{\max}$ case, when adding the next state to the current trial, it has to be checked if it closes a zero-reward SCC which has to be eliminated because it has to be left immediately to reach the goal with minimal reward (`Elim-cycle-min-rew()` in line 24, Alg. 1 and line 27, Alg. 3).

Correctness and optimality proofs for these property types are very similar to the proofs for *MaxProb* and *MinProb* spelled out in the appendices.

3.3 Bounded Properties

Reachability and expected reward properties can be extended by step or reward bounds. $P_{opt}(S_U\ \mathcal{U}_{[l,u]}\ \mathcal{S}_*)$ is the extremal probability of reaching a goal state in $[l,u]$ steps or with accumulated reward in $[l,u]$. Notably, and in contrast to the other properties considered thus far, for such bounded properties, memoryless policies can be outperformed by policies that are aware of the history regarding their past evolution, namely with respect to the number of steps taken or reward accumulated thus far. So, formally, we here work with a definition of policies that deviates form the one in Sect. 2 in that a policy can remember how many steps have already been made or what reward has been accumulated.

Let us first look at step-bounded properties. For those, in standard value iteration, updating all state values synchronously makes it possible to iterate only t times for properties with upper bound t [17] and then to extract a step-dependent policy. In heuristic search algorithms like FRET-LRTDP this is not possible because only the current greedy path is updated. In this case, a straightforward remedy is to encode a step counter in each state and consider all states for which the bounds regarding these counters are exceeded as dead-end. Formally, one works in a derived MDP where states are enriched with counters and where states differing in counter value are different and thus also the policy decision might differ for them (implying history awareness with respect to the original MDP). States which fulfill the reachability property and whose bound-counter lies in the target interval are considered goal states. In our implementation we use the same variants of GLRTDP in combination with FRET like for the unbounded cases above and only add the bound and step counter to the state as described. For reward-bounded properties the basic strategy is the same, except that the counters are now replaced by real-valued variables. If the reward of the current policy exceeds the bound, the current state is considered as a dead-end. In either case (step or reward bounds), the derived MDP can be constructed in such a way that it is guaranteed to be finite-state (which is one of our early assumptions). Expected reward properties with bounds can be solved in a similar way. Since the overall procedures stay the same when adding bounds, the correctness and optimality proofs follow the respective same strategy.

MODYSH is the only tool of the MODEST TOOLSET which fully supports all variants of bounds w.r.t. step/reward bounds and interval types. All other tools do not treat step bounds at all and only support inclusive upper bounds.

4 Empirical Evaluation

Prototypical predecessor versions of MODYSH with less functionality, implemented on a different, less performant code base and with strategies closer to the original version of FRET-LRTDP took part in QComp 2019 and 2020 [13,18], where the approach already showed promising results in comparison to other state-of-the-art model checkers. Since then, the new implementation approach presented in this work and several other optimizations implied a decrease in runtime of MODYSH by a factor of nearly 1/3.

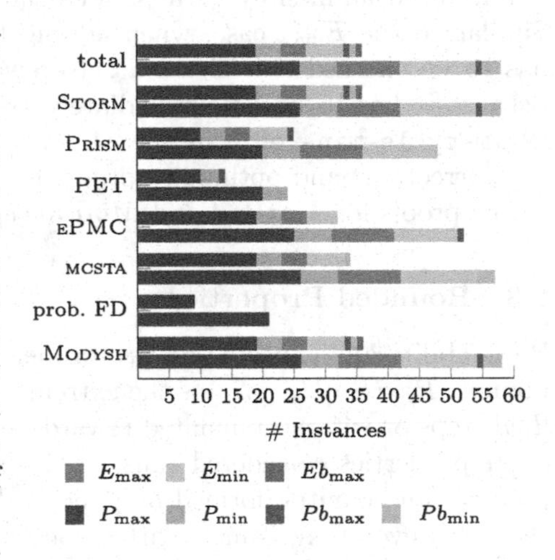

Fig. 1. Number of benchmark instances supported by tools per property type. (upper bars: QComp, lower: additional)

The benchmark set of QComp comprises, appart from other model types, 36 MDP instances. For evaluation purposes, we reran the experiments from QComp 2020 *default often ε-correct track*, i. e., with a precision of $\varepsilon = 10^{-3}$ and a timeout of 30 min, on an Intel Core i7-4790 CPU 3.60GHz with 32 GB RAM. With this setup we are able to show plots which are directly comparable to the evaluation of QComp and the performance improvements of MODYSH are clearly visible. In addition, we added 58 *additional* benchmark instances from the quantitative verification benchmark set QVBS [24] to our case study to enlarge the number of MDP benchmarks and thereby also the number of minimum reach and bounded properties. Furthermore, we wanted to test the tools on both smaller benchmarks, because many tools time out on the difficult QComp instances, as well as on considerably larger instances than in QComp, to demonstrate the capabilities and benefits of MODYSH of only inspecting a fraction of the state space. Therefore, we scaled the models for the *israeli-jalfon*, *philosophers-mdp*, *pnueli-zuck*, *rabin* and *wlan* benchmarks up by parallelizing up to 100 automata for all of them except for wlan, for which 10 processes were already enough such that only MODYSH was able to solve it. The QVBS contains only smaller instances of these benchmarks. For israeli-jalfon, the largest instance results in a state space size of $(2^{100}) - 1$, i. e., $1.268 \cdot 10^{30}$. For 100 dining philosophers the state space grows into the order of 10^{99} and for 100 parallel processes in pnueli-zuck and rabin it is in the order of 10^{100} and 10^{105}, respectively. 10 parallel senders in wlan result in a size of around $7 \cdot 10^{8}$ states. The number of benchmark instances supported by each tool per property type are listed in Figure 1. Since QComp 2020, MODYSH added functionality for bounded properties and some special minimum reach cases.

Not all participating tools of QComp 2020 support MDP benchmarks. Therefore, we were not able to consider MODES [11], STAMINA [34] and DFTRES [35]. But we added new results for Probabilistic FAST DOWNWARD [36], which took part in QComp 2019 but not in 2020. In addition, EPMC [20], MCSTA [16,19,23] of the MODEST TOOLSET [22], PET [10], PRISM [33] and STORM [14] are part of our evaluation. We contacted the authors of all tools and asked for the newest version, i. e., improvements in other tools are also taken into account.

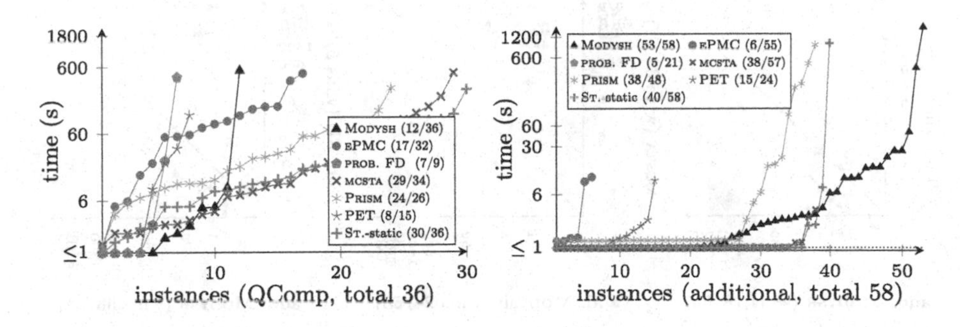

Fig. 2. Quantile plots for default tool versions in often ε-correct track.

In the quantile plots in Fig. 2 a point (x, y) indicates that the runtime of the xth fastest instance of the tool was y seconds. This allows comparing the overall performance of the tools. The benchmark instances are ordered independently for each tool depending on its runtime. The count of correctly solved benchmarks c (no timeout or error) and of supported instances s is given in the label as c/s. MODYSH improved the runtime for many of the QComp instances (Fig. 2, left, contrastable with Fig. 4 bottom right in [13]) in comparison to QComp 2020 such that it is now among the best three tools for a large number of instances. The strength of MODYSH is impressively demonstrated by the results on the additional benchmark set on the right of Fig. 2. It clearly outperforms the other tools on the extremely large scaled benchmarks because only a small fraction of the state space needs to be visited. MODYSH is able to solve 7 benchmarks in less than 30s for which all other tools time out or do not have enough memory. For 5 other models only one other tool is able to solve them. For the largest instances of philosophers, pnueli-zuck, rabin and wlan only a few thousand states have to be visited in MODYSH and only $1.7 \cdot 10^3$ for israeli-jalfon. All these benchmarks have in common that they consist of the parallelization of automata of symmetric structure. The results on both benchmark sets show that MODYSH is clearly able to compete with state-of-the-art model checkers and on certain MDP structures it is even able to quickly solve instances which no other tool is able to handle.

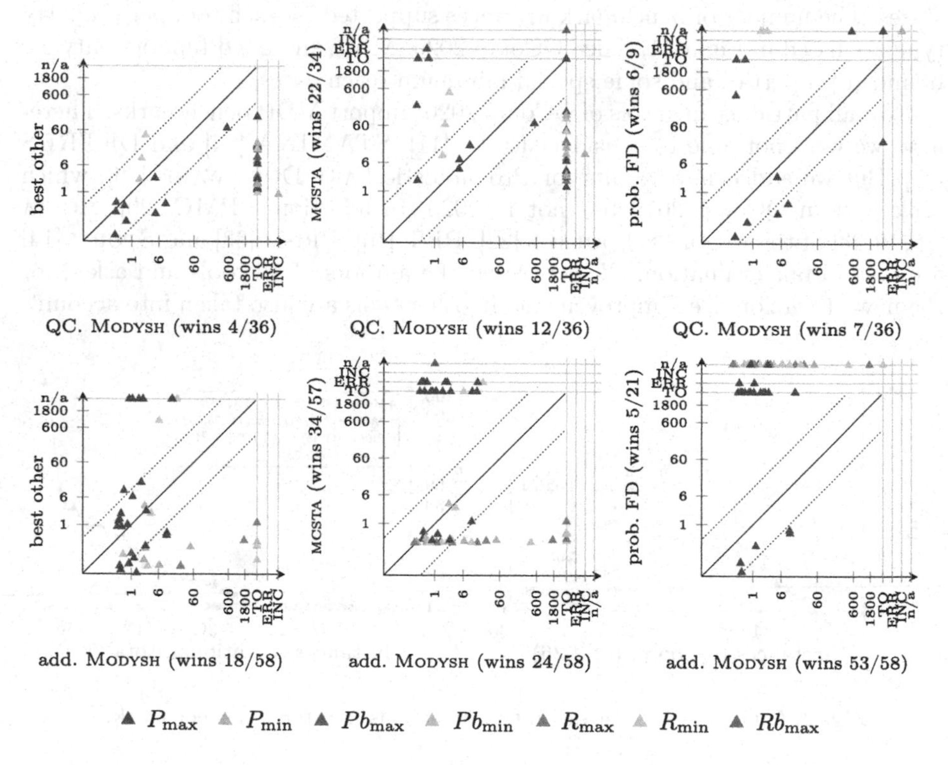

Fig. 3. Scatter plots (row 1: QComp, 2: additional benchmarks).

More detailed results can be inspected in Fig. 3 showing scatter plots comparing individual benchmark instances between two tools or a tool and the best of all other tools. A point (x, y) indicates a runtime of x seconds for the tool on the x-axis and a runtime of y seconds for the tool on the y-axis. This means, if the point lies above the diagonal line, the tool on the x-axis was the fastest. If the point lies above the dotted line, it was more than ten times faster. "TO", "ERR" and "INC" mean timeout, error, e. g., out of memory, and incorrect result, respectively. "n/a" means that the tool is unable to handle the benchmark instance. The number of benchmark instances on which the tool on the x-axis outperformed the tool(s) on the y-axis is given in parenthesis in the label. By the evaluation setup, the upper left plot is a direct update of [13, Fig. 8, middle]. We see that MODYSH is able to compete with the other tools especially on the additional benchmark set for which the results are depicted in the lower row. It solves way more instances and property types than probabilistic FAST DOWNWARD (right column), which is based on the same algorithms. It also supports more properties than MCSTA (upper row, middle), i. e., improves the range of the MODEST TOOLSET and shows better performances on many instances, especially where MCSTA (lower row, middle) or various other tools (lower row, left) are not able to deliver results at all. This demonstrates the potential of the methods implemented in MODYSH because first, it improves the model checking performance of the MODEST TOOLSET in comparison to MCSTA on the same code base. Second, integrating these techniques specifically in STORM looks promising. If MODYSH was dominated by a competitor, e. g., on the QComp benchmarks (upper left), it was often outperformed by STORM. From QComp 2020 it is already known that STORM's code base is highly efficient and the performance is currently out of reach for other model checkers on most of the benchmarks. Implementing our approach in STORM would boost its performance even more.

Interactive result tables which enable a direct runtime comparison across benchmark instances are available online for the QComp benchmarks and for the additional QVBS benchmarks. Furthermore, an artifact enabling the reproduction of all empirical results reported in this paper is available online [29].

5 Related Work

As already described in Sect. 3, our algorithms are generalizations of well-known approaches used in the planning community for the purpose of cost-optimal planning. Of course, ideas behind heuristic search have already been used in model checking. We highlight the parallels but also the differences to our work. *Probabilistic Planning and Heuristic Search.* A variant of FRET-LRTDP is available in the probabilistic version of FAST DOWNWARD [26] which is one of the classical progression planning systems based on heuristic search. It has been extended by Steinmetz et al. [36] for goal probability analysis, i. e., computing the maximal probability to reach a goal. That extension also encompasses several heuristic search algorithms like LRTDP with FRET-π.

The original LRTDP work by Bonet et al. [8] is tailored to $\mathcal{SSP}$ assuming conditions (i′) and (ii′), the second version of Bertsekas [7], with strictly positive action rewards (except at goal states). Kolobov et al. [30,31] instead uses (i) and (ii″) when discussing $\mathcal{GSSP}$ problems. They showed that several MDP problems, including *MaxProb*, can be reduced to this problem class [30,31] and that the respective properties can be solved using FRET with LRTDP. We do not need to assume any of these, but restrict to non-negative reward structures.

Probabilistic Model Checking. This is not the first work to explore probabilistic planning and heuristic search approaches for probabilistic model checking. For instance, heuristic search dynamic programming methods have been applied to MDPs, but for generating probabilistic counterexamples [1]. Closer to our work, Kretinsky et al. developed heuristics for initializing policies in policy iteration such that the computation time to solve long-run average reward properties on MDPs is reduced [32], with specific treatments of SSCs and maximal end components similar to the approach of MODYSH. The PAC tool [3] uses asynchronous bounded value iteration techniques interleaved with guided simulation phases with permanent and transient trap elimination for statistical model checking for reachability analysis on stochastic games. A combination of Bounded Real-Time Dynamic Programming (BRTDP) and Monte Carlo Tree Search has been devised with objectives similar to ours [2]. Technical differences aside, this approach has only been applied to solve *MaxProb* properties.

Machine learning techniques have been exploited [10] to verify reachability properties on MDPs using (1) BRTDP and (2) delayed Q-learning for MDPs with limited information. The techniques are also applicable to arbitrary MDP structures due to special treatments of end components and are implemented in PET (aka. PRISM-TUM), which is part of our evaluation in Sect. 4. In parts, the approach is close to ours for simple reachability properties, but restricted to that, and uses BRTDP instead of FRET-LRTDP. The paper explicitly mentions that so far no attempts have been made to adapt these methods in the context of probabilistic verification. With MODYSH we completely fill this gap.

As became clear in our empirical evaluation, heuristic search can be especially attractive for handling excessively large models. An entirely different approach to attack such problems is the use of external storage to slowly but exhaustively model check problem sizes that otherwise do not fit in memory [23].

6 Conclusion

We introduced a heuristic approach to probabilistic model checking all established property types, except long-run averages, on general MDP structures based on LRTDP combined with FRET. The approach is implemented in MODYSH. We reported on a large empirical evaluation that has demonstrated the competitiveness of MODYSH relative to other state-of-the-art model checking tools. On very large state spaces our tool outperforms its competitors, demonstrating that planning techniques can indeed be used to enhance the performance and capabilities of model checkers.

As a next step we are looking into performance optimizations by exploring the trade-offs between memory usage and runtime. In addition, other heuristics known to work well in the planning community might be worth to implement. Extending the approach to work on other automata types seems also promising.

A Proof for *MinProb*

As announced in Sect. 3.1, this appendix provides a proof that GLRTDP solves *MinProb* properties on general MDP structures correctly by converging to the optimal fixpoint.
To show convergence to the optimal value function from below in case of an admissible initialization, we can argue along the invariant

$$\forall k, \sigma : V_k(s) \leq P_s^\sigma(\diamond G), \text{ where } \sigma \text{ s.t. } P_s^\sigma(\diamond G) = V^*(s)$$

stating that the value function in every iteration is always at most the value under the optimal policy. This means that an initially admissible value function always stays admissible. This is true for the admissible initialization when $k = 0$, because then $V_0(s) = 1$ if $s \in G$ and 0 otherwise. For all other iterations it holds that $V_{k+1}(s) := \sum_{s'} P(s, a, s') \cdot V_k(s')$ for some action a and we can derive that

$$\sum_{s'} P(s, a, s') \cdot V_k(s') \leq \sum_{s'} P(s, a, s') \cdot \min_\sigma P_{s'}^\sigma(\diamond G)$$
$$\leq \sum_{s'} \min_\sigma(P(s, a, s') \cdot P_{s'}^\sigma(\diamond G)) \leq \min_\sigma \sum_{s'} P(s, a, s') \cdot P_{s'}^\sigma(\diamond G).$$

The second inequality holds because σ_{opt} is memoryless and independent of s'. Now assume σ_{opt} is such that $P_s^{\sigma_{opt}}(\diamond G)$ is minimal for all s. Then for action $a = greedy(s, V_k)$ we have for any action b, and in particular for $b = \sigma_{opt}(s)$,

$$\sum_{s'} P(s, a, s') \cdot V_k(s') \leq \sum_{s'} P(s, b, s') \cdot V_k(s').$$

Moreover $V_k(s') \leq P_{s'}^{\sigma_{opt}}(\diamond G)$, which allows us to derive

$$\sum_{s'} P(s, a, s') \cdot V_k(s') \leq \sum_{s'} P(s, \sigma_{opt}(s), s') \cdot P_{s'}^{\sigma_{opt}}(\diamond G) = P_s^{\sigma_{opt}}(\diamond G).$$

Claim: If V_k is a fixpoint for $k \to \infty$ then $P^\sigma(\diamond G) = V_\infty(s_0) \forall \sigma$ greedy in V. (5)
Since $V^*(s) := \min_\sigma P_s^\sigma(\diamond G)$ this means $V^*(s_0) \leq V_\infty(s_0)$ and with the result from above ($\forall k : V_k \leq V^*$) we can conclude $V^*(s_0) = V_\infty(s_0)$.
It remains to show that (5) holds: Let $\sigma_k := greedy(V_k)$, i. e., a greedy policy with respect to the value function V_k and $S_k = \{s | P_{s_0}^{\sigma_k}(\diamond s) > 0\}$, i. e., all states reachable with this greedy policy, then $\max(residual(S_k)) \leq \delta_k$ and for $k \to \infty$ it holds that $\delta_k \to 0$.
To show that δ_k will approach 0 it is enough to argue about the states which will be updated an infinite number of times, i. e., in the end, about the states on optimal policies. These are the states in $S_\infty = \bigcap_{i \geq 0} \bigcup_{k \geq i} S_k$.
Let K be such that $\forall k \geq K : \bigcup_{i \geq k} S_i = S_\infty$, i. e. a step from which on we only consider states which will be infinitely often visited when running GLRTDP infinitely long. Assume we are in a step $j + 1 \geq K$. Let $s \in S_\infty$. We have to distinguish two cases:

- If s has not been updated then $V_{j+1}(s) = V_j(s)$.
- If s is the updated state then $V_{j+1}(s) = \min_\alpha \sum_{s'} P(s, \alpha, s') \cdot V_j(s')$

But this is the same as for simple synchronous value iteration, for which convergence against the optimal fixpoint is proven. For our asynchronous case in GLRTDP we nevertheless have to guarantee fairness among the states in S_∞, i. e., we have to make sure that they are updated infinitely often. This is the case because each possible trial of S_∞ (there are finitely many trials) appears infinitely often, i. e. the states in this trial are updated infinitely often (by construction of GLRTDP when choosing the next greedy action). All other states not in S_∞ can be ignored because they will not influence the greedy policy and optimal values because they are already too large:

For any $s \in S \setminus S_\infty$ it holds that $V_\infty(s) = V_K(s) \leq V^*(s)$ and for any $s \in S_\infty$ by definition of S_∞ and K we know that an action leading again to a state in S_∞ will be chosen, i. e., an $a \in \sigma_\infty$: $V_\infty(s) \leq \sum_{s' \in S_\infty} P(s, \sigma_\infty, s') \cdot V_\infty(s')$ but for every action we choose the greedy one and for any $k \geq K$ it holds that $V_k(s) \leq \sum_{s' \in S} P(s, a, s') \cdot V_k(s') \leq V_\infty(s)$, i. e., the action in σ_∞ must have been the greedy action not leading to $S \setminus S_\infty$. This means that V_∞ defines an optimal strategy on S_∞ for $s_0 \in S_\infty$ which is also an optimal strategy on S because no state $s' \in S \setminus S_\infty$ is visited even with $V_\infty(s') < V^*(s')$. In addition the initial state lies in S_∞ by construction, i. e., $P_{\min}(\diamond G) = V^{\sigma_{opt}}(s_0)$ reaches the fixpoint and is updated infinitely often.

In summary, when running GLRTDP in an infinite number of iterations, the value function for states in S_∞ will approach the optimal values of the minimal probability to reach the goal from below, will never get larger than the optimal value and the difference between V and V^* always becomes strictly smaller for these states. In addition, we can at some point stop updating the value function for parts of the state space because these values will not have an influence on the correct optimal result for the initial state. In our implementation GLRTDP is designed in such a way that it stops when the values on the optimal policy only change by less than ε, which is the same convergence criterion as for simple value iteration.

B Proof for *MaxProb*

Taking up our promise from Sect. 3.1, in the following we will first give an intuition about why the presented combination of GLRTDP and FRET solves *MaxProb* properties on general MDP structures correctly, not only on problems having at least one almost-sure policy as proven in [31], by converging to the optimal fixpoint. Afterwards we sketch a more formal proof.

All greedy policies inspected by GLRTDP at some point end in a goal state or a dead-end state. This could be a real dead-end, i. e., a sink state with only a self-loop or a permanent trap which has been transformed to a dead-end by the cycle elimination of FRET. If it is a permanent trap identified by FRET, the values of all states in it are set to 0. Otherwise, when the sink state is discovered for the first time its value is also directly set to 0. This means we tag these states, do not explore them further and propagate their value back through the graph. Cycling forever is not possible because FRET eventually eliminates all such cycles in greedy policies. With this, we can state that at some point no more states are left to explore in the current GLRTDP trial because all relevant traps are eliminated or a goal or a sink has been found. Then GLRTDP runs until the state values of the current greedy policies are converged up to ε. Even if the greedy policy is not the same in every iteration, at some point it will stay within a set of greedy states which are part of finitely many greedy policies. The values of these states will have converged close enough to the optimal ones such that the algorithm concentrates on these optimal policies. The value function used in GLRTDP is initialized admissibly and therefore can only monotonically decrease and approach the optimal fixpoint from above. When this point is reached (up to ε), the entire procedure (GLRTDP + FRET) terminates. This fixpoint must be the optimal one because the Bellman equation only admits a single fixpoint [7].

To show convergence to the optimal value function from above in case of an admissible initialization, we can argue along the invariant

$$\forall k, \sigma : V_k(s) \geq P_s^\sigma(\diamond G), \text{ where } \sigma \text{ s.t. } P_s^\sigma(\diamond G) = V^*(s)$$

stating that the value function in every iteration is always greater or equal than the optimal value under the optimal policy. This means that an initially admissible value function always stays admissible. This is true for the admissible initialization when $k = 0$, because then $V_0(s) = 0$ if $s \in \mathcal{S}_\perp$ and 1 otherwise. For all other iterations it holds that

$$V_{k+1}(s) := \sum_{s'} P(s, a, s') \cdot V_k(s')$$

for some action a and we can derive that

$$\sum_{s'} P(s, a, s') \cdot V_k(s') \geq \sum_{s'} P(s, a, s') \cdot \max_\sigma P_{s'}^\sigma(\diamond G)$$
$$\geq \sum_{s'} \max_\sigma (P(s, a, s') \cdot P_{s'}^\sigma(\diamond G)) \geq \max_\sigma \sum_{s'} P(s, a, s') \cdot P_{s'}^\sigma(\diamond G)$$

The second inequality holds because σ_{opt} is memoryless and independent of s'.

Now assume σ_{opt} is such that $P_s^{\sigma_{opt}}(\diamond G)$ is maximal for all s. Then for action $a = greedy(s, V_k)$ we have for any action b, and in particular for $b = \sigma_{opt}(s)$,

$$\sum_{s'} P(s, a, s') \cdot V_k(s') \geq \sum_{s'} P(s, b, s') \cdot V_k(s').$$

Moreover $V_k(s') \geq P_{s'}^{\sigma_{opt}}(\diamond G)$ and hence

$$\sum_{s'} P(s, a, s') \cdot V_k(s') \geq \sum_{s'} P(s, \sigma_{opt}(s), s') \cdot P_{s'}^{\sigma_{opt}}(\diamond G) = P_s^{\sigma_{opt}}(\diamond G).$$

Claim: If V_k is a fixpoint for $k \to \infty$ then $P^\sigma(\diamond G) = V_\infty(s_0)$ $\forall \sigma$ greedy in V. (6) Since $V^*(s) := \max_\sigma P_s^\sigma(\diamond G)$ this means $V^*(s_0) \geq V_\infty(s_0)$ and with the result from above ($\forall k : V_k \geq V^*$) we can conclude $V^*(s_0) = V_\infty(s_0)$.

It remains to show that (6) holds: Let $\sigma_k := greedy(V_k)$, i. e., a greedy policy with respect to the value function V_k and $S_k = \{s|P_{s_0}^{\sigma_k}(\diamond s) > 0\}$, i. e., all states reachable with this greedy policy, then $\max(residual(S_k)) \leq \delta_k$ and for $k \rightarrow \infty$ it holds that $\delta_k \rightarrow 0$.
To show that δ_k will approach 0 it is enough to argue about the states which will be updated an infinite number of times, i. e., in the end, about the states on optimal policies. These are the states in $S_\infty = \bigcap_{i \geq 0} \bigcup_{k \geq i} S_k$.
Let K be such that $\forall k \geq K : \bigcup_{i \geq k} S_i = S_\infty$, i. e. a step from which on we only consider states which will be infinitely often visited when running FRET-LRTDP infinitely long. Assume we are in a step $j + 1 \geq K$. Let $s \in S_\infty$. We have to distinguish two cases:

- If s has not been updated then $V_{j+1}(s) = V_j(s)$.
- If s is the updated state then $V_{j+1}(s) = \max_\alpha \sum_{s'} P(s, \alpha, s') \cdot V_j(s')$

This is the same as for simple synchronous value iteration, for which convergence against the optimal fixpoint is proven. For our asynchronous case in GLRTDP we are left with the duty to guarantee fairness among the states in S_∞, i. e., we have to make sure that they are updated infinitely often. This is the case because each possible trial of S_∞ (there are finitely many trials) appears infinitely often, i. e., the states in this trial are updated infinitely often (by construction of GLRTDP when choosing the next greedy action). All other states not in S_∞ can be ignored because they will not influence the greedy policy and optimal values because they are already too large:

For any $s \in S \setminus S_\infty$ it holds that $V_\infty(s) = V_K(s) \geq V^*(s)$ and for any $s \in S_\infty$ by definition of S_∞ and K we know that an action leading again to a state in S_∞ will be chosen, i. e., an $a \in \sigma_\infty$: $V_\infty(s) \geq \sum_{s' \in S_\infty} P(s, \sigma_\infty, s') \cdot V_\infty(s')$ but for every action we choose the greedy one and for any $k \geq K$ it holds that $V_k(s) \geq \sum_{s' \in S} P(s, a, s') \cdot V_k(s') \geq V_\infty(s)$, i. e., the action in σ_∞ must have been the greedy action not leading to $S \setminus S_\infty$.

This means that V_∞ defines an optimal strategy on S_∞ for $s_0 \in S_\infty$ which is also an optimal strategy on S because no state $s' \in S \setminus S_\infty$ is visited even with $V_\infty(s') > V^*(s')$.

In addition the initial state lies in S_∞ by construction, i. e., $P_{\max}(\diamond G) = V^{\sigma_{opt}}(s_0)$ reaches the fixpoint and is updated infinitely often.

Altogether, this shows that when running FRET-LRTDP over an infinite number of iterations, the value function for states in S_∞ will approach the optimal values of the maximal probability to reach the goal from above, will never get smaller than the optimal value and the difference between V and V^* always becomes strictly smaller for these states. In addition, we can at some point stop updating the value function for parts of the state space because these values will not have an influence on the correct optimal result for the initial state. In our implementation FRET-LRTDP is designed in such a way that it stops when the values on the optimal policy only change by less than ε, which is the same convergence criterion as for simple value iteration.

References

1. Aljazzar, H., Leue, S.: Generation of counterexamples for model checking of Markov decision processes. In: QEST 2009, Sixth International Conference on the Quantitative Evaluation of Systems, Budapest, Hungary, 13-16 September 2009. pp. 197–206. IEEE Computer Society (2009). https://doi.org/10.1109/QEST.2009.10, https://ieeexplore.ieee.org/xpl/conhome/5290656/proceeding
2. Ashok, P., Brázdil, T., Kretínský, J., Slámecka, O.: Monte carlo tree search for verifying reachability in Markov decision processes. In: Margaria, T., Steffen, B. (eds.) Leveraging Applications of Formal Methods, Verification and Validation. Verification - 8th International Symposium, ISoLA 2018, Limassol, Cyprus, November 5-9, 2018, Proceedings, Part II. Lecture Notes in Computer Science, vol. 11245, pp. 322–335. Springer (2018). https://doi.org/10.1007/978-3-030-03421-4_21
3. Ashok, P., Kretínský, J., Weininger, M.: PAC statistical model checking for Markov decision processes and stochastic games. In: Dillig and Tasiran [15], pp. 497–519. https://doi.org/10.1007/978-3-030-25540-4_29
4. Barto, A.G., Bradtke, S.J., Singh, S.P.: Learning to act using real-time dynamic programming. Artif. Intell. **72**(1-2), 81–138 (1995). https://doi.org/10.1016/0004-3702(94)00011-O
5. Bellman, R.: A Markovian decision process. Journal of mathematics and mechanics **6**(5), 679–684 (1957)
6. Bertsekas, D.P.: Dynamic Programming and Optimal Control, Vol. 1. Athena Scientific (1995)
7. Bertsekas, D.P.: Dynamic Programming and Optimal Control, Vol. 2. Athena Scientific (1995)
8. Bonet, B., Geffner, H.: Labeled RTDP: improving the convergence of real-time dynamic programming. In: Giunchiglia, E., Muscettola, N., Nau, D.S. (eds.) Proceedings of the Thirteenth International Conference on Automated Planning and Scheduling (ICAPS 2003), June 9-13, 2003, Trento, Italy. pp. 12–21. AAAI (2003), http://www.aaai.org/Library/ICAPS/2003/icaps03-002.php
9. Bonet, B., Geffner, H.: Learning depth-first search: A unified approach to heuristic search in deterministic and non-deterministic settings, and its application to MDPs. In: Long, D., Smith, S.F., Borrajo, D., McCluskey, L. (eds.) Proceedings of the Sixteenth International Conference on Automated Planning and Scheduling, ICAPS 2006, Cumbria, UK, June 6-10, 2006. pp. 142–151. AAAI (2006), http://www.aaai.org/Library/ICAPS/2006/icaps06-015.php
10. Brázdil, T., Chatterjee, K., Chmelik, M., Forejt, V., Kretínský, J., Kwiatkowska, M.Z., Parker, D., Ujma, M.: Verification of Markov decision processes using learning algorithms. In: Cassez, F., Raskin, J. (eds.) Automated Technology for Verification and Analysis - 12th International Symposium, ATVA 2014, Sydney, NSW, Australia, November 3-7, 2014, Proceedings. Lecture Notes in Computer Science, vol. 8837, pp. 98–114. Springer (2014). https://doi.org/10.1007/978-3-319-11936-6_8, https://doi.org/10.1007/978-3-319-11936-6
11. Budde, C.E., D'Argenio, P.R., Hartmanns, A., Sedwards, S.: An efficient statistical model checker for nondeterminism and rare events. Int. J. Softw. Tools Technol. Transf. **22**(6), 759–780 (2020). https://doi.org/10.1007/s10009-020-00563-2
12. Budde, C.E., Dehnert, C., Hahn, E.M., Hartmanns, A., Junges, S., Turrini, A.: JANI: Quantitative model and tool interaction. In: Legay, A., Margaria, T. (eds.) Tools and Algorithms for the Construction and Analysis of Systems - 23rd International Conference, TACAS 2017, Held as Part of the European Joint Conferences

on Theory and Practice of Software, ETAPS 2017, Uppsala, Sweden, April 22-29, 2017, Proceedings, Part II. Lecture Notes in Computer Science, vol. 10206, pp. 151–168 (2017). https://doi.org/10.1007/978-3-662-54580-5_9
13. Budde, C.E., Hartmanns, A., Klauck, M., Kretinsky, J., Parker, D., Quatmann, T., Turrini, A., Zhang, Z.: On Correctness, Precision, and Performance in Quantitative Verification (QComp 2020 Competition Report). In: Proceedings of the 9th International Symposium On Leveraging Applications of Formal Methods, Verification and Validation. Software Verification Tools (2020). https://doi.org/10.1007/978-3-030-83723-5_15
14. Dehnert, C., Junges, S., Katoen, J., Volk, M.: A storm is coming: A modern probabilistic model checker. In: Majumdar, R., Kuncak, V. (eds.) Computer Aided Verification - 29th International Conference, CAV 2017, Heidelberg, Germany, July 24-28, 2017, Proceedings, Part II. Lecture Notes in Computer Science, vol. 10427, pp. 592–600. Springer (2017). https://doi.org/10.1007/978-3-319-63390-9_31
15. Dillig, I., Tasiran, S. (eds.): Computer Aided Verification - 31st International Conference, CAV 2019, New York City, NY, USA, July 15-18, 2019, Proceedings, Part I, Lecture Notes in Computer Science, vol. 11561. Springer (2019). https://doi.org/10.1007/978-3-030-25540-4
16. Hahn, E.M., Hartmanns, A.: A comparison of time- and reward-bounded probabilistic model checking techniques. In: Fränzle, M., Kapur, D., Zhan, N. (eds.) Dependable Software Engineering: Theories, Tools, and Applications - Second International Symposium, SETTA 2016, Beijing, China, November 9-11, 2016, Proceedings. Lecture Notes in Computer Science, vol. 9984, pp. 85–100 (2016). https://doi.org/10.1007/978-3-319-47677-3_6
17. Hahn, E.M., Hartmanns, A.: Efficient algorithms for time- and cost-bounded probabilistic model checking. CoRR **abs/1605.05551** (2016), http://arxiv.org/abs/1605.05551
18. Hahn, E.M., Hartmanns, A., Hensel, C., Klauck, M., Klein, J., Kretínský, J., Parker, D., Quatmann, T., Ruijters, E., Steinmetz, M.: The 2019 comparison of tools for the analysis of quantitative formal models - (QComp 2019 competition report). In: Beyer, D., Huisman, M., Kordon, F., Steffen, B. (eds.) Tools and Algorithms for the Construction and Analysis of Systems - 25 Years of TACAS: TOOLympics, Held as Part of ETAPS 2019, Prague, Czech Republic, April 6-11, 2019, Proceedings, Part III. Lecture Notes in Computer Science, vol. 11429, pp. 69–92. Springer (2019). https://doi.org/10.1007/978-3-030-17502-3_5
19. Hahn, E.M., Hartmanns, A., Hermanns, H.: Reachability and reward checking for stochastic timed automata. Electron. Commun. Eur. Assoc. Softw. Sci. Technol. **70** (2014). https://doi.org/10.14279/tuj.eceasst.70.968
20. Hahn, E.M., Li, Y., Schewe, S., Turrini, A., Zhang, L.: iscasMc: A web-based probabilistic model checker. In: Jones, C.B., Pihlajasaari, P., Sun, J. (eds.) FM 2014: Formal Methods - 19th International Symposium, Singapore, May 12-16, 2014. Proceedings. Lecture Notes in Computer Science, vol. 8442, pp. 312–317. Springer (2014). https://doi.org/10.1007/978-3-319-06410-9_22, https://doi.org/10.1007/978-3-319-06410-9
21. Hansen, E.A., Zilberstein, S.: Lao*: A heuristic search algorithm that finds solutions with loops. Artif. Intell. **129**(1-2), 35–62 (2001). https://doi.org/10.1016/S0004-3702(01)00106-0
22. Hartmanns, A., Hermanns, H.: The Modest Toolset: An integrated environment for quantitative modelling and verification. In: Ábrahám, E., Havelund, K. (eds.) Tools and Algorithms for the Construction and Analysis of Systems - 20th International

Conference, TACAS 2014, Held as Part of the European Joint Conferences on Theory and Practice of Software, ETAPS 2014, Grenoble, France, April 5-13, 2014. Proceedings. Lecture Notes in Computer Science, vol. 8413, pp. 593–598. Springer (2014). https://doi.org/10.1007/978-3-642-54862-8_51
23. Hartmanns, A., Hermanns, H.: Explicit model checking of very large MDP using partitioning and secondary storage. In: Finkbeiner, B., Pu, G., Zhang, L. (eds.) Automated Technology for Verification and Analysis - 13th International Symposium, ATVA 2015, Shanghai, China, October 12-15, 2015, Proceedings. Lecture Notes in Computer Science, vol. 9364, pp. 131–147. Springer (2015). https://doi.org/10.1007/978-3-319-24953-7_10
24. Hartmanns, A., Klauck, M., Parker, D., Quatmann, T., Ruijters, E.: The Quantitative Verification Benchmark Set. In: Vojnar, T., Zhang, L. (eds.) Tools and Algorithms for the Construction and Analysis of Systems - 25th International Conference, TACAS 2019, Held as Part of the European Joint Conferences on Theory and Practice of Software, ETAPS 2019, Prague, Czech Republic, April 6-11, 2019, Proceedings, Part I. Lecture Notes in Computer Science, vol. 11427, pp. 344–350. Springer (2019). https://doi.org/10.1007/978-3-030-17462-0_20
25. Hatefi-Ardakani, H.: Finite horizon analysis of Markov automata. Ph.D. thesis, Saarland University, Germany (2017), http://scidok.sulb.uni-saarland.de/volltexte/2017/6743/
26. Helmert, M.: The fast downward planning system. CoRR **abs/1109.6051** (2011), http://arxiv.org/abs/1109.6051
27. Izadi, M.T.: Sequential decision making under uncertainty. In: Zucker, J., Saitta, L. (eds.) Abstraction, Reformulation and Approximation, 6th International Symposium, SARA 2005, Airth Castle, Scotland, UK, July 26-29, 2005, Proceedings. Lecture Notes in Computer Science, vol. 3607, pp. 360–361. Springer (2005). https://doi.org/10.1007/11527862_33
28. The JANI specification. http://www.jani-spec.org/, accessed on 25/06/2021
29. Klauck, M., Hermanns, H.: Artifact accompanying the paper "A Modest Approach to Dynamic Heuristic Search in Probabilistic Model Checking" (2021), available at http://doi.org/10.5281/zenodo.4922360
30. Kolobov, A.: Scalable methods and expressive models for planning under uncertainty. Ph.D. thesis, University of Washington (2013)
31. Kolobov, A., Mausam, Weld, D.S., Geffner, H.: Heuristic search for generalized stochastic shortest path mdps. In: Bacchus, F., Domshlak, C., Edelkamp, S., Helmert, M. (eds.) Proceedings of the 21st International Conference on Automated Planning and Scheduling, ICAPS 2011, Freiburg, Germany June 11-16, 2011. AAAI (2011), http://aaai.org/ocs/index.php/ICAPS/ICAPS11/paper/view/2682
32. Kretínský, J., Meggendorfer, T.: Efficient strategy iteration for mean payoff in Markov decision processes. In: D'Souza, D., Kumar, K.N. (eds.) Automated Technology for Verification and Analysis - 15th International Symposium, ATVA 2017, Pune, India, October 3-6, 2017, Proceedings. Lecture Notes in Computer Science, vol. 10482, pp. 380–399. Springer (2017). https://doi.org/10.1007/978-3-319-68167-2_25, https://doi.org/10.1007/978-3-319-68167-2
33. Kwiatkowska, M.Z., Norman, G., Parker, D.: PRISM 4.0: Verification of probabilistic real-time systems. In: Gopalakrishnan, G., Qadeer, S. (eds.) Computer Aided Verification - 23rd International Conference, CAV 2011, Snowbird, UT, USA, July 14-20, 2011. Proceedings. Lecture Notes in Computer Science, vol. 6806, pp. 585–591. Springer (2011). https://doi.org/10.1007/978-3-642-22110-1_47, https://doi.org/10.1007/978-3-642-22110-1

34. Neupane, T., Myers, C.J., Madsen, C., Zheng, H., Zhang, Z.: STAMINA: stochastic approximate model-checker for infinite-state analysis. In: Dillig and Tasiran [15], pp. 540–549. https://doi.org/10.1007/978-3-030-25540-4_31
35. Ruijters, E., Reijsbergen, D., de Boer, P., Stoelinga, M.: Rare event simulation for dynamic fault trees. Reliab. Eng. Syst. Saf. **186**, 220–231 (2019). https://doi.org/10.1016/j.ress.2019.02.004
36. Steinmetz, M., Hoffmann, J., Buffet, O.: Goal probability analysis in probabilistic planning: Exploring and enhancing the state of the art. J. Artif. Intell. Res. **57**, 229–271 (2016). https://doi.org/10.1613/jair.5153
37. Tarjan, R.E.: Depth-first search and linear graph algorithms. SIAM J. Comput. **1**(2), 146–160 (1972). https://doi.org/10.1137/0201010

Tweaking the Odds in Probabilistic Timed Automata

Arnd Hartmanns[1(✉)], Joost-Pieter Katoen[1,2], Bram Kohlen[1], and Jip Spel[2]

[1] University of Twente, Enschede, The Netherlands
a.hartmanns@utwente.nl
[2] RWTH Aachen University, Aachen, Germany

QEST Repeatability and Artifact Evaluation 2021 Accepted

Abstract. We consider probabilistic timed automata (PTA) in which probabilities can be parameters, i.e. symbolic constants. They are useful to model randomised real-time systems where exact probabilities are unknown, or where the probability values should be optimised. We prove that existing techniques to transform probabilistic timed automata into equivalent finite-state Markov decision processes (MDPs) remain correct in the parametric setting, using a systematic proof pattern. We implemented two of these parameter-preserving transformations—using digital clocks and backwards reachability—in the MODEST TOOLSET. Using STORM's parameter space partitioning approach, parameter values can be efficiently synthesized in the resulting parametric MDPs. We use several case studies from the literature of varying state and parameter space sizes to experimentally evaluate the performance and scalability of this novel analysis trajectory for parametric PTA.

1 Introduction

Probabilistic timed automata (PTA) [25,45] combine the features of timed automata (TA) [2], to capture hard continuous real-time behaviour with nondeterministic time and choices, with those of Markov decision processes (MDP) [50], to model discrete random decisions. PTA are well-equipped for the study of randomised algorithms interacting with an environment where actions with uncertain outcomes complete after (upper- and lower-bounded) delays. They have been fruitfully applied to verify performance and reliability aspects of communication protocols and networked systems, see e.g. [22–24,39].

Building a PTA model requires knowledge of the precise probabilities of all random events. While unproblematic for a randomised algorithm such as the binary exponential backoff procedure in a CSMA/CA wireless network by itself, uncertainty about the operating environment often means that we do not know, say, the precise probability p of message loss once we decide to send. In such cases, we may turn the verification question around: Instead of computing whether the probability of an eventual successful transmission is above the required threshold

This work was funded by DFG RTG 2236 "UnRAVeL", DFG grant 433044889 PASIWY, NWO grant OCENW.KLEIN.311, and NWO VENI grant 639.021.754.

A. Abate and A. Marin (Eds.): QEST 2021, LNCS 12846, pp. 39–58, 2021.
https://doi.org/10.1007/978-3-030-85172-9_3

given a concrete p, we determine the set of values of p for which the requirement is satisfied. We can then judge whether the network's setup or protocols are sufficiently robust for the intended environments.

In this paper, we focus on *parametric PTA* (pPTA), where the probabilities of some events are unknown and specified as polynomials over a finite set of parameters like p above. We consider the analysis of (time-bounded) probabilistic reachability properties, i.e. statements of the form "is the maximum/minimum probability of eventually/within t time units reaching a goal location above/below $v \in (0,1)$?", but instead of a yes/no answer we want to compute an (approximation of) the set of parameter valuations under which the property is satisfied.

Several approaches have been developed over the past two decades to verify PTA with known probabilities. Most approaches compute a finite abstraction of the continuous-time behaviour, turning the PTA into an equivalent MDP on which the property of interest can be verified using standard MDP model checking [10,11]. This includes using the region graph [45], backwards reachability [46], and digital clocks [34,44]. The latter two are implemented in the PRISM [43] tool while the MODEST TOOLSET's [28] MCSTA model checker uses digital clocks. The stochastic games approach [42], implemented in PRISM, employs games in place of MDP to iteratively refine the abstraction up to the desired precision. UPPAAL SMC [21] applies statistical model checking (SMC) [1] to possibly non-deterministic PTA by interpreting nondeterminism probabilistically. It thus delivers *some* probability between minimum and maximum. Using extensions of lightweight scheduler sampling [47] to PTA [20,31], the MODEST TOOLSET's MODES simulator [15] can deliver upper (lower) bounds on min. (max.) via SMC.

Our contribution is to **extend the reach of model checking to pPTA**: we lift the PTA-to-MDP abstraction techniques to pPTA, then apply *parameter space partitioning* [14,51] on the resulting MDP to approximate the set of satisfying valuations. We **prove** that the abstractions remain correct in the parametric setting (Sect. 3), provide an **implement**ation using digital clocks and backwards reachability in the MODEST TOOLSET followed by parameter space partitioning in STORM [33] (Sect. 4), and use it to experimentally **evaluate** the performance and scalability of our approach (Sect. 5). For the evaluation, we extend all suitable PTA from the Quantitative Verification Benchmark Set (QVBS) [30] with parameters, and additionally study pPTA models of the AODV wireless routing protocol [39]. The latter solves a critical open problem the previous study of the protocol, which had to resort to "testing" by model checking a finite set of selected values for probabilities that are actually unknown.

Related Work. [41] provide a tool to model-check interval PTA. An interval PTA is a special case of pPTA in which no parameter occurs at multiple states. We are not aware of any other work tackling the problem of model-checking pPTA. Instead of parametrising the *probabilities*, however, one may parametrise the *delays* [3]. Where the guard of an edge in a TA with clock c may be the clock constraint $c \geq 3$, it may be $c \geq 2 \cdot p$ in such a *constraint-parametric* TA (cpTA) with p a parameter. For this model, even basic problems like the existence of a parameter valuation under which a goal state is reachable are undecidable [3,4], except in restricted cases such as bounded integer values [37] or L/U TA [35]. Research on cpTA remains active to this day; recent work, for example, proposes a semi-algorithm for liveness [5]. The *inverse method* for cpTA [6] extends one parameter valuation v_0 to a set of valuations $V \supseteq \{ v_0 \}$ such that all $v \in V$ satisfy the same given reachability properties. In its adaptation to cpPTA [8], all $v \in V$ must result in the same reachability probabilities. Parametric interval probabilistic timed automata [7] combine cpPTA with interval Markov chains [36], i.e. compared to cpPTA, the concrete probabilities are replaced by *intervals* of possible probabilities. In contrast to parametric probabilities, intervals cannot express dependencies between the probabilities of different events (such as one event being twice as likely as another). For this model, research currently remains focused on the question of consistency [7].

2 Preliminaries

We now introduce TA, PTA, and pPTA in order, highlighting the differences.

2.1 Timed Automata

A timed automaton [2] is a labelled transition system (LTS) [12] extended with a finite set $\mathscr{X}$ of *clocks* taking non-negative real values and all increasing at rate 1 over time. A *clock valuation* is a function $v \colon \mathscr{X} \to \mathbb{R}_{\geq 0}$. We denote the set of all clock valuations by $\mathbb{R}_{\geq 0}^{\mathscr{X}}$. For $v \in \mathbb{R}_{\geq 0}^{\mathscr{X}}$ and $t \in \mathbb{R}_{\geq 0}$, the valuation $v + t$ is defined by $(v + t)(x) = v(x) + t$ for all $x \in \mathscr{X}$. For $X \subseteq \mathscr{X}$, $v[X := 0]$ is the clock valuation where $v[X := 0](x) = 0$ if $x \in X$ and $v[X := 0](x) = v(x)$ otherwise. Finally, $\mathbf{0}$ is the zero valuation, i.e. $\mathbf{0}(x) = 0$ for all $x \in \mathscr{X}$. The set $CC(\mathscr{X})$ of *clock constraints* over $\mathscr{X}$ contains all expressions defined by

$$\mathcal{X} ::= x < c \mid x \leq c \mid x > c \mid x \geq c \mid \mathcal{X} \wedge \mathcal{X}$$

where $x \in \mathscr{X}$ and $c \in \mathbb{N}$. To keep the presentation simple, w.l.o.g. we omit diagonal and disjunctive clock constraints. Valuation v satisfies clock constraint $\mathcal{X}$, denoted $v \models \mathcal{X}$, iff the expression $\mathcal{X}$ evaluates to true after replacing each $x \in \mathscr{X}$ with $v(x)$. The semantics of clock constraint $\mathcal{X}$ is the *zone* $\zeta_{\mathcal{X}} := \{v \in \mathbb{R}_{\geq 0}^{\mathscr{X}} \mid v \models \mathcal{X}\}$. Let $Zones(\mathscr{X})$ denote the set of zones over the constraints $\mathscr{X}$.

Definition 1. *A* **timed automaton** *(TA) is a tuple* $\mathcal{B} = (Loc, \ell_0, Act, \mathscr{X}, \hookrightarrow, inv)$ *where Loc, Act and* $\mathscr{X}$ *are finite sets of* locations, actions, *and* clocks, *respectively, with initial location* $\ell_0 \in Loc$, *the transition relation is*

$$\hookrightarrow \subseteq Loc \times CC(\mathscr{X}) \times Act \times 2^{\mathscr{X}} \times Loc,$$

and $inv \colon Loc \to CC(\mathscr{X})$ *assigns an* invariant *to each location.*

An edge from ℓ to ℓ' is a tuple $(\ell, g, a, X, \ell') \in \hookrightarrow$ where guard g must be satisfied in order to take the edge, and X contains the clocks to be reset. We assume that every edge is uniquely identified by its source location and action.

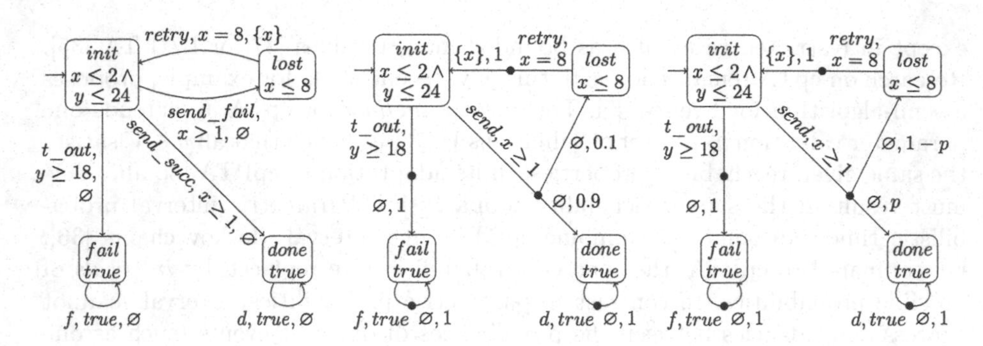

Fig. 1. TA $\mathcal{B}$ **Fig. 2.** PTA $\mathcal{A}$ **Fig. 3.** pPTA $\mathcal{P}$

Example 1. Figure 1 shows TA $\mathcal{B}$ with four locations (labelled by the location name and its invariant) and two clocks x and y. Initially, the system is in location *init*. It can remain there as long as $x \leq 2 \wedge y \leq 24$; after one time unit, the edges to *done* and *lost* become enabled. In *lost*, the system remains as long as $x \leq 8$. The edge back to *init* is enabled when $x = 8$; when taken, clock x is reset. If 18 time units passed, and we are (still or again) in *init*, the edge to *fail* is enabled.

Formally, the semantics of a TA is defined in terms of a *timed* transition system where transitions are labelled with either an action or a time duration.

Definition 2. *The* ***semantics of a TA*** $\mathcal{B}$ *is the timed transition system* $[\![\mathcal{B}]\!]_{ts} = (S, s_0, Act', \rightarrow)$ *where* $S = \{(\ell, v) \in Loc \times \mathbb{R}_{\geq 0}^{\mathcal{X}} \mid v \models inv(\ell)\}$ *is the set of states with initial state* $s_0 = (\ell_0, \mathbf{0})$*, the action labels are in* $Act' = Act \cup \mathbb{R}_{\geq 0}$*, and the transition relation* $\rightarrow$ *is the smallest relation satisfying inference rules*

$$\frac{(\ell, g, a, X, \ell') \in \hookrightarrow \quad v \models g \quad v[X := 0] \models inv(\ell')}{(\ell, v) \xrightarrow{a} (\ell', v[X := 0])} \qquad \frac{v + d \models inv(\ell) \quad d \in \mathbb{R}_{\geq 0}}{(\ell, v) \xrightarrow{d} (\ell, v + d)}$$

We refer to transitions due to the left inference rule as *jumps* with action a and to the others as *delays* of duration d. Observe that $[\![\mathcal{B}]\!]_{ts}$ has uncountably many states in general. Infinite paths of $[\![\mathcal{B}]\!]_{ts}$ are of the form $s_0 \xrightarrow{a_0} s_1 \xrightarrow{a_1} s_2 \cdots$ with the $s_i \in S$ and $a_i \in Act'$. Finite paths are defined similarly. Let $Paths_{inf}^{\mathcal{B}}$ ($Paths_{fin}^{\mathcal{B}}$) denote the set of all infinite (finite) paths of $[\![\mathcal{B}]\!]_{ts}$. W.l.o.g. we can assume that the first transition of a path is a delay and that delays and jumps alternate. $[\![\mathcal{B}]\!]_{ts}$ may contain non-divergent paths [45], i.e. paths along which infinitely many jumps happen in a finite amount of time.

Definition 3. *The path* $\pi \in Paths_{inf}^{\mathcal{B}}$ *is* ***divergent*** *if the sum of the durations of its delays—its* elapsed time*—is* ∞.

2.2 Probabilistic Timed Automata

A *probability distribution* over a countable set X is a function $\mu \colon X \rightarrow [0, 1] \subseteq \mathbb{R}$ with $\sum_{x \in X} \mu(x) = 1$. Let $Distr(X)$ denote the set of distributions on X. Probabilistic timed automata [25, 45] extend TA with probabilistic transitions.

Definition 4. *A* ***probabilistic timed automaton*** *(PTA) is a tuple* $\mathcal{A} = (Loc, \ell_0, Act, \mathscr{X}, prob, inv)$ *as in Definition 1 except that* $prob\colon Loc \times CC(\mathscr{X}) \times Act \to Distr(2^{\mathscr{X}} \times Loc)$ *is the* probabilistic transition function.

If $prob(\ell, g, a)(X, \ell') > 0$ then (ℓ, g, a, X, ℓ') is an *edge* of the PTA.

Example 2. Consider TA $\mathcal{B}$ and PTA $\mathcal{A}$ from Fig. 2. Whereas $\mathcal{B}$ has two send edges (one for successful, one for failed transmissions), $\mathcal{A}$ has one probabilistic edge, where the probability of successful transmission is 0.9.

A TA is a PTA with probability 1 for all edges. We use MDP for their semantics:

Definition 5. *A* ***Markov decision process*** *(MDP)* $\mathcal{M}$ *is a timed transition system with probabilistic transition function* $prob\colon S \times Act' \to Distr(S)$.

If $prob(s, a, \mu)$ with $\mu(s') = p$, we write $prob(s, a, s') = p$. Finite and infinite paths for MDPs are defined as for timed transition systems, with the difference that for path $s_0 \xrightarrow{a_0} s_1 \xrightarrow{a_1} s_2 \cdots$ we require that $prob(s_i, a_i, s_{i+1}) > 0$. To be able to reason about the probabilities of certain events in (timed) MDPs, we use *schedulers.* A scheduler maps finite paths to an available duration if the path has an even number of transitions and to an available action in *Act* otherwise.

Definition 6. *A (history-dependent)* ***scheduler*** *for a timed MDP* $\mathcal{M}$ *is a function* $\sigma\colon Paths_{fin}^{\mathcal{M}} \to (Act \uplus \mathbb{R}_{\geq 0})$ *where, for* $\widehat{\pi} = s_0 \xrightarrow{a_0} s_1 \ldots \xrightarrow{a_{n-1}} s_n$*, we have* $\sigma(\widehat{\pi}) \in Act$ *if* n *is odd and* $\sigma(\widehat{\pi}) \in \mathbb{R}_{\geq 0}$ *otherwise.*

Applying scheduler σ to MDP $\mathcal{M}$ yields the Markov chain $\mathcal{M}^\sigma$. For a fixed scheduler σ and state s, a probability measure $\mathsf{Pr}_s^{\mathcal{M}^\sigma}$ can be defined over the infinite paths starting in s induced by σ using the standard cylinder set construction (see e.g. [12, Ch. 10]). We restrict to *almost-sure divergent* schedulers, which are those schedulers σ where

$$\mathsf{Pr}_s^{\mathcal{M}^\sigma}\{\,\pi \in Paths_{inf}^{\mathcal{M}\sigma} \mid \pi \text{ is divergent}\,\} \;=\; 1.$$

Let $Sched_{div}(\mathcal{M})$ denote the set of almost-sure divergent schedulers of $\mathcal{M}$. The notions of (almost-sure divergent) schedulers can be lifted to PTA $\mathcal{A}$ in a straightforward manner by considering the corresponding notion on the *dense-time* semantics of a PTA, which is an uncountably large MDP:

Definition 7. *The* ***dense-time semantics of a PTA*** $\mathcal{A}$ *is the MDP*

$$[\![\mathcal{A}]\!]_{dense} = (S, s_0, Act', prob')$$

where S*,* s_0*, and* Act' *are as in Definition 2, and if* $prob(l, g, a, \cdot, \cdot)$ *is defined, then* $prob'((\ell, v), a, (\ell', v')) =$

$$\begin{cases} \displaystyle\sum_{\substack{X \subseteq \mathscr{X}:\\ v' = v[X:=0]}} prob(\ell, g, a, X, \ell') & \text{if } a \in Act \wedge v \models g \\ 1 & \text{if } a \in \mathbb{R}_{\geq 0} \wedge \ell = \ell' \wedge v' = v + a \models inv(\ell) \\ 0 & \text{otherwise.} \end{cases}$$

We assume PTA to be *well-formed*: if $prob'((\ell, v), a, (\ell', v')) > 0$ then $v' \models inv(\ell')$, for all reachable (ℓ, v). We lift the notion of a scheduler to PTA $\mathcal{A}$ using the above semantics, i.e. $Sched(\mathcal{A}) := Sched([\![\mathcal{A}]\!]_{dense})$ and $\mathcal{A}^\sigma := [\![\mathcal{A}]\!]^\sigma_{dense}$.

2.3 Parametric Probabilistic Timed Automata

Let V be a set of n real-valued *parameters* (or *variables*) $p_1, \ldots, p_n$. Let $\mathbb{Q}(V)$ denote the set of multivariate polynomials over V. We write $p \in f$ if parameter p occurs in polynomial f. A *parameter instantiation* is a function $u\colon V \to \mathbb{R}$. The *parameter space* for V is the hyper-rectangle over the lower/upper bounds for the parameter in V. A polynomial f can be interpreted as a function $f\colon \mathbb{R}^n \to \mathbb{R}$ where $f(u)$ is obtained by substitution, i.e. in $f(u)$ each occurrence of p_i in f is replaced by $u(p_i)$. To make it clear where substitution occurs, we write $f[u]$ instead of $f(u)$ from now on. Let $Distr_{\mathbb{Q}(V)}(X)$ denote the set of *parametric probability distributions* over X induced by the polynomials in $\mathbb{Q}(V)$, i.e. of functions $\mu\colon X \to \mathbb{Q}(V)$ such that $\sum_{x \in X} \mu(x)[u] = 1$ for all u within the parameter space. A *parametric* PTA is a PTA in which the transitions use parametric distributions to select the clocks to reset and the successor location.

Definition 8. *A* **parametric PTA** *(pPTA) is a tuple* $\mathcal{P} = (Loc, \ell_0, Act, \mathscr{X}, V, prob, inv)$ *where* $Loc, \ell_0, Act, \mathscr{X}$ *and* inv *are as in Definition 1,* V *is a finite set of parameters, and the parametric probabilistic transition function is*

$$prob\colon Loc \times CC(\mathscr{X}) \times Act \to Distr_{\mathbb{Q}(V)}(2^{\mathscr{X}} \times Loc).$$

Applying a parameter instantiation u to the pPTA $\mathcal{P}$ yields the PTA $\mathcal{P}[u]$ by replacing each $f \in \mathbb{Q}(V)$ in the function $prob$ of $\mathcal{P}$ by $f[u]$; we write the resulting function as $prob[u]$. Observe that $prob[u](\ell, g, a, X, \ell') = prob(\ell, g, a, X, \ell')[u]$ for all $\ell, \ell' \in Loc$, guard g, $a \in Act$, $X \subseteq \mathscr{X}$.

Example 3. Consider pPTA $\mathcal{P}$ in Fig. 3 where the probability of successful transmission is p. Applying instantiation $u = \{\, p \mapsto 0.9 \,\}$ to $\mathcal{P}$ yields PTA $\mathcal{A}$ of Fig. 2.

A *region* $R \subseteq \mathbb{R}^n$ is a fragment of the parameter space[1]. A region R is *graph-preserving* for pPTA $\mathcal{P}$ if for all $u \in R$ and polynomials f in $prob$ of $\mathcal{P}$ $f[u] > 0$, that is, none of the edges contains a probability 0. A region is thus graph-preserving if all its valuations induce the same topology. We define parametric MDPs as an extension of MDPs and use them for the semantics of pPTA.

Example 4. Consider pPTA $\mathcal{P}$ again. Region $R = [0.1, 0.9]$ is graph-preserving while $[0, 0.9]$ is not. For both, all probability distributions in $\mathcal{P}$ are well-defined.

Definition 9. *A* **parametric MDP** *(pMDP)* $\mathcal{M}$ *is an MDP over a finite set* V *of variables, and* $prob\colon S \times Act' \to Distr_{\mathbb{Q}(V)}(S)$.

[1] Parameter regions should not be confused with the regions of clock valuations as in the classic region graph construction for a (P)TA.

Definition 10. *The* ***semantics of a pPTA*** *$\mathcal{P}$ over V is the pMDP $[\![\mathcal{P}]\!]_{dense}$ over V defined analogously to Definition 7.*

The definition of $prob'$ includes sums of polynomials, which remain polynomials. Since the transition functions are equivalent, i.e. we have $prob'[u] = prob''$ for $[\![\mathcal{P}]\!]_{dense}[u] = (S, s_0, Act, V, prob'[u])$ and $[\![\mathcal{P}[u]]\!]_{dense} = (S, s_0, Act, V, prob'')$, parameter instantiation and the dense-time semantics commute.

Properties. We consider time-bounded and unbounded reachability properties on PTAs, i.e. the probability of eventually or within a time bound reaching a target set $T \subseteq Loc \times Zones(\mathscr{X})$. Specification $\varphi = \mathbb{P}_{\leq\lambda}(\Diamond T)$ asserts that the probability of eventually reaching T from the initial state $(\ell_0, \mathbf{0})$ is at most λ, where $\lambda \in \mathbb{Q} \cap [0,1]$. That is, for PTA $\mathcal{A}$,

$$\mathcal{A} \models \mathbb{P}_{\sim\lambda}(\Diamond T) \quad \text{iff} \quad \text{for all } \sigma \in Sched(\mathcal{A}) \text{ we have } \mathsf{Pr}^{\mathcal{A}^\sigma}(\Diamond T) \sim \lambda,$$

where $\mathsf{Pr}^{\mathcal{A}^\sigma}(\Diamond T)$ is the probability mass of all infinite paths in $\mathcal{A}$ that start in $(\ell_0, \mathbf{0})$ and visit some pair $(\ell, \zeta) \in T$. To support time-bounded reachability, we assume that each PTA $\mathcal{A}$ has a clock $z \in \mathscr{X}$, which does not occur in any invariant or guard of $\mathcal{A}$ and is never reset. We then check that z does not exceed bound $n \in \mathbb{N}$. We reduce this to the unbounded case and focus on that in the remainder of this paper:

$$\mathcal{A} \models \mathbb{P}_{\sim\lambda}(\Diamond^{\leq n} T) \quad \text{iff} \quad \mathcal{A} \models \mathbb{P}_{\sim\lambda}(\Diamond T^{\leq n})$$

where $T^{\leq n} := \{(\ell, \zeta \cap \zeta_{z\leq n}) \mid (\ell, \zeta) \in T\}$. The definition for strict bounds $<$ is analogous. Negation is defined by

$$\mathcal{A} \models \neg\mathbb{P}_{\sim\lambda}(\Diamond T) \quad \text{iff} \quad \mathcal{A} \models \mathbb{P}_{(\neg\sim)\lambda}(\Diamond T)$$

and similar for $\mathbb{P}_{\sim\lambda}(\Diamond^{\leq n} T)$. These notions are lifted to pPTA by considering properties relative to ranges of parameter values, i.e. regions.

Definition 11. *Given region R and pPTA $\mathcal{P}$, let*

$$\mathcal{P}, R \models \mathbb{P}_{\sim\lambda}(\Diamond T) \quad \textit{iff} \quad \forall\sigma \in Sched(\mathcal{P}[u]),\ u \in R.\ \mathsf{Pr}^{\mathcal{P}[u]^\sigma}(\Diamond T) \sim \lambda \tag{1}$$

$$\mathcal{P}, R \models \neg\mathbb{P}_{\sim\lambda}(\Diamond T) \quad \textit{iff} \quad \forall\sigma \in Sched(\mathcal{P}[u]),\ u \in R.\ \mathsf{Pr}^{\mathcal{P}[u]^\sigma}(\Diamond T)\,(\neg\sim)\,\lambda. \tag{2}$$

We call a region accepting (rejecting), denoted R_a (R_r), if 1 (2) holds. If a region is neither accepting nor rejecting, it is called inconsistent (denoted R_i).

Example 5. Reconsider $\mathcal{P}$ from Fig. 3. Let $\varphi = \mathbb{P}_{\geq 0.75}(\Diamond\{(done, \zeta_{true})\})$. Region $R_r := [0.2, 0.4]$ is rejecting; $R_a := [0.6, 0.8]$ is accepting, and $R_i := [0.4, 0.6]$ is inconsistent, as property φ is violated for $p = 0.4$ but satisfied for $p = 0.6$.

Furthermore, we define the minimal probability in pPTA $\mathcal{P}$ of eventually reaching a state in T on region R as follows:

$$\mathsf{Pr}^{\mathcal{P},R}_{min}(\Diamond T) = \min\{\mathsf{Pr}^{\mathcal{P}[u]^\sigma}(\Diamond T) \mid \forall u \in R, \sigma \in Sched(\mathcal{P})\}.$$

The maximal probability is defined analogously. The definition can be applied to PTA $\mathcal{A}$ with any region, as there are no parameters; we then omit R: $\mathsf{Pr}^{\mathcal{A}}_{min}(\Diamond T)$.

2.4 Problem Statement

This paper focuses on parameter space partitioning [14,51] for pPTA. The key idea is to partition a graph-preserving parameter space into *accepting* and *rejecting* regions w.r.t. a property φ. As obtaining a complete partitioning is practically infeasible, the aim is to cover at least $c\,\%$ of the parameter space.

> *Approximate synthesis problem for pPTA.* Given pPTA $\mathcal{P}$, specification φ, percentage c and region R, partition R into regions R_a, R_r, and R_i such that $\mathcal{P}, R_a \models \varphi$ and $\mathcal{P}, R_r \models \neg\varphi$ where $R_a \cup R_r$ covers at least $c\,\%$ of R.

To solve this problem, we consider techniques to obtain a finite-state pMDP for the semantics of the pPTA in the next section. We then solve the approximate synthesis on the resulting pMDP. Note that this is a computationally hard problem: finding parameter values for a pMDP that satisfy a reachability property is ETR-complete [38][2]. To check whether a region is accepting or rejecting, we apply parameter lifting [51] on this pMDP. Parameter lifting first drops all dependencies between parameters in a pMDP. It then transforms the pMDP into a 2-player stochastic game to obtain upper and lower bounds for the given property φ. It applies to finite-state pMDPs and graph-preserving regions. In Sect. 5 we experimentally show this approach's feasibility.

3 PPTA to pMDP Methods

The main question is now whether existing techniques that verify PTA by obtaining finite-state MDPs carry over to the parametric setting. We will answer this question affirmatively for the digital clocks and backwards reachability techniques. The correctness criterion is, as we will show, that they preserve reachability probabilities as defined on the dense-time semantics. The presentation below is along the lines of [49] adapted to the case with parameters.

3.1 Digital Clocks

The digital clocks approach for TA [9,34] and its adaptation to PTA [44] only consider integer clock valuations, i.e. valuations in $\mathbb{N}^{\mathscr{X}}$, and delays of 1 time unit. By capping the clock valuation for clock x to $\mathbf{k}_x + 1$, where $\mathbf{k}_x$ is the maximal constant to which x is compared in the PTA, digital clocks give rise to a *finite* MDP. To that end, let $(v \oplus t)(x) = \min\{v(x) + t, \mathbf{k}_x + 1\}$ for each $x \in \mathscr{X}$. The digital clock approach requires the PTA to be *closed*, i.e. all clock constraints must only contain non-strict comparisons such as $x \leq c$ and $x \geq c$.

Definition 12. *The* ***digital clocks semantics*** *of a closed pPTA $\mathcal{P}$ is the pMDP $[\![\mathcal{P}]\!]_{dc} = (S, s_0, Act', V, prob')$ with $S = \{\, (\ell, v) \in Loc \times \mathbb{N}^{\mathscr{X}} \mid v \models inv(\ell) \,\}$ and s_0, Act', and $prob'$ are as in Definition 7 (restricted to S), except that for time delays we use*

$$prob'((\ell, v), a, (\ell', v')) = 1 \quad \textit{if } a = 1 \wedge \ell = \ell' \wedge v' = v \oplus 1 \models inv(\ell').$$

[2] Existential Theory of the Reals. ETR problems are between NP and PSPACE, and ETR-hard problems are as hard as finding the roots of a multi-variate polynomial.

Correctness. To prove the correctness of the digital clocks semantics for pPTA we first show that parameter instantiation and digital clocks semantics commute.

Lemma 1. *For pPTA $\mathcal{P}$ and parameter valuation u: $[\![\mathcal{P}[u]]\!]_{dc} = [\![\mathcal{P}]\!]_{dc}[u]$.*

Proof. By Definition 12, we need to prove that the transition functions are equivalent, i.e. $prob'((\ell, v), a, (\ell', v'))[u] = prob'[u]((\ell, v), a, (\ell', v'))$. From Definition 7 this follows for all cases except $a = 1 \wedge \ell = \ell' \wedge v' = v \oplus 1 \models inv(\ell')$. For this case, $prob'((\ell, v), a, (\ell', v')) = 1$. As no parameters occur in this case, equivalence follows trivially. Therefore the transition functions are equivalent. □

Furthermore, similar as for PTA, the following lemma holds:

Lemma 2. *For any closed pPTA $\mathcal{P}$, closed target T, and region R, we have*

$$\mathsf{Pr}_{min}^{\mathcal{P},R}(\Diamond T) \;=\; \mathsf{Pr}_{min}^{[\![\mathcal{P}]\!]_{dc},R}(\Diamond T) \quad \textit{and} \quad \mathsf{Pr}_{max}^{\mathcal{P},R}(\Diamond T) \;=\; \mathsf{Pr}_{max}^{[\![\mathcal{P}]\!]_{dc},R}(\Diamond T).$$

Proof. For minimal reachability, take an arbitrary but fixed $u \in R$. Then $\mathcal{P}[u]$ is a PTA. Thus, $\mathsf{Pr}_{min}^{\mathcal{P}[u]}(\Diamond T) = \mathsf{Pr}_{min}^{[\![\mathcal{P}[u]]\!]_{dc}}(\Diamond T)$ [44]. Using Lemma 1, we conclude $\mathsf{Pr}_{min}^{\mathcal{P}[u]}(\Diamond T) = \mathsf{Pr}_{min}^{[\![\mathcal{P}]\!]_{dc}[u]}(\Diamond T)$. Maximal reachability is proven analogously.

This yields the following result on preserving rejecting and accepting regions:

Theorem 1 (correctness of digital clocks). *For region R, closed target T and closed pPTA $\mathcal{P}$:*

$$\mathcal{P}, R \models \mathbb{P}_{\sim\lambda}(\Diamond T) \iff [\![\mathcal{P}]\!]_{dc}, R \models \mathbb{P}_{\sim\lambda}(\Diamond T).$$
$$\mathcal{P}, R \models \neg\mathbb{P}_{\sim\lambda}(\Diamond T) \iff [\![\mathcal{P}]\!]_{dc}, R \models \neg\mathbb{P}_{\sim\lambda}(\Diamond T)$$

Proof. We show the case of preserving an accepting region for $\lesssim\, \in \{<, \leq\}$:

$$\begin{aligned}
\mathcal{P}, R \models \mathbb{P}_{\lesssim\lambda}(\Diamond T) &\overset{\text{Def. 11}}{\Leftrightarrow} \forall u \in R.\ \mathcal{P}[u] \models \mathbb{P}_{\lesssim\lambda}(\Diamond T)\\
&\Leftrightarrow \mathsf{Pr}_{max}^{\mathcal{P},R}(\Diamond T) \lesssim \lambda\\
&\overset{\text{Lem. 2}}{\Leftrightarrow} \mathsf{Pr}_{max}^{[\![\mathcal{P}]\!]_{dc},R}(\Diamond T) \lesssim \lambda\\
&\Leftrightarrow \forall u \in R.\ [\![\mathcal{P}]\!]_{dc}[u] \models \mathbb{P}_{\lesssim\lambda}(\Diamond T)\\
&\overset{\text{Def. 11}}{\Leftrightarrow} [\![\mathcal{P}]\!]_{dc}, R \models \mathbb{P}_{\lesssim\lambda}(\Diamond T).
\end{aligned}$$

The proofs for preserving rejecting regions for $\lesssim$ and accepting/rejecting regions for $\gtrsim\, \in \{>, \geq\}$ are analogous using $\mathsf{Pr}_{min}^{\mathcal{P},R}(\Diamond T)$ rather than $\mathsf{Pr}_{max}^{\mathcal{P},R}(\Diamond T)$.

From Theorem 1 it follows that accepting/rejecting regions are preserved under the digital clocks semantics. Therefore, the inconsistent regions are preserved, too. Note that region R does not need to be closed. This is only necessary for clock constraints, and they are not influenced by parameters in our setting.

Complexity. An upper bound on the number of states in the digital clocks semantics is $|Loc| \cdot \prod_{x \in \mathscr{X}} (\mathbf{k}_x{+}1)$. This means that the runtime for parameter region verification as used in Theorem 1 is exponential in the number of clocks. In Sect. 5, we will report on the practical feasibility of digital clocks for pPTA.

3.2 Backwards Reachability

To tackle the state space explosion of digital clocks, we consider backwards reachability [46]. Instead of using explicit states (i.e. pairs of locations and valuations), it computes a finite set of *symbolic states*—all that can reach the target T.

As in [46], a *symbolic state* is a pair $(\ell, \zeta) \in Loc \times Zones(\mathscr{X})$. For a set of symbolic states $\mathtt{U} \subseteq Loc \times Zones(\mathscr{X})$, let $\zeta^{\ell}_{\mathtt{U}} = \bigcup\{\zeta \mid (\ell, \zeta) \in \mathtt{U}\}$ be all zones in $\mathtt{U}$ that are paired with location ℓ. To determine the reachable symbolic states, we use time ($tpre_{\mathtt{U}}$) and discrete ($dpre$) predecessor operations. Let $\mathtt{V}$ be the set of symbolic states explored so far; initially $\mathtt{V} = \{T\}$. Then $tpre_{\mathtt{U}}$ determines the symbolic states that can reach a state in $\mathtt{V}$ by delays, all the while staying in $\mathtt{U}$; *dpre* are those that can do so via jumps.

Definition 13. *Given a pPTA $\mathcal{P} = (Loc, \ell_0, Act, \mathscr{X}, V, prob, inv)$, sets of symbolic states $\mathtt{U}$ and $\mathtt{V}$, let:*

$$tpre_{\mathtt{U}}(\mathtt{V}) := \{(\ell, \swarrow_{\zeta^{\ell}_{\mathtt{U}} \cap \zeta_{inv(\ell)}} (\zeta^{\ell}_{\mathtt{V}}) \cap \zeta_{inv(\ell)})\}$$

where $\swarrow_{\zeta'} (\zeta) := \{v \mid \exists t \geq 0.\ (v + t \models \zeta \wedge \forall t' \leq t.\ (v + t' \models \zeta \cup \zeta'))\}$ which denotes the zone that can reach ζ by delays while staying in ζ'.

The function *dpre* is adopted from [46] and is omitted here. Backwards reachability iteratively applies $tpre_{\mathtt{U}}$ and *dpre* to $\mathtt{V}$ until it reaches a fixed point. It returns an initial symbolic state $\mathtt{z}_0$, a set $\mathtt{Z}$ of symbolic states, and a probability function $prob'$ on symbolic states that is based on the probability function *prob* of $\mathcal{P}$. For more details, we refer the interested reader to the *MaxU* algorithm in [46]. Most important for us is the fact that we input a pPTA and a set of symbolic target states and that it returns a pMDP.

Definition 14. *For the initial symbolic state $\mathtt{z}_0$, set $\mathtt{Z}$ of symbolic states and probability function $prob'$, the* **backwards reachability semantics** *of $\mathcal{P} = (Loc, \ell_0, Act, \mathscr{X}, V, prob, inv)$ is the pMDP $[\![\mathcal{P}]\!]_{br(T)} = (\mathtt{Z}, \mathtt{z}_0, Act, V, prob')$.*

Correctness. To prove the correctness of the backwards reachability semantics of a pPTA, we first show that parameter instantiation and the semantics commute similarly to the case for digital clocks.

Lemma 3. *For pPTA $\mathcal{P}$, target T, and valuation u: $[\![\mathcal{P}]\!]_{br(T)}[u] = [\![\mathcal{P}[u]]\!]_{br(T)}$.*

The proof is analogous to the proof of Lemma 1.

Let $T^{\mathsf{C}} = Loc \times Zones(\mathscr{X}) \backslash T$ and let A_T be the set of symbolic states from which there exists a scheduler that almost surely never reaches T, then:

Lemma 4. *For pPTA $\mathcal{P}$, region R, and target T, with $[\![true]\!] = Loc \times Zones(\mathscr{X})$:*

$$\begin{aligned} \mathsf{Pr}^{\mathcal{P},R}_{max}(\Diamond T) &= \mathsf{Pr}^{[\![\mathcal{P}]\!]_{br(T)},R}_{max}(\Diamond tpre_{[\![true]\!]}(T)). \\ \mathsf{Pr}^{\mathcal{P},R}_{min}(\Diamond T) &= 1 - \mathsf{Pr}^{[\![\mathcal{P}]\!]_{br(T)},R}_{max}(\Diamond tpre_{T^{\mathsf{C}}}(A_T)). \end{aligned}$$

The proof for maximal probabilities is analogous to that of Lemma 2, using Lemma 3 and that $\mathsf{Pr}^{\mathcal{A}}_{max}(\Diamond T) = \mathsf{Pr}^{[\![\mathcal{A}]\!]_{br(T)}}_{max}(\Diamond tpre_{[\![true]\!]}(T))$ for PTA $\mathcal{A}$ [46]. The proof cannot generally be applied to minimal probabilities since the backwards reachability semantics only preserves upper-bounded properties. Therefore, we have to convert lower-bounded properties to such. The idea behind the conversion is the following: Instead of calculating whether we reach T, we calculate the opposite, i.e. whether we never reach T. This is achieved by ending up in A_T. However, before reaching A_T, we must not visit T beforehand. This is encoded in the semantics by $MaxU$; paths through T cannot reach A_T. For the correctness of this conversion we refer to [40,46].

Theorem 2 (correctness of backwards reachability). *Given a region R, target T, and pPTA $\mathcal{P}$, we have*

$$\mathcal{P}, R \models \mathbb{P}_{\lesssim\lambda}(\Diamond T) \iff [\![\mathcal{P}]\!]_{br(T)}, R \models \mathbb{P}_{\lesssim\lambda}(\Diamond tpre_{[\![true]\!]}(T)).$$
$$\mathcal{P}, R \models \mathbb{P}_{\gtrsim\lambda}(\Diamond T) \iff [\![\mathcal{P}]\!]_{br(T)}, R \models \mathbb{P}_{\lesssim 1-\lambda}(\Diamond tpre_{T^C}(A_T))$$

The proof is analogous to that of Theorem 1 using Lemma 4 instead of Lemma 2.

From Theorem 2 it follows that accepting regions are preserved under the backwards reachability semantics. The proof for rejecting regions is analogous. Therefore, the inconsistent regions are preserved, too.

Complexity. In the worst case, running the algorithm on a PTA $\mathcal{P}$ generates an MDP in which the set of symbolic states is the set $Loc \times Zones(\mathscr{X})$, which is doubly exponential in the number of clocks for PTA [46]. This is the same for pPTA as parameters do not affect the size of the result. However, case studies have shown that for PTA the state space is significantly smaller than the worst case [46]. It is claimed that the algorithm is feasible for most practical applications. In Sect. 5, we will report on the practical feasibility of backwards reachability for pPTA.

3.3 Other Methods

Digital clocks are only compatible with a limited class of pPTAs and backwards reachability only calculates reachability properties. However, other methods are established for model checking PTAs that do not restrict the PTA and properties. We briefly discuss whether these techniques can be applied to pPTA.

Region Graph. In the *region graph* [45], clock equivalence classes are considered. All clock valuations that satisfy the same constraints are grouped and used to build a *clock region*. This is equivalent to the symbolic states of backwards reachability, where the clock regions are minimal in the most basic variant. This leads to a relatively large state space, although this problem is tackled by other variants, like probabilistic time-abstract bisimulation [17]. The algorithm is applicable to pPTA and the proof is similar to that of digital clocks and

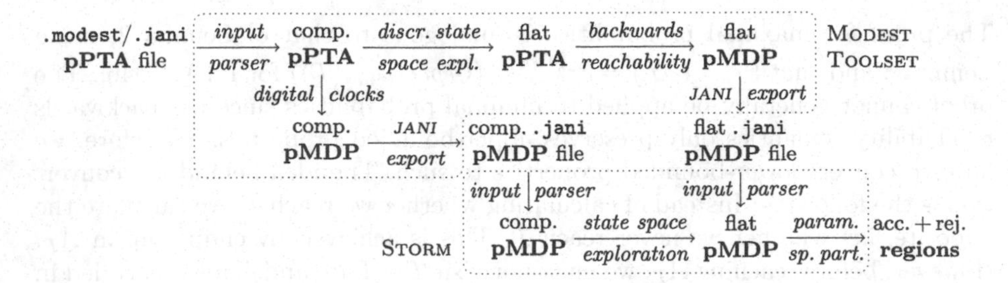

Fig. 4. Our toolchain for pPTA parameter space partitioning

backwards reachability: The substitution of parameters before/after obtaining the region graph semantics is equivalent and minimal/maximal reachability is preserved, which means accepting and rejecting regions are preserved.

Forwards Reachability. As for backwards reachability, *forwards reachability* [45] only considers relevant symbolic states. However, this method performs a forward search from the initial state to the target states. The complexity of forwards reachability is exponential in the number of clocks. However, the forwards algorithm is generally faster than its backwards equivalent [46], but it only provides upper (lower) bounds on the maximal (minimal) probability [22]. This makes forward reachability unsuited for parameter synthesis as regions may be falsely classified as accepting/rejecting. For example, we might have a region that is inconsistent for some upper bounded property. This means that there are both valuations that satisfy and violate the property. As forwards reachability gives bounds on the probability, it might push this probability beyond the bound of the property, resulting in a rejecting region where this is not the case.

Stochastic Games. The stochastic game abstraction [42] transforms the PTA into a 2-player stochastic game. It is usually faster than both digital clocks and backwards reachability in practice [42]. We conjecture that the method can be directly applied on pPTA, resulting in *parametric stochastic games.* However, as parameter synthesis is not implemented for this type of model in Storm or other tools, we would currently not be able to obtain an implementation in the same manner as for the MDP-based approaches.

4 Implementation

We implemented a parameter synthesis pipeline for pPTA by combining the Modest Toolset and the Storm model checker as outlined in Fig. 4. The former has long had support for PTA model checking via digital clocks [27]. Since digital clocks are a syntax-level transformation, the toolset's moconv converter can turn closed PTA models specified in the Modest modelling language [13,26] or the Jani model interchange format [16] into digital clocks MDP, in either of these languages. The implementation is agnostic to the presence of parameters and thus readily works for pPTA, too, exporting pMDP models. Models are usually specified *compositionally* as a network of multiple pPTA extended with discrete (Boolean and bounded integer) variables. The conversion preserves this high-level structure without a blowup in (syntactical) model size. By exporting to JANI, the compositional pMDP can be read back by STORM, which implements parameter space partitioning to deliver the desired set of regions. The actual state space exploration—flattening the composition of variable-extended pMDP into one large pMDP—is performed by STORM in this setup.

For this work, we added an implementation of backwards reachability for maximum probabilities to the MODEST TOOLSET for PTA and pPTA. As shown in the upper branch in Fig. 4, we first turn the compositional pPTA into a single automaton where only clock variables remain. To this we apply backwards reachability using our own implementation of difference-bound matrices. The resulting flat pMDP is then exported to a JANI file, which is usually much larger than the original input. Again, this file can be subsequently be read and regions computed via STORM. For minimum probabilities, backwards reachability would generate non-convex zones, for which we do not have an implementation yet.

5 Evaluation

To evaluate the feasibility of our approach as well as the relative scalability and performance of the two methods, we performed an experimental comparison using our implementation on parametric adaptations of existing PTA benchmarks as well as of the industrial AODV case study.

Table 1. Model characteristics

model	params	clocks	max. $\mathbf{k}_x$
AODV(-n)	1-5	2	4
BRP	1-2	4	16-1657
RM	3	2	5
RH	1	2	5
ZC	1	2	20
FW	1	1	1670

Benchmarks. As we provide the first tool support for pPTA, there are no existing pPTA benchmarks for us to use. We thus turned existing PTA models into pPTA. Among the 9 PTA models in QVBS [30], 5 could be parametrised in a sensible way: those of the bounded retransmission protocol (BRP) [27,32], the repudiation protocol with honest (RH) and malicious receiver (RM) [48], the Zeroconf protocol (ZC) [18], and the IEEE 1394 Firewire protocol (FW) [54]. We consider two variants of BRP: one with equal and one with different loss probabilities for the data and acknowledgement channels. Furthermore, we vary the constants used in the clock constraints to obtain one "small" (S) and one "large" (L) variant of BRP. In RH, we make the probability that a certain message is the last one parametric, and in RM additionally the probability to decode a message. The original repudiation pPTA are not closed and thus only suitable for backwards reachability. We created non-strict (nstr in Fig. 7) variants of these

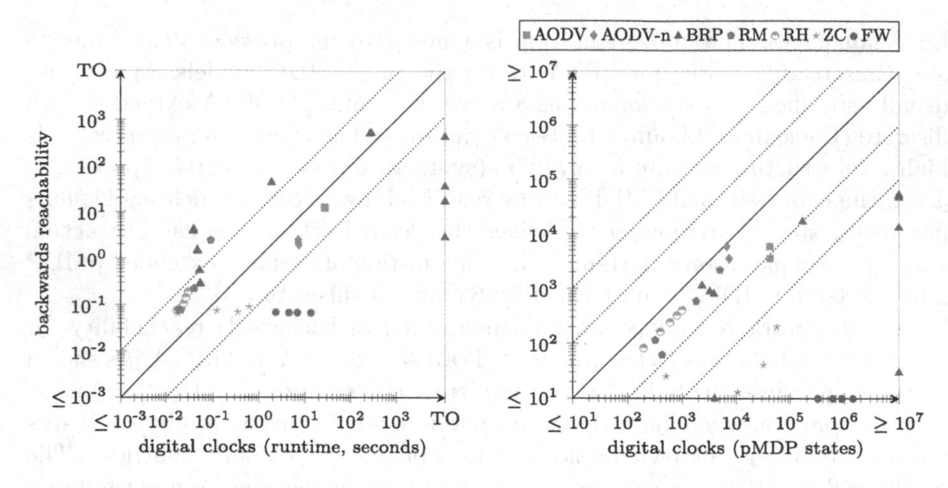

Fig. 5. Runtime up to state space exploration (left) and number of states (right)

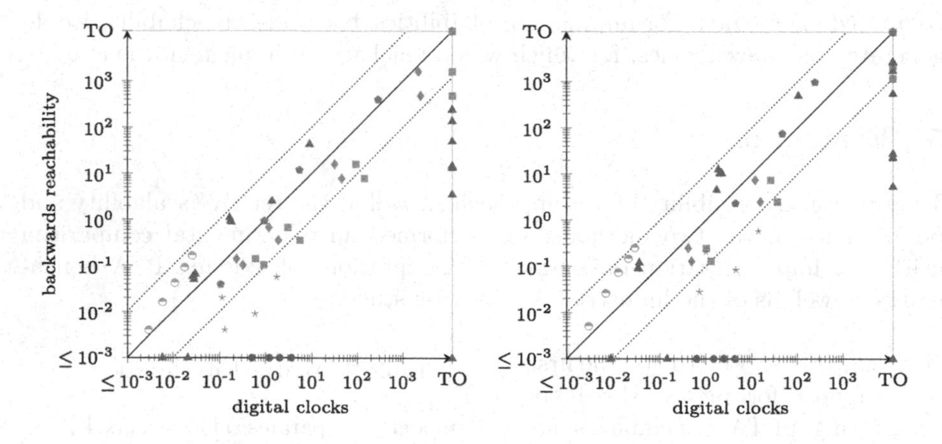

Fig. 6. Parameter space partitioning time (s), 90% (left), 99% (right) coverage

models to enable a comparison. We use two parameters in ZC: the probability to correctly receive a message and the probability that an occupied IP-address is selected. In FW, the parameter is the probability to select between the slow and the fast path.

Additionally, we consider the probabilistic version of the AODV routing protocol [39], which so far had been analysed for selected concrete message loss probabilities only. It comes in two variants, one with routing error (AODV) and one without (AODV-n). For each node, we make the probability to lose an incoming message parametric. By using the same parameter across different subsets of nodes, excluding symmetric cases, we obtain 5 models with 1 (all probabilities are p vs. $1-p$) to 5 (every node has its own parameter p_i) parameters. Table 1 summarises the characteristics of our benchmarks.

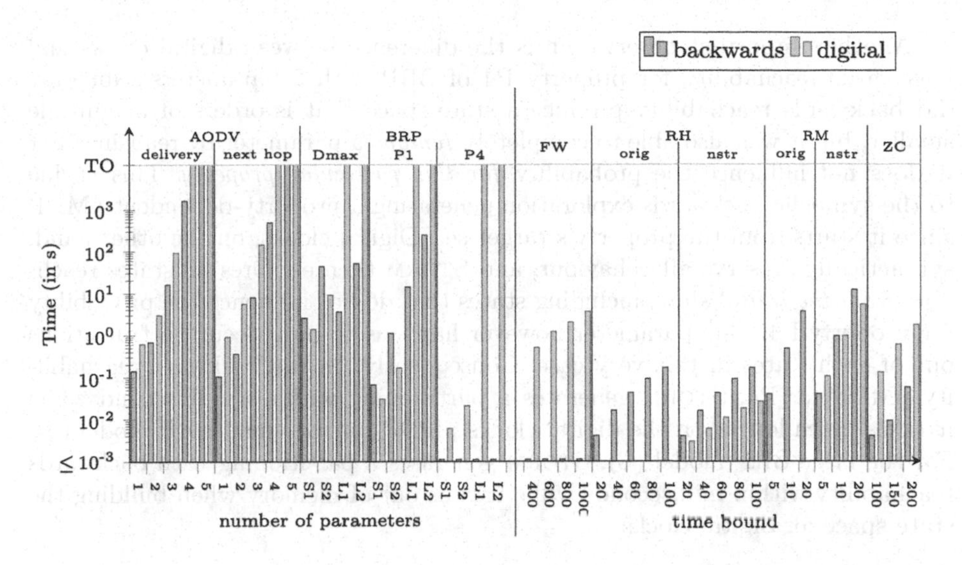

Fig. 7. Runtime for parameter space partitioning with 90% coverage

Setup. We did all experiments on an Intel Core i5-8300H (2.3–4.0 GHz) system with 8 GB of memory running 64-bit Ubuntu 20.04. The timeout was one hour.

Results. In Fig. 5, we compare digital clocks and backwards reachability in terms of the runtime (left) without the parameter space partitioning phase and the number of states of the pMDP (right). Note that for digital clocks, the runtime for the syntactical conversion is negligible. In these scatter plots, a point (x, y) indicates that digital clocks took x seconds or caused x states compared to y seconds/states for backwards reachability, for one specific combination of benchmark, variant (where applicable), and property to check. The dotted lines indicate differences of a factor of 10 and 100. Figure 6 similarly shows the runtime for parameter space partitioning for 90 and 99% coverage.

We observe that digital clocks generate larger state spaces than backwards reachability, just like it does for PTA [46]. We note that the number of transitions per state, however, is often larger with backwards reachability. Ultimately, performance also depends on the topology of the state space, and backwards reachability needs to perform a sometimes expensive symbolic reachability computation, explaining why digital clocks still often manage to be faster as far as obtaining the flat pMDP is concerned (Fig. 5 left). In Fig. 6, however, we can see that backwards reachability is mostly faster in the partitioning phase.

Figure 7 plots the partitioning runtime for those benchmarks where we can vary the number of parameters or the property time bound. On top, we indicate the respective benchmark and property being checked (in case multiple are available). We observe an exponential increase in runtime as we increase the number of parameters for AODV; AODV-n showed similar behaviour. A similar pattern occurs for BRP. Runtime increases mostly linearly with the time bound.

Another important observation is the difference between digital clocks and backwards reachability for property P4 of BRP with 2 parameters. Not only did backwards reachability produce a state space that is orders of magnitude smaller, but it was also able to completely *remove* a parameter by realising that it does not influence the probability *for this particular property*. This is due to the symbolic backwards exploration generating a property-dependent pMDP since it starts from the property's target set. Digital clocks, on the other hand, syntactically preserve all behaviour, and STORM then explores all states reachable *from the initial state*, including states that do not influence the probability for property P4. One parameter however happens to only occur on transitions out of such states in this very case. Consequently, on the backward reachability pMDP for P4, STORM generates a partitioning within a millisecond while it takes much longer on the digital clocks pMDP for the small BRP model (S). For the large BRP model (L), STORM generates a partitioning with backwards reachability within milliseconds while it runs out of memory when building the state space for digital clocks.

6 Conclusion

We presented an approach to tackle the approximate synthesis problem for parametric PTA, that is, to partition the parameter space into accepting and rejecting regions for a given property such that c% of the parameter space is covered. The idea is to first obtain a finite pMDP that is equivalent to the pPTA for the property at hand, and to then apply parameter space partitioning on the pMDP. In the application to AODV, a real-world case study [39], our experiments showed encouraging results, thereby highlighting the usefulness of parametric PTA, our approach, and its implementation.

Beyond this Paper. In this paper, we focused on unbounded and time-bounded probabilistic reachability. The overall approach, however, also works with expected accumulated reachability reward properties when we use digital clocks for the abstraction step [40]. In our case studies, we did not include lower-bounded reachability properties, as those refer to minimal probabilities and are thus affected by divergence. We did not take divergence into account in this paper, but solutions are available in the form of fairness and end-component analysis [40,52]. While backwards reachability calculates probabilities under divergence out of the box, it generates non-convex zones for minimum probabilities, for which we do not (yet) have an efficient implementation. The digital clock and backwards reachability approach can also be used to other parameter synthesis questions for pPTA such as feasibility checking ("does there exist a parameter valuation for which a specification holds?"). The resulting pMDPs can then be analysed using quadratic programming [19].

Outlook. Now that a toolchain for pPTA exists, we would like to study more pPTA case studies; the authors would be happy to assist application experts (e.g. in wireless networks [39]) in modelling and by tuning the tools. Interesting future directions to extend our work are to combine parametrised probabilities with parameters in clock constraints (likely focusing on decidable subclasses as mentioned in Sect. 1), and to consider pPTA in which transition probabilities can depend on clocks [53]. Currently, the stochastic games abstraction is the most competitive technique for PTA. However, parameter space partitioning is not available for parametric stochastic games in Storm or related tools. Developing this would enable the use of the stochastic games abstraction for pPTA.

Data Availability. The tools used and data generated in our experimental evaluation are archived at DOI 10.4121/14910426 [29].

References

1. Agha, G., Palmskog, K.: A survey of statistical model checking. ACM Trans. Model Comput. Simul. **28**(1), 6:1–6:39 (2018). https://doi.org/10.1145/3158668
2. Alur, R., Dill, D.L.: A theory of timed automata. Theor. Comput. Sci. **126**(2), 183–235 (1994)
3. Alur, R., Henzinger, T.A., Vardi, M.Y.: Parametric real-time reasoning. In: STOC, pp. 592–601. ACM (1993). https://doi.org/10.1145/167088.167242
4. André, É.: What's decidable about parametric timed automata? Int. J. Softw. Tools Technol. Transf. **21**(2), 203–219 (2019). https://doi.org/10.1007/s10009-017-0467-0
5. André, É., Arias, J., Petrucci, L., Pol, J.: Iterative bounded synthesis for efficient cycle detection in parametric timed automata. In: TACAS 2021. LNCS, vol. 12651, pp. 311–329. Springer, Cham (2021). https://doi.org/10.1007/978-3-030-72016-2_17
6. André, É., Chatain, T., Fribourg, L., Encrenaz, E.: An inverse method for parametric timed automata. Int. J. Found. Comput. Sci. **20**(5), 819–836 (2009). https://doi.org/10.1142/S0129054109006905
7. André, É., Delahaye, B., Fournier, P.: Consistency in parametric interval probabilistic timed automata. J. Log. Algebraic Methods Program. **110**, 100459 (2020). https://doi.org/10.1016/j.jlamp.2019.04.007
8. André, É., Fribourg, L., Sproston, J.: An extension of the inverse method to probabilistic timed automata. Formal Methods Syst. Des. **42**(2), 119–145 (2013). https://doi.org/10.1007/s10703-012-0169-x
9. Asarin, E., Maler, O., Pnueli, A.: On discretization of delays in timed automata and digital circuits. In: Sangiorgi, D., de Simone, R. (eds.) CONCUR 1998. LNCS, vol. 1466, pp. 470–484. Springer, Heidelberg (1998). https://doi.org/10.1007/BFb0055642
10. Baier, C., de Alfaro, L., Forejt, V., Kwiatkowska, M.: Model checking probabilistic systems. In: Clarke, E., Henzinger, T., Veith, H., Bloem, R. (eds.) Handbook of Model Checking, pp. 963–999. Springer, Cham (2018). https://doi.org/10.1007/978-3-319-10575-8_28
11. Baier, C., Hermanns, H., Katoen, J.-P.: The 10,000 facets of MDP model checking. In: Steffen, B., Woeginger, G. (eds.) Computing and Software Science. LNCS, vol. 10000, pp. 420–451. Springer, Cham (2019). https://doi.org/10.1007/978-3-319-91908-9_21

12. Baier, C., Katoen, J.P.: Principles of Model Checking. MIT Press, Cambridge (2008)
13. Bohnenkamp, H.C., D'Argenio, P.R., Hermanns, H., Katoen, J.P.: MoDeST: a compositional modeling formalism for hard and softly timed systems. IEEE Trans. Software Eng. **32**(10), 812–830 (2006). https://doi.org/10.1109/TSE.2006.104
14. Brim, L., Češka, M., Dražan, S., Šafránek, D.: Exploring parameter space of stochastic biochemical systems using quantitative model checking. In: Sharygina, N., Veith, H. (eds.) CAV 2013. LNCS, vol. 8044, pp. 107–123. Springer, Heidelberg (2013). https://doi.org/10.1007/978-3-642-39799-8_7
15. Budde, C.E., D'Argenio, P.R., Hartmanns, A., Sedwards, S.: An efficient statistical model checker for nondeterminism and rare events. Int. J. Softw. Tools Technol. Transf. **22**(6), 759–780 (2020). https://doi.org/10.1007/s10009-020-00563-2
16. Budde, C.E., Dehnert, C., Hahn, E.M., Hartmanns, A., Junges, S., Turrini, A.: JANI: quantitative model and tool interaction. In: Legay, A., Margaria, T. (eds.) TACAS 2017. LNCS, vol. 10206, pp. 151–168. Springer, Heidelberg (2017). https://doi.org/10.1007/978-3-662-54580-5_9
17. Chen, T., Han, T., Katoen, J.: Time-abstracting bisimulation for probabilistic timed automata. In: Second IEEE/IFIP International Symposium on Theoretical Aspects of Software Engineering, TASE 2008, 17–19 June, 2008, Nanjing, China, pp. 177–184. IEEE Computer Society (2008). https://doi.org/10.1109/TASE.2008.29
18. Cheshire, S., Aboba, B., Guttman, E.: Dynamic configuration of ipv4 link-local addresses. RFC **3927**, 1–33 (2005)
19. Cubuktepe, M., Jansen, N., Junges, S., Katoen, J.-P., Topcu, U.: Synthesis in pMDPs: a tale of 1001 parameters. In: Lahiri, S.K., Wang, C. (eds.) ATVA 2018. LNCS, vol. 11138, pp. 160–176. Springer, Cham (2018). https://doi.org/10.1007/978-3-030-01090-4_10
20. D'Argenio, P.R., Hartmanns, A., Legay, A., Sedwards, S.: Statistical approximation of optimal schedulers for probabilistic timed automata. In: Ábrahám, E., Huisman, M. (eds.) IFM 2016. LNCS, vol. 9681, pp. 99–114. Springer, Cham (2016). https://doi.org/10.1007/978-3-319-33693-0_7
21. David, A., Larsen, K.G., Legay, A., Mikucionis, M., Poulsen, D.B.: Uppaal SMC tutorial. Int. J. Softw. Tools Technol. Transf. **17**(4), 397–415 (2015). https://doi.org/10.1007/s10009-014-0361-y
22. Daws, C., Kwiatkowska, M.Z., Norman, G.: Automatic verification of the IEEE 1394 root contention protocol with KRONOS and PRISM. Int. J. Softw. Tools Technol. Transf. **5**(2–3), 221–236 (2004)
23. Dombrowski, C., Junges, S., Katoen, J., Gross, J.: Model-checking assisted protocol design for ultra-reliable low-latency wireless networks. In: SRDS, pp. 307–316. IEEE Computer Society (2016)
24. Fruth, M.: Probabilistic model checking of contention resolution in the IEEE 802.15.4 low-rate wireless personal area network protocol. In: ISoLA, pp. 290–297. IEEE Computer Society (2006)
25. Gregersen, H., Jensen, H.E.: Formal Design of Reliable Real Time Systems. Master's thesis, Department of Mathematics and Computer Science, Aalborg University (1995)
26. Hahn, E.M., Hartmanns, A., Hermanns, H., Katoen, J.P.: A compositional modelling and analysis framework for stochastic hybrid systems. Formal Methods Syst. Des. **43**(2), 191–232 (2013). https://doi.org/10.1007/s10703-012-0167-z
27. Hartmanns, A., Hermanns, H.: A Modest approach to checking probabilistic timed automata. In: QEST, pp. 187–196. IEEE (2009)

28. Hartmanns, A., Hermanns, H.: The Modest Toolset: an integrated environment for quantitative modelling and verification. In: Ábrahám, E., Havelund, K. (eds.) TACAS 2014. LNCS, vol. 8413, pp. 593–598. Springer, Heidelberg (2014). https://doi.org/10.1007/978-3-642-54862-8_51
29. Hartmanns, A., Katoen, J.P., Kohlen, B., Spel, J.: Tweaking the odds in probabilistic timed automata (artifact). 4TU.Centre for Research Data (2021). https://doi.org/10.4121/14910426
30. Hartmanns, A., Klauck, M., Parker, D., Quatmann, T., Ruijters, E.: The quantitative verification benchmark set. In: Vojnar, T., Zhang, L. (eds.) TACAS 2019. LNCS, vol. 11427, pp. 344–350. Springer, Cham (2019). https://doi.org/10.1007/978-3-030-17462-0_20
31. Hartmanns, A., Sedwards, S., D'Argenio, P.R.: Efficient simulation-based verification of probabilistic timed automata. In: WSC, pp. 1419–1430. IEEE (2017). https://doi.org/10.1109/WSC.2017.8247885
32. Helmink, L., Sellink, M.P.A., Vaandrager, F.W.: Proof-checking a data link protocol. In: Barendregt, H., Nipkow, T. (eds.) TYPES 1993. LNCS, vol. 806, pp. 127–165. Springer, Heidelberg (1994). https://doi.org/10.1007/3-540-58085-9_75
33. Hensel, C., Junges, S., Katoen, J.P., Quatmann, T., Volk, M.: The probabilistic model checker storm. CoRR abs/2002.07080 (2020)
34. Henzinger, T.A., Manna, Z., Pnueli, A.: What good are digital clocks? In: Kuich, W. (ed.) ICALP 1992. LNCS, vol. 623, pp. 545–558. Springer, Heidelberg (1992). https://doi.org/10.1007/3-540-55719-9_103
35. Hune, T., Romijn, J., Stoelinga, M., Vaandrager, F.W.: Linear parametric model checking of timed automata. J. Log. Algebraic Methods Program. **52–53**, 183–220 (2002). https://doi.org/10.1016/S1567-8326(02)00037-1
36. Jonsson, B., Larsen, K.G.: Specification and refinement of probabilistic processes. In: LICS, pp. 266–277. IEEE Computer Society (1991). https://doi.org/10.1109/LICS.1991.151651
37. Jovanovic, A., Lime, D., Roux, O.H.: Integer parameter synthesis for real-time systems. IEEE Trans. Softw. Eng. **41**(5), 445–461 (2015). https://doi.org/10.1109/TSE.2014.2357445
38. Junges, S., Katoen, J., Pérez, G.A., Winkler, T.: The complexity of reachability in parametric Markov decision processes. J. Comput. Syst. Sci. **119**, 183–210 (2021)
39. Kamali, M., Katoen, J.-P.: Probabilistic model checking of AODV. In: Gribaudo, M., Jansen, D.N., Remke, A. (eds.) QEST 2020. LNCS, vol. 12289, pp. 54–73. Springer, Cham (2020). https://doi.org/10.1007/978-3-030-59854-9_6
40. Kohlen, B.: Parameter synthesis in probabilistic timed automata. Master's thesis, RWTH Aachen University, Aachen (2020). https://publications.rwth-aachen.de/record/811856
41. Krause, C., Giese, H.: Model checking probabilistic real-time properties for service-oriented systems with service level agreements. INFINITY. EPTCS, vol. 73, pp. 64–78 (2011)
42. Kwiatkowska, M., Norman, G., Parker, D.: Stochastic games for verification of probabilistic timed automata. In: Ouaknine, J., Vaandrager, F.W. (eds.) FORMATS 2009. LNCS, vol. 5813, pp. 212–227. Springer, Heidelberg (2009). https://doi.org/10.1007/978-3-642-04368-0_17
43. Kwiatkowska, M., Norman, G., Parker, D.: PRISM 4.0: verification of probabilistic real-time systems. In: Gopalakrishnan, G., Qadeer, S. (eds.) CAV 2011. LNCS, vol. 6806, pp. 585–591. Springer, Heidelberg (2011). https://doi.org/10.1007/978-3-642-22110-1_47

44. Kwiatkowska, M.Z., Norman, G., Parker, D., Sproston, J.: Performance analysis of probabilistic timed automata using digital clocks. Formal Methods Syst. Des. **29**(1), 33–78 (2006). https://doi.org/10.1007/s10703-006-0005-2
45. Kwiatkowska, M.Z., Norman, G., Segala, R., Sproston, J.: Automatic verification of real-time systems with discrete probability distributions. Theor. Comput. Sci. **282**(1), 101–150 (2002). https://doi.org/10.1016/S0304-3975(01)00046-9
46. Kwiatkowska, M.Z., Norman, G., Sproston, J., Wang, F.: Symbolic model checking for probabilistic timed automata. Inf. Comput. **205**(7), 1027–1077 (2007)
47. Legay, A., Sedwards, S., Traonouez, L.-M.: Scalable verification of Markov decision processes. In: Canal, C., Idani, A. (eds.) SEFM 2014. LNCS, vol. 8938, pp. 350–362. Springer, Cham (2015). https://doi.org/10.1007/978-3-319-15201-1_23
48. Markowitch, O., Roggeman, Y.: Probabilistic non-repudiation without trusted third party. In: Proceedings 2nd Workshop on Security in Communication Networks (1999)
49. Norman, G., Parker, D., Sproston, J.: Model checking for probabilistic timed automata. Formal Methods Syst. Des. **43**(2), 164–190 (2013)
50. Puterman, M.L.: Markov Decision Processes: Discrete Stochastic Dynamic Programming. Wiley Series in Probability and Mathematical Statistics: Applied Probability and Statistics, John Wiley & Sons Inc., New York (1994)
51. Quatmann, T., Dehnert, C., Jansen, N., Junges, S., Katoen, J.-P.: Parameter synthesis for Markov models: faster than ever. In: Artho, C., Legay, A., Peled, D. (eds.) ATVA 2016. LNCS, vol. 9938, pp. 50–67. Springer, Cham (2016). https://doi.org/10.1007/978-3-319-46520-3_4
52. Sproston, J.: Strict divergence for probabilistic timed automata. In: Bravetti, M., Zavattaro, G. (eds.) CONCUR 2009. LNCS, vol. 5710, pp. 620–636. Springer, Heidelberg (2009). https://doi.org/10.1007/978-3-642-04081-8_41
53. Sproston, J.: Probabilistic timed automata with clock-dependent probabilities. Fundam. Informaticae **178**(1–2), 101–138 (2021)
54. Stoelinga, M., Vaandrager, F.: Root contention in IEEE 1394. In: Katoen, J.-P. (ed.) ARTS 1999. LNCS, vol. 1601, pp. 53–74. Springer, Heidelberg (1999). https://doi.org/10.1007/3-540-48778-6_4

Quantifying Software Reliability via Model-Counting

Samuel Teuber(✉) and Alexander Weigl

Karlsruhe Institute of Technology (KIT), Karslruhe, Germany
samuel@samweb.org, weigl@kit.edu

Abstract. Critical software should be verified. But how to handle the situation when a proof for the functional correctness could not be established? In this case, an assessment of the software is required to estimate the risk of using the software.

In this paper, we contribute to the assessment of critical software with a formal approach to measure the reliability of the software against its functional specification. We support bounded C-programs precisely where the functional specification is given as assumptions and assertions within the source code. We count and categorize the various program runs to compute the reliability as the ratio of failing program runs (violating an assertion) to all terminating runs. Our approach consists of a preparing program translation, the reduction of C-program into SAT instances via software-bounded model-checker (CBMC), and precise or approximate model-counting providing a reliable assessment. We evaluate our prototype implementation on over 24 examples with different model-counters. We show the feasibility of our pipeline and benefits against competitors.

Keywords: Software verification · Software reliability · Model counting

1 Introduction

Formal verified safety, defined as the absence of catastrophic consequences [1], yields a high guarantee on the well-functioning of critical software. But proving safety is a hard and tedious process due to the necessary formalization, verification and (possibly) bug fixing for a given software system. In cases where proof cannot be established, other techniques for the reliability assessment of the software are required. To this end, we want to quantitatively estimate the risk of usage of an assessed software. Traditionally, safety is a qualitative property that a software might or might not fulfill, whereas reliability is often a quantitative, measurable property, e.g., the likelihood of failure or the failure rate. Quantitative analysis is, however, also a valuable addition to formal safety verification. For example, quantitative analyses are useful to assess the strictness of (input)

S. Teuber—This work was supported by funding of the Helmholtz Association (HGF) through the Competence Center for Applied Security Technology (KASTEL).

A. Abate and A. Marin (Eds.): QEST 2021, LNCS 12846, pp. 59–79, 2021.
https://doi.org/10.1007/978-3-030-85172-9_4

assumptions imposed for proving software properties. Those are just some reasons why quantitative and probabilistic software analysis has been identified as an interesting research topic (e.g., [16,23]). In particular, [7,8,13] presented approaches to combine quantitative analysis with symbolic execution.

Contribution. In this work we present a formal approach for the quantification of the violation or adherence to a functional specification through exact and approximate model counting. The approach processes C-programs, where the program specification is given via assertions and assumptions in the procedure bodies. The first step in the approach is a behavior-preserving program translation, which makes the violation or adherence of specification in program runs countable. Afterwards, we use CBMC [5] to convert the transformed C-program into multiple CNF formulae. By using CBMC, our approach is limited to C-programs with a bounded execution and bounded data domain. However, by leveraging CBMC's bit precise semantic, we have wider and more precise support of C-programming language in comparison to previous work. Each model of the CNF formulae represents a possible program run, which we count precisely or approximately with tools like GANAK [17] or APPROXMC [3,15,18]. We present a sophisticated evaluation of our prototype in comparison to its competitor [6] which shows advantages of the bit-precise support and non-determinism. The prototype is publicly available.[1]

Example. Consider the example in Fig. 1 which computes the square root of a non-negative integer in the upper half of variable `a`. The algorithm was later modified to return the actual integer square root `lower`. Note, this modification is flawed which will be discussed in more detail below. There are multiple aspects of this program which can be quantified. Firstly, we can compute the number of inputs `x` for which above mentioned flaw leads to an assertion miss (violation of the assertion condition) before the return-statement. Secondly, we can compute the number of inputs `x` for which the assume-statement fails. While the number of assertion misses is a measure stating to what degree the program at hand is flawed, the number of assumption misses describes how tight the assumptions are under which we try to guarantee correct behavior. In Fig. 1, for example, our assumptions exclude half of the possible input space (namely any negative value). Depending on the use case, this might be considered a very strong assumption which might not match reality.

```
#define TOP2BITS(x) ((x & (3 << 30)) >>
    30)
int sqrt(int x) {
  assume(x>=0);
  int input = x;
  int a = 0, r = 0, e = 0;
  for (int i = 0; i < 32; i++) {
    r = (r << 2) + TOP2BITS(x);
    x <<= 2; a <<= 1;
    e = (a << 1) + 1;
    if (r >= e) { r -= e; a++; }
  }
  int lower = (a>>16);
  unsigned int upper = lower+1;
  assert(lower*lower<=input
    && upper*upper>input);
  return lower;
}
```

Fig. 1. Variation of square root algorithm in [20]

[1] https://github.com/samysweb/counterSharp.

Overview. The formal foundations of our approach, including the definition of reliability, is in Sect. 2. In Fig. 4, we begin building the pipeline, which contains the program transformation, model-counting, and the correctness (Sect. 3). The comparison with [6] and runtime statistics is given in the evaluation (Sect. 4).

2 Foundations of the Pipeline

```
int test1(int x) {
  assert(x!=0);
  return v/x; }
int test2(int x) {
  x = (x<0)? 0 : x;
  int *y = malloc(sizeof(int));
  int div = (*y)*x
  assert(div!=0);
  return v/div;
}
```

Fig. 2. In `test1` the failure only depends on the input. Differently, in `test2` the failure depends on input and non-determinism.

Traditionally, reliability of a program is quantified via the number of inputs for which there exists a program path leading to failure in comparison to the number of inputs for which no such program path exists (as seen for example in [7]). However, this definition draws an incomplete image of the software's reliability for nondeterministic programs.

For example, consider the code in Fig. 2: Both functions are prone to a division by zero failure which is caught by an assertion. The functions differ, because only in `test1` this failure is solely dependent on the input where the division by zero failure *must* happen for exactly one input value (specifically for `x` set to `0`). Meanwhile, the division by zero failure *may* happen for any input of the function `test2` in Fig. 2 while it *must* happen for a negative input value for `x`. We use this semantic of *must* and *may* failures in the following analysis. A failure *must* occur for an input if an assertion miss occurs for every program path. A failure *may* occur for an input if there is both, a program path leading to an assertion miss and a program path leading to an assertion hit.

Partitions of the Input Space. Formally, we consider a program as a relation $P \subseteq \mathcal{S}^2$ between start state $s \in \mathcal{S}$ and the final state $s' \in \mathcal{S}$. We denote this relationship with $s \xrightarrow{P} s'$. We distinguish between the input values $i \in \mathcal{I}$ and output $o \in \mathcal{O}$ of a program: The input values $i = s\lfloor_{\mathcal{I}}$ are part of the start state, whereas the output values $o = s'\lfloor_{\mathcal{O}}$ are part of final state. $s\lfloor_{\Sigma}$ denotes the projection to the variables given from the set Σ. This formalization allows choosing arbitrary values of the non-input variables (e.g., global or local variables) in the start state of a program. Thus, for a given input i, there might be multiple possible start states. And given a single start state s, there might be a set of reachable final states s'.

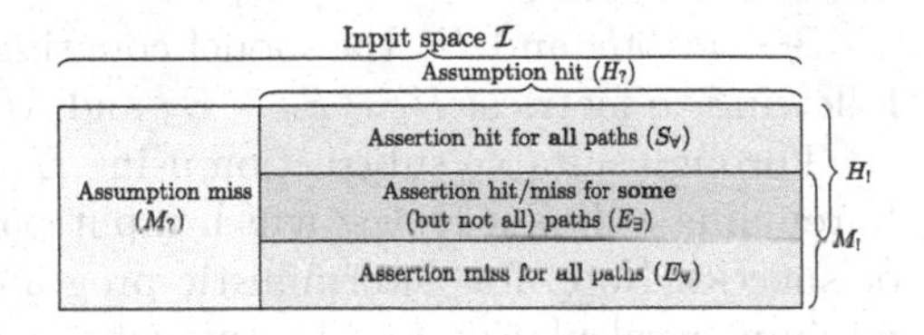

Fig. 3. Partitions $M_?, H_?, E_\forall, E_\exists$, and $S_\forall$ of a programs input space $\mathcal{I}$

In combination, given an input value $i \in \mathcal{I}$, the reachable output values are denoted by $\mathfrak{O}(i) = \{o \mid s\lfloor_{\mathcal{I}} = i \wedge s \xrightarrow{P} s' \wedge o = s'\lfloor_{\mathcal{O}}\} \subseteq \mathcal{O}$. Moreover, we introduce the function $check\colon \mathcal{S} \to \{\checkmark, 🐞, \oslash\}$, which can determine whether the program executed normally ($\checkmark$), adhering all reached assumptions and assertions), or abnormally terminated denoting the first violation of an assertion (🐞) or an assumption ($\oslash$) given the final state s'. We lift *check* to a set of states, i.e., $check(\mathfrak{D}(i)) \subseteq \{\checkmark, 🐞, \oslash\}$ denotes with the possible outcomes for a given input. Later, in Sect. 3.1 we weave the functionality of *check* into the given C-program.

With the definition of *check* we can partition the input space $\mathcal{I}$ of a program into following parts (Fig. 3) under an additional assumption: the given assume-statements within program are only referring to the input variables of the program. The partitions are defined as follows:

$$M_? = \{i \mid check(\mathfrak{D}(i)) = \{\oslash\}\} \tag{1}$$

$$H_? = \{i \mid \oslash \notin check(\mathfrak{D}(i))\} \tag{2}$$

$$E_\forall = \{i \mid check(\mathfrak{D}(i)) = \{🐞\}\} \tag{3}$$

$$E_\exists = \{i \mid check(\mathfrak{D}(i)) = \{\checkmark, 🐞\}\} \tag{4}$$

$$S_\forall = \{i \mid check(\mathfrak{D}(i)) = \{\checkmark\}\} \tag{5}$$

In terms of notation, we denote assumption related variables with ? and assertion related variables with !. M represents *misses* and H represents *hits*, whereas E represents *error* and S represents *success*. Finally, r is used for ratios. Thus, $M_?$ describes the partition of invalid input values according to our assumptions. There can not occur any other observations for these inputs. $H_?$ on the other hand represents all input values which correspond to our assumptions. This can further be split into $E_\forall, E_\exists$ and $S_\forall$. $E_\forall$ represents the input values that always lead to an error, where $E_\exists$ are the input values, where sometimes an error occurred. $S_\forall$ are the input values that always adhere to the assumptions and assertions in the program, regardless of the value of the non-input variables in the start state.

Besides $M_?$ and $H_?$ the model counting approach enables us to measure the following to metrics: $H_! = S_\forall + E_\exists$ and $M_! = E_\forall + E_\exists$

Through suitable subtraction using $H_?$ we can then compute $E_\forall, E_\exists$ and $S_\forall$. It remains a design choice which input partitions of Fig. 3 are treated as error or success. Note, for deterministic programs the partition $E_\exists$ is empty, leading to simpler calculation and less model-counting calls. Depending on the use case we might then be interested in ratios describing how many of the inputs show a particular behavior. To this end, we define the following, exemplary, ratios:

$$r_?^m = \frac{M_?}{\mathcal{I}} \qquad r_\forall^f = \frac{S_\forall}{H_?} \qquad r_\forall^e = \frac{E_\forall}{H_?} \tag{6}$$

3 Pipeline

Figure 4 shows the three pipeline steps of our approach. Firstly, we weave the observation of the *check* function into the given original C-program, by transforming the program flow in such a way that a violation of an assumption or an assertion are explicitly stored in the program state (Sect. 3.1). Also, a violation of either an assumption or an assertion leads to an early termination. The C-program input consists of an entry routine and the (transitively) required routines. Secondly, we use CBMC, a bounded model-checker for C-programs, for the transformation of the program into multiple CNF formulae (Sect. 3.2) where we have one CNF formula for each measured quantity. Thirdly, we run model-counting tools, like APPROXMC or GANAK, on the CNF formulae and thus obtain required metrics (Sect. 3.3) to calculate the reliability.

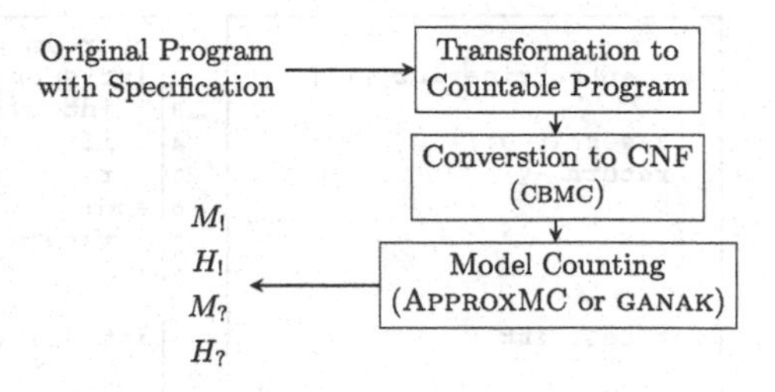

Fig. 4. Overview on the quantification pipeline for the computation of hit and misses of assumption and assertion within the original program.

3.1 Transformation: Make Violation Countable

The goal of the program transformation is to make the assumption or assertion violations explicit in the program state by using dedicated fresh variables. Therefore, the violations become countable by Model Counters. An example of the transformation is presented in Fig. 5. Note, we use the constants `true` and `false` for convenience reason, although they are not defined in standardized C.

Restrictions on C-Programs. The allowed C-subset is restricted by CBMC, which allows both basic datatypes (int, float, char etc.) and complex datatypes (e.g., arrays and structures). Subroutine calls are limited to stand-alone calls and calls assigned to a variable (i.e., no nested subroutine calls and no calls directly within a return-statement). Such calls can easily be transformed into non-nested, non-return calls by introducing appropriate local variables – even automatically if desired. While complex datatypes are supported, they cannot be used as input variables for model counting directly, but must be constructed within the function under evaluation explicitly (e.g., by passing an array's elements into the function explicitly and then constructing the array within the function). We assume that assume-statements solely express conditions imposed on the program's input (and not program internal behavior) and are thus positioned at the beginning of the program before any assertion statements.

Transformation. In the example in Fig. 5, the transformation is applied to the entry function `test`. Required sub-routines are considered for transformation accordingly.

The first step introduces two global variables initialized by `false` (`am` and `as`) which store whether an assertion or assumption was missed in global program state (Line 1). This implies that assume- and assert-statements actually have to modify this variable after checking their specific criterion. Therefore,

```
 1
 2 int subroutine(int y) {
 3
 4   assert(y<0);
 5   return -y;
 6
 7
 8 }
 9
10 int test(int x, int y)
11 {
12
13   assume(x>0);
14   int z = 0;
15   if (y < 0) {
16     z += subroutine(y);
17
18     return z;
19
20
21
22
23
24
25   }
26   assert(z>=0);
27   return z+x;
28
29 }
```

```
 1 char am = false; char as = false;
 2 int subroutine(int y) {
 3   int rv;
 4   if (!(y < 0)) { as = true; goto end; }
 5   rv = -y;
 6 end:
 7   return rv;
 8 }
 9
10 int test(int x, int y)
11 {
12   int rv;
13   if (!(x > 0)) { am = true; goto end; }
14   int z = 0;
15   if (y < 0) {
16     z += subroutine(y);
17     if (as || am) goto end;
18     rv = z;
19 end:
20     assert(am || as); //assertion hit
21     assert(!as);//assertion miss
22     assert(am); //assumption hit
23     assert(!am);//assumption miss
24     return rv;
25   }
26   if (!(z >= 0)) { as = true; goto end; }
27   rv = z + x;
28   goto end;
29 }
```

Fig. 5. Left is the original program, and right the program transformation as described in Sect. 3.1

such statements are transformed into an if-statement with their negated original condition. Missing a condition sets the corresponding variable (am/as) and initiates a jump to the end of the current function. This transformation can be observed in Lines 13 and 26. Note that the same transformation also takes place in subroutines as can be seen in Line 4. In order to support sub-routine calls, every such call triggers a check for assertion- or assumption misses afterwards. If a sub-routine call has violated an assertion, the callee routine directly jumps to its end. This behavior occurs recursively leading to an early termination of the program – similar to exception handling in modern programming languages.

We then unify all return-statements of a routine into a single, labeled return (e.g., Line 18) targeted by multiple goto-statements which replace the return statements in the old program and direct the program flow to the one remaining return statement (e.g., Line 27). Line 18 shows the first return-statement of a routine is kept and labeled with end, while all other return-statements (e.g., Line 27) are replaced by goto-statements. A new value variable rv is introduced which stores the return value across this jump (Lines 12 and 3). If no return-statement exists within a subroutine, we introduce a labeled dummy return-statement at the end of the routine.

Finally, the four assert-statements in Lines 20 to 23 (with conditions as explained later in Sect. 3.2) are added just before the return-statement of the entry routine. Only one of these assert-statements is inserted for the conversion to the CNF formulae in order to generate specific formulae for counting the assumption/assertion hits and misses.

In the remainder of this paper, if we talk about assertion, we mean the assertion given as the specification in the original program, whereas the notion of an assert-statement refers to (one of) the four assertions in the transformed program (Lines 20 to 23).

3.2 Conversion into CNF

For the conversions of C-programs into CNF formulae we use CBMC. CBMC converts the transformed program p and a specification ψ into a CNF formula $\phi = \pi \wedge \neg\psi$ where π is a CNF representation of the unrolled program (w.r.t. a certain loop iteration or recursion depth). Formula ϕ is then satisfiable iff there exists a program path in p leading to a violation of the given specification ψ. Using CBMC with its builtin bit precise semantic, we generate four formulae with varying specification (assert-statements) before the final return-statement: `am`, `!am`, `am || as` and `!as`. Additionally we generate a fifth formula which allows to compute the number of inputs for which the bounds of CBMC were insufficient. The last assert-statement, for instance, is transformed into a formula asserting that `as` is `true` and can thus be used to count the number of models which produce an assertion miss (and thus to compute $M_!$). Note, that enabling only the required assert-statement allows CBMC to reduce the size the generated CNF formula by slicing which improves performance of the model-counters.

3.3 Model Counting in the Pipeline

Preliminaries. For our approach we make use of model counting under projection. Given a CNF formula ϕ over the signature Σ, $|\text{models}(\phi\downharpoonright_\Delta)|$ denotes the number of satisfying assignments (models) of ϕ projected on the variables $\Delta \subseteq \Sigma$, where the formula $\phi\downharpoonright_\Delta$ denotes the projection of ϕ on Δ. Therefore, $\phi\downharpoonright_\Delta$ is the strongest formula over Δ which is entailed by ϕ [12, Logical Foundations] and states the same constraints on the atoms in Δ as ϕ. We define $\text{models}_\Delta(\phi) = \text{models}(\phi\downharpoonright_\Delta)$. For our use case, we distinguish between exact and approximative model-counters. Exact (e.g., sharpSAT [22]) or probabilistic exact (e.g., GANAK [17]) model counters compute the exact number of models t with a certain probability $\delta \leq 1$. Additionally, approximative or (δ, ϵ) model counters (e.g., ApproxMC [3,12,15,18]), return an estimated count $\tilde{c}$ with the guarantee of a relative error ϵ and a maximum uncertainty of δ: $Pr(\tilde{c} \in [t/(1+\epsilon), (1-\epsilon)t]) \geq 1-\delta$

The parameters ϵ and δ are given by the user. We further elaborate on model counting in Appendix A.

Measuring the Reliability. In Sect. 3.2 we obtain formulae which are satisfiable iff there is a start state leading the encoded program to violate the encoded specification. As noted earlier, we encode the inverse of the specification we are interested in and can thus quantify the number of inputs adhering to the specification at hand by using projected model counting with Δ containing exactly the propositional variables corresponding to the program input (a more detailed argument for the correctness of this approach is available in Sect. 3.4). Given the measured counts $\psi_?^m, \psi_?^h, \psi_!^m, \psi_!^h$ (respectively assumption miss and hit, assertion miss and hit), the computation for the case of exact results is relatively straight forward:

$$\begin{aligned} M_? &= \psi_?^m & E_\forall &= \psi_?^h - \psi_!^h \\ H_? &= \psi_?^h & E_\exists &= \psi_?^h - E_\forall - S_\forall = \\ S_\forall &= \psi_?^h - \psi_!^m & &= \psi_!^h + \psi_!^m - \psi_?^h \end{aligned}$$

For the case of approximate model counting using ApproxMC, each of the given counts are burdened with an uncertainty (δ, ϵ). Note, that these uncertainties are further propagated when we compute our ratios in (6). As the ratio $r_?^m$ only depends of the count $M_?$, its error (δ, ϵ) is just propagated towards $r_?^m$. For the ratios $r_\forall^e$, and $r_\forall^f$ the (δ, ϵ) error of each approximated count are multiplied, i.e., the error bound is $(1+\epsilon)^2$ with certainty of $(1-\delta)^2$. We clearly see that the ratios' become less precise.

We consider two numerical examples (for details see Sect. 4) to explore the error bounds.[2] Let us assume ApproxMC's standard parameters $\delta = 0.2$ and $\epsilon = 0.8$. First, consider the case of rangesum03. For this benchmark the following values represent the correct model counts: $H_? = 2^{96}$, $M_! = 2^{64}$, $H_! = 2^{96} - 2^{64}$. By computing the bounds (see Appendix A) we obtain the following error bounds with a probability of 0.64 each:

$$\text{For } r_\forall^e: -2.23 \leq \frac{\psi_?^h - \psi_!^h}{\psi_?^h} \leq 0.69 \quad \text{For } r_\forall^f: \quad 0.99\ldots \leq \frac{\psi_?^h - \psi_!^m}{\psi_?^h} \leq 0.99\ldots$$

We see a strong asymmetry between the error bounds for $r_\forall^e$ and those for $r_\forall^f$ caused by the strong asymmetry of $\psi_!^h$ and $\psi_!^m$. Note that for deterministic benchmarks the ratios can be computed through one minus the other ratio respectively. As a second case, consider the benchmark usqrt-broken for which the correct values are given by: $H_? = 2^{32}$, $M_! = 2^{31}$ and $H_! = 2^{31}$. Here, the split between assertion misses and assertion hits is essentially even. We obtain the following (equal) error bounds both for $r_\forall^e$ and $r_\forall^f$: $-0.62 \leq \frac{\psi_?^h - \phi}{\psi_?^h} \leq 0.89$ for $\phi \in \{\psi_!^h, \psi_!^m\}$.

Note that, firstly, for cases where $\psi_!^m$ and $\psi_!^h$ are expected or found to be of similar magnitude, a stricter value for ϵ should be considered. Secondly, in all cases it is worthwhile to examine, both, the ratios and all the absolute numbers: While a success ratio $r_\forall^f \geq 0.99$ seems good, it might still be the case that a large number of inputs yield an assertion miss if the input space $\mathcal{I}$ and $H_?$ are sufficiently large (as seen for case rangesum03 in Sect. 4).

[2] For conciseness, we mostly round the given bounds to two decimal places.

3.4 Correctness of the Pipeline

The pipeline we described in Sect. 3 is correct. By correctness, we mean that we get correct counts $\psi_?^m, \psi_?^h, \psi_!^m$ and $\psi_!^h$ for $M_?, H_?, M_!$ and $H_!$ when running the pipeline with an exact model counter. The correctness depends on three elementary properties: First, the program transformation preserves the behavior of the original program. Of course, a new program flow is established, but in cases without a violation of an assumption or assertion the transformed program behaves equally. Secondly, every hit or miss of an assumption or assertion is captured faithfully by a violated assert-statement (Lines 20 to 23). Thirdly, every (projected) model of the generated CNF formulae is indeed a representative for a violating or valid program path. In the Appendix B we elaborate the correctness deeply.

4 Evaluation

In order to evaluate our approach, we ran our prototypically-implemented pipeline on a number of C-programs from various benchmark families with the objective of showing its strengths, weaknesses and limits. Our experiments aimed to answer the following questions:

(Q1) Does our pipeline admit the quantification of more complex programs in comparison to [6]?
(Q2) How large can programs and input spaces become for given resource limits?
(Q3) Where does our bit-precise semantic help in obtaining more precise results and where is it too costly in comparison to [6]?
(Q4) Can our pipeline accurately quantify non-deterministic program behavior?

Implementation. For our experiments we implemented COUNTERSHARP[3] as a tool which transforms input C-programs into countable CNF formulae using CBMC. We apply the transformation (Sect. 3.1) on the input C-program's abstract syntax tree. Afterwards, the modified abstract syntax tree is converted back into C-code which is automatically passed to CBMC producing a total of five CNF formulae: Four checking for assertion/assumption hits and misses and another formulae to check for how many inputs (if any) the given bound is insufficient (i.e., for which inputs there are paths which need a deeper unroll of loops). The formulae provided by CBMC are adjusted to contain model counting projection instructions. We extract the propositional variables which represent the inputs variables in the C-program. Finally, our tool returns the five formulae containing projection instructions, which can then be processed by a suitable model counter of the user's choice. This approach both leaves the freedom to make use of other model counters and allows the trivial parallelization of this last quantification step. Our tool is publicly available.[4]

[3] COUNTERSHARP counts (thus sharp) counterexamples (thus counter) for a given specification.

[4] https://github.com/samysweb/counterSharp.

Experiment Setup. All experiments were run on a 4 core Intel Core i5-6500 processor and 16 GB of RAM. COUNTERSHARP was run with a timeout of 15 min. while the model counters had 5 min per instance. In order to achieve a fair comparison, the tool by [6] was given a timeout of 40 min to account for the multiple model counter runs in the case of COUNTERSHARP. All runs had a restricted memory of 2 GB. The execution of the tools was monitored using the *runlim* utility [2]. All scripts and raw results are available online [21]. The stated performance data is given as the median of 5 runs. APPROXMC was used in default configuration (though we used activated sparse hashing) and varying seeds across the runs, analogue for GANAK.

In the remaining, we compare varying setups: The setup `dim` denotes the analyzer of [6]. The setup `cS-gan` consists of COUNTERSHARP with GANAK for model counting for determinstic programs. The runtime is given as the time of COUNTERSHARP c added by the minimum runtime of (parallel) counting assertion hits $g_{H_!}$ and misses $g_{M_!}$: $c + \min(g_{H_!}, g_{M_!})$. Analogue, COUNTERSHARP with APPROXMC (`cS-app`), where the runtime is given as $c + \max(a_{H_!}, a_{M_!})$. Both setups `cS-gan` and `cS-app` simulate a run of COUNTERSHARP followed by a parallel run of GANAK or APPROXMC to compute assertion hit/miss counts. Additionally, we defined two setups used for non-deterministic cases named `cSn-gan` and `cSn-app`, where the runtime is given as $c{+}\max(\min(g_{H_?}, g_{M_?}), g_{H_!}, g_{M_!})$ and $c{+}\max(a_{H_?}, a_{M_?}, a_{H_!}, a_{M_!})$. Here, including extra counts to obtain reliable results for assumption hits, assertion hits and assertion misses—a necessity because of the possible overlap between assertion misses and hits in the case of non-determinism. GANAK can stop this computation once one of the two counters for assumption hits and misses returns, as the complement can be computed through subtraction. On the other hand, APPROXMC has to count both due to approximation errors (hence the max).

Benchmarks by [6]. In a first step we compared `cS-gan` and `cs-app` directly with `dim` using the benchmark set presented in their original work where each benchmark provided a version with $\mathcal{I}$ of size 10 and 1000. The comparison in the first part of Table 1 clearly shows that neither `cS-gan` nor `cS-app` can win against `dim` on the paper's original benchmark set. Indeed, it seems `dim` is very well suited to answer quantification questions on their benchmark set. To illustrate the reasons for its success, it is worth having a look at the considered benchmarks. One of the benchmarks where `cS-app` is particularly bad in comparison to `dim` is the count_up_down benchmark in Fig. 6. It is clear that a bounded model checker approach is worse at solving instances like this one due to the large number of loop unrolls necessary. An abstract interpretation approach like `dim`, seems to handle this kind of task a lot better—and independent of the loop size. This difference becomes even clearer for the case of the benchmark Mono3_1 which contains a loop repeated one million times.

```
int testfun(int n) {
  assume(n>=0&&n<=999);
  int x=n, y=0;
  while(x>0) { x--; y++;
    }
  assert(y==n);
}
```

Fig. 6. Benchmark count_up_down for input size 1000

Table 1. Evaluation of dim [6], cS-gan, cS-app, cSn-gan, cSn-app on the applicable benchmarks: Runtimes in seconds (median of 5 runs). (MO = out-of-memory, TO = timeout)

Deterministic Benchmarks: Comparison to dim[6]

Benchmark	Source	dim				cS-gan				cS-app			
Input Size		10		1000		10		1000		10		1000	
		time	exact	time	exact	time	exact	time	exact	time	exact	time	exact
bwd_loop1a	[6]	**0.51**	✓	**0.55**	✓	1.68	✓	2.82	✓	4.98	✓	8.68	≈
bwd_loop2	[6]	**0.67**	✓	**0.41**	≈	1.76	✓	1.35	✓	2.67	✓	4.24	≈
count_up_down	[6]	**0.54**	✓	**0.27**	✓	1.16	✓	8.75	✓	1.19	✓	142.45	≈
example1a	[6]	**0.64**	✓	**0.46**	✓	1.59	✓	1.83	✓	5.18	✓	10.02	≈
example7a	[6]	**0.54**	✓	**0.82**	✓	1.07	✓	TO	–	1.76	✓	TO	–
gsv2008	[6]	**0.43**	✓	**0.62**	✓	1.6	✓	8.89	✓	3.03	✓	79.41	≈
hhk2008	[6]	**0.44**	✓	**0.68**	✓	0.96	✓	TO	–	1.56	✓	TO	–
Log	[6]	**0.57**	≈	**0.69**	≈	1.45	✓	TO	–	1.90	✓	TO	–
Mono3_1	[6]	**0.48**	≈	**0.74**	≈	TO	–	TO	–	TO	–	TO	–
Waldkirch	[6]	**0.4**	✓	**0.37**	✓	1.35	✓	8.17	✓	1.56	✓	146.56	≈

Complex benchmarks

Benchmark	Source	dim	cS-gan		cS-app	
		reason for failure	time	exact	time	exact
floor-broken	[20]	float	TO	–	**11.13**	≈
floor	[20]	float	**1.07**	✓	8.47	✓
overflow	crafted	incorrect (overflow)	**1.41**	✓	1.69	≈
Problem10_16	[19]/[10]	timeout	TO	–	**723.73**	≈
Problem13_4	[19]/[10]	timeout	TO	–	TO	–
rangesum03	[19]/[4]	arrays	**0.51**	✓	2.0	≈
rangesum05	[19]/[4]	arrays	TO	–	TO	–
usqrt-broken	[20]	bit arithmetic	TO	–	**17.06**	✓
usqrt	[20]	bit arithmetic	TO	–	**105.83**	✓

Nondeterministic Benchmarks[a]: Comparison to dim[6]

Benchmark	Source	dim	cSn-gan	cSn-app
		time	time	time
bwd_loop10-2	[6]	**0.39**	MO	MO
bwd_loop10	[6]	**0.65**	MO	MO
bwd_loop7-2	[6]	**0.55**	230.11	6.41
bwd_loop7	[6]	**0.88**	6.36	6.41
example7b-2	[6]	**0.45**	TO	TO
example7b	[6]	**0.17**	4.18	3.63
for_bounded-2	[6]	**0.25**	MO	135.51
for_bounded-1	[6]	**0.67**	2.91	2.17
nondet	crafted	**0.87**	TO	1.83

[a] Exactness classification is not applicable for nondeterministic programs.

As a matter of fact, every time cS-gan or cS-app take more than 15 s or time out on the benchmark set in the first part of Table 1, it is due to a benchmark which requires an unrolling of a loop with depth larger or equal to 500.

The only exception to this rule is the benchmark gsv2008 which takes longer than 15 s for a run with unroll depth 101. Conversely, nearly all benchmarks executed within 15 s require a less deep loop-unrolling. It thus seems that loop depth is the main bottleneck for our approach on this benchmark set. At the same time, the program logic of all benchmarks in the set is comparatively simple: subtraction and addition paired with while, for or if-statements. Towards answering (Q3) we observe that the approach proposed by Dimovski and Legay [6] works very well on this type of benchmark. However, the question remains whether more complex benchmarks might require a more precise semantic than the one available in `dim`.

Complex Programs. To this end, we looked at a number of other benchmarks from the SV-Competition [19] as well as the C-Snippets [20] code repository. We collected a number of benchmarks which allowed some form of quantification through input variables and either had a wider range of interesting, representative program constructs (such as arrays, floats, or bit arithmetic) or represented suitable candidates to test the scalability of our approach. These benchmarks were manually modified to allow quantification on them (addition of suitable assertion- and assumption-statements) and to compare different sizes (e.g., comparison of 3 element and 5 element array input). Additionally, we crafted one benchmark showing behavior we were particularly interested in. All of the instances mentioned in the second part of Table 1 show behavior which cannot be analyzed by `dim`: The tool was either unable to analyze the programs due to the use of arrays and bit arithmetic, ran into a timeout due to instance size or even produced wrong results due to the neglection of overflows. We discuss these benchmarks as well as why we believe they are difficult and what our results show.

The first two benchmarks in the table represent a broken and correct algorithm to compute the floor function of a floating point number. For the experiment we assumed a positive, non-infinite, non-NaN 128 bit `long double` input which the program is supposed to round down into another `long double` number using bit arithmetic. The final assertion checks whether the computed number is smaller than the original number. We added a broken version of the program which, according to `cS-app`, breaks the specification for approx. $1.5 * 2^{95}$ of the 2^{127} positive inputs. The `cS-app` tool also returns that 2^{127} inputs comply to the specification. Here we see the reason why it is necessary to compute both counts (assertion miss and assertion hit): The ratio between assertion hits and misses is approx. $1.75 * 10^{-10}$ and thus the entirety of assertion misses well lies within the error bounds of the assertion hit measurement. Therefore, it seems good advice to always inspect the numbers and their relations manually in order to spot such approximation errors during data interpretation.

The float instances are followed by the only instance which `dim` was able to compute faster than our tools and which is a crafted benchmark. However, `dim` returns faulty results due to an integer overflow which remains undetected in polyhedra: The benchmarks assumes an input `x` strictly larger than `INT_MIN/2`. If `x` is negative, `|INT_MIN/2|` is subtracted while `|INT_MIN/2|` is added if `x` is positive. The final assertion requires that `x` be negative. `dim` returns that the assertion is met by one third of the inputs (i.e., by all inputs which are negative) and this is of course what a polyhedra tool would have to return due to its abstract domain. However, closer examination shows that the addition in the positive case leads to an overflow for any number larger than `|INT_MIN/2|`. Thus, the assertion is actually hit for two thirds of all inputs (i.e., all negative inputs and all sufficiently large positive inputs). This benchmark was of course crafted to show the drawbacks in the use of polyhedra for quantification. While one might argue that this is an unfair comparison, we merely want to draw attention to the fact that such mistakes are not uncommon in programming and can be better quantified with a tool with bit-precise semantic, like COUNTERSHARP.

The following two benchmarks stem from the RERS Challenge 2018 [10] and concern the reachability problem for linear time logic (LTL) formulae. For each instance a number of error states are defined which should not be reached. The states reached by the program depend on the input variables. `Prob10` is then run for 16 time steps converting the program in a way that it takes 16 char variables as inputs, while `Prob13` was only supposed to be run for 4 time steps. As only values from 1 to 5 (instance 10) or 1 to 10 (instance 13) are allowed, suitable input restrictions were introduced for both `dim` and `cS-*`. The original RERS challenge had 3 problem levels of which `Prob10` was level *small* and `Prob13` was level *medium*. `Prob10` consists of approx. 1.3k LoC while `Prob13` contains around 114k LoC. While `cS-app` works relatively well on the small instance, COUNTERSHARP runs into a timeout for the medium size instance when trying to construct the CNF formulae. On the other hand, `cS-gan` cannot compute either of the two benchmarks. This is probably due to the fact that the number of clauses (and variables) in the instance are at least one order of magnitude larger than the numbers in other benchmarks solvable by `cS-gan`. It turned out that `dim` seems to be unable to solve both instances due to a memout. We believe the reason for this is the scale of the benchmark instances which is much larger than instances previously considered in the evaluation of `dim`.[5]

The following two rangesum examples implement the computation of a rangesum for an array of size 3 and 5. This benchmark can, again, only be evaluated for the `cS-*` tools as it requires the ability to handle arrays. We modified the benchmark in such a way that the elements inside the array are given as input parameters to the main function thus allowing an easy quantification across all possible array values. Both, `cS-app` and `cS-gan` are able to quickly solve the 3 element instance. We also see that `cS-gan` is particularly efficient in cases where one of the two counts (here the assertion hits) can be computed particularly fast, as the complementary count (which in this case would have timed out) can be computed exactly. At the same time, both tools still fail to solve the 5 element instance. This can be explained by the vastly larger input space for an array of 5 integer elements which considers an input space 2^{64} times larger than the input space for 3 elements.

[5] Given the choice of a relatively low memout initially, we reran `dim` on Problem_10 with 8 GB of memory available. However, the program ran into a timeout after 2400 s using 4.5 GB of memory.

The final two benchmarks stem from Fig. 1 which represents the computation of an integer square root. As we already mentioned in the introduction, the code in the listing is flawed (this corresponds to usqrt-broken). Namely, the shift in Line 12 is a signed right shift which introduced flawed results for any input `x` larger than 2^{30}. Indeed, `cS-app` correctly returns in all 5 runs that the assertion succeeds (resp. fails) for 2^{30} (resp. also 2^{30}) inputs while 2^{31} inputs already miss the assumption. For the case of the fixed version (usqrt) the main challenge for `cS-app` (or its underlying SAT solver to be precise) is showing the unsatisfiability of the assertion miss formula which takes $104.93s$ in comparison to $3.01s$ for counting all 2^{31} assertion hits.

Concerning questions (Q1) and (Q2) we see that there is a category of benchmarks which the tool by Dimovski and Legay [6] fails to analyze and in all fairness was probably never meant to analyze given the polyhedral domain approach. On the other hand our tools `cS-gan` and `cS-app` are able to solve a number of these (in our opinion) interesting benchmarks. In particular our tools admit the quantification of larger programs (e.g., Problem_10) and more complex programs (i.e., arrays or advanced arithmetic). Concerning (Q3) we showed how a bit precise semantic helps in correctly quantifying overflow errors.

Non-deterministic Programs. Turning to the question of non-deterministic programs (Q4), we can observe in the last part of Table 1 that all benchmarks considered are solved the fastest by `dim`. We observe, once again, that the cases where `cSn-gan` or `cSn-app` yield a timeout or memout are cases with very deep loops with the notable exception of the benchmark nondet which does not involve loops. While `dim` is the fastest tool, it turns out that `cSn-gan` and `cSn-app` can sometimes produce more precise results than the approximate result yielded by `dim`. In particular, the benchmark nondet which corresponds to the code in `test2` in Fig. 2 demonstrates this behavior. As previously discussed, the assertion can fail for any input due to the non-deterministic value of `y`, but it must fail for any negative input `x`. Accordingly, `cSn-app` reports a possible assertion miss for 2^{32} cases and a possible assertion hit for 2^{31} cases. This corresponds to a probability of success between 0 and 50% and a probability of violation between 50 and 100% for a uniform input distribution. While `dim` is faster in reporting its result, it reports that both success and violation probability lie within the range 0 to 100% which is a lot less precise than `cSn-app`. The reason for this behavior is quite likely the multiplication in the program which is difficult to handle for the polyhedra abstract domain. Equally, the complexity of the formula due to multiplication might be a reason why the exact counter `cSn-gan` fails to quantify the benchmark. Concerning (Q4) we find that our tools admit the accurate quantification of non-deterministic benchmarks. Just as for deterministic benchmarks our approach is limited by loop depth. Once again certain cases can profit from the bit precise semantic allowing for a more precise quantification (as in the case of nondet).

Analysis of Bottlenecks. CBMC and the program transformation take most of the required time for instances with a small input space. As the input space size grows, the counting step either becomes infeasible (especially for GANAK) or requires a lot more computation time in comparison to CBMC (especially for APPROXMC). Since counting can take significant amounts of time and GANAK and APPROXMC showed drastic variations in performance on some benchmarks (e.g. Waldkirch where GANAK solves the instance a magnitude faster than APPROXMC or float where the opposite behavior can be observed), an approach using a portfolio of model counters could reduce computation time.

5 Related Work

Klebanov et al. [11] present the idea of using CBMC to generate CNF formulae for model counting to compute the information leakage of a program.

Geldenhuys et al. [8] proposed an approach using symbolic execution for the extraction of path probabilities. To this end, a classic symbolic execution methodology is modified such that it computes probabilities instead of path conditions: At every branching point the algorithm computes the likelihood of descending into the subprocedures at hand under the assumption of a uniform distribution of input variables using the polytope utility LattE [14]. Path probabilities are then computed through multiplication of branch probabilities. Lui and Zhang [13] propose a similar approach which does not use formula slicing or memoization. Filieri et al. [7] applies symbolic execution on Java-programs to extract path conditions for every possible execution path. The paths are then labeled as success, failure or "gray" paths with unknown success status. The number of models for these path conditions are, again, counted with LattE, though they may be constrained by usage profiles describing an input distribution. This allows to compute the probability of failure (failure paths) and the confidence in the result (gray paths) under the given usage profile. The approach furthermore supports multi-threading. In contrast, our approach differs in both: the programming language under evaluation and the counting technique approach allowing more complex behavior than the linear constraints described by polytopes.

The approach of Dimovski and Legay [6] allows the computation of assertion hit and miss probabilities under a uniform input distribution for programs written in a subset of C. The approach uses abstract interpretation to describe the program behavior using the polyhedra domain, which encodes linear constraints between program variables. In a first step, a forward analysis of the program is performed computing an invariant which must hold after the program execution. The forward analysis is based on intervals which have to be defined for the input variables of interest. The forward analysis works by applying sound (but overapproximating) transfer functions on the specified input domain. Afterwards, in a backward pass, conditions specifying an assertion hit/miss are propagated backward using appropriate transfer functions. This approach produces two overapproximating linear constraint preconditions: One for assertion hits and one for assertion misses. The number of inputs corresponding to the precondition at hand are then again computed using LattE [14]. Both results (assertion hit and miss) represent an upper-bound for the number of inputs leading to the specified behavior. By calculating the complement of each count, the tool can further provide a lower bound for each of the two cases. Additionally, the analyzer provides a semantic that also allows a quantification of non-deterministic programs. While the tool of [6] is similar, it only supports a more restrictive subset of the C programming language in comparison to our approach: The subset supports no complex data types and only integers as basic data types while our work supports both arrays and floats. Additionally, the use of non-linear operators as well as bit operators is greatly restricted. As our tool relies on bounded model checking and SAT model counting, we can avoid this limitation by making use of a bit-precise semantic which also takes overflows into account. On the other hand, our tool has stricter limitations on the loops that can be evaluated as the bounded model checker must unroll such loops up to a sufficient depth for exact results, while Dimovski and Legay [6] can analyze such loops in considerably less time using abstract interpretation if the loop conditions are sufficiently simple.

6 Conclusion

In this work we presented a formal approach allowing the precise quantification of software properties given as assumptions and assertions. The resulting pipeline contains three steps: (1) program transformation, (2) conversion into CNF formulae and (3) model counting. We implemented the pipeline prototypically and undertook an extensive, qualitative evaluation of the prototype. We show that our quantification approach is both, feasible, even a program with 1300 LoC was still analyzable with the given resources, and useful, sometimes yielding results beyond the precision of previous approaches. We further introduced a semantic allowing the quantitative analysis of non-deterministic programs and showed that the pipeline is capable of analyzing such instances, too. In our comparison to the tool in [6], we showed strengths and weaknesses of both approaches: While their approach works very well on programs with high loop depth, our approach allows the quantification of programs using more complex arithmetical operations such as bit operators or multiplication. Equally, we presented a case in which the bit precise semantic helps in returning more precise results for the quantification of non-deterministic programs.

To summarize, we believe that a SAT counting based approach is an interesting addition to the tool set available for software quantification. While some benchmarks also remain out of reach for our approach (either due to the input space explosion with growth in input variables or due to the sheer problem size) and approximation errors need to be considered when analyzing, the capabilities of our approach will scale with advances in model counting in the same way the capabilities of bounded model checkers grow with the advances in SAT solving.

Future Work. The current pipeline can only serve as an estimator of probability of failure for the case of a uniform input distribution. An extension to non-uniform input distributions could further increase the utility of the tool.

To this end, we identify a need for *projected* weighted model counters (approximate or exact). Additionally, CBMC allows the use of complex data types such as structs. However, the current setup only allows basic data types as input variables. A methodology allowing a quantification over a set of complex data structure instances as input might be an interesting addition to the tool in its current form.

A Model Counting

Model counting is the counting of satisfying assignments (models) of a given (propositional) formula. We use model counting to quantify the number of inputs for which a given variable corresponds to our specification. This section gives a brief overview over the notions of model counting and recent advances in the field before it explains how we harness propositional model counting techniques in our pipeline.

Preliminaries. Assuming a CNF formula ϕ over the signature Σ, an assignment is the interpretation of the propositional atoms $I\colon \Sigma \to \{t, f\}$. There are $2^{|\Sigma|}$ possible assignments for the variables in ϕ. We can then define $\text{models}(\phi) := \{I \mid I \models \phi\}$ as the set of all assignments satisfying ϕ. A model counter thus computes the cardinality $|\text{models}(\phi)|$. In our case, we are only interested in the assignments of the program's input variables. For this we need the projection on propositional formulae. Let $\Delta \subseteq \Sigma$ be a signature, then $\phi{\downharpoonright_\Delta}$ denotes the strongest formula over Δ which is entailed by ϕ [12, Logical Foundations]. Note, $\phi{\downharpoonright_\Delta}$ states the same constraints on the atoms in Δ as ϕ. We denote with $\text{models}_\Delta(\phi) := \{I{\downharpoonright_\Delta} \mid I \models \phi{\downharpoonright_\Delta}\}$ the set of Δ-models of the Δ-projection of ϕ. There exist exact and approximative model-counters.

Exact Model Counting. In our work we use the probabilistic exact model counter GANAK [17] which is freely available and represents an improved version of sharpSAT [22]. For a given parameter δ the model counter returns the exact model count with probability $1 - \delta$. In practice the tool returned exact results for all benchmarks considered by [17] when setting $\delta = 0.05$. The model counter also recently won a first place in the Model Counting Competition 2020 [9] as part of a portfolio.

Approximate Model Counting. Alternatively, we can use the approximate model counter ApproxMC [3,12,15,18] which uses a *probably approximate correct* (or PAC) algorithm. This means the algorithm provides a theoretical guarantee on the relation between the correct model count ψ and the result of ApproxMC $\psi_\mathcal{A}$ which can be configured by some (ϵ, δ):

$$\Pr\left[\frac{\psi}{1+\epsilon} \leq \psi_\mathcal{A} \leq (1+\epsilon)\psi\right] \geq 1 - \delta \tag{7}$$

It is worth noting that two runs of ApproxMC can be considered independent random variables as the correctness of the result depends on the random variables within the algorithm and not its input value.

Bounds for Ratios. For the ratios $r^e_\forall$, and $r^f_\forall$ we obtain error bounds through the following corollary:

Corollary 1. *Let ϕ_1, ϕ_2 be two model counts measured according to (7) and let further ϕ_1^*, ϕ_2^* be the correct model counts, then:*

$$\Pr\left[1-(1+\epsilon)^2\frac{\phi_2^*}{\phi_1^*} \leq \frac{\phi_1-\phi_2}{\phi_1} \leq 1-\frac{\phi_2^*}{(1+\epsilon)^2\phi_1^*}\right] \geq (1-\delta)^2$$

For both ratios ϕ_1 in Corollary 1 corresponds to $\psi^h_?$ (consequently ϕ_1^* represents $H_?$). ϕ_2 then either corresponds to $\psi^h_!$ or $\psi^m_!$ respectively yielding the measurement for $r^e_\forall$ or $r^f_\forall$ (hence ϕ_2^* represents $H_!$ or $M_!$). We clearly see that the ratios' error bounds become less precise in comparison to the single measure case.

B Correctness of the pipeline

In this section we will argue why the pipeline we described in Sect. 3 is correct. By correctness, we mean that we get correct counts $\psi^m_?, \psi^h_?, \psi^m_!$ and $\psi^h_!$ for $M_?, H_?, M_!$ and $H_!$ when running the pipeline with an exact model counter. The correctness depends on three elementary properties: First, the program transformation preserves the behavior of the original program. Of course, a new program flow is established, but in cases without a violation of an assumption or assertion the transformed program behaves equally. Secondly, every hit or miss of an assumption or assertion is captured faithfully by a violated assert-statement (Lines 20 to 23). Thirdly, every (projected) model of the generated CNF formulae is indeed a representative for a violating or valid program path.

Program Transformation Preserves Behavior. We need to consider that unifying return statements, and early execution abortion do not alter the program flow unfaithfully, especially, no program behavior is introduced which violates an assumption or assertion.

Considering the unifying of return statements: When in the old program a return statement is reached, the sub-routine is immediately returned to the callee. In the transformed program, we store the return value if necessary, and jump to the last remaining return statement under the end-label. This takes an extra step, but this step does not modify the value of any program variable.

The early abortion is triggered only when an assumption or assertion violation occurs, as the variables `am` and `as` are not in the original program. Therefore, for any input without a specification violation, the transformed program behaves exactly the same, terminating with `am` and `as` equal to `false`.

Capturing Hits and Misses. We need to consider four cases (assumption/assert hit/miss), but due to symmetry reason, we only argue in the two critical cases: the violation of an assumption or assertion. Please reconsider, that we encode the required property with a negation into the program, because a negation on the property is added by CBMC. As a summary, we count the inputs which violate an assert-statement.

Let us consider the case of assumption misses ($M_?$). As previously explained, the resulting program should contain exactly one assert-statement which is violated iff the given input violates one of the specified assumptions. This assert-statement (see Line 23 in Fig. 5) checks the program variable `am`. The program execution, just as in the original program, begins with the subroutine under consideration and the checks of the assumptions. If at least one assumption is violated for a specific input, the variable `am` becomes `true`, and the program is aborted early. For the entry routine, we directly jump to the assert-statement which will be violated. Therefore, any violation of user-defined assertion is ignored in the further program flow.

Next, we consider the case of assumption hit ($H_?$) with an occurring assertion miss ($M_!$). The original (and transformed) program starts again in the entry function, but as the assumptions are met, the remaining body is executed (`am` stays `false` for the complete remaining execution). Afterwards, if an assertion in is violated somewhere during the program execution, the variable `as` becomes and stays `true`, and also the program flow of the current routine and its callees is aborted recursively until one of the assert-statements (Lines 20 to 23) is reached. In detail, only the assert-statements in Line 21 (assertion miss) and in Line 22 (assumption hit) are violated. If no assertion had been violated, the assert-statement in Line 20 would be violated.

Model-Counting. Our projected model counter computes the number of distinguishable assignments projected to a variable set Δ which satisfy the given formula. If we want to obtain correct results for our metrics, we thus need to choose our formulae and projection set in such a way that the model counter returns correct counts. Our bounded model checker CBMC returns formulae which are satisfiable iff a specified assert-statement can be missed for some program path (within the defined bound). The formula returned by CBMC contains propositional variables for all bits of our program's input. We thus choose exactly those input representatives as set Δ. If we are able to construct a program containing exactly one assert-statement which is missed iff the given input has a program path that implies an assertion hit, we can then compute $H_!$ by constructing this program and subsequently passing the transformed program to CBMC and later on to the model counter projecting on the input variables. Correspondingly, we can compute $M_!, H_?$ and $M_?$ if we can construct a program containing exactly one assert-statement which is missed iff the conditions for the corresponding variable are met.

References

1. Avizienis, A., Laprie, J.C., Randell, B., Landwehr, C.: Basic concepts and taxonomy of dependable and secure computing. Inst. Syst. Res. **0114**, 10 (2001). https://doi.org/10.1016/S0005-1098(00)00082-0, http://drum.lib.umd.edu/handle/1903/5952
2. Biere, A.: runlim. Website (2016). http://fmv.jku.at/runlim
3. Chakraborty, S., Meel, K.S., Vardi, M.Y.: Algorithmic improvements in approximate counting for probabilistic inference: from linear to logarithmic SAT calls. In: Kambhampati, S. (ed.) Proceedings of the Twenty-Fifth International Joint Conference on Artificial Intelligence, IJCAI 2016, New York, NY, USA, 9–15 July 2016, pp. 3569–3576. IJCAI/AAAI Press (2016). http://www.ijcai.org/Abstract/16/503
4. Chen, Y.-F., Hong, C.-D., Sinha, N., Wang, B.-Y.: Commutativity of reducers. In: Baier, C., Tinelli, C. (eds.) TACAS 2015. LNCS, vol. 9035, pp. 131–146. Springer, Heidelberg (2015). https://doi.org/10.1007/978-3-662-46681-0_9
5. Clarke, E., Kroening, D., Lerda, F.: A tool for checking ANSI-C programs. In: Jensen, K., Podelski, A. (eds.) TACAS 2004. LNCS, vol. 2988, pp. 168–176. Springer, Heidelberg (2004). https://doi.org/10.1007/978-3-540-24730-2_15
6. Dimovski, A.S., Legay, A.: Computing program reliability using forward-backward precondition analysis and model counting. In: FASE 2020. LNCS, vol. 12076, pp. 182–202. Springer, Cham (2020). https://doi.org/10.1007/978-3-030-45234-6_9
7. Filieri, A., Pasareanu, C.S., Visser, W.: Reliability analysis in symbolic PathFinder. In: Proceedings - International Conference on Software Engineering, pp. 622–631 (2013). https://doi.org/10.1109/ICSE.2013.6606608
8. Geldenhuys, J., Dwyer, M.B., Visser, W.: Probabilistic symbolic execution. In: Proceedings of the 2012 International Symposium on Software Testing and Analysis, pp. 166–176. ACM (2012). https://doi.org/10.1145/2338965.2336773
9. Hechter, M., Fichter, J.K.: Model Counting Competition 2020. Website (2020). https://mccompetition.org/. Accessed 5 Dec 2020
10. Jasper, M., Mues, M., Schlüter, M., Steffen, B., Howar, F.: RERS 2018: CTL, LTL, and reachability. In: Margaria, T., Steffen, B. (eds.) ISoLA 2018. LNCS, vol. 11245, pp. 433–447. Springer, Cham (2018). https://doi.org/10.1007/978-3-030-03421-4_27
11. Klebanov, V., Manthey, N., Muise, C.: SAT-based analysis and quantification of information flow in programs. In: Joshi, K., Siegle, M., Stoelinga, M., D'Argenio, P.R. (eds.) QEST 2013. LNCS, vol. 8054, pp. 177–192. Springer, Heidelberg (2013). https://doi.org/10.1007/978-3-642-40196-1_16
12. Klebanov, V., Weigl, A., Weisbarth, J.: Sound probabilistic #SAT with projection. In: Electronic Proceedings in Theoretical Computer Science, EPTCS, vol. 227, pp. 15–29 (2016). https://doi.org/10.4204/EPTCS.227.2
13. Liu, S., Zhang, J.: Program analysis: From qualitative analysis to quantitative analysis. In: Proceedings - International Conference on Software Engineering, pp. 956–959 (2011). https://doi.org/10.1145/1985793.1985957
14. Loera, J.A.D., Hemmecke, R., Tauzer, J., Yoshida, R.: Effective lattice point counting in rational convex polytopes. J. Symb. Comput. **38**(4), 1273–1302 (2004). https://doi.org/10.1016/j.jsc.2003.04.003

15. Meel, K.S., Akshay, S.: Sparse hashing for scalable approximate model counting: theory and practice. In: Hermanns, H., Zhang, L., Kobayashi, N., Miller, D. (eds.) LICS 2020: 35th Annual ACM/IEEE Symposium on Logic in Computer Science, Saarbrücken, Germany, 8–11 July 2020, pp. 728–741. ACM (2020). https://doi.org/10.1145/3373718.3394809
16. Orso, A., Rothermel, G.: Software testing: a research travelogue (2000–2014). In: Future of Software Engineering, FOSE 2014 - Proceedings, pp. 117–132 (2014). https://doi.org/10.1145/2593882.2593885
17. Sharma, S., Roy, S., Soos, M., Meel, K.S.: GANAK: a scalable probabilistic exact model counter. In: Kraus, S. (ed.) Proceedings of the Twenty-Eighth International Joint Conference on Artificial Intelligence, IJCAI 2019, Macao, China, 10–16 August 2019, pp. 1169–1176. ijcai.org (2019). https://doi.org/10.24963/ijcai.2019/163
18. Soos, M., Meel, K.S.: BIRD: engineering an efficient CNF-XOR SAT solver and its applications to approximate model counting. In: The Thirty-Third AAAI Conference on Artificial Intelligence, AAAI 2019, The Thirty-First Innovative Applications of Artificial Intelligence Conference, IAAI 2019, The Ninth AAAI Symposium on Educational Advances in Artificial Intelligence, EAAI 2019, Honolulu, Hawaii, USA, 27 January–1 February 2019, pp. 1592–1599. AAAI Press (2019). https://doi.org/10.1609/aaai.v33i01.33011592
19. SoSy-Lab LMU: SV-Benchmarks (2020). https://github.com/sosy-lab/sv-benchmarks
20. Stout, B.: C Snippets (2009). http://web.archive.org/web/20101204075132/c.snippets.org/
21. Teuber, S., Weigl, A.: Evaluated artifact for "quantifying software reliability via model-counting" (2021). https://doi.org/10.5445/IR/1000134169
22. Thurley, M.: sharpSAT – counting models with advanced component caching and implicit BCP. In: Biere, A., Gomes, C.P. (eds.) SAT 2006. LNCS, vol. 4121, pp. 424–429. Springer, Heidelberg (2006). https://doi.org/10.1007/11814948_38
23. Visser, W., Bjørner, N., Shankar, N.: Software engineering and automated deduction. In: Herbsleb, J.D., Dwyer, M.B. (eds.) Proceedings of the on Future of Software Engineering, FOSE 2014, Hyderabad, India, 31 May–7 June 2014, pp. 155–166. ACM (2014). https://doi.org/10.1145/2593882.2593899

Quantitative Models and Metamodels: Analysis and Validation

Compositional Safe Approximation of Response Time Distribution of Complex Workflows

Laura Carnevali[1], Marco Paolieri[2], Riccardo Reali[1(✉)], and Enrico Vicario[1]

[1] Department of Information Engineering, University of Florence, Florence, Italy
{laura.carnevali,riccardo.reali,enrico.vicario}@unifi.it
[2] Department of Computer Science, University of Southern California, Los Angeles, USA
paolieri@usc.edu

Abstract. We propose a compositional technique for efficient evaluation of the cumulative distribution function of the response time of complex workflows, consisting of activities with generally distributed stochastic durations composed through sequence, choice/merge, split/join, and repetition blocks, with unbalanced split and join constructs that break the structure of well-formed nesting. Workflows are specified using a formalism defined in terms of stochastic Petri nets, that permits decomposition of the model into a hierarchy of sub-workflows with positively correlated response times, which guarantees a stochastically ordered approximation of the end-to-end response time when intermediate results are approximated by stochastically ordered distributions and when dependencies are simplified by replicating activities appearing in multiple sub-workflows. This opens the way to an efficient hierarchical solution that manages complex models by recursive application of Markov regenerative analysis and numerical composition of monovariate distributions.

Keywords: Stochastic workflow · Response time distribution · Structured model · Compositional evaluation · Stochastic ordering

1 Introduction

A workflow is an orchestration of concurrent and sequential activities, mainly shaped by constructs of split/join, choice/merge, and sequence, occasionally including dependencies that break well-formed nesting and repetitions that produce transient cycles [29]. This abstraction fits a large variety of material and digital processes, in multiple contexts such as supply chain management [18], administration [1], composite web services [12], cloud "functions as a service" [33].

When the model is associated with a measure of probability, quantitative evaluation may provide measures of interest for different stages of development and operation [28,8,14,7], supporting the achievement of a tradeoff between some of them, such as the average response time, the subtask dispersion, and the energy consumption [26]. In particular, the Cumulative Distribution Function (CDF) of the end-to-end response time is relevant for the evaluation of the

A. Abate and A. Marin (Eds.): QEST 2021, LNCS 12846, pp. 83–104, 2021.
https://doi.org/10.1007/978-3-030-85172-9_5

expected reward under a Service Level Agreement (SLA) with soft deadlines and penalty functions [16,27]. In this case, a stochastically ordered approximation of the CDF produces a safe approximation of the expected reward.

However, practical feasibility faces a difficult combination of recurring complexities: activity durations follow general (i.e., non-Exponential) probability distributions (GEN), often supported within firm bounds enforced by design or by contract, which cast the problem in the class of non-Markovian processes [11]; the interleaving of actions in concurrent sub-workflows leads to explosion of the state space and complex dependencies due to the overlap among concurrent activities with GEN duration [6].

Compositional solution can address both complexities by avoiding explicit representation of interleavings, limiting state space explosion and simplifying the structure of underlying stochastic processes of individual components [6]. In [36], mean time and standard deviation of the completion time of a workflow of activities with GEN duration composed by fork/join, sequence, and repetition are derived through an efficient bottom-up calculus, which is extended in [22] for the special case of Continuous Phase (CPH) durations. In [2], the completion time of an acyclic attack tree with CPH delays is evaluated by repeatedly composing CPH distributions in a bottom-up approach, with possible approximation to compress their representation to maintain a bounded number of phases. In [10], the response time of an acyclic workflow obtained by well-formed nesting of activities with GEN durations is evaluated in a two-step approach, first applying Markov regenerative analysis to nested sub-workflows identified so as to have a limited degree of concurrency, and then repeatedly composing the resulting monovariate distributions bottom-up to obtain the workflow response time.

In this paper, we propose a compositional technique for efficient evaluation of the response time CDF of complex workflows consisting of activities with GEN durations composed through sequence, choice/merge, split/join, and *repetition blocks*, with *unbalanced split and join constructs that may break the structure of well-formed nesting*. Workflows are specified using a higher level formalism defined in terms of stochastic Petri nets, that permits decomposition of the model into a hierarchy of sub-workflows with positively correlated response times. This guarantees a stochastically ordered approximation of the end-to-end response time when completion times of intermediate sub-workflows are approximated by a stochastically ordered fitting distribution and when activities shared among multiple sub-workflows are replicated as independent random variables so as to reduce dependencies. These approximations allow an efficient hierarchical solution approach that manages complex models by recursive application of Markov regenerative analysis and numerical composition of monovariate distributions.

The rest of the paper is organized in four sections, addressing: specification of models and their representation as a *structure tree* (Sect. 2); decomposition of models into a hierarchy of sub-workflows identified in the structure tree by heuristics trading approximation for complexity, and re-composition of the end-to-end response time distribution (Sect. 3); results of numerical experimentation (Sect. 4); and conclusions (Sect. 5). Theorem proofs and other details are deferred to the Appendix (Sect. 6).

2 Modeling Workflows with Structured STPNs

We specify workflows with stochastic activity durations using a formalism that constrains expressivity of a class of stochastic Petri nets (Sects. 2.1 and 2.2) so as to ensure positive correlation among their response times and to enable derivation of a structured representation of the model (Sect. 2.3).

2.1 Stochastic Time Petri Nets (STPNs)

Stochastic Time Petri Nets (STPNs) model concurrent timed systems: transitions represent the execution of activities, tokens within places account for the system logical state, and directed arcs from input places to transitions, and from transitions to output places, define precedence relations among activities [15]. A transition is enabled if each of its input places contains at least one token; at newly enabling, a transition samples a time-to-fire from a CDF with support between an Earliest Firing Time (EFT) and a Latest Firing Time (LFT); upon firing, it removes a token from each input place and adds one token to each output place. The choice among transitions with equal time-to-fire is solved by a random switch determined by probabilistic weights.

As shown in Fig. 1a, places are represented as circles, tokens as dots inside places, immediate (IMM) transitions (i.e., with zero time-to-fire) as thin bars (e.g., `as1`), deterministic (DET) transitions (i.e., with nonzero deterministic time-to-fire) as gray bars (e.g., `z1`), EXP transitions as white bars (e.g., `v1`), and other GEN transition as black bars (e.g., `q`). Where necessary, labels indicate rates, firing intervals $[EFT, LFT]$, CDF types (e.g., *uniform* or *expolynomial*).

2.2 STPN Blocks

Workflows with stochastic durations can be specified using a fragment of the expressivity of STPNs, which is sufficient to represent a variety of workflow con-

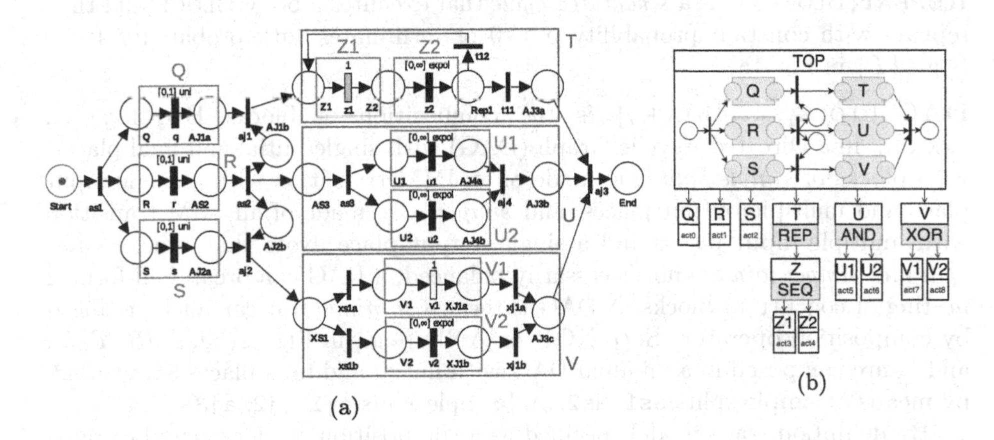

Fig. 1. (a) Workflow STPN and (b) its structure tree (composite blocks in blue). (Color figure online)

trol patterns [29,37] while making explicit a structure of composition enabling efficient timed analysis. To this end, we specify workflows by recursive composition of *blocks*, each defined as an STPN with a single *initial place* and a single *final place*. The execution of a block starts when a token is added to the initial place, and it eventually terminates, with probability 1 (w.p.1), when a token reaches the final place. Blocks compose elementary STPN transitions through nested constructs of concurrent (split/join) and sequential (sequence, choice/merge, repeat) behavior, or by acyclic compositions that break well-formed nesting through unbalanced fork and join operations (simple split, simple join):

$$\begin{aligned}\textsc{Block} := \textsc{Act} \mid \textsc{Seq}\{\textsc{Block}_1, \ldots, \textsc{Block}_n\} \mid \text{AND}\{\textsc{Block}_1, \ldots, \textsc{Block}_n\} \\ \mid \text{XOR}\{\textsc{Block}_1, \ldots, \textsc{Block}_n, p_1, \ldots, p_n\} \mid \textsc{Repeat}\{\textsc{Block}, p\} \\ \mid \text{DAG}\{\textsc{Block}_1, \ldots, \textsc{Block}_n\} \qquad (1)\end{aligned}$$

Act is an elementary *activity* represented by an STPN with a single transition with GEN duration connecting the initial and final places (e.g., Q in Fig. 1a).

Seq$\{\textsc{Block}_1, \ldots, \textsc{Block}_n\}$ is the *sequence* of n blocks $\textsc{Block}_1$, ..., $\textsc{Block}_n$ (e.g., Z in Fig. 1a).

XOR$\{\textsc{Block}_1, \ldots, \textsc{Block}_n, p_1, \ldots, p_n\}$ is made of an initial immediate random *exclusive choice*, with probabilities $p_1, \ldots, p_n$, among n alternative blocks $\textsc{Block}_1$, ..., $\textsc{Block}_n$, each connected to a final IMM *simple merge* transition (e.g., V in Fig. 1a). XOR is said to be *balanced*, meaning that all the concurrent paths started at the initial split are terminated at the final join.

AND$\{\textsc{Block}_1, \ldots, \textsc{Block}_n\}$ is a balanced split-join made of an initial IMM *parallel split* transition that forks execution along n concurrent blocks $\textsc{Block}_1$, ..., $\textsc{Block}_n$ and a final IMM *synchronization* transition that terminates the block (e.g., U in Fig. 1a).

Repeat$\{\textsc{Block}, p\}$ is a *structured cycle* that executes a body $\textsc{Block}$ and then repeats with constant probability $p > 0$ or terminates with probability $1 - p$ (e.g., $\{T\}$ in Fig. 1a).

DAG$\{\textsc{Block}_1, \ldots, \textsc{Block}_n\}$ is the composition of blocks $\textsc{Block}_1$, ..., $\textsc{Block}_n$ in a Directed Acyclic Graph (DAG) with single initial and final places, with blocks of *simple split* [29], made of an IMM transition with a single input place and multiple output places, and *simple join*, made of an IMM transition with multiple input places and a single output place. Note that, since *simple split* and *simple join* are not necessarily balanced, a DAG can break well-formed nesting of concurrent blocks. A DAG is termed *minimal* if it cannot be reduced by composition operators Seq, XOR or AND (e.g., in Fig. 1a, Q, R, S, T, U, and V are composed in a minimal DAG with initial and final places `Start`, `End`, by means of simple splits `as1`, `as2`, and simple joins `aj1`, `aj2`, `aj3`).

By definition, each model specified as a composition of blocks can be translated into a unique STPN. Conversely, blocks do not cover all the expressivity of STPNs. In particular, since choices are expressed only by IMM transitions within balanced XOR or REPEAT blocks, a model cannot represent a race selection where a choice is determined by values sampled by concurrent activities, e.g., early preemption of a timed activity that may occur in a timeout mechanism. As a positive consequence, this restriction also rules out *anomalies* where the early completion of some intermediate step can result in a longer workflow duration, providing the basis to prove positive correlation among completion times of different intermediate points in the workflow.

2.3 Structure Tree

Based on the grammar of Eq. (1), a workflow can be decomposed as a *structure tree* $S = \langle N, E, n_0 \rangle$: N is the set of nodes (blocks); E is the set of directed edges (connecting each block with its component blocks); n_0 is the root node. In Fig. 1b, a block is represented as a box labeled with the block name; the box is labeled also with the activity name (for ACT blocks) or block type (for SEQ, AND, XOR, REPEAT blocks), or contains places and transitions connecting the component blocks (for DAG blocks). Hierarchical graphs with single-entry single-exit blocks are inspired by *program structure trees* [17] and *process structure trees* [34]: similarly, we ensure that the structure tree is unique and robust with respect to local changes (i.e., modifying a sub-workflow in the STPN affects only its subtree in the structure tree) by using maximal blocks (e.g., SEQ blocks with as many components as possible) and by matching DAG blocks with lowest priority (i.e., SEQ or AND are used instead of DAG nodes, if possible).

3 Compositional Evaluation of Workflows Response Time

We evaluate the end-to-end response time CDF of a workflow by composition of the results of separate analyses of a hierarchy of sub-workflows. In so doing, we repeatedly apply both straight *numerical combination* of monovariate CDFs (Sect. 6.2 in the Appendix) and *Markov regenerative transient analysis* (Sect. 3.1) so as to leverage their different strengths. Numerical combination turns out to be efficient in the composition of independent sub-workflows through well-nested operators (AND, XOR, SEQ, REPEAT), but it is not feasible for sub-workflows with common dependencies (DAG). Markov regenerative analysis suffers from the degree of concurrency among activities with GEN durations, and its efficient implementation requires that sub-workflow durations be represented in analytic form, which may require approximated fitting of numerical results.

The structure tree is used to aggregate model components and to select solution techniques according to heuristics that trade approximation for complexity (Sects. 3.3 and 3.2) while ensuring that the final result is a stochastic upper bound of the exact CDF of the end-to-end workflow response time (Sect. 3.4).

3.1 Regenerative Transient Analysis

The structure of a workflow naturally leads to concurrent execution of multiple activities, which in most practical cases are generally distributed (GEN), often within a bounded support. In this setting, the marking process $\{ M(t), t \geqslant 0 \}$ of the STPN representing a sub-workflow is a Markov Regenerative Process (MRP) if a new *regeneration point* (i.e., a time instant where the Markov property is satisfied) is eventually reached w.p.1 from any state [20]. In this case, transient probabilities $P_{ij}(t)$ of each marking j and initial regeneration i (including the enabling time of each GEN transition [25]) can be evaluated by numerical integration of Generalized Markov Renewal equations $P_{ij}(t) = L_{ij}(t) + \sum_{k \in \mathcal{R}} \int_0^t dG_{ik}(u)\, P_{kj}(t-u)$ for all i in the set of reachable regenerations $\mathcal{R}$ and for all j in the set of markings $\mathcal{M}$, where the *global kernel* $G_{ik}(t) := P\{X_1 = k, T_1 \leqslant t \mid X_0 = i\}$ characterizes the next regeneration point $T_1 \geqslant 0$ and regeneration $X_1 \in \mathcal{R}$, while the *local kernel* $L_{ij}(t) := P\{M(t) = j, T_1 > t \mid X_0 = i\}$ defines transient probabilities of the process until the next regeneration point.

In turn, kernels can be evaluated numerically if at most one GEN transition is enabled in each state [6], or if multiple GEN transitions are enabled concurrently but the number of firings between regeneration points is bounded [5]. For this larger class of MRPs, kernels can be evaluated using *stochastic state classes* [15], which encode the marking, joint Probability Density Function (PDF), and support of the times-to-fire of enabled transitions after each sequence of firings between any two regeneration points. This joint PDF is continuous, with piecewise representation over Difference Bounds Matrix (DBM) zones [9], and can be evaluated in closed-form by the ORIS tool (or the Sirio Java library) [24] provided that each transition has expolynomial PDF [32].

The complexity of regenerative transient analysis of an STPN can be efficiently estimated by *nondeterministic analysis* of the underlying TPN, which is sufficient to identify the set of feasible behaviors of the model while avoiding the complexity of evaluation of their measure of probability. This is obtained by enumeration of a *state class graph*, encoding the continuous set of executions of the STPN into a discrete representation [4,35]: each vertex is a *state class* $S = \langle m, D \rangle$ including a marking m and a *DBM zone* D [13], i.e., a continuous set of values for the times-to-fire of enabled transitions. For each transition t that can fire first, the graph includes a directed edge (S, t, S') from S to the state class $S' = \langle m', D' \rangle$ with marking m' after the firing and the zone D' of reachable times-to-fire. The graph is finite under fairly general conditions (requiring that the number of reachable markings be finite and the earliest and latest firing times of transitions be rational values [35]), it permits detection of regeneration points (as classes where each GEN transition is newly enabled, disabled, or enabled since a deterministic time), and it makes explicit the number of transitions between regeneration points and the degree of concurrency among GEN timers.

3.2 Complexity Heuristics

Complexity Factors. Regenerative transient analysis incurs different factors of complexity in the enumeration of stochastic state classes, derivation of the local and global kernels, and solution of the Markov renewal equations:

- The results of [30] show that the number of stochastic state classes depends on the number of concurrently enabled GEN transitions, the number of firings

after which a GEN transition is persistent (i.e., continuously enabled), and the number of expmonomial terms of the PDFs of the GEN transitions.
In our approach, the number of stochastic state classes is kept limited by decomposing a workflow into sub-workflows that are separately analyzed.

- The results of [30] also show that the number of DBM zones and the number of expmonomial terms of the joint PDF of each stochastic state class depend on the same factors as the number of stochastic state classes. In particular, at each firing, the number of DBM zones increases polynomially with the number of persistent transitions, the number of expmonomial terms increases linearly with the polynomial degree of the joint PDF, and, in turn, if the analytical form of the joint PDF contains no EXP factor, the polynomial degree increases linearly with number of fired/disabled transitions.
In our approach, the factors of complexity of stochastic state classes are kept limited not only by decomposing a workflow into sub-workflows, but also by approximating the numerical form of the response time distribution of a sub-workflow with a piecewise PDF made of EXP terms (the approximation is needed when a sub-workflow is analyzed in isolation, either through regenerative transient analysis or through numerical analysis, and then regenerative transient analysis of a higher-level sub-workflow has to be performed).
- According to [15], the derivation of the kernels and the solution of the Markov renewal equations have linear complexity in the number of stochastic state classes and in the number of DBM zones and expmonomial terms of the joint PDF of each stochastic state class. Moreover, the derivation of the kernels and the solution of the Markov renewal equations also have linear and quadratic complexity, respectively, in the number of time points, i.e., the number of times the time step is contained in the analysis time limit. In our approach, the derivation of the local kernel is limited to the evaluation of $L_{if}(t)$, where i is the initial regeneration and f is the final absorbing marking.

Complexity Measures. To estimate the complexity of regenerative transient analysis of an STPN, we use the state class graph of the underlying TPN to compute the maximum number c of GEN transitions concurrently enabled in a state class (*concurrency degree*) and the maximum length r of paths of a regeneration epoch (*epoch length*), i.e., paths between regenerative state classes. However, the state class graph may be huge for complex blocks with high values of c and r, and would not tell how much of the complexity depends on the structure of the block itself and how much on the blocks it contains, which instead is relevant to decide how to decompose the block. To cope with both aspects, for each block b, we compute upper bounds C and R on c and r, respectively, and we also compute the concurrency degree $\bar{c}$ and the epoch length $\bar{r}$ of a *simplified* block $\bar{b}$, obtained by replacing each composite block of b with an (elementary) activity block. To this end, we perform a bottom-up visit of the structure tree:

- At the bottom level, we perform nondeterministic analysis of the TPN of each composite block, computing the tuple $\langle c, r, t_{\min}, t_{\max} \rangle$, where $t_{\min}$ and $t_{\max}$ are the minimum and the maximum execution time of the block, respectively, and can be derived as the minimum and the maximum duration, respectively, of paths between the initial and the final state class [35].

- At the next higher level, we perform nondeterministic analysis of the TPN of the simplified block $\bar{b}$, obtained by replacing each composite block b' of b with an activity block with duration interval equal to the min-max execution interval of b' (computed at the previous step). Evaluation of the complexity measures on the resulting state class graph $\bar{\Gamma}$ yields the tuple $\langle \bar{c}, \bar{r}, t_{\min}, t_{\max} \rangle$. Upper bounds on c and r are $C = \max_{S \in \Omega}\{\sum_{t \in E_S} C_t\}$ and $R = \max_{\rho \in \Psi_r}\{\sum_{e \in G_\rho} R_e\}$, respectively, where Ω is the set of state classes of $\bar{\Gamma}$, E_S is the set of transitions enabled in state class S, C_t is the concurrency degree upper bound of the block corresponding to transition t (1 for activity blocks), Ψ_r is the set of paths of $\bar{\Gamma}$ between regenerative state classes, G_ρ is the set of edges of path ρ, and R_e is the epoch length upper bound of the block corresponding to the transition of edge e (1 for activity blocks).
- Evaluation is repeated until the tuple $\langle \bar{c}, \bar{r}, C, R, t_{\min}, t_{\max} \rangle$ is computed for each composite block of the structure tree (thus also for the root block).

Then, we define the *complexity heuristics*: a block b (a simplified block $\bar{b}$) is *easy* to analyze if both C and R ($\bar{c}$ and $\bar{r}$) are not larger than some thresholds Θ_c and Θ_r, respectively, and *complex* otherwise, e.g., for the workflow STPN of Fig. 1, R is infinite, due to the concurrency between the REPEAT block T and blocks S, U, and V (in fact, for the simplified STPN obtained replacing T and V with an activity block each, we have $\bar{c} = 3$ and $\bar{r} = 7$). The goal of our compositional analysis is to reduce the workflow complexity, e.g., by analyzing block T in isolation and replacing it with a transition approximating its duration.

3.3 Analysis Heuristics

In the evaluation of the response time CDF $\Phi_b(t)$ of a block b, we consider four *actions*, each introducing a different approximation.

Action 1 (Numerical Analysis): If b is well-structured, i.e., neither it is a DAG or REPEAT block nor it contains DAG or REPEAT blocks (e.g., U in Fig. 1), evaluate $\Phi_b(t)$ by numerical analysis.

Action 2 (Regenerative Transient Analysis): If the execution time CDF of each activity block in b is expressed in analytical form (e.g., T in Fig. 1), evaluate $\Phi_b(t)$ through regenerative transient analysis of the STPN of b, otherwise (i.e., if b has some activity CDF expressed in numerical form, e.g., the workflow of Fig. 1 after the response time CDF of T is evaluated separately) replace each numerical CDF with the analytical form of a stochastic upper bound CDF (by Lemma 2), and evaluate a stochastic upper bound on $\Phi_b(t)$ through regenerative transient analysis of the STPN of the resulting block (by Lemma 3).

Action 3 (Inner Block Analysis): Evaluate the response time CDF of a composite block c (see Fig. 6a) contained in block b (through some action α_1), replace c with an activity block having the computed CDF as execution time

CDF (see Fig. 6b), and compute the response time CDF of the obtained block b' (through some action α_2). Note that α_1 and α_2 may yield $\Phi_b(t)$ (e.g., if both are action 1) or a stochastic upper bound on $\Phi_b(t)$ (e.g., if α_1 is action 1 and α_2 is action 2).

Action 4 (Inner Block Replication): If b is a DAG block, replicate some predecessors of a block to evaluate its response time independently of the rest of the DAG (replicated blocks are identical). Specifically, let $G = (V, E, v_I, v_F)$ be the DAG where V is the set of vertices (i.e., blocks of b) plus (fictitious) zero-duration initial vertex v_I and final vertex v_F (not shown in Fig. 1), and E is the set of edges (i.e., precedence relations between blocks): identify vertex $v \in V \setminus \{v_I, v_F\}$ (e.g., block T in Fig. 1); let the set K of vertices in $V \setminus \{v_I, v_F\}$ that are predecessors both of v and of some node $u \in V$ not predecessor of v (i.e., $K = \{R\}$); replicate the vertices in K and the edges to/from nodes in K (i.e., add R' to V; add $v_I \to R'$ and $R' \to T$ to E); evaluate the response time CDF of v (by some action, see Fig. 6c) and replace it with an activity block with the computed CDF as execution time CDF (see Fig. 6d); and, evaluate the response time CDF of the obtained block (by some action), which is a stochastic upper bound on the response time CDF $\Phi_b(t)$ of the original block b (by Lemma 4).

We define *analysis heuristics* to visit the structure tree and repeatedly select an action until a safe approximation of the workflow response time is evaluated. To exploit our complexity heuristics, which characterizes both the complexity of the structure of a workflow and the complexity of the blocks that it contains, we consider three analysis heuristics that perform a *top-down* visit of the structure tree. Nevertheless, the approach is open to the definition of different heuristics.

Overall, if a block is, or can be reduced to, a well-structured composition of independent sub-workflows, then numerical analysis guarantees efficient and accurate evaluation. Conversely, REPEAT blocks can be evaluated in isolation through regenerative transient analysis when their parallel composition with other blocks prevents the occurrence of regenerations. Finally, DAG blocks, representing dependent sub-workflows, can be evaluated through regenerative transient analysis, operating some simplification if the DAG is too complex to analyze (i.e., analyzing some block or some sub-workflow in isolation).

Specifically, **analysis heuristics 1** operates as follows on a visited block b:

1. If b is well-structured, then select action 1 (numerical analysis).
2. If b is a SEQ or an AND or an XOR block, and contains DAG or REPEAT blocks at some hierarchy level, then select action 3 (inner block analysis) as many times as the number of composite blocks of b (which are replaced with an activity block each) and then select action 1 (numerical analysis).
3. If b is a REPEAT block, use the analysis heuristics to select the next action:
   (a) if b is easy to analyze, select action 2 (regenerative transient analysis);
   (b) otherwise, select action 3 (inner block analysis) to compute (through some action) the response time CDF of the block repeated by the loop.
4. If b is a DAG block, use the analysis heuristics to select the next action:

(a) if both b and the simplified block $\bar{b}$ are easy to analyze, then select action 2 (regenerative transient analysis);
(b) otherwise, until block b becomes easy to analyze, repeatedly select action 4 (inner block replication), each time analyzing in isolation one of the sub-workflows AND-joined by the final IMM transition of b.

For instance, the DAG in Fig. 1 is too complex to analyze: heuristics 1 performs regenerative transient analysis of the sub-workflow $\{Q, R, T\}$ (by replicating R), and then performs regenerative transient analysis of the obtained block.

To evaluate how approximating intermediate numerical CDFs impacts result accuracy and computational complexity with respect to decoupling dependent sub-workflows, we consider **analysis heuristics 2**, a variant of heuristics 1 that manages complex DAG blocks (point 4a) by repeatedly performing first action 3 (inner block analysis, replacing complex composite blocks with activity blocks), and then action 4, until the DAG is easy to analyze.

Finally, to show the efficacy of numerical analysis in the evaluation of well-structured workflows, we consider **analysis heuristics 3**, another variant of heuristics 1 that performs regenerative transient analysis in cases where heuristics 1 would perform numerical analysis.

3.4 Approximation Safety

We now ensure that our compositional analysis method is safe when workflows are used to guarantee soft deadlines of SLAs. Our proofs (in the Appendix) hinge on the idea of *stochastic order* and on the following lemma on the order of independent replication of positively correlated random variables (r.v.s) [3].

Definition 1 (Stochastic order). *Given two random vectors* $\mathbf{X}_1$ *and* $\mathbf{X}_2$*, we say that "*$\mathbf{X}_1$ *is smaller than* $\mathbf{X}_2$*" (*$\mathbf{X}_1 \leqslant_{st} \mathbf{X}_2$*), if* $E[f(\mathbf{X}_1)] \leqslant E[f(\mathbf{X}_2)]$ *for all monotone nondecreasing functions* f*. For scalar* X_1 *and* X_2 *with CDFs* $F_1(x)$ *and* $F_2(x)$*, respectively, this is equivalent to* $F_1(x) \geqslant F_2(x)$ *for all* x*.*

Lemma 1 (Order of independent replicas under positive correlation). *Let* $\mathbf{X} = (X_1, \ldots, X_n)$ *be a vector of positively correlated r.v.s, i.e.,* $\mathrm{Cov}[f(\mathbf{X}), g(\mathbf{X})] \geqslant 0$ *holds for all monotone nondecreasing* $f, g\colon \mathbb{R}^n \to \mathbb{R}$*. Then,* $\overline{\mathbf{X}} \geqslant_{st} \mathbf{X}$*, where* $\overline{\mathbf{X}}$ *is a vector of independent r.v.s with* $\overline{X}_i \sim X_i$ *for all* i*.*

In our regenerative analysis, numerical CDFs are replaced with analytical stochastic upper bound CDFs (which guarantee stochastic order for each known point of the numerical CDFs). The next lemma proves that such bound can be a piecewise CDF combining a shifted truncated EXP (*body*) and a shifted EXP (*tail*). The approximant accuracy could be improved by considering multiple pieces for the body (e.g., to better approximate multimodel CDFs).

Lemma 2 (Stochastic upper bound CDF). *Given a r.v.* X *with numerical CDF* $F(x)$ *with* $x \in D = \{a, a+\delta, \ldots, a+(L-1)\,\delta\}$*,* $a \in \mathbb{R}_{\geqslant 0}$*,* $\delta \in \mathbb{R}_{>0}$*, and* $L \in \mathbb{N}$*, let* $\hat{X}$ *be the r.v. with CDF* $\hat{F}(x)$ *s.t.* $\hat{F}(x) = 0\ \forall x < d$*,* $\hat{F}(x) = 0.75\,(1 -$

$e^{-\lambda_b(x-d)})/(1 - e^{-\lambda_b(q_3-d)})$ $\forall x \in [d, q_3]$, $\hat{F}(x) = 0.25(1 - e^{-\lambda_t(x-q_3)}) + 0.75$ $\forall x \in [q_3, \infty)$, *where d is equal to a if $F(x)$ starts with downward concavity and equal to the abscissa of the intersection of the x-axis with the line tangent to the inflection point of $F(x)$ if $F(x)$ starts with upward concavity, q_3 is the third quartile of X, λ_b is the minimum of the values that maximize $\hat{F}(x)$ and satisfy $\hat{F}(x) \leqslant F(x)$ $\forall x \in D \cap [d, q_3]$, and λ_t is the minimum of the values that maximize $\hat{F}(x)$ and satisfy $\hat{F}(x) \leqslant F(x)$ $\forall x \in D \cap (q_3, \infty)$. Then, $\hat{X} \geqslant_{st} X$.*

In our inner block analysis, a node n in the structure tree is replaced with an activity block with duration stochastically larger than the response time of n. The next lemma proves that, after this approximation, the response time of the obtained workflow is stochastically larger than the actual response time.

Lemma 3 (Stochastic order of inner block analysis). *Let $S = (N, E, n_0)$ be the structure tree of a workflow with root node $n_0 \in N$, and let $T(n)$ be the response time of the subtree rooted in $n \in N$. If n is replaced with n' s.t. $T(n) \leqslant_{st} T(n')$, yielding the new structure tree $S' = (N', E', n'_0)$, then $T(n_0) \leqslant_{st} T(n'_0)$.*

In our inner block replication, ancestors of a vertex v are replicated in a DAG block to evaluate the response time of v independently of the rest of the DAG. The next lemma proves that, also after this approximation, the response time of the obtained workflow is stochastically larger than the actual response time.

Lemma 4 (Stochastic order of inner block replication). *Given a DAG block $G = (V, E, v_I, v_F)$ and a vertex $v \in V$, let $T(v)$ be the response time of v, let K be the set of vertices in $V \setminus \{v_I, v_F\}$ that are predecessors both of v and of some node $u \in V$ not predecessor of v, let F be the set of edges in E to/from a node in K, and let $G' = (V', E', v'_I, v'_F)$ be the DAG s.t. V' includes all vertices in V plus a new node k' with $T(k') \sim T(k)$ $\forall k \in K$, and E' includes all edges in E plus an edge to/from each new node k' for each edge to/from the corresponding node $k \in K$. Then, $T(v'_F) \geqslant_{st} T(v_F)$.*

4 Experimentation

In this section, we answer the following questions on our proposed approach:

Q1. Is the approach feasible for the considered concurrency structures?
Q2. Is the approach accurate with respect to a simulated ground truth?
Q3. Does the approach obtain accurate results in reasonable times?

To this end, we consider eight models combining four structures that gradually increase the workflow complexity, evaluating how the approximation of intermediate numerical results and the replication of dependent events affect result accuracy and computational complexity. For each model, we compare a ground truth with the results of our analysis heuristics and simulation. For our complexity heuristics, we consider thresholds $\Theta_c = 3$ and $\Theta_r = 10$ on the concurrency degree and the epoch length, respectively, and we consider activity durations with uniform CDF over $[0, 1]$. Experiments are performed using a single core of an Intel Xeon Gold 5120 CPU (2.20 GHz) equipped with 32 GB of RAM.

4.1 Experimentation Models

Figure 2 shows the four structures used to build the experimentation models:

Simple DAG (Fig. 2a) has concurrency degree 3 and epoch length 8. Thus, it can be efficiently analyzed by regenerative transient analysis.

Complex DAG (Fig. 2b) has concurrency degree 5. It cannot be analyzed as a whole and needs to be decomposed by one of the analysis heuristics.

Complex AND (Fig. 2c) is a well-structured tree, with two instances of simple DAG as leaves. Once the latter are analyzed by regenerative transient analysis, the resulting tree can be analyzed numerically (analysis heuristics 1 and 2), or, given that it has concurrency degree 4, regenerative transient analysis can be applied to blocks A and F, and then to the resulting model (analysis heuristics 3).

Nested Repetitions (Fig. 2d) has two nested REPEAT blocks, with an instance of (simple or complex) DAG and of complex AND as leaves: once the latter are

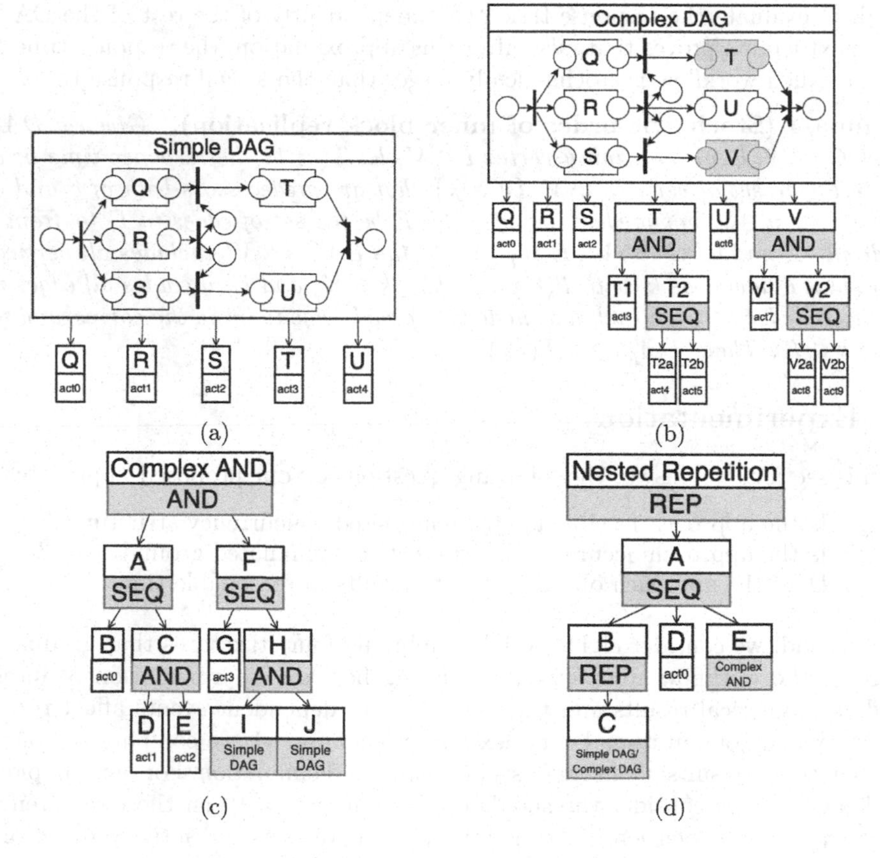

Fig. 2. Structure elements used to compose experimentation models.

analyzed in isolation, the resulting model has concurrency degree 1 and path length 3, and can be analyzed by regenerative transient analysis (all heuristics).

Figure 3 shows the models obtained combining the structure elements of Fig. 2.

Models 1a and 2b (Fig. 3a) are well-structured trees with two instances of (simple and complex, respectively) DAG and an instance of complex AND as leaves: once the latter composite blocks are analyzed, the resulting tree can be analyzed numerically (analysis heuristics 1 and 2), or, given that it has concurrency degree 4, regenerative transient analysis can be applied first to blocks A and K, and then to the resulting model (analysis heuristics 3).

Models 2a and 2b (Fig. 3b) are well-structured trees with an instance of (simple and complex, respectively) DAG and of Nested Repetitions as leaves: as for models 1 and 2, once the latter composite blocks are analyzed, the resulting tree can be analyzed numerically (analysis heuristics 1 and 2), or, given that it

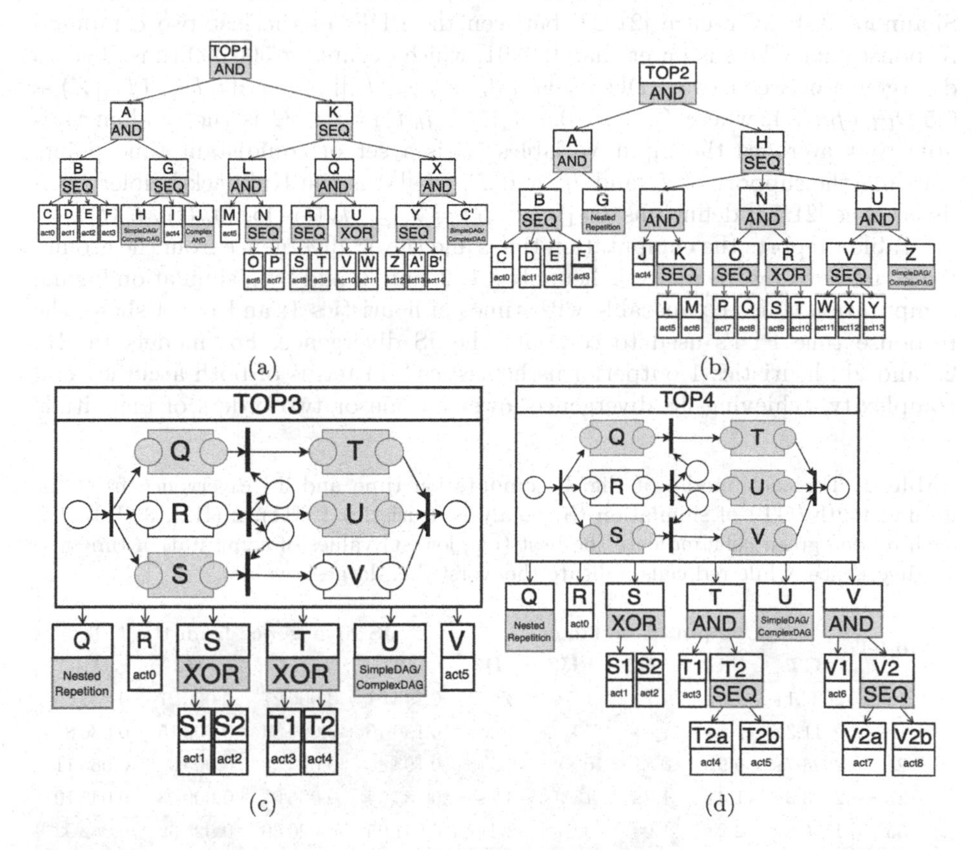

Fig. 3. Four models used for experimentation. Each model combines some of the structures defined in Fig. 2 and is tested in a simple variant, using instances of Simple DAG, and a complex variant, using instances of Complex DAG.

has concurrency degree 4, regenerative transient analysis can be applied first to blocks A and H, and then to the resulting model (analysis heuristics 3).

Models 3a and 3b (Fig. 3c) consist of a top DAG made of an instance of (simple and complex, respectively) DAG, one of Nested Repetitions, and two well-structured sub-trees. Once the instance of Nested Repetitions is evaluated, the model is still too complex and must be decomposed.

Models 4a and 4b (Fig. 3d) are variants of models 3a and 3b, respectively, with more complex well-structured trees. Once the instance of Nested Repetitions is evaluated, the model is still too complex and must be decomposed.

4.2 Experimentation Results

For each workflow of Fig. 3, a ground truth is computed by performing simulation of the corresponding STPN through the SIRIO library of the ORIS tool [24,31], increasing the number of simulation runs by 100 000 at a time until the Jensen-Shannon (JS) divergence [21,23] between the PDFs of the last two computed response time CDFs is lower than 0.0001, which occurs for 500 000 runs. The JS divergence between two PDFs f_a and f_b is $D_{\mathrm{JS}}(f_a \,||\, f_b) := 0.5\, D_{\mathrm{KL}}(f_a \,||\, Z) + 0.5\, D_{\mathrm{KL}}(f_b \,||\, Z)$, where $Z(t) := 0.5\,(f_a(t) + f_b(t))\ \forall t \in \Omega$ is the random variable that averages the input variables, Ω is a set of equidistant time points covering the support of f_a and f_b, and $D_{\mathrm{KL}}(\cdot \,||\, \cdot)$ is the Kullback-Leibler (KL) divergence [21,23] defined as $D_{\mathrm{KL}}(f_a \,||\, f_b) = \sum_{t\in\Omega} f_a(t) \cdot \log(f_b(t)\,/\,f_a(t))$.

Table 1 reports the computation times and the JS divergence from the ground truth achieved by the analysis heuristics 1, 2, and 3, and by a simulation having computation times comparable with times of heuristics 1, and Fig. 4 shows the response time PDFs used to compute the JS divergence. For models 1a, 1b, 2a and 2b, heuristics 1 outperforms heuristics 3 in terms of both accuracy and complexity, achieving JS divergence lower by one or two orders of magnitude

Table 1. For each model of Fig. 3, computation time and JS divergence from the ground truth (GT) of simulation (S), analysis heuristics 1 (H1), 2 (H2), 3 (H3). For each model, green cells indicate the best (i.e., lowest) values of computation time and JS divergence, while red cells indicate the worst (i.e., largest) values.

Model	Computation Times					JS divergence from the GT			
	GT	S	H1	H2	H3	S	H1	H2	H3
1a	2027.1 s	1.5 s	1.5 s	1.5 s	32.7 s	0.196 91	0.000 27	0.000 27	0.033 26
1b	2711.3 s	1.0 s	1.0 s	13.8 s	10.6 s	0.190 46	0.000 84	0.003 15	0.036 81
2a	2028.7 s	0.9 s	0.8 s	0.8 s	3.2 s	0.193 85	0.003 63	0.003 63	0.060 11
2b	2723.3 s	1.4 s	1.2 s	13.9 s	3.2 s	0.184 64	0.006 62	0.009 43	0.043 10
3a	1566.8 s	2.2 s	1.9 s	3.8 s	2.4 s	0.041 05	0.026 89	0.025 69	0.082 92
3b	2232.2 s	4.2 s	3.9 s	16.8 s	2.5 s	0.081 02	0.026 60	0.025 48	0.081 73
4a	1803.9 s	1.8 s	1.7 s	72.6 s	5.9 s	0.194 71	0.019 72	0.037 30	0.075 77
4b	2534.5 s	2.6 s	2.6 s	72.0 s	6.0 s	0.202 84	0.021 98	0.039 41	0.076 71

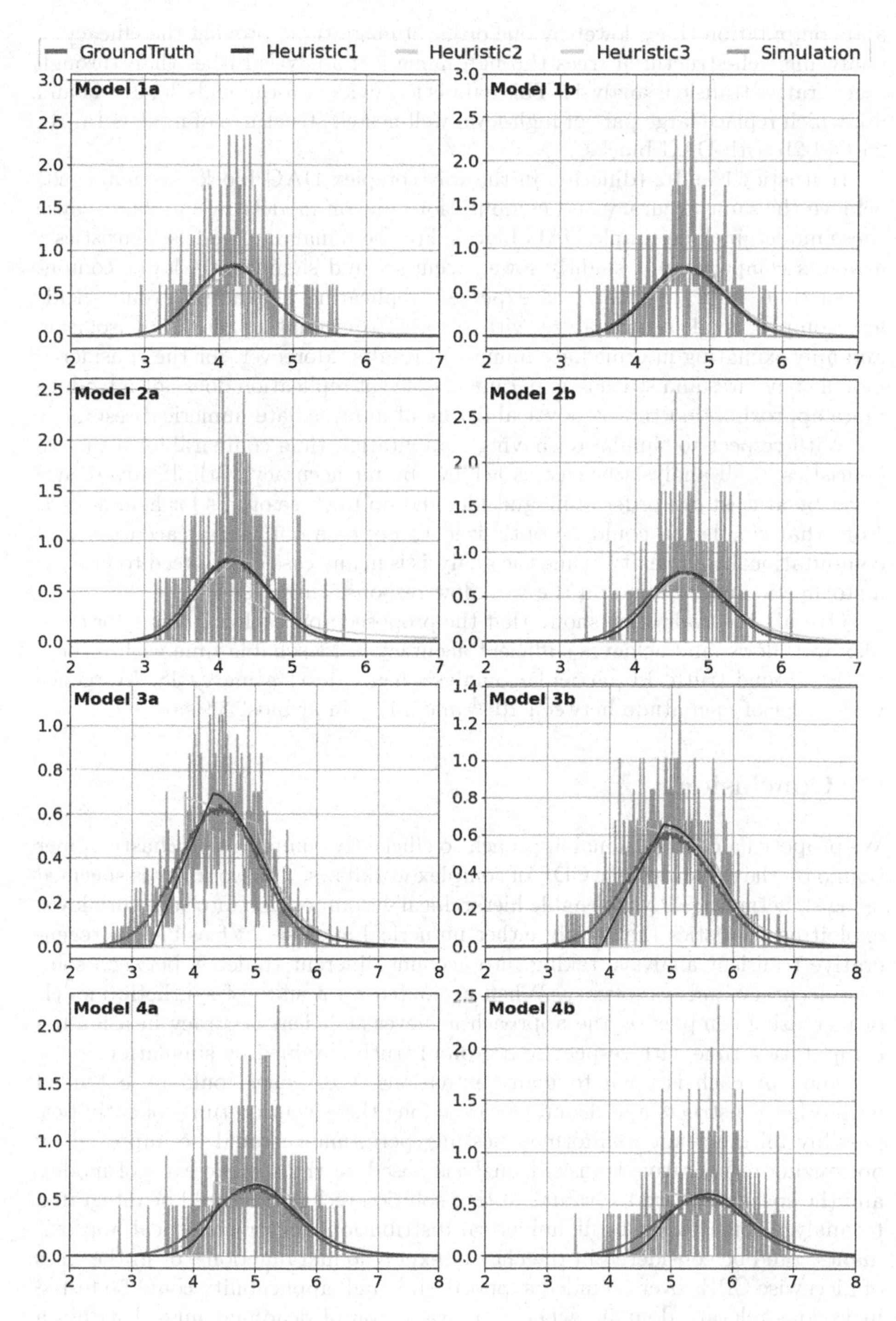

Fig. 4. Response time PDFs of the workflow models of Fig. 3.

and computation times lower by one order of magnitude, proving the efficacy of analyzing well-structured trees through numerical analysis rather than through regenerative transient analysis. This gain is less evident for models 3a, 3b, 4a and 4b, which replace large part of high-level well-nested structures of models 1a, 1b, 2a and 2b with DAG blocks.

Heuristics 1 and 2 (differing in the way complex DAG blocks are managed) achieve the same accuracy and computation time on models 1a and 2a, because these models include simple DAG blocks. For the remaining models, heuristics 1 achieves comparable or slightly lower accuracy and significantly lower computation time, indicating that, as expected, replicating dependent events yields less complex models to analyze with respect to evaluating blocks in isolation and approximating intermediate numerical results. Moreover, for the considered model structures and stochastic parameters, event replication does not introduce more approximation than analytical fitting of intermediate numerical results.

With respect to simulation having computation time comparable to that of heuristics 1, all analysis heuristics achieve better accuracy with JS divergence lower by at least one order of magnitude, and up to three orders for heuristics 1. Note that simulation could be optimized to improve both result accuracy and computational complexity, while the analysis is in any case guaranteed to provide a stochastic upper bound on the workflow response time CDF.

Overall, the evaluation shows that the proposed approach is feasible for complex workflows, and achieves sufficient accuracy in reasonable time with respect to the ground truth. In particular, analysis heuristics 1 achieves JS divergence with order of magnitude between 10^{-4} and 10^{-2}, in at most 3.9 s.

5 Conclusions

We proposed a compositional approach to efficiently compute a stochastic upper bound on the response time CDF of complex workflows. The workflow is specified using structured STPNs to enable hierarchical decomposition into sub-workflows, exploiting heuristics that apply either numerical analysis (if feasible) or regenerative transient analysis, taking into account different tradeoffs between solution accuracy and complexity. When evaluated on a suite of synthetic models of increasing complexity, the approach achieves sufficient accuracy in a limited computation time with respect to a ground truth obtained by simulation.

The approach is open to many extensions: the model could be extended with other constructs and dependencies among the execution times of activities, possibly affecting not well-formed nesting; performance could be improved by optimizing regenerative transient analysis based on the specific class of models and the specific reward to evaluate; other solution techniques could be integrated to analyze some block; to fit numerical distributions, other analytical approximants could be considered in the class of expolynomial functions, or in the class of piecewise CPHs over bounded supports [19]; and, applicability could be tested in various relevant domains where the evaluation of deadlines missed within a given time requires the computation of the response time CDF.

6 Appendix

We recall syntax and semantics of STPNs (Sect. 6.1) and numerical analysis of well-structured workflows (Sect. 6.2), and we report theorem proofs (Sect. 6.3) and a graphical representation of analysis heuristics (Sect. 6.4).

6.1 Formal Syntax and Semantics of STPNs

Syntax. An STPN is a tuple $\langle P, T, A^-, A^+, EFT, LFT, F, W, Z\rangle$ where: P and T are disjoint sets of places and transitions, respectively; $A^- \subseteq P \times T$ and $A^+ \subseteq T \times P$ are pre-condition and post-condition relations, respectively; EFT and LFT associate each transition $t \in T$ with an earliest firing time $EFT(t) \in \mathbb{Q}_{\geqslant 0}$ and a latest firing time $LFT(t) \in \mathbb{Q}_{\geqslant 0} \cup \{\infty\}$ such that $EFT(t) \leqslant LFT(t)$; F, W, and Z associate each transition $t \in T$ with a Cumulative Distribution Function (CDF) F_t for its duration $\tau(t) \in [EFT(t), LFT(t)]$ (i.e., $F_t(x) = P\{\tau(t) \leqslant x\}$, with $F_t(x) = 0$ for $x < EFT(t)$ and $F_t(x) = 1$ for $x > LFT(t)$), a weight $W(t) \in \mathbb{R}_{>0}$, and a priority $Z(t) \in \mathbb{N}$, respectively.

As usual in stochastic Petri nets, a transition t is called *immediate* (IMM) if $EFT(t) = LFT(t) = 0$ and *timed* otherwise; a timed transition t is called *exponential* (EXP) if $F_t(x) = 1 - \exp(-\lambda x)$ for some rate $\lambda \in \mathbb{R}_{>0}$, or *general* (GEN) otherwise. For each GEN transition t, we assume that F_t can be expressed as the integral function of a probability density function (PDF) f_t, i.e., $F_t(x) = \int_0^x f_t(y)\,dy$. Similarly, an IMM transition $t \in T$ is associated with a generalized PDF represented by the Dirac impulse function $f_t(y) = \delta(y - \overline{y})$ with $\overline{y} = EFT(t) = LFT(t)$. A place $p \in P$ is called an *input* or *output* place for a transition $t \in T$ if $(p, t) \in A^-$ or $(t, p) \in A^+$, respectively.

Semantics. The state of an STPN is a pair $s = \langle m, \tau\rangle$, where $m : P \to \mathbb{N}$ is a *marking* assigning a number of tokens to each place and $\tau : T \to \mathbb{R}_{\geqslant 0}$ associates each transition with a *time-to-fire*. A transition is *enabled* by a marking if each of its input places contains at least one token. The next transition t to fire in a state s is selected from the set E of enabled transitions having time-to-fire equal to zero and maximum priority with probability $W(t)/\sum_{t_i \in E} W(t_i)$. When t fires, s is replaced with $s' = \langle m', \tau'\rangle$, where: m' is derived from m by removing a token from each input place of t, yielding an intermediate marking m_{tmp}, and adding a token to each output place of t; τ' is derived from τ by: *i*) reducing the time-to-fire of each *persistent* transition (i.e., enabled by m, m_{tmp} and m') by the time elapsed in s; *ii*) sampling the time-to-fire of each *newly-enabled* transition t_n (i.e., enabled by m' but not by m_{tmp}) according to F_{t_n}; and, *iii*) removing the time-to-fire of each *disabled* transition (i.e., enabled by m but not by m').

6.2 Numerical Analysis of Well-Structured Workflows

We derive the numerical form of the response time CDF of a block by combining bottom-up the numerical forms of the response time CDFs of the blocks that it contains, provided that the block has a well-formed structure, i.e., it is not a DAG block and it does not contain DAG blocks, and that it does not contain REPEAT blocks. Specifically, given n blocks $b_1, \dots, b_n$ with response time CDF $\Phi_1(t), \dots, \Phi_n(t)$ and PDF $\phi_1(t), \dots, \phi_n(t)$, respectively:

- the response time CDF $\Phi_{\text{seq}}(t)$ of a SEQ block made of $b_1, \dots, b_n$ is derived by performing subsequent convolutions of $\phi_1(t), \dots, \phi_n(t)$ $\forall t \in [0, t_{\max}]$:

$$\Phi_{\text{seq}}(t) = \Phi^{1,n}(t)$$

$$\Phi^{1,i}(t) = \int_0^t \int_0^\tau \phi^{1,i-1}(x)\, \phi_i(\tau - x)\, dx\, d\tau \;\forall i \in \{2, \dots, n\} \tag{2}$$

$$\phi^{1,i-1}(t) = \frac{d}{dt}\Phi^{1,i-1}(t) \;\forall i \in \{2, \dots, n-1\}, \quad \phi^{1,1}(t) = \phi_1(t)$$

- the response time CDF $\Phi_{\text{and}}(t)$ of an AND block made of $b_1, \dots, b_n$ is the CDF of the maximum among the response times of $b_1, \dots, b_n$, which is derived as the product of $\Phi_1(t), \dots, \Phi_n(t)$ $\forall t \in [0, t_{\max}]$ due to the fact that the response times of $b_1, \dots, b_n$ are independent random variables:

$$\Phi_{\text{and}}(t) = \Phi_1(t) \cdot \ldots \cdot \Phi_n(t) \tag{3}$$

- the response time CDF $\Phi_{\text{xor}}(t)$ of an XOR block made of $b_1, \dots, b_n$ is derived as the weighted sum of $\Phi_1(t), \dots, \Phi_n(t)$ $\forall t \in [0, t_{\max}]$:

$$\Phi_{\text{xor}}(t) = p_1\, \Phi_1(t) + \ldots + p_n\, \Phi_n(t) \tag{4}$$

6.3 Theorem Proofs

Proof of Lemma 2. By construction, $\hat{F}(x) \leqslant F(x)\; \forall x \in D \cap [d, \infty)$, and $\hat{F}(x) = 0$ $\forall x < d$. Starting from the point with abscissa a, the PDF of $F(x)$ may be either increasing or decreasing: If the PDF is increasing, then it will reach a maximum point that comprises an inflection point for the CDF $F(x)$, where the concavity changes from upward to downward (see Fig. 5b). By construction, the tangent line to the inflection point intersects the x-axis in a point whose abscissa d is larger than a, otherwise the CDF should have downward concavity before the inflection point, which contradicts the hypothesis. Otherwise (i.e., if the PDF is decreasing), $F(x)$ has downward concavity (see Fig. 5a), d is selected equal to a, and stochastic order is verified for $\forall x \in D \cap [a, \infty)$. Therefore, $\hat{X} \geqslant_{st} X$. □

Proof of Lemma 3. The duration of the sub-workflow associated with any node m (SEQ, AND, XOR, REPEAT, DAG) is a monotone nondecreasing function of the durations of the sub-workflows associated with its children; respectively, the sum (SEQ), max (AND), random mixture (XOR), series (REPEAT), max over all paths from the initial to the final node (DAG). By definition of stochastic order, if a child n is replaced with n' s.t. $T(n) \leqslant_{st} T(n')$, then $T(m) \leqslant_{st} T(m')$ for the new node m'. By recursion, $T(n_0) \leqslant_{st} T(n_0')$ for the new root n_0'. □

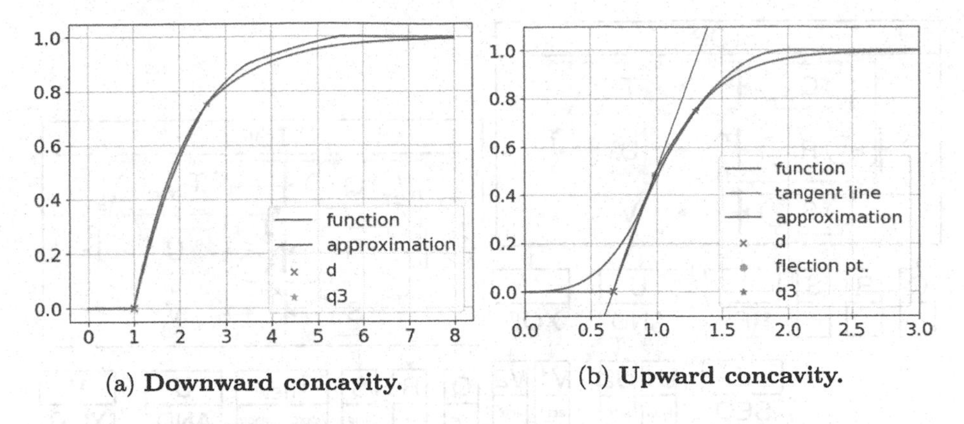

(a) **Downward concavity.** (b) **Upward concavity.**

Fig. 5. Stochastic upper bound CDF: concavity of approximated CDF.

Proof of Lemma 4. Since DAG edges denote AND-join dependencies, the response time of a vertex v is $T(v) = D(v) + \max(T(k_1), \ldots, T(k_n))$ where $D(v)$ is the duration of the block associated with v and $T(k_1), \ldots, T(k_n)$ are the response times of its predecessors. By visiting the vertices of G in topological order, we can evaluate the response time $T(v_F)$ of the DAG as an expression combining nonnegative block durations $D(v)$ $\forall v \in V$ through monotone nondecreasing operators (i.e., summation and maximum). The intermediate values of this expression obtained during the visit are the response times $T(\cdot)$ of the nodes of G, which, by construction, are positively correlated. In the evaluation of $T(v'_F)$ in G', the random variable $T(k)$ of each node $k \in K$ is replaced with the independent replica $T(k') \sim T(k)$. Then, by Lemma 1, we obtain $T(v'_F) \geqslant_{st} T(v_F)$. □

6.4 Analysis Actions

Figure 6 illustrates the application of the sequence of actions 3 and 4 on the structure tree presented in Fig. 1b. In particular, in Fig. 6a, a red and a green box identify two sub-structures on which actions 3 and 4 are applied, respectively:

- Action 3: is applied as follows: some action (depending on the considered analysis heuristic) is used to evaluate the response time of the sub-structure in the red box, which is then replaced with an activity block T_{new} associated with the evaluated response time CDF (see Fig. 6b).
- Action 4 evaluates the sub-structure in the green box independently of the rest of the DAG: the blocks that are shared with the rest of the DAG (i.e., block R) are replicated (i.e., block R_{bis} is added), also adding two fictitious zero-duration nodes v_I and v_F (see Fig. 6c); then, this sub-structure is evaluated though some action (depending on the considered analysis heuristic); finally, the sub-structure is replaced with an activity block QRT_{new} associated with the evaluated response time CDF (see Fig. 6d).

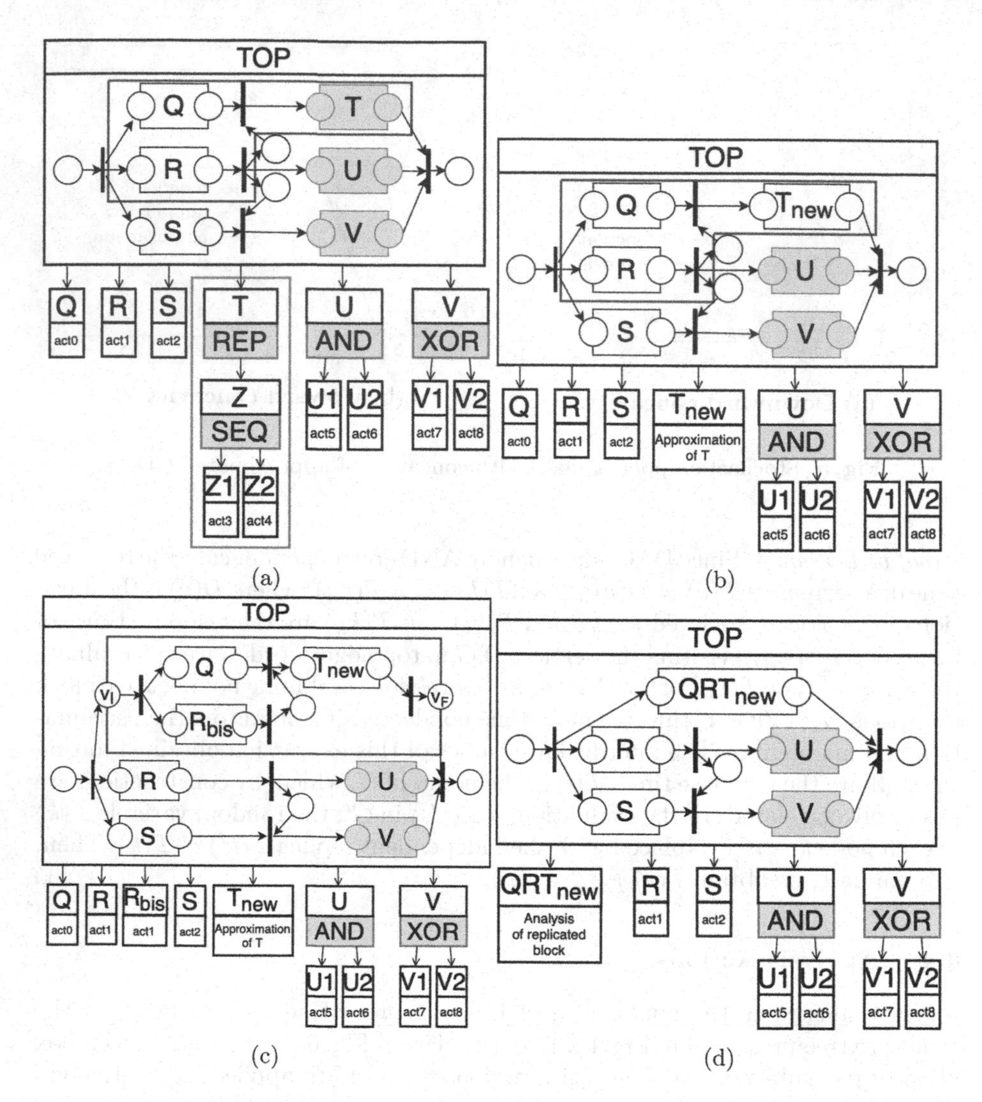

Fig. 6. A graphical representation for actions 3 and 4. (a) The structure tree presented in Fig. 1b, where the red and the green boxes indicate the sub-structures subject to actions 3 and 4, respectively. (b) The structure tree after the application of action 3, which replaces the sub-structure in the red box with an activity block. (c) The structure tree after the replication of block R during action 4. (d) The structure tree after the execution of action 4, which replaces the sub-structure in the green box with an activity block. (Color figure online)

References

1. Van der Aalst, W.M.: The application of Petri nets to workflow management. J. Circ. Syst. Comput. **8**(01), 21–66 (1998)

2. Arnold, F., Hermanns, H., Pulungan, R., Stoelinga, M.: Time-dependent analysis of attacks. In: Abadi, M., Kremer, S. (eds.) POST 2014. LNCS, vol. 8414, pp. 285–305. Springer, Heidelberg (2014). https://doi.org/10.1007/978-3-642-54792-8_16
3. Baccelli, F., Makowski, A.M.: Multidimensional stochastic ordering and associated random variables. Oper. Res. **37**(3), 478–487 (1989)
4. Berthomieu, B., Diaz, M.: Modeling and verification of time dependent systems using time Petri nets. IEEE Trans. Softw. Eng. **17**(3), 259–273 (1991)
5. Biagi, M., Carnevali, L., Paolieri, M., Papini, T., Vicario, E.: Exploiting non-deterministic analysis in the integration of transient solution techniques for Markov regenerative processes. In: Bertrand, N., Bortolussi, L. (eds.) QEST 2017. LNCS, vol. 10503, pp. 20–35. Springer, Cham (2017). https://doi.org/10.1007/978-3-319-66335-7_2
6. Bobbio, A., Telek, M.: Markov regenerative SPN with non-overlapping activity cycles. In: Proceedings of International Computer Performance and Dependability Symposium, pp. 124–133 (1995)
7. Bruneo, D., Distefano, S., Longo, F., Scarpa, M.: QoS assessment of WS-BPEL processes through non-Markovian stochastic Petri nets. In: Proceedings of IPDPS, pp. 1–12. IEEE (2010)
8. Canfora, G., Di Penta, M., Esposito, R., Villani, M.L.: QoS-aware replanning of composite web services. In: Proceedings of IEEE International Conference on Web Services, pp. 121–129. IEEE (2005)
9. Carnevali, L., Grassi, L., Vicario, E.: State-density functions over DBM domains in the analysis of non-Markovian models. IEEE Trans. Soft. Eng. **35**(2), 178–194 (2009)
10. Carnevali, L., Reali, R., Vicario, E.: Compositional evaluation of stochastic workflows for response time analysis of composite web services. In: Proceedings of the ACM/SPEC International Conference on Performance Engineering, pp. 177–188 (2021)
11. Ciardo, G., German, R., Lindemann, C.: A characterization of the stochastic process underlying a stochastic Petri net. IEEE Trans. Soft. Eng. **20**(7), 506–515 (1994)
12. Curbera, F., et al.: Business process execution language for web services (2002)
13. Dill, D.L.: Timing assumptions and verification of finite-state concurrent systems. In: Sifakis, J. (ed.) CAV 1989. LNCS, vol. 407, pp. 197–212. Springer, Heidelberg (1990). https://doi.org/10.1007/3-540-52148-8_17
14. Gias, A.U., van Hoorn, A., Zhu, L., Casale, G., Düllmann, T.F., Wurster, M.: Performance engineering for microservices and serverless applications: the radon approach. In: Companion of the ACM/SPEC International Conference on Performance Engineering, pp. 46–49 (2020)
15. Horváth, A., Paolieri, M., Ridi, L., Vicario, E.: Transient analysis of non-Markovian models using stochastic state classes. Perf. Eval. **69**(7–8), 315–335 (2012)
16. Jensen, E.D., Locke, C.D., Tokuda, H.: A time-driven scheduling model for real-time operating systems. In: Rtss, vol. 85, pp. 112–122 (1985)
17. Johnson, R., Pearson, D., Pingali, K.: The program structure tree: computing control regions in linear time. In: ACM Conference on Programming Language Design and Implementation (PLDI), pp. 171–185. ACM (1994)
18. de Kok, T.G., Fransoo, J.C.: Planning supply chain operations: definition and comparison of planning concepts. Handb. Oper. Res. Manage. Sci. **11**, 597–675 (2003)

19. Korenčiak, L., Krčál, J., Řehák, V.: Dealing with zero density using piecewise phase-type approximation. In: Horváth, A., Wolter, K. (eds.) EPEW 2014. LNCS, vol. 8721, pp. 119–134. Springer, Cham (2014). https://doi.org/10.1007/978-3-319-10885-8_9
20. Kulkarni, V.: Modeling and Analysis of Stochastic Systems. Chapman & Hall (1995)
21. Lin, J.: Divergence measures based on the Shannon entropy. IEEE Trans. Inf. Theory **37**(1), 145–151 (1991)
22. Liu, Y., Zheng, Z., Zhang, J.: Markov model of web services for their performance based on phase-type expansion. In: Proceedings of DASC-PICOM-CBDCOM-CYBERSCITECH, pp. 699–704. IEEE (2019)
23. Nielsen, F.: On a generalization of the Jensen-Shannon divergence and the JS-symmetrization of distances relying on abstract means. arXiv preprint arXiv:1904.04017 (2019)
24. Paolieri, M., Biagi, M., Carnevali, L., Vicario, E.: The ORIS tool: quantitative evaluation of non-Markovian systems. IEEE Trans. Soft. Eng. **47**, 1211–1225 (2021)
25. Paolieri, M., Horváth, A., Vicario, E.: Probabilistic model checking of regenerative concurrent systems. IEEE Trans. Softw. Eng. **42**(2), 153–169 (2016)
26. Pesu, T., Kettunen, J., Knottenbelt, W.J., Wolter, K.: Three-way optimisation of response time, subtask dispersion and energy consumption in split-merge systems. In: Proceedings of VALUETOOLS 2017, pp. 244–251. ACM (2017)
27. Rahman, J., Lama, P.: Predicting the end-to-end tail latency of containerized microservices in the cloud. In: 2019 IEEE International Conference on Cloud Engineering (IC2E), pp. 200–210. IEEE (2019)
28. Rogge-Solti, A., Weske, M.: Prediction of business process durations using non-Markovian stochastic Petri nets. Inf. Syst. **54**, 1–14 (2015)
29. Russell, N., Ter Hofstede, A.H., Van Der Aalst, W.M., Mulyar, N.: Workflow control-flow patterns: a revised view. BPM Center Report BPM-06-22, pp. 06–22. BPMcenter.org (2006)
30. Sassoli, L., Vicario, E.: Close form derivation of state-density functions over DBM domains in the analysis of non-Markovian models. In: Proceedings of International Conference on Quantitative Evaluation of Systems. pp. 59–68. IEEE (2007)
31. SIRIO Library (2020). https://github.com/oris-tool/sirio
32. Trivedi, K.S., Sahner, R.: Sharpe at the age of twenty two. ACM SIGMETRICS Perform. Eval. Rev. **36**(4), 52–57 (2009)
33. Van Eyk, E., Iosup, A., Abad, C.L., Grohmann, J., Eismann, S.: A SPEC RG cloud group's vision on the performance challenges of FaaS cloud architectures. In: Companion of the 2018 ACM/SPEC International Conference on Performance Engineering, pp. 21–24 (2018)
34. Vanhatalo, J., Völzer, H., Koehler, J.: The refined process structure tree. Data Knowl. Eng. **68**(9), 793–818 (2009)
35. Vicario, E.: Static analysis and dynamic steering of time-dependent systems. IEEE Trans. Softw. Eng. **27**(8), 728–748 (2001)
36. Zhang, Y., Zheng, Z., Lyu, M.R.: WSPred: a time-aware personalized QoS prediction framework for Web services. In: IEEE International Symposium on Software Reliability Engineering, pp. 210–219. IEEE (2011)
37. Zheng, Z., Trivedi, K.S., Qiu, K., Xia, R.: Semi-Markov models of composite web services for their performance, reliability and bottlenecks. IEEE Trans. Serv. Comput. **10**(3), 448–460 (2015)

Transient Analysis of Hierarchical Semi-Markov Process Models with Tool Support in Stateflow

Stefan Kaalen(✉), Mattias Nyberg, and Olle Mattsson

Department of Machine Design, KTH Royal Institute of Technology, Stockholm, Sweden
{kaalen,matny,ollemat}@kth.se

Abstract. Semi-Markov process (SMP) models can not always accurately model real-world systems. To help the situation the paper proposes an hierarchical extension to SMP-models, called Hierarchical SMP-models (HSMP-models) and as the first contribution present an algorithm for performing transient analysis of HSMP-models where the CDF of each transition is an expolynomial. As the second contribution, a numerical method is presented for HSMP-models with an underlying stochastic process not satisfying bounded regeneration. The third, and final, contribution is tool support based on Stateflow for transient analysis of HSMP-models. Furthermore, an industrial case study for transient analysis of HSMP-models with unbounded regeneration representing a battery management system is presented.

1 Introduction

As the complexity of *safety-critical systems* increases, the need for model-based *transient analysis*, such as quantifying the *reliability* and *availability*, becomes increasingly important [1–3]. However, even for relatively simple systems, transient analysis can be difficult.

Solutions for performing these analyses have been presented for different families of models. One of the most influential being *Markov processes*, which has a long history of use in transient analysis of systems [4]. However, real-world systems can often not realistically be modeled to satisfy the Markov property [5, 6]. *Semi-Markov Processes* (SMPs), introduced in [7] and [8], generalizes Markov process and thereby allow for more realistic models of real-world systems. In SMPs the Markov property holds only directly after a transition has occurred. For the subclass of SMPs where the Cumulative Distribution Function (CDF) of each transition time is *expolynomial*, methods for performing analytical transient analysis has been presented by e.g. [9]. The problem of accurately modeling real-world systems does however remain also for SMPs [2,10], which can also be seen

The authors acknowledge the following agencies and projects for general and financial support: the European H2020 - ECSEL PRYSTINE project and Vinnova FFI, through the AVerT project, and the Wallenberg AI, Autonomous Systems and Software Program (WASP) funded by Knut and Alice Wallenberg Foundation.

A. Abate and A. Marin (Eds.): QEST 2021, LNCS 12846, pp. 105–126, 2021.
https://doi.org/10.1007/978-3-030-85172-9_6

in the case study of the present paper. The main problem is the lack of support for *hierarchy*, i.e. states containing internal SMPs, and *parallelism*.

As a step in the direction of increasing the amount of real-world systems that can be realistically modeled and analyzed, the present paper focuses on a hierarchical extension of SMP-models. This class of models will be referred to as *Hierarchical Semi-Markov Process* (HSMP) models and as the first contribution, an algorithm for performing transient analyses for the subclass of HSMP-models where the CDF of each transition is expolynomial is presented. The results extends the previous work in [11,12] with the addition of hierarchy.

The underlying process of HSMP-models is a subclass of a *Markov Regenerative Processes* MRGPs [2], which unlike SMPs only satisfy the Markov property directly after each transition into a subset of states known as *regeneration states*. For HSMP-models, this corresponds to the states that are not internal states of a super-state. MRGPs can be partitioned into those for which the maximum number of transitions that can occur between entering two regeneration states is bounded, referred to as having *bounded regeneration*, and the rest having *unbounded regeneration*. Performing transient analysis of HSMP-models is straight-forward using the proposed algorithm when bounded regeneration holds. As the second contribution, a method for utilizing the algorithm also when the model of interest has *unbounded regeneration* is presented. The problem of performing transient analysis of MRGPs with unbounded regeneration has also been approached in [13]. However the method in [13], unlike the method proposed here, has a complexity that directly correlates with the expected number of transitions between two regeneration points. This complexity implies that the method in [13] can become intractable when the number of expected transitions is high.

There are other modeling languages capable of modeling processes with the same underlying processes as HSMP-models. One such language is stochastic *petri nets* [14,15], used in the process of performing transient analysis in [13]. However, HSMP-models, are built upon *Statecharts* developed in [16] which has a much wider spread within industry, as is evident by their central role in the standard Unified Modeling Language (UML) [17]. Furthermore, many industrial tools have adopted Statecharts. One prominent example is *Stateflow* [18], a product within Matlab [19]. Matlab has more than 5 million users worldwide [20] and Matlab is widely used in industry around the world, an example being the automotive industry [21]. Therefore, to make the theory presented in the present paper easily accessible by industry, as the third contribution, the Matlab tool *SMP-tool* presented in [11,12] is extended with support for transient analysis of HSMP-models modeled in Stateflow.

Several previous approaches have been presented to model and analyze systems specified with stochastic extensions of Statecharts [22–24]. However, the analyses are either based on simulations, which may take an unfeasible long time when proving high levels of reliability, or only allow for modeling where each transition CDF has an exponential or fixed delay distribution.

The paper is outlined as follows. In Sect. 2, some preliminaries about SMPs, the *Laplace-Stieltjes transform* (L-S transform), and expolynomials are presented. In Sect. 3, the first contribution, i.e. the algorithm for transient analysis of HSMP-models is presented together with the definition of HSMP-models.

In Sect. 4, the second contribution, i.e. the method for utilizing the algorithm on HSMP-models with unbounded regeneration is presented. Finally, in Sect. 5 the third contribution, i.e. expanding the Matlab tool SMP-tool with support for transient analysis of HSMP-models, is presented and applied to a case study based on a real industrial system for electric trucks provided by the heavy-vehicle manufacturer Scania.

2 Preliminaries

2.1 Semi-Markov Process

Semi-Markov Processes (SMPs) will here be defined following [10]. Consider a system that evolves randomly over time and which at each point of time is in a state belonging to a countable state space $\mathcal{S}$. Moreover, let $\{X_n,\ n \in \mathbb{N}_0\}$ be the discrete-time stochastic process where X_n describes the state of the system directly after n state transitions has occurred. We let U_n denote the *sojourn time* in state X_n, i.e. the time that is spent in state X_n before transitioning into state X_{n+1}. Furthermore, assume that the first state of the stochastic process X_0 is entered at time $t = 0$. Consider now the continuous-time stochastic process $\{Z(t),\ t \geq 0\}$, with state space $\mathcal{S}$, where $Z(t) = X_{N(t)}$ and

$$N(t) = \begin{cases} 0 & \text{if } U_0 > t \\ \sup\{n \in \mathbb{N}_+ : U_0 + ... + U_{n-1} \leq t\} & \text{if } U_0 \leq t\ , \end{cases}$$

i.e. where $Z(t)$ describes the state of the system at time t. Note that $Z(t)$ is right-continuous. The process $\{Z(t),\ t \geq 0\}$ is a *Semi-Markov Process* if

$$P(X_{n+1} = j, U_n \leq t \mid X_n = i, U_{n-1}, ..., X_0, U_0) = P(X_{n+1} = j, U_n \leq t \mid X_n = i).$$

Let the matrix $\mathbf{Q}(t)$ be the *semi-Markov kernel* associated with an SMP $\{Z(t),\ t \geq 0\}$. The elements of the semi-Markov kernel [2] are given by

$$Q_{ij}(t) = P(X_{n+1} = j,\ U_n \leq t \mid X_n = i)\ , \quad i, j \in \mathcal{S},\ n \geq 0\ . \tag{1}$$

Note that we allow transitions from a state to itself represented by non-zero elements on the diagonal. An SMP is characterized by its semi-Markov kernel and its initial distribution $\mathbf{p}_0 = (P(Z(0) = i)),\ i \in \mathcal{S}$.

SMPs are often modelled as state transition diagrams [2]. Following [11], consider such models and let each transition have one source state and several possible target states with an associated probability vector. Furthermore, assign each transition a timer with a CDF. When a state i is entered, the timers of each transition with i as source state starts counting down from a time given by the corresponding CDF. As the first timer reaches zero, the corresponding transition is taken. Let $\mathcal{T}_i$ be the set of all transitions with i as source state. Moreover, let $\mathcal{T}_{ij}$ denote the set of all transitions with i as source state and j as possible target state. Furthermore, let $F_k(t)$ denote the CDF of transition $k \in \mathcal{T}_i$ and let $\mathbf{p}_k(j)$

denote the probability that transition k has state j as target. Assuming that no two transitions from the same state can occur at the same time instant, the semi-Markov kernel of an SMP can, following [11], be found through the relation

$$Q_{ij}(t) = \sum_{k \in \mathcal{T}_{ij}} \mathbf{p}_k(j) \int_0^t \left(\prod_{l \in \mathcal{T}_i \setminus \{k\}} (1 - F_l(u)) \right) dF_k(u) \,. \tag{2}$$

The above integral, and every integral on the form $\int_a^b f(t)dg(t)$ should be interpreted as a Lebesgue-Stieltjes integral on the interval $[a, b]$ unless otherwise stated.

Transient analysis of SMPs will now be presented following among others [25]. The Stieltjes convolution of two matrix-valued functions $\mathbf{A}(t)$ and $\mathbf{B}(t)$ of dimension $m \times m$ is given by $\mathbf{C}(t)$ where $C_{ij}(t) = \sum_{k=1}^{m} A_{ik}(t) \star B_{kj}(t)$ and

$$A_{ik}(t) \star B_{kj}(t) = \begin{cases} \int_0^t A_{ik}(t-y) dB_{kj}(y) & \text{if } t \geq 0 \\ 0 & \text{if } t < 0 \,. \end{cases}$$

Consider a matrix-valued function $\mathbf{A}(t)$. The Stieltjes-convolution inverse of $\mathbf{A}(t)$ is given by $(\mathbf{I} - \mathbf{A})^{(-1)}(t) = \sum_{n=0}^{\infty} \mathbf{A}^{(n)}(t)$, where $\mathbf{A}^{(n)}(t) = \mathbf{A}^{(n-1)}(t) \star \mathbf{A}(t)$, for $n > 0$, and $\mathbf{A}^{(0)}(t) = \boldsymbol{\Theta}(t)$, where $\boldsymbol{\Theta}(t)$ is the diagonal matrix with the Heaviside step function, $\Theta(t)$, on its diagonal. Moreover, let $\mathbf{H}(t)$ be a diagonal matrix where each element on the diagonal is the sum of the corresponding row of $\mathbf{Q}(t)$. At time $t \geq 0$ it holds that

$$P_{ij}(t) = \sum_{k \in \mathcal{S}} \left[(\mathbf{I} - \mathbf{Q}(t))^{(-1)} \right]_{ik} \star (\mathbf{I} - \mathbf{H}(t))_{kj} \,, \tag{3}$$

where $P_{ij}(t)$ is the probability of being in state j at time t given that state i is the initial state.

2.2 Laplace-Stieltjes Transform

The Laplace-Stieltjes (L-S) transform [26] of a function $g(t)$ is defined by

$$\mathcal{L}^*\{g(t)\} = \int_{-\infty}^{\infty} e^{-st}\, dg(t) = \lim_{R_1 \to \infty} \lim_{R_2 \to \infty} \int_{(-R_1, R_2)} e^{-st}\, dg(t) \qquad \text{where } s \in \mathbb{C} \,,$$

whenever this Lebesgue-Stieltjes integral exists.

2.3 Expolynomials

Define *expolynomial* as a sum of terms on the form $\Theta(t-b)t^u \boldsymbol{\alpha}^T e^{\mathbf{A}t} \boldsymbol{\beta}$, where b is a non-negative real number, u is a non-negative integer, A is an $N \times N$-matrix of real numbers where N is some positive integer, and $\boldsymbol{\alpha}$ and $\boldsymbol{\beta}$ are real column vectors of length N. The definition differ somewhat from the definition in e.g. [27] in order to more align with the phase-type distribution. Expolynomials are, as proven by Proposition 1 in Appendix A, closed under Stieltjes convolution.

3 Transient Analysis of Hierarchical Semi-Markov Processes

The first contribution, i.e. the algorithm for transient analysis of *HSMP-models* will here be presented. But first, the concept of HSMP-models, which is essentially a subclass of StoCharts presented in [22], will be formalized. The greatest restriction of HSMP-models compared to StoCharts is that HSMP-models does not allow for parallelism as StoCharts do with the "AND"-types of their nodes.

3.1 HSMP-models

Let $\mathcal{P}(\mathcal{A})$ denote the power set of a set $\mathcal{A}$, i.e. the set of all subsets of $\mathcal{A}$.

Definition 1. *A* Hierarchical Semi-Markov Process (HSMP)-model *is a tuple* $(\mathcal{S}, \mathcal{S}^{down}, \mathbf{p}_0, \mathcal{F}, \mathcal{T}, \text{ch})$ *where*

- $\mathcal{S}$ *is a finite, non-empty set of states.*
- $\mathcal{S}^{down} \subseteq \mathcal{S}$ *is a set of states (representing system failure).*
- $\mathbf{p}_0$ *is a probability vector representing the initial distribution of the states in* $\mathcal{S}$.
- $\mathcal{F}$ *is a set of Cumulative Distribution Functions (CDFs),* $F(t)$, *each satisfying* $\lim_{t\to 0^-} F(t) = 0$.
- $\mathcal{T} \subseteq \mathcal{S} \times \mathcal{F} \times (\mathcal{S} \to [0,1])$ *is a set of transitions, each containing a source state, a CDF, and a probability vector.*
- $\text{ch} : \mathcal{S} \to \mathcal{P}(\mathcal{S})$ *associates each state with a possible empty set of children states such that* $\mathcal{S}$ *is organized as a tree.*

An example of an HSMP-model is illustrated in Fig. 1a. For the remainder of the paper, the following notation will be used for a given HSMP-model, $\mathcal{H} = (\mathcal{S}, \mathcal{S}^{down}, \mathbf{p}_0, \mathcal{F}, \mathcal{T}, \text{ch})$:

- $\mathcal{S}^0$ is the set of all possible initial states of $\mathcal{H}$, i.e. all states corresponding to a nonzero element in $\mathbf{p}_0$.
- $\text{ch}_{\text{in}}(i) \subseteq \text{ch}(i)$ is the set of all internal initial states of state i, i.e. all children states of state $i \in \mathcal{S}$ that are possible target states of at least one transition with source not in $\text{ch}(i)$.

Inspired by [16], we say that state j is an *external state* of a state i if $j \notin (\text{ch}(i) \cup \{i\})$. Moreover, a state i is a *super-state* if $\text{ch}(i) \neq \emptyset$.

The semantics of HSMP-models, influenced by [22], is similar to that of SMP modeled with timers as described in Sect. 2.1; when a state i is entered, one timer for each of the transitions with i as source state starts counting down from a time given by their CDFs and the first to reach zero triggers the corresponding transition. The target state of the transition is then chosen according to its probability vector. Where HSMP-models differ from SMP-models is that the timers of a super-state j starts counting down when any state in $\text{ch}(j)$ is entered, they then keep counting down without resetting until one reaches zero, causing the corresponding transition, or until a state $k \notin \text{ch}(j)$ is reached by a transition from a state in $\text{ch}(j)$. Each time-point when a state i, which is not a child state of any super-state of an HSMP-model, is entered, is a regeneration point, i.e. the Markov property holds in these time points. The underlying process of an HSMP-model is thereby a MRGP [2].

For the algorithm, it is assumed that there is only one level of hierarchy, i.e. if a state $j \in \text{ch}(i)$ for some state i then $\text{ch}(j) = \emptyset$. Note that this, together with the tree structure of the states in an HSMP-model, assures that no two super-states can share the same child state. How to generalize to more levels of hierarchy is discussed later in Sect. 3.3. Furthermore, it is assumed that no state $i \in \mathcal{S}^{down} \cup \mathcal{S}^0$ satisfies $i \in \text{ch}(j)$ for any state $j \in \mathcal{S}$, an assumption without loss of generality for the task of computing the reliability. Without loss of generality, it is further assumed that: (1) each state $i \in \mathcal{S}^{down} \cup \mathcal{S}^0$ satisfies $\text{ch}(i) = \emptyset$, (2) there are no transitions with a super-state as possible target state, (3) no super-state has more than one internal initial state, i.e. for all the transitions from states not in a super-state i, the probability of a state $j \in \text{ch}(i)$ being the target state is zero except possibly for the internal initial state, and (4) no transition has a state $i \in \mathcal{S}^{down}$ as source state. If any of these four assumptions does not hold, the HSMP-model can with small effort be transformed to an HSMP-model satisfying these conditions. More discussion about this is provided in Sect. 3.3.

3.2 Algorithm for Transient Analysis of HSMP-models

As in Sec. 2.1, let $\mathcal{T}_i$ denote the set of all transitions with a state i as source state. The following algorithm is focused on computing the reliability. However, the same algorithm can with small adjustments be used for computing the availability as is discussed in Sect. 3.3. The reliability is the probability that no down-state has been reached after a certain system life-time. The reliability of HSMP-models where all transitions have expolynomial distributions can be found utilizing Algorithm 1. The algorithm contains four enumerated steps, all described below.

Step 1. In this case, all transitions in $\mathcal{T}_i$ have exponential CDFs. It follows directly from the memoryless property of the exponential distribution that each of the transitions in $\mathcal{T}_i$ can be replaced by $|\text{ch}(i)|$ new transitions which is each identical to the original transition with the only difference that each of the new transitions has a unique element in $\text{ch}(i)$ as source. After performing this procedure for all transitions $\mathcal{T}_i$, state i can be removed without affecting the reliability since i will no longer be the source (or possible target) of any transition.

Step 2. The goal of this step is to transform $\mathcal{H}$ in such a way that $\text{ch}(i) = \emptyset$ without affecting the reliability. It is done as follows.

Firstly, all transitions with a state $j \notin \text{ch}(i)$ as source and with the internal initial state $j \in \text{ch}(i)$ of state i as a possible target is substituted for a transition

Algorithm 1: Transient analysis of HSMP-models

Input: An HSMP-model $\mathcal{H} = (\mathcal{S}, \mathcal{S}^{down}, \mathbf{p}_0, \mathcal{T}, \mathcal{F}, ch)$ and a vector $\boldsymbol{\tau}$ of points in time.
Output: Reliability of $\mathcal{H}$ for the time-points $\boldsymbol{\tau}$.
foreach $i \in \mathcal{S}$ **do**
  if ch$(i) \neq \emptyset$ **then**
    if *All transitions in* $\mathcal{T}_i$ *have exponential distributions* **then**
1   Flatten out i by moving all transitions in $\mathcal{T}_i$ to the inner states.
    else
2   Flatten out i by translating the transitions with a state $j \in$ ch(i) as source (possible target) state to transitions with i as source (possible target) state.
    end
  end
end
3 Find the semi-Markov kernel in the L-S domain of $\mathcal{H}$.
4 Calculate the system failure distribution in the L-S domain, $f_T(s)$, of $\mathcal{H}$ and inverse Laplace transform $f_T(s)/s$ in order to find the reliability, $R(t)$, of $\mathcal{H}$ for the time points $\boldsymbol{\tau}$.

with same source state and transition timer but where the possible target state j has been changed to i with the same probability of occurring.

Secondly, consider the HSMP-model $\mathcal{H}_i$ made up by all states in ch(i) in union with all states that can be reached by one transition from a state in ch(i), all transitions from $\mathcal{H}$ for which the source state is in ch(i), and with the internal initial state j of i as the initial state. Figure 1b illustrates $\mathcal{H}_2$ for the HSMP-model illustrated in Fig. 1a. Let $\mathcal{S}_i$ denote the state space of $\mathcal{H}_i$. For $\mathcal{H}_i$, find the probability, $P^i_{jl}(t)$, of having reached state l as a function of the time for each state $l \neq$ ch(i). Let $\mathbf{Q}^i(t)$ denote the semi-Markov kernel of $\mathcal{H}_i$ and $\mathbf{H}^i(t)$ denote the diagonal matrix where each diagonal element contains the sum of the corresponding row in $\mathbf{Q}^i(t)$. The quantities can utilizing Eq. (3) be expressed through

$$P^i_{jl}(t) = \sum_{m \in \mathcal{S}_i} \left[\left(\mathbf{I} - \mathbf{Q}^i(t) \right)^{(-1)} \right]_{jm} \star \left(\mathbf{I} - \mathbf{H}^i(t) \right)_{ml} . \quad (4)$$

Once this expression is found for all states $l \in (\mathcal{S}_i \setminus \text{ch}(i))$ of $\mathcal{H}_i$ we again consider $\mathcal{H}$. All transitions with a state belonging to ch(i) as source are now removed and new transitions, one for each $l \in (\mathcal{S}_i \setminus \text{ch}(i))$ are instead added with i as source state and l as only possible target state. Each of these new transitions are assigned (possibly defective [28]) CDFs according to their target state l given by $P^i_{jl}(t)$. Note that this may result in an HSMP-model $\mathcal{H}$ with several transitions from state i that have the same possible target states.

It now holds that no internal state $j \in$ ch(i) of i is the source state of any transition with a state $l \notin$ ch(i) as possible target state. Furthermore, no internal state $j \in \text{ch}(i)$ of i is a possible target state of any transition with a state $l \notin \text{ch}(i)$ as source state. Therefore all internal states $j \in \text{ch}(i)$ of i can be removed together with the transitions between them without affecting the reliability.

A simple HSMP-model is illustrated before and after applying steps 1–2 in Fig. 1a 1c. The new CDF $F_{2,1}$ is yielded from applying step 2 of the algorithm and depend on $F_{3,4}$, $F_{4,3}$, and $F_{4.1}$.

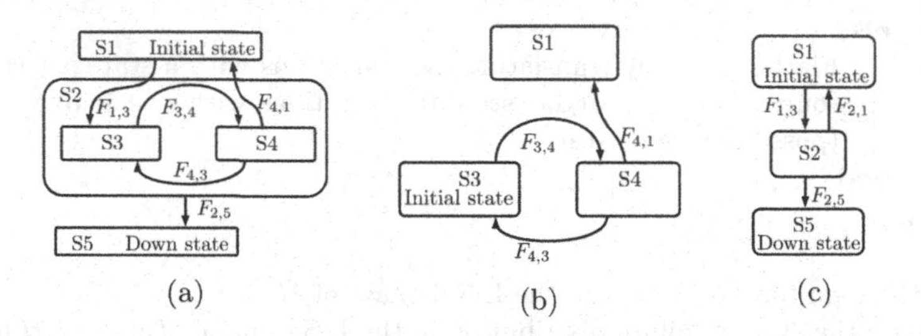

Fig. 1. a) An HSMP-model to be analyzed where $F_{i,j}$ denotes the CDF of the corresponding timer. It is assumed that $F_{2,5}$ is not an exponential. b) SMP utilized to find $F_{2,1}$ needed for the flattened model. c) Model of the resulting HSMP-model yielded by applying steps 1–2 of Algorithm 1 to the HSMP-model in Fig. 1a.

Solving Eq. (4) analytically is in general a difficult task given the infinite sum of convolutions the Stieltjes convolution inverse induces. One specific situation is when P^i_{jl} not only can be found analytically but the expolynomial shape is also preserved. One case where this holds is when all transitions with a state $k \in \text{ch}(i)$ as source has a CDF given by an exponential distribution. In this case, each searched element P^i_{jl} is simply the CDF of the time to absorption of the Markov process $\mathcal{H}_i$ and is thereby by the definition in [29] easily represented as a phase-type distribution.

Finding Transient Probability Under Bounded Regeneration. We consider another case where P^i_{jl} both can be found analytically and preserves the expolynomial shape. The case is equivalent to what is referred to as *bounded regeneration* in [30] and that terminology will be imported here. Let t_i denote the possibly infinite maximal time that can be spent in one visit of state i before leaving. Bounded regeneration occurs when the maximum number of transitions N_i that can occur between states in ch(i) in the time t_i without entering a state $m \notin \text{ch}(i)$ is finite. Following [31] and [5], it holds for the Stieltjes convolution inverse part of Eq. (4) that $(\mathbf{I} - \mathbf{Q}^i(t))^{(-1)} = \sum_{m=0}^{\infty} \mathbf{Q}^i(t)^{(m)}$. Now, bounded regeneration with the limit N_i in state i directly implies that the maximum number of transitions that can occur in $\mathbf{H}_i$ within the time t_i is $N_i + 1$, i.e. N_i transitions between the states in ch(i) and thereafter one transition leaving the set ch(i). This in turn directly implies that $\mathbf{Q}^i(t)^{(m)} = \mathbf{0}$ for all $m \geq N_i + 2$ and for all $t < t_i$. It now follows directly from Proposition 1 that $P^i_{jl}(t)$ is a sum of expolynomial expressions which can be found following the derivations in Appendix A. For the case when state i does not satisfy the restriction of bounded regeneration, the reader is referred to Sect. 4 where a numerical method is utilized to find $P^i_{jl}(t)$.

Step 3. Since there are no remaining super states in $\mathcal{H}$ after steps 1 and 2, each time a transition occurs constitutes a regeneration point and $\mathcal{H}$ is thereby a SMP. In step 3 the goal is to find all elements $q_{ij}(s)$ of the L-S transform of the semi-Markov kernel, $\mathbf{Q}(t)$, for $\mathcal{H}$. The matrix $\mathbf{Q}(t)$ is equivalent to the global kernel of the underlying MRGP. Let $\mathcal{T}_{ij}$ denote the subset of $\mathcal{T}_i$ with transitions having j as a possible target state. Moreover, let $F_k(t)$ and $\mathbf{p}_k$ denote the CDF and the target state probability vector of the transition k for any $k \in \mathcal{T}$. By applying the L-S transform on Eq. (2) and utilizing Leibniz's integral rule and Th. 6.1.7 in [42] it now holds for all non-zero elements of $q_{ij}(s)$ that

$$q_{ij}(s) = \sum_{k \in \mathcal{T}_{ij}} \mathbf{p}_k(j) \int_{0-}^{\infty} e^{-st} \prod_{l \in (\mathcal{T}_i) \setminus \{k\}} (1 - F_l(t)) dF_k(t) \; . \tag{5}$$

In the case of bounded regeneration, $q_{ij}(s)$ can be found symbolically for all states $i, j \in \mathcal{S}$ in the manner that is presented in Appendix B. If bounded regeneration is not fulfilled, the reader is referred to Sect. 4 where a numerical method for finding all element of $q_{ij}(s)$ is presented.

Step 4. Following Eq. (3), the L-S transform, $f_T(s)$ of the system failure distribution, i.e. the probability that a down-state has occurred as a function of the time, is given by $f_T(s) = \sum_{i \in \mathcal{S}^0} \sum_{j \in \mathcal{S}^{down}} \mathbf{p}_0(i)((\mathbf{I} - \mathbf{q}(s))^{-1})_{ij}$. For the case when all super-states of $\mathcal{H}$ has bounded regeneration, $f_T(s)$ is found completely symbolically. Utilizing the relation that $f_T(s) = \mathcal{L}^\star\{F_T(t)\} = s\mathcal{L}\{F_T(t)\}$, the reliability for τ can be computed from $f_T(s)$ by first finding the system failure probability, $F_T(t)$, using a numerical inverse Laplace transform of $f_T(s)/s$ and then utilizing $R(t) = 1 - F_T(t)$.

3.3 Discussion

When there are several levels of hierarchy, the algorithm can be generalized in the following way assuming that all of the super-states, that are themselves children of a state, satisfy either bounded regeneration or only have children states with exponential CDFs. Steps 1 and 2 are performed first for the super-states furthest away from the root for each branch in the tree of states. This will result in some of the leaves of the tree being deleted, creating a set of new set of super-states furthest away from the root. For this new tree again, perform steps 1 and 2 of Algorithm 1. Repeat this procedure until there are no more super-states when steps 3 and 4 will be applied.

In the end of Sect. 3.1, four assumptions was mentioned for which it was stated that an HSMP-model that does not satisfy these assumptions can be transformed into one that does. The perhaps most intricate of these transformations is for the case when a super-state i has several internal initial states $j, k, ..., l$. For these models, step 2 of the algorithm is done a little differently; instead of flattening state i into one state it will be flattened into several states $i_j, i_k, ..., i_l$. To get the CDFs of each of the new transitions that will be created for each of the source states $j, k, ..., l$, the HSMP-model $\mathcal{H}_{i_m}$ where $m \in \{j, k, ...l\}$ is considered where a unique internal initial state of i is chosen for each m. Furthermore, each transition with an internal initial state n of i as possible target state will be transformed so that the new possible target state with the same probability will be the one of the state $i_j, i_k, ..., i_l$ for which state m were chosen.

For the HSMP-models considered, the reliability and availability are equivalent. However, if there exists down-states being source states of transitions, the availability is given by applying the algorithm directly while the reliability is given by first eliminating all transitions with a down-state as source state and then applying the algorithm.

4 Transient Analysis of HSMP-models with Unbounded Regeneration

The case when at least one super-state of an HSMP-model, $\mathcal{H} = (\mathcal{S}, \mathcal{S}^{down}, \mathbf{p}_0, \mathcal{F}, \mathcal{T}, \mathrm{ch})$, has unbounded regeneration is now considered. The solution requires that the integral in Eq. (5) can be solved numerically. One straightforward case is when all super-states of the HSMP-model is the source of at least one transition m satisfying that $F_m(t) = 1$ for some time $t < \infty$. In this case the integral in Eq. (5) is always over a finite interval for $\mathcal{H}_i$. Step 1 of the algorithm is performed in the same manner as for HSMP-models of bounded generation. As for the step 2, problems arise in finding the elements $P^k_{jl}(t)$ from Eq. (4). This problem is worked around by instead of trying to find a symbolic expression of $P^k_{jl}(t)$, a symbolic expression is found for its L-S transform $p^k_{jl}(s) = \mathcal{L}^\star\{P^k_{jl}(t)\}$. Applying the transform on Eq. (4) yields

$$p^k_{jl}(s) = \sum_{m \in \mathcal{S}_k} \left[(\mathbf{I} - \mathbf{q}^k(s))^{-1}\right]_{jm} (\mathbf{I} - \mathbf{h}^k(s))_{ml} \,. \tag{6}$$

Note that in Eq. (6), the Stieltjes convolution and Stieltjes convolution inverse of Eq. (4) has been transformed into a product and a matrix inverse respectively. Since all CDFs in $\mathcal{F}$ have expolynomial distributions, so will all CDFs of $\mathcal{F}^k \subseteq \mathcal{F}$. It follows from this that both $\mathbf{Q}^k(t)$ and $\mathbf{H}^k(t)$ has expolynomial expressions which in turn implies that $\mathbf{q}^k(s)$ and $\mathbf{h}^k(s)$ can be found symbolically. A symbolical expression of $p^k_{jl}(s)$ can thereby be found from Eq. (6). Note that the CDFs of the transitions with k as source being added to $\mathcal{H}$ in step 2 will be unknown but the L-S transforms of the these CDFs will be known symbolically.

We now remember that the goal of step 3 of Algorithm 1 is to find all elements $q_{ij}(s)$ for $\mathcal{H}$. Recall Eq. (5) presenting how $q_{ij}(s)$ depends on the transitions of $\mathcal{H}$. The problem with solving Eq. (5) analytically is that the CDF of some elements in $\mathcal{H}$ is unknown. However, since the L-S transform of these CDFs are known, a numerical inverse Laplace transform can be utilized to find the values of these CDFs for some specified values in time. Since this is the case, when the integral cannot be found analytically, numerical integration can be used to numerically compute a symbolic expression of $q_{ij}(s)$. To enable the numerical integration, each term in each of the elements $F_k(t)$ can, by the product rule [39], be transformed into two parts. The first part contains only a heaviside step function, for which the integral can be solved analytically, and the second part is differentiable implying that its derivative can be found in the same manner as for the CDFs. Once all elements $q_{ij}(s)$ has been found, step 4 of the algorithm is performed as explained in Sect. 3.

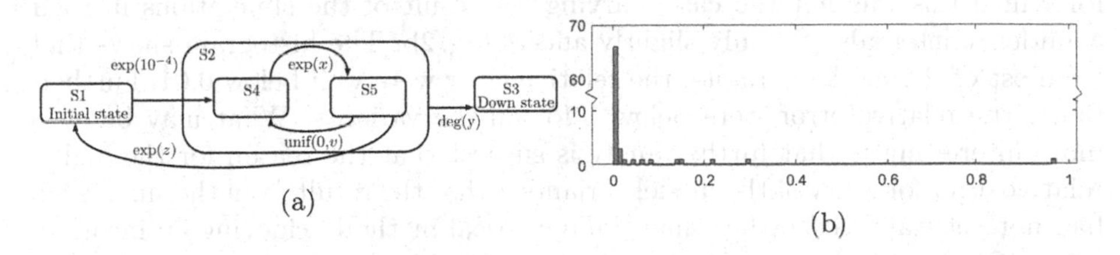

Fig. 2. a) A Simple model with unbounded regeneration used for evaluation. b) Histogram over for how many of the model variants of the model in 2a, a certain relative error of the transient probability after 45000 h were reached when using the solution method presented in Sect. 4.

4.1 Evaluation of Numerical Performance

Since the method presented earlier in this section utilizes numerical steps, an evaluation of the result of the algorithm will here be presented in order to indicate its performance. A computer with a 2.60 GHz, 6 core CPU and 16 GB RAM running Windows 10 Enterprise has been used for all computing analyses in this and upcoming sections. The evaluation is done by applying the numerical method to several variants of a simple model and comparing the results with the results achieved through simulation. The model is visualized in Fig. 2a. The model has been chosen since it is a minimal HSMP-model with unbounded regeneration. The values of the parameters x, y, z, and v are varied through the different variants of the model. The parameter for the transition from state S1 to state S4 is unchanged through the variants since it does not affect the approximations made in performing the transient analysis. For each of the varied parameters, three possible values are chosen and the model is analyzed for all combinations of parameter values resulting in a total of 81 model variants. The values chosen for each of the parameters are $x \in \{5 \cdot 10^{-1}, 10^{-2}, 10^{-3}\}$, $y \in \{150, 25, 10\}$, $z \in \{10, 10^{-1}, 5 \cdot 10^{-2}\}$, and $v \in \{200, 15, 2\}$ with units h or h^{-1} depending on the distributions. The possible values of the parameters have been chosen such that a down-state is reached in most of the simulations. For each of the variants of the model, the probability of having reached the down state S3 after 45000 h is computed. The numerical method were performed as it is implemented in SMP-tool 2.0 which is discussed in Sect. 5 and the analysis took a total time of 7 min. For the simulation, each variant of the model were simulated 10^6 times and it took a total time of 17 h.

For three of the model variants, S3 were not reached for any of the simulations given the low probability of it occurring. These variants were therefore removed. Let $F_T^{num}(45000)$ denote the system failure probability at time 45000 h yielded by the numerical method and let $F_T^{sim}(45000)$ denote the corresponding quantity yielded from the simulation approach. The result of how many of the model variants had what relative error, $|F_T^{sim}(45000) - F_T^{num}(45000)|/F_T^{sim}(45000)$, is presented in the histogram illustrated in Fig. 2b. For most of the model variants, the confidence interval of the simulations were small and for the very few for which this was not the case, varying the result of the simulations in their confidence intervals still only slightly affects Fig. 2b. The histogram shows that for most of the model variants, the relative error were well below 0.01. Furthermore, the relative error were below 1 for all the variants. What may be even more interesting is that further analysis suggest that the reason for the higher relative error for a few of the model variants is that the result from the simulation had not yet stabilized rather than the numerical method achieving an incorrect result. Given this, the numerical method validated here proved accurate for each of the 78 variants of the model. The simulations took 17 h to finalize

5 Tool Support and Case Study

The third contribution, tool support for transient analysis of HSMP-models will here be presented and applied to an industrial case study.

5.1 Tool Support

The tool named SMP-tool has in an earlier version previously been presented in [12] and [11] as a tool for performing steady-state, transient, and sensitivity analyses of SMPs. In SMP-tool 2.0 [32], support for transient analyses of HSMP-models both with bounded and unbounded regeneration following Sect. 3 and 4 has been added. Stateflow, within the Matlab and more specifically Simulink product family, provides a graphical language for state transition diagrams [18] and is widely used within the automotive industry [33]. Based on this, the tool support given by SMP-tool is based on Stateflow models. However, in order to include all stochastic behavior of HSMP-model, the models used by SMP-tool do while they are specified in Stateflow differ some from the models that can be analyzed directly in Stateflow.

SMP-tool allow the user to specify which inversion method that is used for the numerical inverse Laplace transform utilized in step 4 and in case of unbounded regeneration also in step 3 of Algorithm 1, the default method being the CME [34] method. As for the numerical integral solutions found in step 3 in the case of unbounded regeneration, the numerical method used is the *trapezoidal* method implemented the Matlab function "trapz()". Furthermore, the step-length of each integral is set as $0.1h$ as default but may be changed by the user. The main user interface for creating models is Stateflow and will be discussed further in Sect. 5.2.

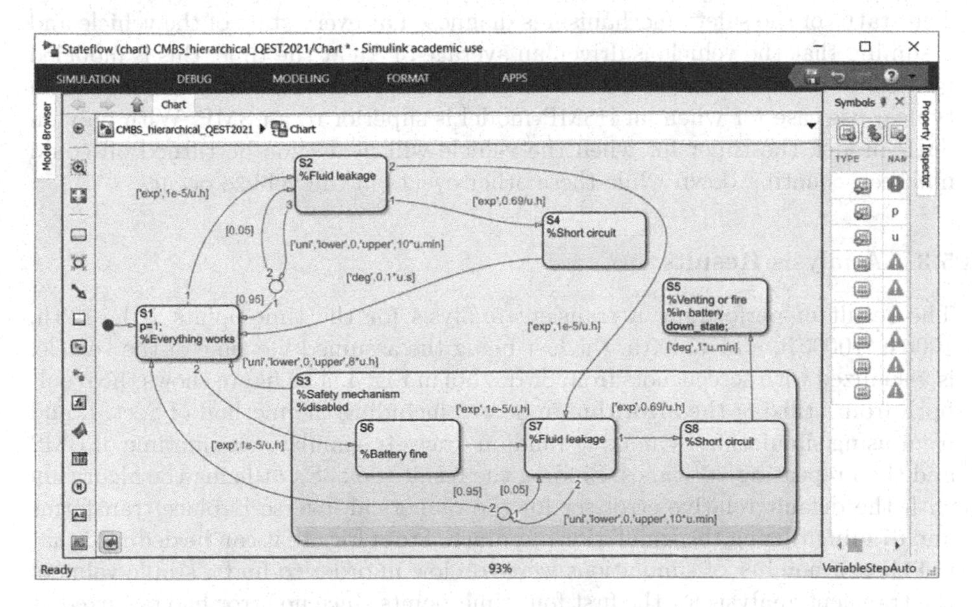

Fig. 3. HSMP-model of BMS case study including nine states in total, four of which contained in the super-state S3.

5.2 Case Study

The case study is based on a real industrial system from the heavy-vehicles manufacturer Scania and concerns a Battery Management System (BMS) in electric trucks and is visualized as an HSMP-model with unbounded regeneration modeled in Stateflow in Fig. 3. Model parameters have been chosen according to the real system but some have been adjusted for confidentiality reasons. Each arrow from a state represents a transition and is labeled by the CDF associated to the transition. Moreover, when there are several possible destination states, this is modeled by adding a junction (circle) from which several arrows origins, each of these arrows are labeled by the probability that the transition takes the corresponding route.

In the battery, there can be a fluid leakage corresponding to states S2 and S7. Fluid leakage is diagnosed periodically with a certain *diagnostic coverage* causing either a transition back to itself or to S1 if the fault is discovered and repaired. When a fluid leakage is left unattended, eventually it will lead to a short circuit of the battery corresponding to state S4 and S8. As long as the safety mechanism, which is a pyro-fuse, is enabled, corresponding to not being in state S3, the current will be shut down quickly when a short circuit happens. After this the truck will be repaired in a workshop before heading out on the road fully functioning again corresponding to state S1. However, if the safety mechanism is disabled as a fluid leakage leads to short circuit, this will eventually lead to the battery venting harmful gas and possibly cause a fire corresponding to state S5. The status of the safety mechanism is diagnosed at every start of the vehicle and assuming that the vehicle is driven an average of 8 h at the time, this is modeled by a uniform distribution from state S3 to state S1. The model illustrates a typical use case for when an HSMP-model is superior to an SMP. With only an SMP model, the timer for when the vehicle will next time be turned off could not keep counting down while these other events in the vehicle occur.

5.3 Analysis Results

The result of performing a transient analysis for the time-points 10 h, 100 h, 1000 h, 10000 h, and 45000 h, the last being the assumed life-time of the vehicle, is visualized with screenshots from SMP-tool in Fig. 4. The figure shows the result both from utilizing the algorithm in Sect. 3 including the method of Sect. 4, and from using simulation by making random draws to simulate one lifetime of BMS and then repeating 10^7 times. Finding the result took 38 s utilizing the algorithm with the default relative error set for the numerical inverse Laplace transforms and 67 min utilizing the simulation approach. From Fig. 4b it can be deduced that the chosen number of simulations were too low in order to find a stable value of the transient analysis for the first four time-points since no error had occurred at those times for any of the simulations. Looking at the life-time of the system, the relative error of the proposed algorithm, assuming that the simulation result is the correct value, is given by $|1.03 \cdot 10^{-6} - 8.00 \cdot 10^{-7}|/8.00 \cdot 10^{-7} = 0.29$. For the purpose of safety- and dependability-analysis, it is most often the magnitude and not the error that is of greatest important, implying that the proposed algorithm found an accurate enough result in a fraction of the time needed for the simulation. Furthermore, the result from the proposed algorithm is well inside the 95% confidence interval of the simulation visualized by its lower and upper confidence bounds in Fig. 4b. Note that the confidence interval is present for each time point but since no system failures had yet occurred, the lower confidence bound is 0 for some of the time-points. While this is merely the result of one example it indicates that the proposed algorithm can provide an answer with satisfying accuracy for real-world systems.

	Method	Time [hours]	Reliability	System failure	System failure / hour	Availability	1 - Availability
1	CME	10	0.999999999832949	1.670512617124587e-10	1.670512617124586e-11	0.999999999832949	1.670512617124587e-10
2	CME	100	0.999999997766440	2.233560225661790e-09	2.233560225661790e-11	0.999999997766440	2.233560225661790e-09
3	CME	1000	0.999999977099406	2.290059419873813e-08	2.290059419873813e-11	0.999999977099406	2.290059419873813e-08
4	CME	10000	0.999999770490072	2.295099283955437e-07	2.295099283955437e-11	0.999999770490072	2.295099283955437e-07
5	CME	45000	0.999998967903988	1.032096012432149e-06	2.293546694293664e-11	0.999998967903988	1.032096012432149e-06

(a)

	Method	Time [hours]	Reliability	System failure	Lower confidence bound	Upper confidence bound	System failure / hour
1	Simulation	10	1.000000000000000	0	0	3.668011163370465e-07	0
2	Simulation	100	1.000000000000000	0	0	3.668011163370465e-07	0
3	Simulation	1000	1.000000000000000	0	0	3.668011163370465e-07	0
4	Simulation	10000	1.000000000000000	0	0	3.668011163370465e-07	0
5	Simulation	45000	0.999999200000000	8.000000000230045e-07	3.444691268207785e-07	1.576399436681288e-06	1.777777777828899e-11

(b)

Fig. 4. a) Screenshot from SMP-tool for the result of transient analysis of BMS using the proposed algorithm. b) Screenshot from SMP-tool for the result of transient analysis of BMS using simulation.

6 Related Work

There are several tools capable of analyzing models with underlying non semi-Markov processes, such as ORIS [35], SHARPE [9], TimeNET [36], and GreatSPN [37]. However, along with SMP-tool, ORIS is the only one which can perform non-simulation based transient analysis of processes that do not satisfy the enabling restriction [35], i.e. where there can be several transitions with non-exponential CDFs enabled at the same time. The related work will therefore focus on the ORIS tool.

The ORIS tool allow for transient analysis of stochastic Time Petri Nets (sTPN) with unbounded regeneration through the *regenerative engine* and the *forward engine* in ORIS Tool 2.3.0 [35]. We compare these two ORIS engines with SMP-tool by applying them to the case study presented in Sect. 5.2. An sTPN-model corresponding to the HSMP-model in Fig. 3 is illustrated in Fig. 5. The error allowed was set to 10^{-6}, implying a maximum relative error of about 1, in order to capture the correct magnitude of the probability of reaching the down-state. For the forward engine, the tool was still enumerating stochastic state classes after 12 h. The computation was at that point aborted. For the regenerative engine, using a discretization step of 10 h, the probability of reaching the down-state after 45000 h was found to be $1.02 \cdot 10^{-6}$ with computation time 18 s. The result shows that the tools were comparable both when it comes to analysis time and the analysis result. However, the performance of the ORIS Tool depends on the discretization step both considering the analysis result and analysis time.

Following the description in [13], ORIS Tool works by what in the modeling framework of HSMP-models could be described as approximating the model of unbounded regeneration as having bounded regeneration; after a number of transitions within hierarchical state, as the probability of more transitions occurring becomes low enough in relation to the chosen error bound, it is assumed that no

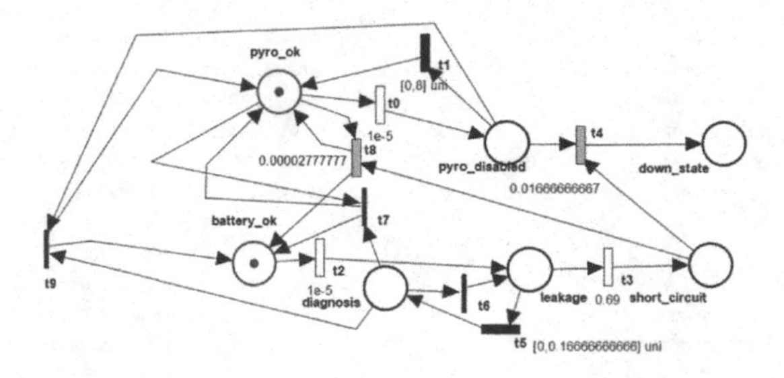

Fig. 5. BMS modeled as a timed stochastic petri net in ORIS tool.

more transition can occur within the hierarchical state. This implies that when the expected number of transitions within an hierarchical state is high, the analysis time of ORIS can become intractable. In contrast, SMP-tool approximates the model of unbounded regeneration directly, without treating it as having bounded regeneration and therefore lacks the same issue with the number of transitions made. To exemplify this, both SMP-tool and the regenerative engine of ORIS have been applied to the variant of the model in Fig. 2a with $x = 1$, $y = 10^3$, $z = 5 \cdot 10^{-3}$, and $v = 1$. In 12 s, SMP-tool managed to compute that the probability of reaching the down-state after 45000 h is 0.55, which corresponds well to the result of 0.56 yielded by simulating the model 1000 times. The same model were remodeled into an sTPN and analyzed with discretization step 5000 and error 0.1. The regenerative engine had still not found a result after 12 h when the computation were interrupted.

7 Conclusions and Future Work

When developing safety-critical systems, model-based transient analysis of reliability and availability is of utmost importance. Furthermore, it is imperative to make modeling and analysis easily accessible by the industry through tool support. This is with advantage achieved by basing the tools on what is already used in the industry. Examples are the modeling language Statecharts, which is a part of UML, and the tool Stateflow, which is a part of Matlab.

In order to achieve this result, the present paper has made the following contributions. As the first contribution an algorithm for transient analysis of HSMP-models, a stochastic extension of Statecharts, has been presented with a method for applying the algorithm for the subclass of models having bounded regeneration. Moreover, since it is important to be able to model as many real-world systems as possible, as the second contribution, a method for applying the algorithm for models with unbounded regeneration has been presented. Finally, as the third contribution, the Matlab app SMP-tool has been extended with support for HSMP-models. SMP-tool has also successfully been applied to an industrial case study provided by the heavy-vehicle manufacturer Scania.

Future work includes generalizing the approach and tool to support also parallel hierarchical states.

Appendix

A Convolution of Expolynomials

Proposition 1. *Expolynomials are closed under Stieltjes convolution.*

Proof. Consider two expolynomial expressions $G_1(t) = \Theta(t-b)t^u \boldsymbol{\alpha} e^{\mathbf{A}t}\boldsymbol{\beta}$ and $G_2(t) = \Theta(t-b')t^{u'}\boldsymbol{\alpha}' e^{\mathbf{A}'t}\boldsymbol{\beta}'$. By performing a Jordan decomposition of the square matrices A and A' into UJU^{-1} and $U'J'U'^{-1}$ where J is the Jordan normal (canonical) form, it following [38] holds that $G_1(t) = \Theta(t-b)t^u\boldsymbol{\alpha}Ue^{\mathbf{J}t}U^{-1}\boldsymbol{\beta}$ and $G_2(t) = \Theta(t-b')t^{u'}\boldsymbol{\alpha}'U'e^{\mathbf{J}'t}U'^{-1}\boldsymbol{\beta}'$. Furthermore it holds that all Jordan blocks created by a decomposition of a matrix $\mathbf{B}$, satisfies $\mathbf{J}_i = \mathbf{I}_i\lambda_i + \mathbf{N}_i$ where λ_i is an eigenvalue of B, $\mathbf{I}$ is a unit matrix, and $\mathbf{N}_i$ has value one on each element directly above the diagonal and zero everywhere else. Now following [38], it holds that $e^{\mathbf{J}t}$ is a block diagonal matrix with $e^{\mathbf{J}_1t}$, $e^{\mathbf{J}_2t}$, ..., $e^{\mathbf{J}_mt}$ on the diagonal and that $e^{\mathbf{J}_it} = e^{\lambda_it}\left(\mathbf{I}_i + t\mathbf{N}_i + \ldots + t^{m_i-1}\mathbf{N}_i^{m_i-1}/(m_i-1)!\right)$, where m_i is the size of the square matrix $\mathbf{J}_i$. Utilizing this result, it holds that $G_1(t)$ and $G_2(t)$ can be rewritten as $G_1(t) = \sum_{m\in\mathcal{M}}\Theta(t-b)t^{u_m}c_me^{a_mt}$ and $G_2(t) = \sum_{n\in\mathcal{M}'}\Theta(t-b')t^{u'_n}c'_ne^{a'_nt}$, for some real numbers c_m, c'_n, a_m and a'_n, for some non-negative integers u_m and u'_m, for some sets $\mathcal{M} = \{1, 2, ..., N\}$ and $\mathcal{M}' = \{1, 2, ..., N'\}$, and for some nonnegative integers N and N'

We will now consider the Stieltjes convolution of two expolynomial terms on the scalar form $K_1(t) = \Theta(t-b)t^uce^{at}$ and $K_2(t) = \Theta(t-b')t^{u'}c'e^{a't}$. Utilizing the product rule [39] it holds that

$$\begin{aligned}
(K_1 \star K_2)(t) &= \int_{[0,t]} \Theta(t-v-b)(t-v)^u ce^{a(t-v)} d\left(\Theta(v-b')v^{u'}c'_ne^{a'v}\right) \\
&= \Theta(t-b')\Theta(t-b'-b)(t-b')^u ce^{a(t-b')}b'^{u'}c'_ne^{a'b'} \\
&\quad + \Theta(t-b-b')\int_{[b',t-b]}(t-v)^u ce^{a(t-v)}u'v^{u'-1}c'_ne^{a'v}dv \\
&\quad + \Theta(t-b-b')\int_{[b',t-b]}(t-v)^u ce^{a(t-v)}v^{u'}c'_na'e^{a'v}dv\ .
\end{aligned}$$

Now utilizing the binomial theorem it holds

$$\begin{aligned}
(K_1 \star K_2)(t) &= \Theta(t-b'-b)\sum_{k=0}^{u}\binom{u}{k}t^k(-b')^{u-k}ce^{at}e^{-ab'}b'^{u'}c'_ne^{a'b'} \\
&+ \Theta(t-b-b')\sum_{l=0}^{u}\binom{u}{l}t^{n-l}(-1)^l ce^{at}u'c'_n\int_{[b',t-b]}v^{l+u'-1}e^{(a'-a)v}dv \\
&+ \Theta(t-b-b')\sum_{m=0}^{u}\binom{u}{m}t^{n-m}(-1)^m ce^{at}c'_na'\int_{[b',t-b]}v^{u'+m}e^{(a'-a)v}dv\ .
\end{aligned}$$

Now, it holds for any non-negative integer n and for any non-zero real number a that

$$\int x^ne^{ax}\,dx = \frac{e^{ax}}{a^{n+1}}\sum_{k=0}^{n}(-1)^k(ax)^{n-k}\frac{n!}{(n-k)!}\ .$$

Utilizing this result for the remaining two integrals in the convolution of $K_1(t)$ and $K_2(t)$ finally yields that Proposition 1 holds and yields an expression for $(K_1 \star K_2)(t)$.

B Semi-Markov Kernel of Expolynomials

In order to find the shape of $\mathbf{q}_{ij}(s)$ when each transition has an expolynomial CDF, some propositions are needed. The proof of Proposition 2 is straightforward and based upon substitution.

Proposition 2. *Let $m\%n$ denote m modulo n and let $\lceil x \rceil$ denote the ceiling function. The product $f(t)$ of two expolynomial expressions $\sum_{m\in\mathcal{M}} \Theta(t - b_m)t^{u_m}\boldsymbol{\alpha}_m^T e^{\mathbf{A}_m t}\boldsymbol{\beta}_m$ and $\sum_{n\in\mathcal{M}'} \Theta(t-b'_n)t^{u'_n}\boldsymbol{\alpha}_n'^T e^{\mathbf{A}'_n t}\boldsymbol{\beta}'_n$ where $\mathcal{M} = \{1, 2, ..., N\}$ and $\mathcal{M}' = \{1, 2, ..., N'\}$ satisfies*

$$f(t) = \sum_{m\in\mathcal{M}^*} \Theta(t - b_m^*)t^{u_m^*}\boldsymbol{\alpha}_m^{*T} e^{\mathbf{A}_m^* t}\boldsymbol{\beta}_m^* ,$$

where $u_m^* = u_{m\%N} + u'_{\lceil m/N \rceil}$, $\boldsymbol{\alpha}_m^{*T} = \boldsymbol{\alpha}_{m\%N}^T \otimes \boldsymbol{\alpha}'^T_{\lceil m/N \rceil}$, $\mathbf{A}_m^* = \mathbf{A}_{m\%N} \oplus \mathbf{A}'_{\lceil m/N \rceil}$, $\boldsymbol{\beta}_m^* = \boldsymbol{\beta}_{m\%N} \otimes \boldsymbol{\beta}'_{\lceil m/N \rceil}$, and $\otimes$ and $\oplus$ denote the kronecker product *and* kronecker sum *respectively[40].*

Proposition 3. *Consider any expolynomial $\sum_{m\in\mathcal{M}} \Theta(t - b_m)t^{u_m}\boldsymbol{\alpha}_m^T e^{\mathbf{A}_m t}\boldsymbol{\beta}_m$ where $b_m \in \mathbb{R}_+$, $u_m \in \{0, 1, 2, ...\}$, $\boldsymbol{\alpha}_m^T \in \mathbb{R}^{1\times p_m}$, $\mathbf{A}_m \in \mathbb{R}^{p_m\times p_m}$, $\boldsymbol{\beta}_m \in \mathbb{R}^{p_m\times 1}$, $p_m \in \{1, 2, ...\}$, and $\mathcal{M} = \{1, 2, ..., N\}$ for some $N \in \mathbb{N}_0$. It holds that*

$$\mathcal{L}\left\{\sum_{m\in\mathcal{M}} \Theta(t - b_m)t^{u_m}\boldsymbol{\alpha}_m^T e^{\mathbf{A}_m t}\boldsymbol{\beta}_m\right\}$$
$$= \sum_{m\in\mathcal{M}}\sum_{v=0}^{u_m} \frac{u_m!}{(u_m - v)!} b_m^{u_m-v}\boldsymbol{\alpha}_m^T e^{-b_m(s\mathbf{I}-\mathbf{A}_m)}(s\mathbf{I} - \mathbf{A}_m)^{-(v+1)}\boldsymbol{\beta}_m .$$

Proof. By use of well known properties of the Laplace transform presented in among others [41], it holds for any expolynomial that

$$\mathcal{L}\left\{\sum_{m\in\mathcal{M}}\Theta(t - b_m)t^{u_m}\boldsymbol{\alpha}_m^T e^{\mathbf{A}_m t}\boldsymbol{\beta}_m\right\} = \sum_{m\in\mathcal{M}}\boldsymbol{\alpha}_m^T \mathcal{L}\{\Theta(t - b_m)t^{u_m}\}(s_*)\boldsymbol{\beta}_m\Big|_{s_*=(s\mathbf{I}-\mathbf{A}_m)}$$
$$= \sum_{m\in\mathcal{M}} e^{-b_m s_*}\boldsymbol{\alpha}_m^T \mathcal{L}\{(t + b_m)^{u_m}\}(s_*)\boldsymbol{\beta}_m\Big|_{s_*=(s\mathbf{I}-\mathbf{A}_m)}$$
$$= \sum_{m\in\mathcal{M}}\sum_{v=0}^{u_m} \binom{u_m}{v} b_m^{u_m-v} e^{-b_m s_*}\boldsymbol{\alpha}_m^T \left(\frac{v!}{s_*^{v+1}}\right)\boldsymbol{\beta}_m\Big|_{s_*=(s\mathbf{I}-\mathbf{A}_m)}$$
$$= \sum_{m\in\mathcal{M}}\sum_{v=0}^{u_m} \frac{u_m!}{(u_m - v)!} b_m^{u_m-v}\boldsymbol{\alpha}_m^T e^{-b_m(s\mathbf{I}-\mathbf{A}_m)}(s\mathbf{I} - \mathbf{A}_m)^{-(v+1)}\boldsymbol{\beta}_m .$$

□

Computing the L-S Transform of the Semi-Markov Kernel. The expression for an element $q_{ij}(s)$ of $\mathbf{q}(s)$ for an SMP can now be found by applying the L-S transform on Eq. (2):

$$\begin{aligned} q_{ij}(s) &= \int_{-\infty}^{\infty} e^{-st} d \left[\sum_{k \in \mathcal{T}_{ij}} \left(\mathbf{p}_k(j) \int_0^t \prod_{l \in \mathcal{T}_i \setminus \{k\}} (1 - F_l(u))\, dF_k(u) \right) \right] \\ &= \sum_{k \in \mathcal{T}_{ij}} \mathbf{p}_k(j) \int_{-\infty}^{\infty} e^{-st} \prod_{l \in \mathcal{T}_i \setminus \{k\}} (1 - F_l(t))\, dF_k(t) \,, \end{aligned}$$

Write the CDF of a transition as $F_k(t) = \sum_{m \in \mathcal{M}} \Theta(t - b_{km}) t^{u_{km}} \boldsymbol{\alpha}_{km}^T e^{\mathbf{A}_{km} t} \boldsymbol{\beta}_{km}$ where $\mathcal{M} = \{1, 2, ..., N_k\}$. Without loss of generality we assume that $b_{km} \geq 0$ for each $k \in \mathcal{T}$, and $m \in \mathcal{M}$. Each element $q_{ij}(s)$ of $\mathbf{q}(s)$ now takes the form

$$\begin{aligned} q_{ij}(s) = \sum_{k \in \mathcal{T}_{ij}} \mathbf{p}_k(j) \int_{-\infty}^{\infty} e^{-st} \prod_{l \in \mathcal{T}_i \setminus \{k\}} \left(1 - \sum_{m \in \mathcal{M}} \Theta(t - b_{lm}) t^{u_{lm}} \boldsymbol{\alpha}_{lm}^T e^{\mathbf{A}_{lm} t} \boldsymbol{\beta}_{lm} \right) \\ d \left(\sum_{n \in \mathcal{M}'} \Theta(t - b'_{kn}) t^{u'_{kn}} \boldsymbol{\alpha}_{kn}'^T e^{\mathbf{A}'_{kn} t} \boldsymbol{\beta}'_{kn} \right) \,, \end{aligned}$$

where $\mathcal{M} = \{1, 2, ..., N_l\}$ and $\mathcal{M}' = \{1, 2, ..., N_k\}$. Now, for each $m \in \mathcal{M}$ and $l \in \mathcal{T}_i \setminus \{k\}$ define $b_{lm}^* = -b_{lm}$, $u_{lm}^* = -u_{lm}$, $\boldsymbol{\alpha}_{lm}^{*T} = -\boldsymbol{\alpha}_{lm}^T$, $\mathbf{A}_{lm}^* = -\mathbf{A}_{lm}$, and $\boldsymbol{\beta}_{lm}^* = -\boldsymbol{\beta}_{lm}$. The element $q_{ij}(s)$ can now be rewritten as

$$\begin{aligned} q_{ij}(s) = \sum_{k \in \mathcal{T}_{ij}} \mathbf{p}_k(j) \int_{-\infty}^{\infty} e^{-st} \prod_{l \in \mathcal{T}_i \setminus \{k\}} \left(\sum_{m \in \mathcal{M}^*} \Theta(t - b_{lm}^*) t^{u_{lm}^*} \boldsymbol{\alpha}_{lm}^{*T} e^{\mathbf{A}_{lm}^* t} \boldsymbol{\beta}_{lm}^* \right) \\ d \left(\sum_{n \in \mathcal{M}'} \Theta(t - b'_{kn}) t^{u'_{kn}} \boldsymbol{\alpha}_{kn}'^T e^{\mathbf{A}'_{kn} t} \boldsymbol{\beta}'_{kn} \right) \,, \end{aligned}$$

where for each $l \in \mathcal{T}_i \setminus \{k\}$, $\mathcal{M}^* = \{1, ..., N_l^* = N_l + 1\}$, $b_{l(n+1)}^* = -\infty$, $u_{l(n+1)}^* = 0$, $\boldsymbol{\alpha}_{l(n+1)}^{*T} = 1$, $\mathbf{A}_{l(n+1)}^* = 0$, and $\boldsymbol{\beta}_{l(n+1)}^* = -1$. By the theorems in [42] of the Lebegue-Stieltjes integral, $q_{ij}(s)$ satisfies

$$\begin{aligned} q_{ij}(s) = & \sum_{k \in \mathcal{T}_{ij}} \mathbf{p}_k(j) \sum_{n \in \mathcal{M}'} e^{-s b'_{kn}} \\ & \prod_{l \in \mathcal{T}_i \setminus \{k\}} \left(\sum_{m \in \mathcal{M}^*} \Theta(b'_{kn} - b_{lm}^*) b_{kn}'^{u_{lm}^*} \boldsymbol{\alpha}_{lm}^{*T} e^{\mathbf{A}_{lm}^* b'_{kn}} \boldsymbol{\beta}_{lm}^* \right) b_{kn}'^{u'_{kn}} \boldsymbol{\alpha}_{kn}'^T e^{\mathbf{A}'_{kn} b'_{kn}} \boldsymbol{\beta}'_{kn} \\ & + \sum_{k \in \mathcal{T}_{ij}} \mathbf{p}_k(j) \int_{-\infty}^{\infty} e^{-st} \\ & \prod_{l \in \mathcal{T}_i \setminus \{k\}} \left(\sum_{m \in \mathcal{M}^*} \Theta(t - b_{lm}^*) t^{u_{lm}^*} \boldsymbol{\alpha}_{lm}^{*T} e^{\mathbf{A}_{lm}^* t} \boldsymbol{\beta}_{lm}^* \right) \sum_{n \in \mathcal{M}^\star} \Theta(t - b_{kn}^\star) t^{u_{kn}^\star} \boldsymbol{\alpha}_{kn}^{\star T} e^{\mathbf{A}_{kn}^\star t} \boldsymbol{\beta}_{kn}^\star dt \,, \end{aligned}$$

where $\mathcal{M}^\star = \{1, 2, ..., N_k^\star = 2N_k\}$ and where where for each $n \in \{1, 2, ..., N_k\}$ $b_n^\star = b'_n$, $u_n^\star = u'_n$, $\boldsymbol{\alpha}_n^{\star T} = \boldsymbol{\alpha}_n'^T$, $\mathbf{A}_n^\star = \mathbf{A}'_n$, and $\boldsymbol{\beta}_n^\star = \mathbf{A}_n \boldsymbol{\beta}'_n$, and where for each $n \in \{N_k + 1, ..., N_k^\star\}$ $b_n^\star = b'_{n-c_k}$, $u_n^\star = u'_{n-c_k-1}$, $\boldsymbol{\alpha}_n^{\star T} = \boldsymbol{\alpha}'^T_{n-c_k}$, $\mathbf{A}_n^\star = \mathbf{A}'_{n-c_k}$, and $\boldsymbol{\beta}_n^\star = \boldsymbol{\beta}'_{n-c_k} u_{n-c_k}$. Note that since $b'_{kn} \geq 0$ for each $k \in \mathcal{T}$ and $n \in \mathcal{M}$, it follows directly that $b_{kn}^\star \geq 0$ for each $k \in \mathcal{T}$ and $n \in \mathcal{M}^\star$. Now let h denote $|\mathcal{T}_i \backslash \{k\}|$ and let $l^1, l^2, ..., l^h$ denote the elements in $\mathcal{T}_i \backslash \{k\}$. By Proposition 2 $q_{ij}(s)$ now satisfies

$$
\begin{aligned}
q_{ij}(s) = &\sum_{k \in \mathcal{T}_{ij}} \mathbf{p}_k(j) \sum_{n \in \mathcal{M}'} e^{-sb'_{km}} \\
&\prod_{l \in \mathcal{T}_i \backslash \{k\}} \left(\sum_{m \in \mathcal{M}^*} \Theta(b'_{kn} - b^*_{lm}) b_{kn}'^{u^*_{lm}} \boldsymbol{\alpha}^{*T}_{lm} e^{\mathbf{A}^*_{lm} b'_{kn}} \boldsymbol{\beta}^*_{lm} \right) b_{kn}'^{u'_{kn}} \boldsymbol{\alpha}'^T_{kn} e^{\mathbf{A}'_{kn} b'_{kn}} \boldsymbol{\beta}'_{kn} \\
&+ \sum_{k \in \mathcal{T}_{ij}} \mathbf{p}_k(j) \int_{-\infty}^{\infty} e^{-st} \sum_{m \in \mathcal{M}^\dagger} \Theta(t - b^\dagger_m) t^{u^\dagger_m} \boldsymbol{\alpha}^{\dagger T}_m e^{\mathbf{A}^\dagger_m t} \boldsymbol{\beta}^\dagger_m dt \,,
\end{aligned}
$$

where $\mathcal{M}^\dagger = \{1, 2, ..., \prod_{l \in \mathcal{T}_i \backslash \{k\}} (N_l^* + 1) N_k^\star\}$ and

$$
\begin{aligned}
b^\dagger_m &= \max(b^\star_{k,m\%N_k^\star}, b^*_{l^1 \lceil m/N_k^\star \rceil}, b^*_{l^2 \lceil m/N_k^\star N_{l^1}^\star \rceil}, ..., b^*_{l^h \lceil m/N_k^\star \prod_{\gamma=1}^{h-1} N_{l^\gamma}^\star \rceil}) \\
u^\dagger_m &= u^\star_{k,m\%N_k^\star} + u^*_{l^1 \lceil m/N_k^\star \rceil} + u^*_{l^2 \lceil m/N_k^\star N_{l^1}^\star \rceil} + ... + u^*_{l^h \lceil m/N_k^\star \prod_{\gamma=1}^{h-1} N_{l^\gamma}^\star \rceil} \\
\boldsymbol{\alpha}^{\dagger T}_m &= \boldsymbol{\alpha}^{\star T}_{k,m\%N_k^\star} \otimes \boldsymbol{\alpha}^{*T}_{l^1 \lceil m/N_k^\star \rceil} \otimes \boldsymbol{\alpha}^{*T}_{l^2 \lceil m/N_k^\star N_{l^1}^\star \rceil} \otimes ... \otimes \boldsymbol{\alpha}^{*T}_{l^h \lceil m/N_k^\star \prod_{\gamma=1}^{h-1} N_{l^\gamma}^\star \rceil} \\
\mathbf{A}^\dagger_m &= \mathbf{A}^\star_{k,m\%N_k^\star} \oplus \mathbf{A}^*_{l^1 \lceil m/N_k^\star \rceil} \oplus \mathbf{A}^*_{l^2 \lceil m/N_k^\star N_{l^1}^\star \rceil} \oplus ... \oplus \mathbf{A}^*_{l^h \lceil m/N_k^\star \prod_{\gamma=1}^{h-1} N_{l^\gamma}^\star \rceil} \\
\boldsymbol{\beta}^\dagger_m &= \boldsymbol{\beta}^\star_{k,m\%N_k^\star} \otimes \boldsymbol{\beta}^*_{l^1 \lceil m/N_k^\star \rceil} \otimes \boldsymbol{\beta}^*_{l^2 \lceil m/N_k^\star N_{l^1}^\star \rceil} \otimes ... \otimes \boldsymbol{\beta}^*_{l^h \lceil m/N_k^\star \prod_{\gamma=1}^{h-1} N_{l^\gamma}^\star \rceil} .
\end{aligned}
$$

Finally, utilizing Proposition 3 yields

$$
\begin{aligned}
q_{ij}(s) = &\sum_{k \in \mathcal{K}_{ij}} \mathbf{p}_k(j) \sum_{n \in \mathcal{M}'} e^{-sb'_{kn}} \\
&\prod_{l \in \mathcal{T}_i \backslash \{k\}} \left(\sum_{m \in \mathcal{M}^*} \Theta(b'_{kn} - b^*_{lm}) b_{kn}'^{u^*_{lm}} \boldsymbol{\alpha}^{*T}_{lm} e^{\mathbf{A}^*_{lm} b'_{kn}} \boldsymbol{\beta}^*_{lm} \right) b_{kn}'^{u'_{kn}} \boldsymbol{\alpha}'^T_{kn} e^{\mathbf{A}'_{kn} b'_{kn}} \boldsymbol{\beta}'_{kn} \\
&+ \sum_{k \in \mathcal{T}_{ij}} \mathbf{p}_k(j) \sum_{m \in \mathcal{M}^\dagger} \sum_{v=0}^{u^\dagger_m} \frac{u^\dagger_m!}{(u^\dagger_m - v)!} b_m^{\dagger u^\dagger_m - v} e^{-b^\dagger_m s} \boldsymbol{\alpha}^{\dagger T}_m e^{b^\dagger_m \mathbf{A}^\dagger_m} (s\mathbf{I} - \mathbf{A}^\dagger_m)^{-(v+1)} \boldsymbol{\beta}^\dagger_m \,.
\end{aligned}
\tag{7}
$$

To summerize, utilizing the definitions of the parameters, which can be found as described above, Eq. (7) finally give the expression of the semi-Markov kernel of any SMP where each transition has an expolynomial CDF.

References

1. Rausand, M., Høyland, A.: System Reliability Theory: Models, Statistical Methods, and Applications, 2nd edn. Wiley, Hoboken (2004)

2. Trivedi, K.S., Bobbio, A.: Reliability and Availability Engineering: Modeling, Analysis, and Applications. Cambridge University Press, Cambridge (2017)
3. Limnios, N., Oprişan, G.: Semi-Markov Processes and Reliability. Springer, New York (2001). https://doi.org/10.1007/978-1-4612-0161-8
4. International Electrotechnical Commission: Functional Safety of Electrical/Electronic/Programmable Electronic Safety-related Systems (IEC61508) (2010)
5. Limnios, N.: Dependability analysis of semi-Markov systems. Reliab. Eng. Syst. Saf. **55**, 203–207 (1997)
6. Marsan, M.A.: Stochastic petri nets: an elementary introduction. In: Rozenberg, G. (ed.) APN 1988. LNCS, vol. 424, pp. 1–29. Springer, Heidelberg (1990). https://doi.org/10.1007/3-540-52494-0_23
7. Levy, P.: Processus semi-Markoviens. In: Proceedings of the International Congress of Mathematicians, Amsterdam, pp. 416–426 (1954)
8. Smith, W.: Regenerative stochastic processes. Proc. R. Soc. Lond. Ser. A Math. Phys. Sci. **232**(1188), 6–31 (1955)
9. Trivedi, K.S., Sahner, R.: SHARPE at the age of twenty two. SIGMETRICS Perform. Eval. Rev. **36**(4), 52–57 (2009)
10. Kulkarni, V.G.: Modeling and Analysis of Stochastic Systems, 3rd edn. Taylor & Francis Group, LLC, Boca Raton (2017)
11. Kaalen, S., Nyberg, M.: Branching transitions for semi-Markov processes with application to safety-critical systems. In: Zeller, M., Höfig, K. (eds.) IMBSA 2020. LNCS, vol. 12297, pp. 68–82. Springer, Cham (2020). https://doi.org/10.1007/978-3-030-58920-2_5
12. Kaalen, S., et al.: Tool-supported dependability analysis of semi-markov processes with application to autonomous driving. In: 4th International Conference on System Reliability and Safety (ICSRS), pp. 126–135 (2019)
13. Horvath, A., et al.: Transient analysis of non-Markovian models using stochastic state classes. Perform. Eval. **68**, 315–335 (2012)
14. Petri, C.A.: Communication with automata, Ph.D. thesis (1966)
15. Vicario, E.: Using stochastic state classes in quantitative evaluation of dense-time reactive systems. IEEE Trans. Softw. Eng. **26**(5), 703–719 (2009)
16. Harel, D.: Statecharts: a visual formalism for complex systems. Sci. Comput. Program. **8**, 231–274 (1987)
17. UML homepage. https://www.uml.org/. Accessed 2 July 2021
18. Stateflow Homepage. https://se.mathworks.com/products/stateflow.html. Accessed 2 July 2021
19. Matlab homepage. https://se.mathworks.com/products/matlab.html. Accessed 2 July 2021
20. Matlab company page. https://se.mathworks.com/company.html. Accessed 2 July 2021
21. Matlab Company Factsheet. https://se.mathworks.com/content/dam/mathworks/handout/2020-company-factsheet-8-5x11-8282v20.pdf. Accessed 26 Apr 2021
22. Janssen, D.N.: Extensions of Statecharts with probability, time, and stochastic timing. Ph.D. thesis (2003)
23. Lindemann, C., et al.: Performance analysis of time-enhanced UML diagrams based on stochastic processes. In: Proceedings of the 3rd International Workshop on Software and Performance (WOSP), pp. 25–34 (2002)

24. Hermanns, H., et al.: From StoCharts to MoDeST: a comparative reliability analysis of train radio communications. In: Proceedings of the 5th International Workshop on Software and Performance (WOSP), pp. 13–23 (2005)
25. Pyke, R.: Markov renewal processes with finitely many states. Ann. Math. Stat. **32**(4), 1243–1259 (1961)
26. Grimmet, G., Stirzaker, D.: Probability and Random Processes, 3rd edn. Oxford University Press Inc., Oxford (2001)
27. Horváth, A., et al.: Transient analysis of generalised semi-Markov processes using transient stochastic state classes. In: International Conference on the Quantitative Evaluation of Systems (QEST), Williamsburg (2010)
28. Beaumont, G.P.: Probability and random variables. International Publishers in Science and Technology, West Sussex (2005)
29. Neuts, M.F.: Matrix-Geometric Solutions in Stochastic Models: An Algorithmic Approach. Dover Publications, Inc., New York (1981)
30. Biagi, M., Carnevali, L., Paolieri, M., Papini, T., Vicario, E.: Exploiting non-deterministic analysis in the integration of transient solution techniques for Markov regenerative processes. In: Bertrand, N., Bortolussi, L. (eds.) QEST 2017. LNCS, vol. 10503, pp. 20–35. Springer, Cham (2017). https://doi.org/10.1007/978-3-319-66335-7_2
31. Çinlar, E.: Introduction to Stochastic Processes. Dover Publications Inc., New York (1975)
32. SMP-tool homepage. https://www.kth.se/itm/smptool. Accessed 2 July 2021
33. Matlab automotive solutions page. https://se.mathworks.com/solutions/automotive.html. Accessed 2 July 2021
34. Horváth, I., et al.: An optimal inverse Laplace transform method without positive and negative overshoot - an integral based interpretation. Electron. Notes Theoret. Comput. Sci. **337**, 87–104 (2018)
35. Paolieri, M., et al.: The ORIS tool: quantitative evaluation of non-Markovian systems. IEEE Trans. Softw. Eng. **47**(6), 1211–1225 (2019)
36. Zimmermann, A.: Modelling and performance evaluation with TimeNET 4.4. In: Bertrand, N., Bortolussi, L. (eds.) QEST 2017. LNCS, vol. 10503, pp. 300–303. Springer, Cham (2017). https://doi.org/10.1007/978-3-319-66335-7_19
37. Amparore, E.G., Balbo, G., Beccuti, M., Donatelli, S., Franceschinis, G.: 30 years of GreatSPN. In: Fiondella, L., Puliafito, A. (eds.) Principles of Performance and Reliability Modeling and Evaluation. SSRE, pp. 227–254. Springer, Cham (2016). https://doi.org/10.1007/978-3-319-30599-8_9
38. Notes on the Matrix Exponential. http://www.ctr.maths.lu.se/media11/MATM14/2012vt12/exp_.pdf. Accessed 25 Mar 2021
39. Adams, R.A., Essex, C.: Calculus: A Complete Course, 8th edn. Pearson Canada Inc., Toronto (2014)
40. Liesen, J., Mehrmann, V.: Linear Algebra. Springer, Cham (2015)
41. Oberhettinger, F., Badii, L.: Tables of Laplace Transforms. Springer, Heidelberg (1973). https://doi.org/10.1007/978-3-642-65645-3
42. Carter, M., van Brunt, B.: The Lebesgue-Stieltjes Integral: A Practical Introduction. Springer, New York (2000). https://doi.org/10.1007/978-1-4612-1174-7

Evaluating the Effectiveness of Metamodeling in Emulating Quantitative Models

Michael Rausch[1](✉) and William H. Sanders[2]

[1] University of Illinois at Urbana-Champaign, Urbana, IL, USA
mjrausc2@illinois.edu

[2] Carnegie Mellon University, Pittsburgh, PA, USA
sanders@cmu.edu

Abstract. It is often prohibitively time-consuming to do sensitivity analysis, uncertainty quantification, and optimization with complex state-based quantitative models because each model execution or solution takes so long to complete, and many such executions are necessary to complete the analysis. One way to approach this problem is to use metamodels that emulate the behavior of the base model but run much faster. These metamodels may be automatically constructed using machine learning techniques, and then the relevant analysis may be conducted on the fast-running metamodel in place of the slow-running model.

In this work, we evaluate the effectiveness of several different types of metamodels in emulating seven publicly available PRISM and Möbius models. In our evaluation, we found that the metamodels are reasonably accurate and are several thousand times faster than the corresponding models they emulate. Furthermore, we find that stacking-based metamodels are significantly more accurate than state-of-the-practice metamodels. We show that metamodeling is a powerful and practical tool for modelers interested in understanding the behavior of their models, because it makes feasible analysis techniques that would otherwise take too long to run on the original models.

Keywords: Metamodels · Surrogate models · Emulators · Security models · Reliability models · Sensitivity analysis · Uncertainty quantification · Optimization

1 Introduction

Many realistic state-based quantitative models have a large number of input variables. The precise values of the input variables are often unknown in practice. Alternatively, some of the input variable values may be under the direct control of the modeler, and the modeler must choose values that will maximize the system's performance or utility. Sensitivity analysis (SA) and uncertainty quantification (UQ) can help a modeler understand and manage the resulting

A. Abate and A. Marin (Eds.): QEST 2021, LNCS 12846, pp. 127–145, 2021.
https://doi.org/10.1007/978-3-030-85172-9_7

uncertainties, while optimization can help a modeler determine which combination of input variable values may be best to achieve a particular goal. SA, UQ, and optimization techniques are all of vital importance to modelers. However, applying SA, UQ, or optimization techniques directly to the model can be impractical for models with long execution times, since the techniques usually mandate that the model be solved for many different combinations of values for the input parameters.

Alternatively, a modeler can perform SA, UQ, or optimization indirectly through the use of a metamodel. A metamodel is a model of a model. The metamodel (usually imperfectly) emulates the behavior of the model (which we shall refer to as the *base model* in this paper). The metamodel generally runs significantly faster than the base model, trading a great gain in speed for slightly less accuracy. It is unnecessary to construct metamodels manually; metamodels can be constructed automatically using machine learning techniques. Once a metamodel has been built, SA, UQ, or optimization techniques can be applied to it as a "stand-in" for the base model and completed in a reasonable amount of time. Using metamodels, modelers can indirectly perform SA, UQ, and optimization on models that run so slowly that it would be impractical to evaluate them directly.

In recent work, a stacking-based metamodel architecture was applied to the analysis of a botnet simulation model [9] and an Advanced Metering Infrastructure (AMI) simulation model [10]. The resulting metamodels ran hundreds to thousands of times faster than the original base models and were more accurate than traditional metamodels in emulating their behavior. While the results were impressive, it was unknown whether the approach would generalize and produce similarly good results for other models. If the metamodeling techniques were shown to work on a wider variety of test cases, modelers would gain greater confidence in the efficacy of metamodels for quantitative modeling.

In this work, we use seven previously published models (six PRISM models and one Möbius model) as test cases to evaluate the speed and accuracy of metamodels built via machine learning. We evaluate state-of-the-practice metamodel architectures, the recently proposed stacking-based metamodel architecture, and several variants. We show that the metamodels we built run thousands of times faster than the corresponding base models, and all are reasonably accurate. In addition, based on the results of our experiments, we give advice on which metamodel architectures to use for those who wish to apply the techniques to their own models. Our work gives confidence that metamodels can be a useful tool for performing SA, UQ, and optimization on models with many uncertain input parameters that would normally take too long to execute.

The rest of this paper is organized as follows. In Sect. 2, we describe the seven models we use as test cases. In Sect. 3, we briefly explain the metamodeling approach in general, the current state-of-the-practice, the recently developed stacking-based approach, and variants to the stacking-based approach. In Sect. 4, we show the speed and accuracy of the metamodels on the test case models. We discuss key insights, use cases, and limitations in Sect. 5. We briefly review related work in Sect. 6, and conclude in Sect. 7. The appendix contains details of the initial conditions of our test case models.

2 Test Cases

We evaluate seven different test cases. We used the botnet model that was previously used as a test case in a metamodeling investigation [9], plus six published PRISM [4] models that, to the best of our knowledge, have never been used as subjects of metamodeling. We examined the publicly available PRISM case studies[1]. Since we are especially interested in the use of metamodeling for sensitivity analysis, uncertainty quantification, and optimization, which are commonly used when a model has many uncertain input variables, we selected case studies that had many model input variables. We also tried to select models that were non-trivial. We chose to use published models, rather than create synthetic models, to help evaluate the real-world effectiveness of the metamodeling techniques. We are also pleased that these models cover a variety of domains: cybersecurity, reliability, chemistry, and biology. We shall briefly describe each test case in turn.

- **Botnet** The Botnet model is a Möbius model that is described in detail in [12] and was used as the sole test case in [9]. The model can be used to study the growth of a botnet over the course of a week given certain conditions, such as the probability that an uninfected computer will become infected, and the rate of removal of bots from the net. The values of eleven input variables are uncertain in this model. The input variables, and the values that they assume in our evaluation, can be found in the Appendix in Table 7.
- **Circadian** The Circadian Clock model[2], based on the abstract model found in [1] and [15], is a CTMC PRISM model. Given rates for transcription, translation, binding, release, and degradation, for activator and repressor genes and mRNA the model calculates the amount of activator protein at a given time. There are 17 input variables in this model. The input variables, and the range of values that they assume in our evaluation, can be found in the Appendix in Table 8.
- **Cluster** The Workstation Cluster model[3] is a PRISM model taken from [3]. It can be used to calculate the quality of service (QoS) of a workstation cluster arranged in a star topology. The components of the cluster fail and are repaired at specific rates. The values of thirteen input variables are uncertain in this model. The majority of these uncertain input variables are various failure and repair rates. The input variables, and the range of values that they assume in our evaluation, can be found in the Appendix in Table 9.
- **Cyclin** The Cyclin model[4] is a CTMC PRISM model based on a formal specification from [5]. It models cell cycle control in eukaryotes, given a specific quantity of various molecules and various base reaction rates. This model contains 14 input variables. The input variables, and the range of values that they assume in our evaluation, can be found in the Appendix in Table 10.

---

[1] Available here: https://www.prismmodelchecker.org/casestudies/.
[2] Model code here: https://www.prismmodelchecker.org/tutorial/circadian.php.
[3] Model code here: https://www.prismmodelchecker.org/casestudies/cluster.php.
[4] Model code here: https://www.prismmodelchecker.org/casestudies/cyclin.php.

- **Embedded** The Embedded System model[5] is a CTMC PRISM model of an embedded control system based on a model description from [7]. The embedded system consists of three sensors, a sensor input processor, a main processor, an output processor, two actuators, and a bus that connects the processors. The components have a probability of failing, either permanently or in a transient manner. Some failures can be repaired by a system reboot. This model can be used to calculate reliability and availability metrics. There are six input variables in this model. The input variables, and the range of values that they assume in our evaluation, can be found in the Appendix in Table 11.
- **Kanban** The Kanban Manufacturing System model[6] is a CTMC PRISM model based on a model found in [2]. The model can be used to estimate the throughput of a manufacturing system. There are fourteen input variables in this model. The input variables, and the range of values that they assume in our evaluation, can be found in the Appendix in Table 12.
- **Molecules** The Simple Molecular Reactions model[7] is a CTMC PRISM model. Given a particular amount of Na, Cl, and K, various reaction rates, and length of time, it can calculate the expected percentage of Na/K molecules. There are nine input variables in this model. The input variables, and the range of values that they assume in our evaluation, can be found in the Appendix in Table 13.

3 Approach

A *metamodel* or *emulator* or model surrogate is a model of a model. Recall that we refer to the model that the metamodel emulates as the *base model*. The metamodel will attempt to produce the same output as the base model, given the same vector of input variable values. In general, a metamodel will not emulate the base model perfectly, but will run much faster than the base model. It is possible to construct metamodels using machine learning (ML) by following this general process:

1. Collect training and testing data by generating a number of different inputs, running the base model with those inputs, and recording each input vector and the corresponding modeling result.
2. Train a machine learning regressor to emulate the base model using the training data.
3. Test the trained regressor with the testing data. This is done by running each test input vector through the metamodel and observing the difference between the metamodel output and the previously-recorded base model output.

[5] Model code here: https://www.prismmodelchecker.org/casestudies/embedded.php.
[6] Model code here: https://www.prismmodelchecker.org/casestudies/kanban.php.
[7] Model code here: https://www.prismmodelchecker.org/casestudies/molecules.php.

In this work, we generated all training and testing data inputs uniformly at random for ease of analysis and coding, though it is possible to use more sophisticated sampling methods (such as Latin Hypercube Sampling or Sobol Sequence Sampling) to collect the training data, which may produce slightly more accurate metamodels [9].

The choice of the kind of regressor to train is important. For example, someone may create a metamodel by training a random forest, multilayer perceptron, Gaussian Process regressor, or something more exotic. Some types of regressors may be substantially more accurate than others, and it is difficult to know *a priori* which will perform well. In much of the related work, the author chose regressors based on intuition, familiarity and comfort with a particular technique, or certain properties of the data. In other papers, the authors evaluate several different regressor types, and chose the best-performing regressor (the *Best of Many* approach). In recent work [9,10], a stacking-based metamodel approach was shown to significantly outperform the Best of Many approach. However, it was only shown to work on two test cases, and no variants of the metamodel architecture were explored.

In the remainder of the section, we shall briefly review the stacking-based metamodel approach for SA and UQ introduced in [9], and then describe the variants of the architecture that we developed and evaluate in this paper.

3.1 Stacking Review and Variants

In machine learning, an ensemble is a collection of learning algorithms that are used together with the goal of creating a stronger overall learner than any of the constituent learning algorithms alone. One of the simplest ensemble techniques one could imagine is to create a voting committee of heterogeneous regressors (for example, one multilayer perceptron, one random forest, one KNN regressor, etc.), each trained separately on the training data. The output of the committee would be the average of the regressor predictions. However, better performance could be achieved if the predictions of more accurate regressors were given more weight in the vote than inaccurate regressors. In particular, it may be that one regressor may perform relatively well in one region of the input space, and relatively poorly in another region of the input space. The appropriate weighting is not obvious, but could be learned using a separate regressor or committee of regressors. This is the core idea of stacking. With stacking, the predictions of the regressors in the first committee are appended to the training data, and then a second committee of regressors is trained with the augmented training datasets [16]. The stacking approach was successfully used to win Kaggle ML competitions [11].

To build a stacked metamodel, one would first train multiple regressors with the training dataset, to form the first level committee. Next, every regressor in the first level committee would be run with every input vector from the training dataset to obtain the corresponding regressor prediction for that input vector, and then the regressor prediction would be appended to the input vector as another feature. At the end of the process, the modeler will have an augmented

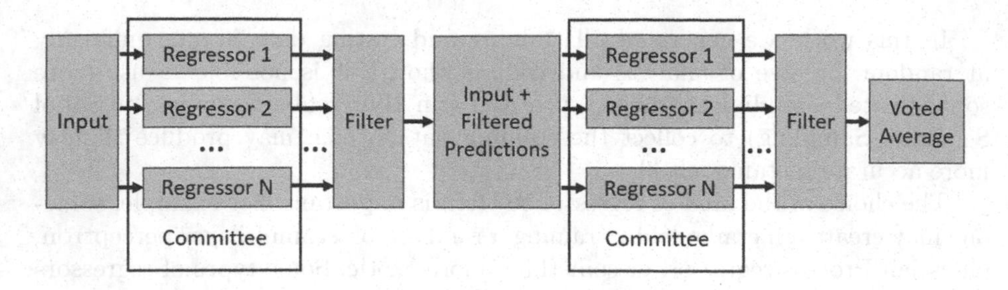

Fig. 1. Overview of the metamodel architecture.

training dataset that contains all of the original input vectors, but each input vector will include additional features: the predictions of the first level regressors. The augmented dataset would then be used train another set of regressors to form the second level committee. The regressors in the second level will have the benefit of additional information in their training dataset to help them make better predictions: the "recommendations" of the first level regressors.

See Fig. 1 for a diagram of the flow of execution in the metamodel architecture. Once the metamodel has been built, it may be executed. To execute the metamodel, an input vector is given to each regressor in the first level committee, and each regressor calculates its prediction. We have experimentally found that it is useful to filter predictions from inaccurate regressors (later in this section we discuss details of the filter), but predictions from accurate regressors are appended to the input vector. This augmented input vector is then given to each of the regressors in the second level committee, which in turn make their predictions. The predictions are filtered again, and the predictions that make it through the filter are averaged to calculate the final metamodel prediction.

In general, one may choose any set of regressors as members of the two committees. However, one must of course choose a specific set of regressors. This is an important choice, since the overall accuracy of the metamodel heavily depends on which regressors are members of the two committees. Only one kind of committee was evaluated in [9] and [10]. We go further than those works by evaluating five different committees with a variety of properties.

Similarly, appropriate filters can help improve the performance of the metamodel. In general, one may use the predictions of each regressor in the committee, regardless of the accuracy of the regressor. However, we have experimentally found that it can be useful to filter predictions from inaccurate regressors. Only one kind of filter was evaluated in [9] and [10], however, filtering can be accomplished in a variety of ways. We build upon previous work by evaluating variants on two different types of filters.

We shall describe the different committee and filter types we evaluate in this work.

Committees. We evaluate five different committees in this work. We hypothesize that larger committees would outperform smaller committees, and committees with more heterogeneity would outperform those with less, and wish to test the hypothesis. To that end, we evaluate committees of different sizes, and with varying degrees of heterogeneity in membership. We evaluate a large committee with many heterogeneous regressors (Committee 0), a couple of small committees with relatively homogeneous members (Committees 1 and 2, which only contain different kinds of random forests and multilayer perceptrons, respectively, as members), a medium-sized committee with heterogeneous members (Committee 3), and a small committee with heterogeneous members (Committee 4). What follows is a description of each committee. A summary of the properties of the committees can be found in Table 1.

0. Committee 0: The regressors in this committee are the same as the regressors used in the committees in [9]. It is also the largest committee we evaluate, with 25 members. The members are 1 random forest regressor, 8 different multilayer perceptrons (each with a different combination of solvers and activation functions), 4 different Gradient Boosting Regressors (each with a different loss function), a RidgeCV regressor, 10 different KNN regressors (each vary by the number of neighbors and whether the neighbor votes are weighted uniformly or by distance), a Gaussian Process regressor, and a stochastic gradient descent regressor. Hyperparameter tuning is less of an issue in this large committee compared to the smaller committees, because one can simply include many variants of the same "base" algorithm with different hyperparameters without being particularly concerned with finding the "best" set of hyperparameters. In effect, the stacking algorithm will indirectly perform the work of finding the "best" set of hyperparameters in the given set for the modeler in the course of its normal operation.
1. Committee 1: All of the regressors in this committee are variants of the random forest architecture with different hyperparameters. There are 4 regressors in this committee, and each varies by the number of trees in the forest (either 10 or 100) and by the criterion used (either *mean squared error* or *mean absolute error*).
2. Committee 2: All of the regressors in this committee are variants of the multilayer perceptron architecture with different hyperparameters. There are 6 regressors in this committee, and each has a unique combination of solver (either *adam*, *sgd*, or *lbfgs*) and activation function (either *logistic* or *tanh*).
3. Committee 3: This committee is our representative example of a committee with a moderate number of regressors. There are six regressors in this committee: one random forest, one multilayer perceptron, one support vector machine, one Gaussian process regressor, one KNN regressor, one Gradient Boosting regressor, and one stochastic gradient descent regressor.
4. Committee 4: This committee is our representative example of a small committee of just three regressors, each having a different architecture. This committee contains a random forest, a multilayer perceptron, and an Gaussian process regressor.

Table 1. Properties of committees.

Committee index	Regressor mix (heterogeneous or homogeneous)	Size of committee (# members)
0	Heterogeneous	Largest (25)
1	Homogeneous	Small (4)
2	Homogeneous	Moderate (6)
3	Heterogeneous	Moderate (6)
4	Heterogeneous	Small (3)

Filters. We evaluate two different types of filters in this work: *Top k* and *Within n Percent.* The *Top k* filter only allows the predictions from the k most accurate regressors to pass through. In this work we consider $k = \{1, 2, 3\}$. The *Within n Percent* filter will allow predictions from all regressors through as long as the regressor's error is no worse than n percent worse than the best performing regressor. In this work we consider $n = \{10\%, 25\%, 50\%, 100\%\}$. With this filter, it is not known *a priori* how many regressors will be filtered - theoretically, all of the regressor predictions could be filtered (except the predictions from the best regressor), or all of them could pass through the filter. As a control, we also construct metamodels with no filters. The *Within 25%* filter was used in the architecture of the metamodel found in [9].

4 Evaluation of Accuracy and Speed

In this evaluation, we first determine the most effective combination of committee and filter to use for the metamodel, and rank the committees and filter types by their effectiveness. Then, we calculate the accuracy of the metamodel on each of the test cases. We also investigate by how much the accuracy of the metamodels increase with more training data. Finally, we evaluate speed of the metamodel compared to the base models, as well as the time it takes to train the metamodels.

Before we begin the evaluation, it is important to explain how the accuracy of the metamodels is calculated. For each of the test case base models we created a dataset of one thousand randomly generated input vectors and executed the base model with those vectors to obtain the corresponding outputs to create a "ground truth" test dataset. The test dataset was distinct from the training dataset. The trained metamodel is executed with all of the input vectors from the test dataset, and the metamodel prediction for each input vector is recorded. We then calculate the absolute error for each vector by taking the absolute value of the difference between the metamodel's prediction and the ground truth model output. We sum all of the absolute errors and then divide by the number of input vectors in the test dataset to obtain the mean absolute error. Finally, to facilitate comparison between the different test cases (whose outputs have very

Table 2. Most accurate committee/filter combination for each test case. *Note: Description of committee by given index found in Sect.* 3.

Base model	Committee index	Filter index
Botnet	0	*Within n Percent*, n = 50%
Circadian	1	No filter
Cluster	0	*Within n Percent*, n = 25%
Cyclin	1	*Within n Percent*, n = 25%
Embedded	0	*Within n Percent*, n = 10%
Kanban	0	*Within n Percent*, n = 50%
Molecules	0	*Within n Percent*, n = 10%

different ranges of values), we normalize the errors by dividing the mean absolute error by the range of the base model's outputs in the test dataset.

We wrote the code to build the metamodels in Python with the aid of the scikit-learn package [8].

4.1 Accuracy Given Different Committee Compositions and Filters

As we described in Sect. 3, we evaluate five different committee types and eight different filters, so there are a total of $8 \times 5 = 40$ different combinations of committee and filter. The most accurate stacked metamodel committee/filter combination for each test case we consider is given in Table 2. We see that Committee 0, the largest and most diverse of the committees, is the best performing committee for five of the seven test cases, and that Committee 1, which is composed of four different random forest regressors, is the best performing committee for the remaining two test cases. The best performing filters were all variants of the *Within n Percent* filter, with the exception of the Circadian test case, where not having a filter outperformed all of the other filters.

It would be useful to know which committee/filter combination works best in general. Table 2 shows that the best committee/filter combination for one base model will not be the best combination for another base model. One may of course try all forty combinations and select the best for their particular model. It does not take long to train a metamodel, so it is feasible to try many different variants. However, it is illuminating to know which committee and filter combinations work well across many different kinds of models. To determine this, we ranked each committee/filter combination for each of our seven test cases from 1 to 40, from most accurate to least. We then calculated the *average rank* for each committee/filter combination by summing all of the individual ranks and then dividing by the number of test cases. The committee/filter combination with the lowest average rank would be the most accurate metamodel across the test cases. When we performed that calculation, we found that Committee 0 (the largest and most heterogeneous committee) paired with the *Within n Percent, n=10%* filter was the most accurate combination across all of the test case models.

It would also be illuminative to rank the committees (each paired with the filter that maximizes its performance) and filters (each paired with the committee that maximizes its performance) across the test models, respectively. We shall describe the process to determine the rank of a committee. A similar process can be used to rank a filter. To rank the committees, we first calculate the average rank of each combination of the 40 combinations of committee/filter as described in the paragraph above, and then sort this list from the smallest value to the largest. We then start at the top of this list and if we see an entry with a committee we have seen before, we delete that entry. When we reach the end of the list we will have a ranked list of committees. By ranking in this way, we compare each committee when paired with the filter that maximizes its performance.

The ranking of the committees and filters can be found in Table 3. From the table it is clear that the stacking approach benefits from having committees with many heterogeneous members. Committees with more homogeneity (e.g. Committees 1 and 2) and small committees (e.g. Committee 4) do not perform as well in general (though, as we saw previously in Table 2, they may perform better than the alternatives on specific models). The results validate the intuition that an ensemble with many diverse members will outperform an smaller ensemble with fewer, more homogeneous members.

Table 3 also provides striking results for the filter. Clearly, the *Within n Percent* filter dramatically outperformed the *Top k* filter. In fact, the *Top k* filter was worse than having no filter at all. It is interesting that the more restrictive *Within n Percent* filters outperformed the less restrictive filters of the same type, while the less restrictive *Top k* filters outperformed the more restrictive. We hypothesize that the *Top k* filters were too restrictive to allow for the diversity necessary for a well-performing ensemble.

4.2 Metamodel Accuracy: Naive vs. Best of Many vs. Stacked

Up to this point, we have only evaluated the accuracy of different variants of the stacked metamodel relative to one another. It is also important to know (a) the accuracy of the stacked metamodel compared to other metamodels, and (b) the absolute accuracy of the model.

First, it is important to know whether sophisticated ML techniques perform better than very naive methods. To help make this comparison, we created a *Naive metamodel*, which consists of a regressor that predicts that the output will be the average of the outputs in the training data, regardless of the input. More sophisticated metamodels must show that they are substantially better than this naive metamodel to demonstrate utility.

Table 4 reports the errors of the Naive, Best of Many, and Stacked metamodels. The errors we report are the mean absolute error normalized by the range of observed test values. We normalize the errors to make it possible to compare the accuracy of the metamodels across test cases. We chose to use the stacked metamodel architecture that was most suited for each individual test case (e.g. the metamodel variants listed in Table 2), rather than use the same

Table 3. Committees and filters ranked by accuracy. Description of committee by given index found in Sect. 3.

Rank	Committee Index	Committee Description
1	0	Large
2	3	Medium
3	1	4 RF
4	4	MLP, RF, GPR
5	2	6 MLP

Rank	Filter Description
1	*Within n Percent*, n = 10%
2	*Within n Percent*, n = 25%
3	*Within n Percent*, n = 50%
4	No filter
5	*Within n Percent*, n = 100%
6	*Top k*, k = 3
7	*Top k*, k = 2
8	*Top k*, k = 1

committee/filter combination for every test case. All metamodels were trained with a datasets that contained 1000 random inputs each.

By examining the table we see that the Best of Many and Stacked metamodels always substantially outperform the Naive metamodel, as we had hoped. Further, for five of the seven test cases, the stacked metamodel was more than 5% more accurate than the Best of Many metamodel, and it was never less accurate. The stacked metamodels were, on average, 8.2% more accurate than the base models. These numbers demonstrate the effectiveness of the stacked approach compared to the current state-of-the-practice Best of Many approach, and validates its general applicability.

It is obvious that if a metamodel is not accurate enough it will not be a practical tool for modelers. However, it is difficult to know how accurate a metamodel must be to be a good emulator. One of the primary challenges in determining an acceptable accuracy threshold is that a modeler may use metamodels for different reasons, and different use cases may require more accuracy than others. We believe that the stacked metamodels are likely accurate enough to be useful for a variety of applications given the errors reported in Table 4, perhaps with the exception of the Circadian test case. We leave to future work an investigation to determine the acceptable accuracy threshold for a variety of common applications of metamodels.

4.3 Accuracy Given Different Training Sample Dataset Sizes

It can be time consuming to collect the training data because the base model runs slowly. To be time efficient it would be best to collect as little training data as possible while maintaining reasonable metamodel accuracy. For this reason we investigated the impact of the size of the training set on the accuracy of the metamodel. We trained stacked metamodels with Committee 0 (the largest and most heterogeneous committee) and the *Within n percent, n = 10%* filter with datasets that contained 250, 500, 750, and 1000 input vectors, respectively,

Table 4. Average metamodel prediction error. The error is the mean absolute error normalized by the range of test values.

Metamodel type	Naive metamodel error	Best of many metamodel error	Stacked metamodel error	Error reduction stacked vs. best of many
Botnet	0.00161	0.00111	0.00100	10.4%
Circadian	0.08461	0.05658	0.05648	0.2%
Cluster	0.03322	0.01181	0.01141	3.4%
Cyclin	0.14129	0.02369	0.02111	10.9%
Embedded	0.03726	0.00459	0.00432	5.9%
Kanban	0.17832	0.02798	0.02508	10.3%
Molecules	0.06610	0.03899	0.03262	16.3%

Table 5. Mean absolute error normalized by range of stacked metamodel trained with training datasets of different sizes.

Metamodel type	250 training samples	500 training samples	750 training samples	1000 training samples
Botnet	0.03817	0.03227	0.01816	0.01626
Circadian	0.07011	0.064872	0.06065	0.05707
Cluster	0.02151	0.01631	0.01562	0.01349
Cyclin	0.05285	0.04281	0.03048	0.02606
Embedded	0.02445	0.01547	0.01271	0.01126
Kanban	0.03956	0.03222	0.02956	0.02555
Molecules	0.05672	0.05688	0.05072	0.04750

and evaluated their accuracy. The results of our investigation are contained in Table 5. As expected, the metamodels trained with more training data are more accurate than those that were trained with less. However, it is encouraging to see that even those metamodels that were trained with just 250 training samples are often reasonably accurate compared to the metamodels trained with 1000 training samples.

4.4 Speed Comparison

We performed these speed experiments on an Ubuntu VM running on a laptop with an Intel i7-7500U processor and 8 GB of RAM. For each test case, both the base model and the metamodel were run with the same 200 randomly generated inputs, and the time it took to execute with those inputs was recorded. In addition, we recorded the time it took to train the metamodel for each case. We used the same committee/filter combination for each metamodel in these experiments to make it easier to compare across test cases. Committee 0 paired with the *Within n Percent, n = 10%* filter was the combination we chose, for two reasons. First, we had previously determined that combination was, on average,

Table 6. Base model vs. metamodel execution speed comparison and metamodel training time (all in seconds).

Name	Base model execution (seconds)	Metamodel execution (seconds)	Metamodel training (seconds)
Botnet	1115.5	0.3	137.4
Circadian	18291.6	0.5	204.7
Cluster	2034.9	0.6	66.4
Cyclin	2668.2	0.5	76.9
Embedded	684.7	0.2	60.6
Kanban	5737.6	0.3	64.5
Molecules	3714.5	1.1	67.1

the most accurate across all the test cases. Second, Committee 0 is the largest committee, so it will likely take longer to train compared to the other committees, so it can be considered a "worst-case" training time. More members will take longer to train, since each member of the committee is trained separately in sequence. None of the code used in these experiments was parallelized, though regressor training could be easily parallelized.

Table 6 shows that the metamodel runs faster than the base model by several orders of magnitude, thousands of times faster. On average it took an hour and twenty minutes to run the base model with the dataset containing 200 inputs, while it took the metamodels on average half a second to run the same dataset. The metamodels are almost ten thousand times faster than their corresponding base model on average. The table also shows that training the metamodel with the training dataset can be done reasonably quickly.

5 Discussion and Recommendations

Our analysis can help modelers who are interested in using metamodeling for their own models. At a high level, it appears that reasonably accurate metamodels can be created for a variety of different kinds of models using a variety of different machine learning algorithms. Even relatively slow running complicated ensemble methods run thousands of times faster than the base model, making feasible analyses that would otherwise be unfeasible if performed directly on the base model. We recommend that modelers consider the use of metamodels to help with analyses involving slow-running models that have many uncertain input variables.

A stacking-based ensemble approach appears to produce more accurate metamodels than approaches that use a single regressor, even if that regressor is the best among many candidate regressors. Our analysis of the seven test cases shows that the best stacking metamodels were never worse than those metamodels produced by using the Best of Many approach, and were often significantly better.

The accuracy of the metamodel was impacted by the size and heterogeneity of the constituent committees: more accurate metamodels had larger and more heterogeneous committees, while less accurate metamodels had smaller and more homogeneous committees. Judicious use of filters can increase the accuracy of the metamodel, with relatively restrictive versions of the *Within n Percent* filter being the best we evaluated. Our evaluation shows that it does not take long to train a metamodel if one already has the training dataset, so it is possible to try a number of different metamodel variants to see which would work best for that particular dataset. We recommend that modelers using metamodeling should (1) use stacking rather than the more common Best of Many approach, (2) use committees that are large and heterogeneous in their stacked metamodels, (3) use a filter to remove predictions of under performing regressors, and (4) take advantage of how quickly metamodels can be trained by trying a number of different metamodel variants to find one that works particularly well for the dataset.

We found that the accuracy of the metamodel depends in part on the size of the training dataset: the larger the training dataset, the more accurate the metamodel. However, the stacked metamodels were reasonably accurate even with little training data. This is encouraging, because our original motivation for using metamodels was as a replacement for models that run slowly. Collecting training data can be the most time consuming stage of the ML-based metamodeling, so it is encouraging that metamodels do not require onerously large training datasets to be accurate. We recommend that modelers obtain as much training data as is practical, but to not be discouraged if only a small training dataset can be obtained.

A modeler should be aware of the limitation of the approach. If the model being considered runs too slowly it may not be feasible to collect enough training data to build a reasonably accurate metamodel. However, if the model being considered runs quickly, it would be better to do the analysis (whether SA, UQ, optimization, or something else) directly on the model instead of doing the analysis indirectly with the use of a metamodel, because metamodels usually don't perfectly emulate the base model and can introduce errors into the analysis. For this reason, in some cases it may be beneficial to do a hybrid approach in which a broad metamodel-based analysis is paired with (and perhaps even guides) a more limited and narrowly focused analysis on the base model.

6 Related Work

To the best of our knowledge, no other published research exists that evaluates the effectiveness (accuracy and speed) of metamodeling on multiple state-based quantitative models. We also know of no other work that evaluates as many different kinds of metamodels.

As mentioned previously, the stacking metamodel approach applied to quantitative state-based models was first demonstrated in [9] and [10], though only one test case was used in each of those papers, and only one committee/filter combination was considered. We, however, demonstrate the effectiveness of our approach with numerous test cases and a variety of committee/filter combinations. Many other kinds of metamodel approaches were used before stacking was introduced as a means to build a metamodel; consult [6] and [14] for a broad overview of the topic of metamodeling in general. Eisenhower et al. performed a metamodel-based analysis of a building energy model with hundreds of parameters, though the metamodel was a simple support vector machine with a Gaussian kernel, not an ensemble. Our work demonstrates that an ensemble can significantly outperform a single regressor. The approach shown in [13] uses multiple metamodels, but not as an ensemble, and [17] uses an ensemble but with a recursive arithmetic average method to combine the predictions of multiple regressors, while we use stacking. Stacking has been shown to be effective in winning machine learning competitions [11].

7 Conclusion

We have evaluated the effectiveness of metamodeling on seven real-world published models. We showed that the metamodels we created can be quite accurate in emulating the behavior of the base model, and run almost ten thousand times faster on average. Since the metamodels are so fast relative to the base model, analyses that could take too long to complete on the base model directly (such as sensitivity analysis, uncertainty quantification, and optimization) can be done indirectly with the aid of the metamodel instead in a reasonable amount of time. Metamodels can be constructed automatically with the aid of machine learning, minimizing the manual efforts of the modeler. We evaluated a variety of metamodels, and found that stacked metamodels performed better than the more commonly used Best of Many metamodels. Of the stacked metamodels, those that had larger and more diverse committees were more accurate than those that had smaller and more homogeneous committees. Appropriate filters, particularly the *Within n Percent* filters increase metamodel accuracy. We believe that ML metamodels, like those shown in this work, are an underused and relatively-unknown tool in the modeler's toolbox. We hope that our descriptions and evaluations will encourage their future use in exploring slow-running state-based quantitative models with many uncertain input variables. Future work should evaluate models with dozens or hundreds of input variables to evaluate the scalability of the approach.

Acknowledgements. The authors would like to thank Jenny Applequist and the reviewers for their feedback on the paper. This material is based upon work supported by the Maryland Procurement Office under Contract No. H98230-18-D-0007. Any opinions, findings and conclusions or recommendations expressed in this material are those of the author(s) and do not necessarily reflect the views of the Maryland Procurement Office.

A Appendix: Values of Input Variables for Test Case Models

In this appendix we list the input variables for each of the seven models we use as test cases in our evaluation, along with the corresponding domain. When the domain was not specified by the original authors, we estimated it, informed by the original paper and subject matter expertise.

Table 7. List of inputs used in the Botnet test case.

Variable name	Domain
ProbConnectToPeers	[0.25, 1]
ProbPropagationBot	[0.05, 0.15]
ProbInstallInitialInfection	[0.05, 0.15]
Prob2ndInjctnSuccessful	[0.25, 1]
RateConnectBotToPeers	[0.0166, 6]
RateOfAttack	[0.0166, 6]
RateSecondaryInjection	[0.0166, 6]
RateBotSleeps	[0.05, 0.15]
RateBotWakens	[0.0005, 0.0015]
RateActiveBotRemoved	[0.05, 0.15]
RateInactiveBotRemoved	[0.00005, 0.00015]

Table 8. List of inputs used in the Circadian test case.

Variable name	Domain
T	[0, 200]
transc_da	[45, 55]
transc_da_a	[450, 550]
transc_dr	[0.009, 0.011]
transc_dr_a	[45, 55]
transl_a	[45, 55]
transl_r	[4.5, 5.5]
bind_a	[0.9, 1.1]
bind_r	[0.9, 1.1]
deactivate	[1.8, 2.2]
rel_a	[45, 55]
rel_r	[90, 110]
deg_a	[0.9, 1.1]
deg_c	[0.9, 1.1]
deg_r	[0.18, 0.22]
deg_ma	[9, 11]
deg_mr	[0.45, 0.55]

Table 9. List of inputs used in the Cluster test case.

Variable name	Domain
ws_fail	[0.0002, 0.02]
switch_fail	[0.000025, 0.0025]
line_fail	[0.00002, 0.002]
startLeft	[5, 15]
startRight	[5, 15]
startToLeft	[5, 15]
startToRight	[5, 15]
startLine	[5, 15]
repairLeft	[1, 3]
repairRight	[1, 3]
repairToLeft	[0.125, 0.375]
repairToRight	[0.125, 0.375]
repairLine	[0.0625, 0.1875]

Table 10. List of inputs used in the Cyclin test case.

Variable name	Domain
N	[2, 4]
t	[0, 60]
k	[0, 4]
R1	[0.0045, 0.055]
R2	[0.0009, 0.0011]
R3	[0.0027, 0.0033]
R4	[0.45, 0.55]
R5	[0.27, 0.33]
R6	[0.0045, 0.0055]
R7	[0.0081, 0.0099]
R8	[0.0081, 0.0099]
R9	[0.009, 0.011]
R10	[0.0153, 0.0187]
R11	[0.018, 0.022]

Table 11. List of inputs used in the Embedded test case.

Variable name	Domain
lambda_p	[1/(2 * 365 * 24 * 60 * 60), 1/(30 * 24 * 60 * 60)]
lambda_s	[1/(90 * 24 * 60 * 60), 1/(7 * 24 * 60 * 60)]
lambda_a	[1/(4 * 30 * 24 * 60 * 60), 1/(7 * 24 * 60 * 60)]
tau	[1/90, 1/30]
delta_f	[1/(2 * 24 * 60), 1/(8 * 60 * 60)]
delta_r	[1/(5 * 60), 1]

Table 12. List of inputs used in the Kanban test case.

Variable name	Domain
t	[1, 5]
in1	[0.5, 1.5]
out4	[0.45, 1.35]
synch123	[0.2, 0.6]
synch234	[0.25, 0.75]
back	[0.15, 0.45]
redo1	[0.18, 0.54]
redo2	[0.21, 0.63]
redo3	[0.195, 0.585]
redo4	[0.165, 0.495]
ok1	[0.42, 1.26]
ok2	[0.46, 1.38]
ok3	[0.455, 1.365]
ok4	[0.385, 1.155]

Table 13. List of inputs used in the Molecules test case.

Variable name	Domain
T	[0, 0.003]
i	[0, 10]
N1	[10, 100]
N2	[10, 100]
N3	[10, 100]
e1rate	[90, 110]
e2rate	[9, 11]
e3rate	[27, 33]
e4rate	[18, 22]

References

1. Barkai, N., Leibler, S.: Biological rhythms: circadian clocks limited by noise. Nature **403**, 267–268 (2000)
2. Ciardo, G., Tilgner, M.: On the use of Kronecker operators for the solution of generalized stocastic Petri nets. ICASE Report 96–35, Institute for Computer Applications in Science and Engineering (1996)
3. Haverkort, B., Hermanns, H., Katoen, J.P.: On the use of model checking techniques for dependability evaluation. In: Proceedings of 19th IEEE Symposium on Reliable Distributed Systems (SRDS 2000), Erlangen, Germany, pp. 228–237, October 2000
4. Kwiatkowska, M., Norman, G., Parker, D.: PRISM: probabilistic symbolic model checker. In: Field, T., Harrison, P.G., Bradley, J., Harder, U. (eds.) TOOLS 2002. LNCS, vol. 2324, pp. 200–204. Springer, Heidelberg (2002). https://doi.org/10.1007/3-540-46029-2_13
5. Lecca, P., Priami, C.: Cell cycle control in eukaryotes: a BioSpi model. In: Proceedings of Workshop on Concurrent Models in Molecular Biology (BioConcur 2003). Electronic Notes in Theoretical Computer Science (2003)
6. Liu, H., Ong, Y.S., Cai, J.: A survey of adaptive sampling for global metamodeling in support of simulation-based complex engineering design. Struct. Multidiscip. Optim. **57**(1), 393–416 (2018)
7. Muppala, J., Ciardo, G., Trivedi, K.: Stochastic reward nets for reliability prediction. Commun. Reliab. Maintain. Serviceabil. **1**(2), 9–20 (1994)
8. Pedregosa, F., et al.: Scikit-learn: machine learning in Python. J. Mach. Learn. Res. **12**, 2825–2830 (2011)
9. Rausch, M., Sanders, W.H.: Sensitivity analysis and uncertainty quantification of statde-based discrete-event simulation models through a stacked ensemble of metamodels. In: Gribaudo, M., Jansen, D.N., Remke, A. (eds.) QEST 2020. LNCS, vol. 12289, pp. 276–293. Springer, Cham (2020). https://doi.org/10.1007/978-3-030-59854-9_20
10. Rausch, M., Sanders, W.H.: Stacked metamodels for sensitivity analysis and uncertainty quantification of AMI models. In: Proceedings of the 2020 IEEE International Conference on Communications, Control, and Computing Technologies for Smart Grids (SmartGridComm), pp. 1–7 (2020)
11. Risdal, M.: Stacking made easy: an introduction to StackNet by competitions grandmaster Marios Michailidis (KazAnova). http://blog.kaggle.com/2017/06/15/stacking-made-easy-an-introduction-to-stacknet-by-competitions-grandmaster-marios-michailidis-kazanova/. Accessed 13 Dec 2019
12. Ruitenbeek, E.V., Sanders, W.H.: Modeling peer-to-peer botnets. In: Proceedings of 2008 Fifth International Conference on Quantitative Evaluation of Systems, pp. 307–316, September 2008
13. Tenne, Y.: An optimization algorithm employing multiple metamodels and optimizers. Int. J. Autom. Comput. **10**(3), 227–241 (2013)
14. Viana, F., Gogu, C., Haftka, R.: Making the most out of surrogate models: tricks of the trade. In: Proceedings of the ASME Design Engineering Technical Conference, vol. 1, pp. 587–598 (2010)
15. Vilar, J., Kueh, H.Y., Barkai, N., Leibler, S.: Mechanisms of noise-resistance in genetic oscillators. Proc. Natl. Acad. Sci. U.S.A. **99**(9), 5988–5992 (2002)
16. Wolpert, D.H.: Stacked generalization. Neural Netw. **5**(2), 241–259 (1992)
17. Zhou, X.J., Ma, Y.Z., Li, X.F.: Ensemble of surrogates with recursive arithmetic average. Struct. Multidiscip. Optim. **44**(5), 651–671 (2011)

Queueing Systems

Network Calculus for Bounding Delays in Feedforward Networks of FIFO Queueing Systems

Alexander Scheffler(✉) and Steffen Bondorf

Distributed and Networks Systems Group, Faulty of Mathematics,
Ruhr University Bochum, Bochum, Germany
alexander.scheffler@rub.de

Abstract. Networks for safety-critical operation must guarantee deterministic bounds on the end-to-end delay of data transmission despite the usually many data flows that all share the available data forwarding resources. Queueing is inevitable and the queueing delay becomes the important impact factor for communication delays. Network Calculus can calculate verifiable delay bounds in networks of such queues and the tighter the bounds are, the less over-provisioning is required when they are used for the design of safety-critical networked systems.

Tightening delay bounds is an important objective of Network Calculus research. In this paper, we focus on the improvement of the overall analysis algorithm bounding delays in feedforward networks. FIFO queueing is widespread in practice, yet, considering it to model the fraction any data flow gets of the forwarding resource turned out to be complex with Network Calculus. The currently only analysis with practically usable performance was developed for tandem topologies. On the other hand, there are sophisticated algorithms for the feedforward analysis without considering the FIFO property. Here, big gains in tightness were achieved by properly extending the algorithms for tandem topologies. We aim at bringing these gains to the FIFO analysis – and the FIFO analysis to feedforward networks. We provide a thorough integration of both – theoretically and with novel tool support. Our new analysis shows a considerable tightness improvement over the feedforward analysis without FIFO considerations as well as a straightforward extension of the FIFO analysis.

1 Introduction

Overview. Guaranteed performance in safety-critical networked systems is often achieved by expensive overprovisioning. A formal verification methodology that derives accurate performance metrics, even for high network utilizations, is essential for identifying inefficient designs. Network Calculus (NC) can compute bounds on the end-to-end delay of data flows crossing a network of deterministic queueing systems. The more characteristics of the real system we can consider in the model and the analysis, the tighter the NC bounds. Therefore, tightening NC delay bounds is usually achieved in at least one of multiple different ways: *i)* with

A. Abate and A. Marin (Eds.): QEST 2021, LNCS 12846, pp. 149–167, 2021.
https://doi.org/10.1007/978-3-030-85172-9_8

more detailed system models (availability of the forwarding resource at servers and the demand thereof by data flows), *ii)* creation of more sophisticated models capturing the resource allocation among different flows queued at a server or *iii)* advancing NC algorithms that convert these models into a network analysis.

The second research topic in NC, modeling of resource sharing, has seen quite some treatment in the context of Time-Sensitive Networking (TSN) lately [32–34]. These results focus on resource sharing for complex FIFO per-class schemes, computing residual service and delay bounds at individual servers. The network analysis is enabled by a simple NC result. The bound on data put into a server by a flow is described by a function. The bound after crossing a server can be computed by shifting this function by the respective server's delay bound. This new bound on the flow's data to forward is then used for computations at the subsequent server on its path. While this enables for a feedforward (FF) network analysis, the computed delay bounds will lack tightness.

For *iii)*, more advanced models for resource sharing across sequences of servers, so-called tandems, exist in NC. Additionally, there are algorithms to convert them into the analysis of feedforward networks – already revealing some potential sources of overestimation on the bounds that should be avoided. For the analysis of FIFO queueing systems, the Least Upper Delay Bounds (LUDB) [1,3] was developed as the currently only algorithm that extends single-server FIFO queueing results in a non-trivial, tightness-impeding way and still shows practically usable performance. Competing analysis approaches that convert the entire network into a mixed-integer linear program (MILP) [11,14] can compute better delay bounds but lack computational performance. LUDB was, however, developed as a tandem analysis only. In this paper, we aim at bringing the ideas behind the LUDB tandem analysis to feedforward networks. We make use of the knowledge that has been created since LUDB was developed, but for the analysis that does not make use of the FIFO property at all [4,7]. Tailoring these results to consider the unique properties of LUDB and FIFO analysis gives us the tightest delay bounds in networks without abandoning the classic NC for MILP optimization. We provide a novel method called LUDB-FF based on [6] and benchmark our delay bounds against those of available tools and approaches: the Tandem Matching Analysis that does not consider FIFO [4], the LUDB tandem tool extended to a network analysis by the simple curve-shift used in TSN research, and the current state of feedforward analysis tool support in [6]. LUDB-FF delay bounds outperform these three alternatives considerably across different network sizes and utilizations as we show in our numerical evaluation.

Background and Problem Statement. Commonly taken assumptions in NC are that there are no cyclic dependencies between flows (feedforward property), the data within flows remains unchanged by forwarding (i.e., it remains FIFO per individual data flow), the amount of data remains unchanged, and data in a server's queue is forwarded in FIFO order. There is work that alleviates these assumptions, e.g., on data scaling [30], non-FIFO systems [28] and cyclic dependencies [10]. But most research on NC had been dedicated to different assumptions on how multiple flows may multiplex into a common server's queue. The two main streams of research are on *arbitrary multiplexing* that assumes anything can happen and *FIFO multiplexing* that allows for end-to-end FIFO assumptions. The task of an NC analysis translates to the questions how to derive the least (non-zero) residual forwarding service for a flow of interest (foi) given its crossflow interference. Arbitrary multiplexing turns out to be a convenient assumption. Without any further knowledge, the foi only receives service after its crossflows have been fully served – the computation only depends on the crossflows and is independent of the foi's actually sent data. While this gives a valid bound for FIFO multiplexing, too, bounds can of course be improved when accounting for the FIFO property of the system. This becomes complex as the foi and its crossflows are interdependent. Put simple, without knowing the data sent by the foi, we cannot derive the FIFO residual service. Partially restoring the convenience of arbitrary multiplexing, an NC result exists that computes a residual service curve with a free parameter θ that captures the FIFO effects between foi and crossflows. The θ can be set later, depending on the foi and thus deferring the decision. It allows for reusing the NC analysis formula to test differently shaped, alternative foi configurations. However, the analysis of a network of FIFO queueing systems usually creates a multitude of interdependent θ parameters. Setting them to obtain the best FIFO delay bound remains a highly complex task as before, just encoded differently. Moreover, complexity in the network analysis is increased by analysis aspects such as maximum aggregation of flows that reduces the amount of θ parameters. We provide a thorough integration of LUDB into the NC feedforward analysis that considers these aspects for best delay bounds.

Paper Outline. This paper is organized as follows: Section 2 surveys related work and provides NC background. Section 3 contributes LUDB-FF, our extension of the LUDB FIFO analysis to a modern feedforward network analysis. Numerical results are presented in Sect. 4 and Sect. 5 concludes the paper and outlines potential future research directions based on LUDB-FF.

2 Network Calculus Basics and Related Work

2.1 Network Calculus Ressource Models

Extensive depictions of the algebraic Network Calculus (NC) can be found in [9, 12]. We restrict our presentation to briefly listing the basics used in our paper.

Definition 1 (Arrival Curve). *Let A be the function cumulatively counting the data created by a flow f. By convention, $\forall t \leq 0 : A(t) = 0$. Then we call α an arrival curve for f iff*

$$\forall 0 \leq d \leq t : A(t) - A(t-d) \leq \alpha(d). \tag{1}$$

Definition 2 (Service Curve). *Suppose A is the input to server s and A' is cumulatively counting output from s. We say that β is a service curve for $\mathcal{S}$ iff*

$$\forall t \geq 0 : A'(t) \geq \inf_{0 \leq s \leq t}\{A(t-s) + \beta(s)\} =: A \otimes \beta(t). \tag{2}$$

Theorem 1 (Performance Bounds). *Consider a server s that offers a service curve β. Assume flow f has arrival curve α. Then we get the following bounds:*

$$\text{Output Bound: } \alpha'(t) = \alpha \oslash \beta(t) := \sup_{u \geq 0}\{\alpha(t+u) - \beta(u)\} \quad (3)$$

$$\text{Delay Bound: } hdev(\alpha, \beta) = \inf\{d \geq 0 : (\alpha \oslash \beta)(-d) \leq 0\} \quad (4)$$

where α' is a bound on A', the output from s (see Definition 2), and the horizontal deviation hdev between arrival curve and service curve bounds the delay experienced by f at s.

Commonly found shapes of NC curves are token buckets $\gamma_{\sigma,\rho}(d) = \rho \cdot d + \sigma$ for arrival curves and rate latencies $\beta_{T,R}(d) = R \cdot \max\{0, d - T\}$ for service curves, both from the set of pseudoaffine functions [15]. Note that σ, ρ, T, R are positive rationals including 0. T is called the latency of curve $\beta_{T,R}$.

2.2 Network Calculus Analyses and Tool Support

The oldest analysis in NC that can compute bounds in FIFO systems is the so-called Total Flow Analysis (TFA) [19]. Although it is known to be inferior to more modern analyses when analyzing entire networks, it is still widely used today due to its simplicity. Foremost, the idea to shift arrival curves bounding the data put into a server by the server's delay bound is known from TFA and still used in papers focusing on modeling single servers, e.g., in Time-Sensitive Networking research [33]. The Separated Flow Analysis (SFA) [9] has superseeded TFA in modern analysis algorithms. Whereas TFA aggregates all data flows at a server, SFA computes a lower bound on the residual service a flow gets.

Theorem 2 (FIFO Residual Service Curve). *Consider a server s that offers a service curve β. Assume flows f_1 and f_2 multiplex in the queue of server s. Let α_1 and α_2 be the arrival curves of the two flows, respectively. Assuming FIFO multiplexing, the residual (or left-over l.o.) service curve denoting a lower bound on the forwarding of f_1 is:*

$$\beta_{f_1}^{l.o.}(t, \theta) = [\beta(t) - \alpha_2(t - \theta)]^{\uparrow} \cdot \mathbf{1}_{\{t > \theta\}} =: \beta \ominus_{\theta} \alpha_2, \forall \theta \geq 0 \quad (5)$$

$\mathbf{1}_{\{condition\}}$ *is the indicator function (1 if the condition is true, 0 otherwise) and* $[g(x)]^{\uparrow} = \max\{0, \sup_{0 \leq z \leq x} g(z)\}$.

By aggregation of all flows, TFA does not compute the residual service and hence, it is not creating any free θ parameter. The Separated Flow Analysis (SFA) computes the per-server, per-flow residual service. It therefore needs to set the θ parameter locally at every server (see Appendix A). This analysis, setting the parameter as suggested in [20], is implemented in the open-source NetworkCalculus.org Deterministic Network Calculator (NCorg DNC) [6][1] and note that a general overview on NC-tools can be found for example in [35].

[1] Available at http://dnc.networkcalculus.org.

Fixing the θ parameter for an analysis also allows for a closed-form calculation of a residual service on sequences of servers (tandems) [20]. Yet, it was shown that this is not optimal for bounding end-to-end delays. The Least Upper Delay Bound Analysis (LUDB) was derived to allow for a more advanced setting of θ parameters on tandems [1,3] that considers interdependencies between subsequent servers of a tandem. A reference implementation called DElay BOund Rating AlgoritHm (DEBORAH) [2] exists[2]. For instance, LUDB was used in network-on-chip analysis [16,23,24] and sensor network analysis [31]. Still, all these network analyses are in effect tandem analyses.

A branch of NC that prevents setting θ parameters is the optimization-based analysis. Results for tandems [13] and for feedforward analysis [14] exist. Unfortunately, they do not scale well and current efforts try to replace optimization constraints with knowledge from some of the above analyses [11]. Scalability results in larger feedforward networks are currently not available, neither is a tool.

In this paper, we take the most advanced NC analysis to set the θs, the LUDB, whose design and tool support was focused on the analysis of tandems. We have implemented the LUDB tandem analysis and combined it with the arrival bounding backtracking scheme in the feedforward analysis [8] to derive residual curves instead of delays bounds directly. Our efforts to bring this approach to the analysis of feedforward networks results in a new method, LUDB-FF, that beats all alternatives (arbitrary and FIFO) for feedforward analysis by considerable margins.

2.3 The LUDB Analysis and the DEBORAH Tool

Since LUDB [1,3] is one of the integral parts of LUDB-FF, we delve into its delay bound computation and the challenges for an application to feedforward networks. On tandems, the LUDB aims at implementing two main principles of the NC analysis: Pay Bursts Only Once (PBOO) [9] and Pay Multiplexing Only Once (PMOO) [29]. Both aim at the fact that flows do not wait at a subsequent server on a tandem for their worst-case burstiness to build up (again). PBOO targets implementation of this property for the foi, PMOO additionally for its crossflows. Neither is implemented in TFA, constituting its inferiority to modern analyses. SFA implements the former principle only whereas LUDB implements the former fully as well as the latter one for specific types of interference patterns. Intuitively, the PMOO can be implemented by first combining servers into single systems and then subtracting crossflows. This is easily possible for nested tandems[3] but not for non-nested ones. Non-nested tandems need to be cut into a sequence of nested tandems, see Fig. 1.

Also note that LUDB is restricted to a specific set of arrival and (input) service curves, namely token bucket and rate latency curves respectively. LUDB

[2] Available at http://cng1.iet.unipi.it/wiki/index.php/Deborah.

[3] A tandem has nested interference iff for every pair of flows either both flows do not have common servers or the path of one flow is included in the other flow's path.

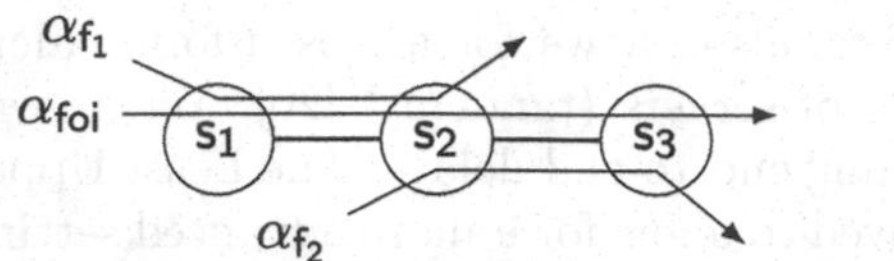

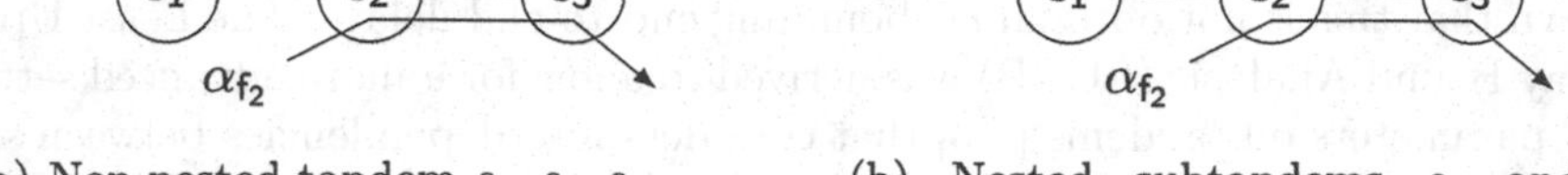

(a) Non-nested tandem s_1, s_2, s_3.

(b) Nested subtandems s_1 and s_2, s_3.

Fig. 1. Converting the non-nested tandem (a) into a sequence of nested subtandems by cutting between servers $s1$ and $s2$ (b). Note, that an arrival curve for the cut flow f_1 after the cutting location, α_{f_1}, needs to be computed.

describes the residual service curves and delay bounds symbolically. The non-linearity of the objective function "bound the delay" is tackled by splitting the problem into linear pieces (simplexes), defining linear programs (LPs). This results in $O(m!)$ LPs to solve for nested tandems with m crossflows. For each LP, the optimal objective value gives one valid delay bound. The least delay bound from the many created LPs is the LUDB result. Even though the LUDB uses optimization as the later NC approaches and it is similarly faced with a potentially vast number of LPs, its first reference implementation DEBORAH [2] already showed practically useable performance on tandems. Therefore, we opt to extend LUDB to a tightest possible network analysis. The complementary approach, trading off tightness of the comprehensive optimization analysis for tractable network analysis can be found in [11].

3 Bringing the LUDB Analysis to Feedforward Networks

This section's contribution is LUDB-FF, the extension of the LUDB tandem analysis to an analysis of feedforward networks that considers all the complexities arising from this step such as the tradeoffs between flow aggregation [1,7] and flow segregation [5] within the analysis as well as the differentiation between delay and output bounding during the analysis procedure. It was shown that the majority of the feedforward analysis effort stems from bounding the arrivals of crosstraffic at their respective location of interference with the foi [7]. A generic arrival bounding procedure for the arbitrary multiplexing analysis was presented in [4]. It is a backtracking scheme with multiplexing-specific pruning that we adapt to FIFO multiplexing and the LUDB to create LUDB-FF. As a comparison that does not invest heavily in arrival bounding, we combined the LUDB tool DEBORAH with the TFA output bounding from delay. We implemented both analyses as extensions to the NCorg DNC[4].

[4] The DEBORAH tool is licensed under GPL while the NCorg DNC is licensed under LGPL. Thus, we cannot redistribute DEBORAH and we opted for a new implementation in the NCorg DNC.

Section 3.1 presents the arrival bounding procedure stripped from arbitrary multiplexing assumptions and adapted to FIFO, Sect. 3.2 details the LUDB-FF and Sect. 3.3 presents the aforementioned straightforward DEBORAH integration into the DNC.

3.1 Arrival Bounding Procedure for FIFO

For the arrival bounding, we have adapted the method presented in [4], see the algorithmic description Algorithm 1. The difference compared to arbitrary multiplexing is as follows: with arbitrary multiplexing assumptions it is possible to take into account the foi when bounding the crossflows in the sense that we do not have to take into account the interference with foi (again). For an illustration of the analysis proceedings, take this sample application to Fig. 2:

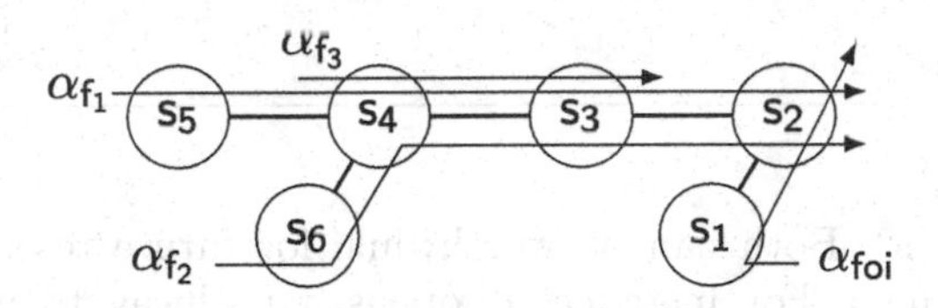

Fig. 2. Feedforward Sample Network.

We aim to bound the delay of the flow of interest (foi), subject to its crossflow interference. We require a bound on the arrivals of f_1 and f_2 at server s_2, hence we set $F = \{f_1, f_2\}$ and $s = s_2$. From the flows in F we compute one arrival curve per incoming link and sum them up. In our example, all flows in F take the same link, namely $l = (s_3, s_2)$. I.e., we bound the two flows in aggregate. With GETDIVERGINGSERVER we backtrack from the source of this link (dest $= s_3$) to the server where the flows diverge (start $= s_4$). At the found server the arrival curve will be computed recursively. With this information, a FIFO analysis on the path start $\rightsquigarrow$ dest is started – resulting in a residual service curve for the aggregate of flows f_1 and f_2, subtracting interference of f_3. The *true* (in Algorithm 1) indicates that this is done in an output minimizing instead of delay bound minimizing way. Last, LUDB-FF computes the output of f_1 and f_2, their arrivals at server s_2, by the deconvolution-based output bound computation (Theorem 1).

3.2 Parameter θ and Residual Service Computation of LUDB-FF

Working with NC curves allows for an algebraic NC feedforward analysis (see Sect. 2). The DEBORAH tool for tandems, in contrast, directly returns the delay bound without revealing the actual LP computing it or the setting of θs found

Algorithm 1. Arrival Bounding Algorithm

Input (F, s) Flows to bound at server s
Output α_F^s Output arrival curve at s for flows F

1: **procedure** COMPUTEARRIVALCURVE(F, s)
2: $\alpha_F^s \leftarrow \gamma_{0,0}$
3: **for** $l \in$ IngoingLinks(s) **do**
4: $F_l \leftarrow F \cap$ Flows(l)
5: **if** $F_l \neq \emptyset$ **then**
6: dest $\leftarrow$ Source(l)
7: start $\leftarrow$ GETDIVERGINGSERVER(F_l, dest)
8: $\alpha_{F_l}^{\text{start}} \leftarrow$ COMPUTEARRIVALCURVE(F_l, start)
9: $\beta_{F_l}^{\text{l.o.}} \leftarrow$ COMPUTESERVICECURVE($F_l, \alpha_{F_l}^{\text{start}}$, start $\rightsquigarrow$ dest, *true*)
10: $\alpha_F^s \leftarrow \alpha_F^s + \alpha_{F_l}^{\text{start}} \oslash \beta_{F_l}^{\text{l.o.}}$
11: **end if**
12: **end for**
13: return α_F^s
14: **end procedure**

by solving the LPs[5]. Both can be worthwhile for future research and we will grant access to them. For instance, it opens a pathway to machine learning (ML) by providing θ settings to learn from an ML-assisted NC analysis like [21]. Our design decision to increase modularity also allows for use of existing LP solvers. We opted for IBM CPLEX whereas DEBORAH has a custom internal solver implementation. From the θ solution vector, we can easily compute the residual service curve for further use in the NCorg DNC, see Algorithm 1. Last, the tight integration into the DNC tool lets LUDB-FF benefit from more recent features in the arrival bounding such as caching of intermediate results [4] and parallelization [27].

Before we present the details of our LUDB-FF analysis, we detail the notation. Figure 2 serves as visualizing sample network. Set the path $p := s_4 \rightsquigarrow s_2$ from server 4 to 2, i.e., $(4, 3, 2)$. $\text{Path}(f_2)_{|p} = p$ since f_2 crosses the entire path p. $p_1 \subseteq p_2$ holds if p_1 is a subpath of p_2, e.g., $(4, 3) \subseteq p$. $P_x := \text{Paths}(F_x)_{|p} = \cup_{f_x \in F_x} \text{Path}(f_x)_{|p}$ is the set of all subpaths on p for flows in F_x. For simplicity we assume that a flow does not rejoin a given path otherwise, for sake of notation, we just split it up and give it two distinct identifiers. For a nested tandem, a crossflow with path p has level 1 ($l = 1$) if there is no other crossflow whose path entirely includes p. With $\beta_f^{\text{l.o.}}$ we denote the residual service curve for flow f.

First, we define a simple, local bound on the θ parameter.

[5] DEBORAH works with the pseudoaffine curve framework. Although it outputs variables s_i that relate to the variables θ_i from the FIFO residual service curve in Theorem 2, it is not trivial to infer θ_i from s_i: for a crossflow indexed with i, we can compute $\theta_i = hdev(\alpha_i, \beta_i) + s_i$ from its residual service curve β_i and the s_i. However, this requires to additionally create a tree-structure that captures the relative paths of crossflows to the foi on the analyzed tandem, the so-called nesting tree, first.

Definition 3 (Lower bound on θ). *Given an arrival curve $\alpha_f := \gamma_{\sigma,\rho}$ and residual service curve $\beta_f^{l.o.}$, we define $\underline{\theta}_f(\beta_f^{l.o.}, \alpha_f) := \inf\{t \geq 0 : \beta_f^{l.o.}(t) \geq \sigma\}$.*

This can be seen as the θ setting which results in a residual service curve with lowest latency and is thus output minimizing – however only locally as the following Lemma shows. Also note that $\underline{\theta}_f(\beta_f^{l.o.}, \alpha_f) = hdev(\alpha_f, \beta_f^{l.o.})$.

Next, we make use of a bound on the output derived in [1] which holds for nested tandems that we transform into the corresponding residual service curve.

Lemma 1 (Output Bound for FIFO $\beta^{l.o.}$ [1]). *Consider a nested tandem T with crossflows F_x and a foi. Further assume w.l.o.g. that all flows in F_x have distinct paths. Then the following residual service curve minimizes*[6] *the output bound on a foi*

$$\left[\bigotimes_{i \in I} \beta_{T_i,R_i}\right] \otimes \left[\bigotimes_{f \in F_{x|l=1}} [\beta_f^{l.o.}(t) - \alpha_f^{Source(f)}(t - \theta_f)]^{\uparrow} \cdot 1_{\{t > \theta_f\}}\right] \tag{6}$$

with $I = \{i \in T : \nexists f \in F_x : Path(f) \cap i \neq \emptyset\}$ and $\theta_f = \underline{\theta}_f(\beta_f^{l.o.}, \alpha_f^{Source(f)})$. $\beta_f^{l.o.}$ is the residual service curve of crossflow f computed by LUDB—for this a nested tandem analysis with foi=f and crossflows $\{f_x \in F_x : Path(f_x) \subsetneq Path(f)\}$ has to be executed. Assume foi has arrival curve $\gamma_{\sigma,\rho}$. The respective output bound is then given by $\gamma_{\sigma',\rho'}$ with $\rho' = \rho$ and

$$\sigma' = \sigma + \left(\sum_{f \in F_{x|l=1}} \theta_f + \sum_{i \in I} T_i\right) \cdot \rho \tag{7}$$

Now we give the algorithm to compute the residual service curve for the flows of interest. The LUDB-FF delay analysis for the foi is initialized by calling $\text{COMPUTESERVICECURVE}(\{\text{foi}\}, \alpha_{\text{foi}}, \text{Path(foi)}, false)$.

The algorithm first finds the crossflows on path p and aggregates them based on their subpaths on p. It then either does a nested or non-nested analysis while also distinguishing different cases to tighten the bounds.

NESTEDANALYSIS $F_{\text{foi}}, F_{\text{foi}} \cup F_x, A_{P_x}, \alpha_{F_{\text{foi}}}, p, o, agg$ For the nested analysis where we want to find the best output bound we make use of Lemma 1 which gives us the respective residual service curve for the flows of interest. For the delay bound case it is important to aggregate the flows of interest with crossflows that also have p as subpath (in this case *agg* holds) – from the respective residual service curve we have to subtract those crossflows again but do so such that the delay bound stays the same (as with the crossflow aggregate). In detail:

- *Case (Output bound)* We proceed as discussed in Lemma 1.
- *Case (Delay bound)* Here we need to differentiate whether *agg* holds or not, i.e., whether there are crossflows having the same path p as subpath (see also

[6] For the resulting output bound $\gamma_{\sigma',\rho'} = \min_{\theta_1,\ldots,\theta_{|F_x|}} (\gamma_{\sigma,\rho} \oslash \beta_{\text{foi}}^{\text{l.o.}}(\theta_1, \ldots, \theta_{|F_x|}))$ holds.

Algorithm 2. Compute Service Curve Algorithm (LUDB-FF)

Input $(F, \alpha_F^{\text{Source}(p)}, p, o)$ Flows for which to compute the (aggregate) residual service on path p either optimizing for the output (o) or delay bound
Output Residual service curve for flow aggregate F
Note F_{all} denotes all flows of the network

```
1: procedure COMPUTESERVICECURVE(F, α_F^Source(p), p, o)
2:     F_x ← {f ∈ F_all : Path(f) ∩ p ≠ ∅}\F
3:     P_x ← Paths(F_x)_|p
4:     nested ← determine if F ∪ F_x forms a nested (tandem) interference on p
5:     agg ← determine if P_x ∩ p ≠ ∅ holds
6:     if nested then
7:         A_{P_x} ← ∅
8:         for p_x ∈ P_x do
9:             α_{p_x} ← COMPUTEARRIVALCURVE({f ∈ F_x : Path(f)_|p = p_x},
           Source(p_x))
10:            A_{P_x} ← A_{P_x} ∪ α_{p_x}
11:        end for
12:        return NESTEDANALYSIS(F, F ∪ F_x, A_{P_x}, α_F^Source(p), p, o, agg)
13:    else
14:        return NONNESTEDANALYSIS(F, F ∪ F_x, α_F^Source(p), p, o, agg)
15:    end if
16: end procedure
```

Algorithm 2). In case there are none, we compute the LUDB with foi $= F_{\text{foi}}$ and crossflows F_x. If there are certain crossflows with this property, first define $F_{x_{\text{foi}}} := \{f \in F_x : \text{Path}(f)_{|p} = p\}$. We set foi $= F_{\text{foi}} \cup F_{x_{\text{foi}}}$ and crossflows $F_x \backslash$foi and compute the LUDB. From the returned service curve $\beta_{\text{foi}}^{\text{l.o.}}$ we need to subtract $F_{x_{\text{foi}}}$, i.e., $\beta_{F_{\text{foi}}}^{\text{l.o.}} := \left[\beta_{\text{foi}}^{\text{l.o.}}(t) - \alpha_{F_{x_{\text{foi}}}}(t - \theta_{F_{x_{\text{foi}}}})\right]^{\uparrow} \cdot 1_{\{t > \theta_{F_{x_{\text{foi}}}}\}}$ with $\theta_{F_{x_{\text{foi}}}} = \underline{\theta}_{F_{x_{\text{foi}}}}(\beta_{\text{foi}}^{\text{l.o.}}, \alpha_{\text{foi}})$. Note that $hdev(\alpha_{\text{foi}}, \beta_{\text{foi}}^{\text{l.o.}}) = hdev(\alpha_{F_{\text{foi}}}, \beta_{F_{\text{foi}}}^{\text{l.o.}})$.

NonNestedAnalysis $F_{\text{foi}}, F_{\text{foi}} \cup F_x, \alpha_{F_{\text{foi}}}, p, o, agg$ The non-nested tandem gets cut at specific servers to get a nested one. Only reduced sets of cuts [3] will be considered and are denoted by S, i.e., once a set of cuts is found which will result in a nested tandem no further cuts will be added to this set of cuts. For example for the network in Fig. 1a we could either cut before server s_2 or s_3, i.e., $S = \{\{2\}, \{3\}\}$. For each possibility we need to compute the output arrival bounds of the crossflows at the cuts. In case *agg* holds we propose not to cut the crossflows which go through the entire path p to further tighten the bounds. We choose a set of cuts which either yields the lowest delay bound or the lowest output bound depending on o. Moreover, note that in contrast to the NestedAnalysis we have not computed the arrival curves for the crossflow-aggregates because due to the cutting, arrival curves have to be computed at each cut we consider. More precisely:

- *Case (not agg: foi can't be aggregated with crossflows)* For a $C \in S$ with $C = \{c_1, c_2, ..., c_m\}$ we define $c_0 :=$ Source$(p), c_{m+1} :=$ Sink(p) and assume w.l.o.g. $c_{i-1} < c_i \forall i \in \{1, ..., m+1\}$. Then, for each $c_i \in C \cup \{c_{m+1}\}$ we consider for every $p_x \in \text{Paths}(F_x)_{|c_{i-1} \rightsquigarrow c_i}$ the crossflow-aggregate $F_{p_x} := \{f \in F_x : \text{Path}(f)_{|c_{i-1} \rightsquigarrow c_i} = p_x\}$ for which we compute the arrival curve at $\text{Source}(p_x)$, i.e., $\alpha_{F_{p_x}} := \text{COMPUTEARRIVALCURVE}(F_{p_x}, \text{Source}(p_x))$. In other words, for each subtandem defined by the path $c_{i-1} \rightsquigarrow c_i$ we compute the aggregate arrival bounds for crossflows having the same path p_x within this subtandem. For each aggregate within a subtandem we create a dedicated flow and add it to F'_x (this set is related to the current cut set C). Choose foi $= F_{\text{foi}}$ and crossflows F'_x and do a nested analysis, i.e., $\text{NESTEDANALYSIS}(F_{\text{foi}}, F_{foi} \cup F'_x, A_{F'_x}, \alpha_{F_{\text{foi}}}, p, o, false)$.

- *Case (agg: foi can be aggregated with crossflows)* Let $F_{x_p} := \{f \in F_x : \text{Path}(f)_{|p} = p\}$. In this case we first only consider the crossflow subset $F_{x_r} := F_x \backslash F_{x_P}$ and compute $A_{F'_{x_r}}$ analog to the previous case. Next, we compute the arrival curve, $\alpha_{F_{x_p}}^{\text{Source}(P)}$, for F_{x_P} at $\text{Source}(P)$ and do a nested analysis, i.e., $\text{NESTEDANALYSIS}(F_{\text{foi}}, F_{\text{foi}} \cup F'_{x_r} \cup F_{x_p}, A_{F'_{x_r}} \cup \alpha_{F_{x_p}}^{\text{Source}(P)}, \alpha_{F_{\text{foi}}}, p, o, true)$.

After considering all $C \in S$ we pick the service curve $\beta^{\text{l.o.}}_{F_{\text{foi}},C}$ which either has the minimal latency (if o holds) or delay bound $hdev(\alpha_{F_{\text{foi}}}, \beta^{\text{l.o.}}_{F_{\text{foi}},C})$.

Apart from the extension to FIFO feedforward networks, the novelty of LUDB-FF lies in the use of residual service curves during the arrival bounding process and the differentiation of several cases to tighten the bounds, revealing the interplay of flow aggregation, delay and output bounding. The latter brought about the insight to not cut crossflows in the non-nested analysis that cross the entire analyzed tandem of servers (instead of always cutting all crossflows as [3] proposes).

3.3 DEBORAH-Integration by TFA's Output from Delay

Since there is the DEBORAH tool which implements the LUDB on tandems we can, in principle, use its delay bounds to compute the output bound required in a feedforward analysis:

Theorem 3. *(Output From Delay [19]) Consider a server s that offers a service curve β. Assume flow f with arrival curve α traverses the server and experiences a delay bounded by d. Then $\alpha'(t) = \alpha(t+d)$ is an output bound for flow f.*

We pair this idea with the arrival bounding procedure described by Algorithm 1. In order to make this work, the following adaptions have to be applied. After line 7 in Algorithm 1, we compute F_p as the set of flows that have $p := \text{start} \rightsquigarrow \text{dest}$ as subpath. More formally, $F_p = \{f \in F : \text{Path}(f)_{|p} = p\}$. Similarly, we compute $\alpha_{F_p}^{\text{start}} = \text{COMPUTEARRIVALCURVE}(F_p, \text{start})$ and then call DEBORAH to retrieve the delay bound $d^p_{F_p}$ – note that DEBORAH demands that the crossflows with the same subpath in p have to be aggregated before as well as the aggregate for the foi, F_p. From $d^p_{F_p}$ we compute the arrival curve for F_p at s by using Theorem 3, i.e., $\alpha^s_{F_p}(t) = \alpha^{\text{start}}_{F_p}(t + d^p_{F_p})$ which holds for F_l at s as well since $F_l \subseteq F_p$.

Regarding tightness of the bounds it must be noted that Theorem 3 gives a rather coarse bound in general. More importantly, note that in case $F_l \neq F_p$, more flows than actually present are considered, thus worsening the bounds and potentially violating the stability constraint. This constraint signifies that the long-term service must exceed arrivals, more formally for each server i, $\sum_{j \in \text{Flows}(i)} \rho_j \leq R_i$ has to hold. Overall, we call this analysis DEBORAH-Integration. It provides us with a competitor that implements LUDB on tandems, developed into a feedforward analysis with a well-known and often applied output bounding computation.

4 Numerical Evaluation

Setup. For our numerical evaluation, we use a subset of the publicly available networks from [4] that are part of the NCorg DNC. They have been created to mimic Internet topologies, following the General Linear Preference (GLP) model [17]. The networks have been created to test the Tandem Matching Analysis (TMA) analysis that does not make use of the FIFO multiplexing assumption but assumes worst case multiplexing in the analyzed flow's point of view, called arbitrary multiplexing. TMA results are therefore valid for FIFO queueing systems, too, and we aim at giving some insight into the inherent untightness of delay bounds this lack of the FIFO assumption causes.

Secondly, we adapted the GLP networks to be more in line with existing evaluations of FIFO multiplexing NC analyses. In the original networks, the flow arrival curves and server service curves were fixed. The server utilization varies between the servers, as this is caused solely by the presence of flows. These are routed on the shortest path between randomly chosen source servers and sink servers, such that their amount varies between servers. The difference between such arbitrary multiplexing analyses and FIFO analyses is known to become significant with increasing network utilization. We assume the same holds to at least some degree between different FIFO analyses, too. Therefore, we have set equal utilization at every server in the network by fixing the flow arrival curves α and adapting servers' service curves β according to the desired utilization u:

$$\alpha(d) = \gamma_{\sigma,\rho}(d) = \gamma_{5,5}(d) = 5d + 5 \tag{8}$$

$$\beta(d) = \beta_{0,R_i}(d) = R_i \cdot d \text{ with } R_i = \frac{\sum_{j \in \text{Flows}(i)} \rho_j}{u} \tag{9}$$

Our dataset for our evaluation includes the aforementioned GLP networks with size 20, 40, 100 and 200 devices. Note, that a transformation of the network takes place that transforms the devices' output ports to server [18], i.e., we assume output queueing like in TSN for example. Overall, our dataset consists of 5040 data flows across the four networks sizes. For these four sizes, we created networks with homogeneous utilizations of 70%, 90% and 99%. In the resulting twelve network configurations each data flow's end-to-end delay bound was then computed with the NCorg DNC's most advanced FIFO analysis SFA-FIFO (see

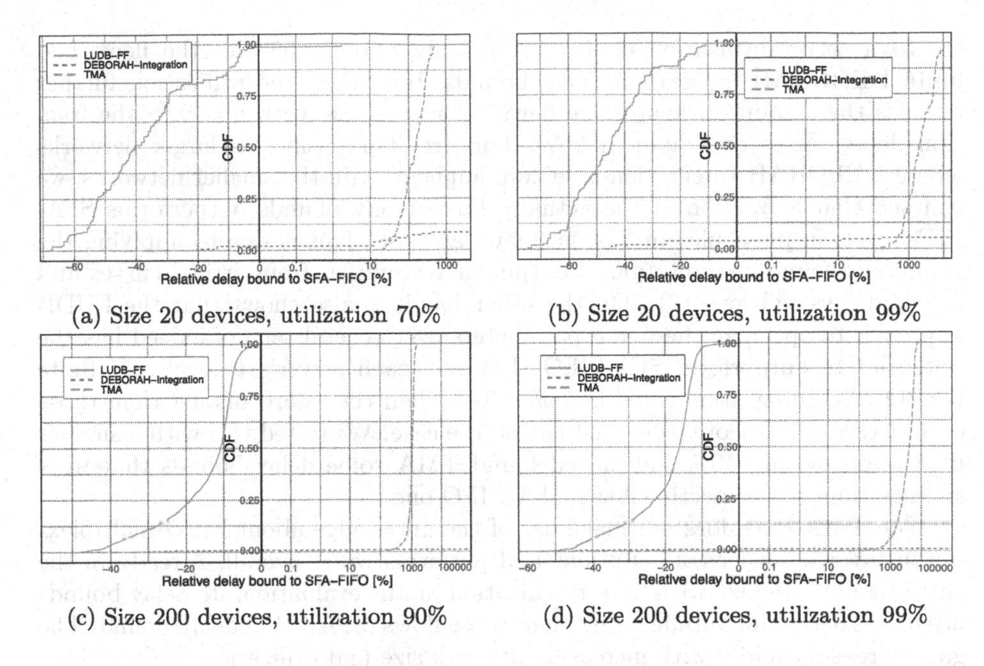

(a) Size 20 devices, utilization 70%
(b) Size 20 devices, utilization 99%
(c) Size 200 devices, utilization 90%
(d) Size 200 devices, utilization 99%

Fig. 3. Cumulative distribution of relative delay bounds w.r.t. SFA-FIFO, i.e., $delay^{\text{other}}_{\text{SFA-FIFO}}$ [%], for different network sizes and utilizations.

Appendix A), with the simple extension of the DEBORAH analysis DEBORAH-Integration, with the arbitrary multiplexing TMA, and with our new LUDB-FF. Our full data set is available online[7].

The measure of interest in our evaluation is the deviation of some analysis' per-flow delay bounds from a reference analysis, the baseline. I.e., for every analyzed flow in one specific configuration, we compute

$$delay^{\text{other}}_{\text{baseline}} = \frac{delay^{\text{other}} - delay^{\text{baseline}}}{delay^{\text{baseline}}}. \tag{10}$$

We present results relative to SFA-FIFO delay bounds, i.e., to the currently available tool support, first. Revealing that our LUDB-FF is performing best, we will switch to it as the new de-facto reference for NC FIFO analysis benchmarks.

Observations. We first depict the relative delay bound w.r.t. SFA-FIFO across different network sizes and utilizations, shown in Fig. 3. The low fraction of DEBORAH-Integration shown in the various configuration is immediately visible. Due to the problems of this approach to extend DEBORAH to an analysis for feedforward networks (violation of the stability constraint even at actual utilizations below 1, see Sect. 3.3), the majority of delay bounds computed by DEBORAH-Integration were infinite. For example, in the network of 20 devices,

[7] https://github.com/alexscheffler/dataset-qest2021.

set to a server utilization of $u = 70\%$, only about 8.55% of the flows had finite DEBORAH-Integration delay bounds. Increasing the utilization further reduces the amount of analyzable flows – for example, with $u = 99\%$ the fraction decreases to only about 5.26%. This trend intensifies in larger networks where DEBORAH-Integration is seldom applicable. In the smaller networks, we can see that a fraction of the actually finite delay bounds outperforms SFA-FIFO, yet, in particular in Fig. 3a, many are not. This is due to applying the commonly found computation of output arrival curves from arrival curves and server delays (Theorem 3). On the other hand, it also shows that the LUDB approach to optimize the free θ parameters over several servers indeed has the potential to outperform SFA-FIFO that sets each server's θ locally. If finite DEBORAH-Integration could be computed, then they were smaller than those of TMA. Yet, in theory, this need not be the case. We tested this with a smaller utilization of only 50% and indeed found TMA some delay bounds that were smaller than the respective flow's SFA-FIFO one.

The TMA that does not make use of the knowledge about FIFO multiplexing in favor of easier computations and performs worst overall. Already in the smallest network size with lowest utilization in our evaluation, its delay bounds are one order of magnitude larger than those flows' SFA-FIFO delay bound. The gap increases rapidly with increasing network size and utilization.

Our new LUDB-FF in contrast almost always beats the alternative analyses. When it comes to DEBORAH-Integration it must be noted that LUDB-FF's delay bounds are never worse. The maximum improvement over SFA-FIFO tends to decrease with the network size, while staying at over 50%, and increases with the utilization per network size. Our extensive evaluation also reveals that there are some considerable exceptions that require further investigation, though. In Fig. 3a, there are two flows where SFA-FIFO slightly outperforms LUDB-FF and in the configurations partially not shown in Fig. 3, we can further report:

- $n = 100$, $u = 70\%$: 11 instances
- $n = 100$, $u = 90\%$: 4 instances
- $n = 100$, $u = 99\%$: 0 instances
- $n = 200$, $u = 70\%$: 9 instances
- $n = 200$, $u = 90\%$: 1 instance
- $n = 200$, $u = 99\%$: 1 instance

In particular in the network with 100 devices, we can see the dependence on the utilization. By successively increasing the utilization, some flows drop from the list where SFA-FIFO outperforms LUDB-FF. Interestingly, we can observe a different trend in the network with 200 devices. At utilizations of 90% and 99%, the same flow constitutes the single instance. Yet, it was not in the set of flows at utilizations 70%. These observations show that there is potential for further improvement of LUDB-FF. In particular since SFA-FIFO means that we make a cut at each server, it could be studied if additional cuts in the LUDB-FF method can improve the bounds. However, this would then come with additional costs since it is not clear in which cases this can be beneficial. But overall, we only observed 28 instances across our 15120 analyzed flows, i.e., 0.1852% of delay bounds that could be improved by SFA-FIFO.

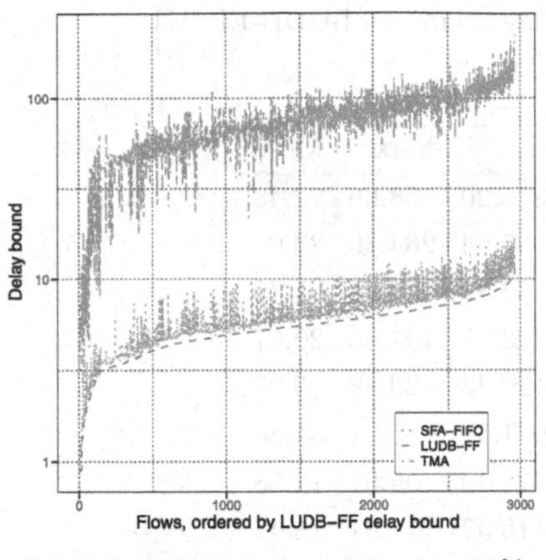

(a) Size 200 devices, utilization 90%

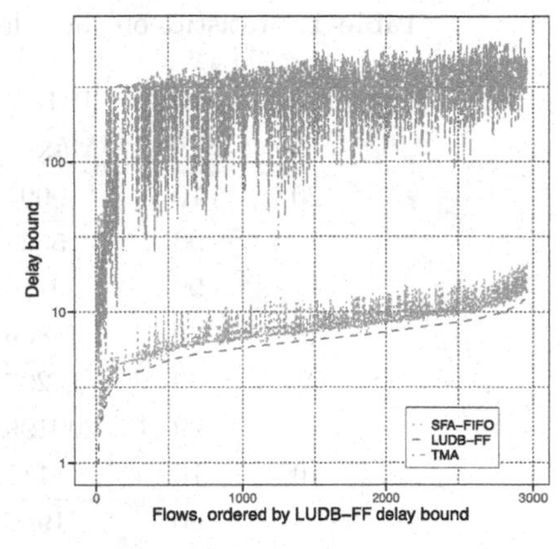

(b) Size 200 devices, utilization 99%

Fig. 4. Absolute delay bounds computed by different DNC analyses for different network sizes and utilizations.

Secondly, Fig. 4 presents the absolute delay bounds as computed by the analyses in two sample network configurations: size 200 devices and utilizations of 90% and 99%. We omitted DEBORAH-Integration bounds due to their low number – for example for the networks of size 200 and the studied utilizations, for more than 99% of the flows we obtain infinite bounds. Representative observations of interest are that in Fig. 4a and Fig. 4b those flows where the TMA delay bound outperforms SFA-FIFO or where our LUDB-FF delay bound is outperformed must be on the far left end. I.e., the bounds are actually low, meaning that the fraction of the feedforward network crossed by those flows with an impact must be small. Gains by our competitors are thus relatively small. Another observation to be taken in Fig. 4 is that with increasing utilization, TMA delay bounds not only become larger but they increase by different factors – despite a uniform utilization at servers. This is visualized by large oscillations of the lines connecting TMA data points. In contrast, the two FIFO analyses SFA-FIFO and LUDB-FF are considerably more resilient to increasing utilization of a network. I.e., ranking alternative network designs w.r.t. some flows' delay bounds would yield vastly different results if TMA was used.

Given that our new LUDB-FF analysis shows most promising results, we set it as the new baseline for a comparison to the other main alternatives TMA and SFA-FIFO. Table 1 provides statistics about how much these other alternatives overestimate worst-case delay bounds w.r.t. LUDB-FF. As already obvious from our evaluation, using TMA instead of a FIFO analysis results in vast overestimation of delay bounds across all analyzed network sizes and utilizations. SFA-FIFO, in comparison, is more precise than TMA but the delay bound of some potentially important data flows might still be overestimated more than 100%.

Table 1. Statistics on the relative delay bound to LUDB-FF [%]

GLP n	Util u	TMA		SFA-FIFO	
		Max	Mean	Max	Mean
20	70	997.92	348.20	208.66	72.43
	90	5568.44	1507.54	264.40	83.0
	99	19710.72	5752.82	293.23	95.30
40	70	866.76	428.37	107.90	20.67
	90	12297.87	3120.43	120.58	25.97
	99	200199.03	35041.52	136.81	34.86
100	70	1112.25	478.40	109.51	14.58
	90	1987.59	1115.57	113.81	18.47
	99	9018.02	4084.48	122.06	26.06
200	70	1289.94	537.57	108.09	15.90
	90	2169.07	1234.69	124.04	19.61
	99	11460.48	4850.10	133.83	26.43

5 Conclusion

In this paper we present LUDB-FF, an advanced Network Calculus analysis for deriving delay bounds in networks of FIFO queueing systems. LUDB-FF combines the ideas manifested in the so-called Least Upper Delay Bound analysis with the NC framework for feedforward analysis that was developed for use with analyses not considering FIFO but arbitrary multiplexing. We thoroughly investigated the potential gains in delay bound tightness by flow aggregation [1,7] during the complex analysis proceedings. For this many cases have to be taken care of w.r.t. on how to proceed in the analysis in order to compute the tightest-possible bounds, e.g., when during the analysis a non-nested interference on a tandem occurs, it is advisable to not cut crossflows that travel along the entire tandem's path which has not been discussed before. We implemented our findings into the open-source NCorg DNC tool that only provided support for the FIFO analysis with SFA-FIFO until now. Our numerical evaluation shows that LUDB-FF outperforms alternative approaches to bound flow delays in feedforward networks of FIFO queueing systems, namely SFA-FIFO, a straightforward extension of the reference tool for LUDB that can only work with tandem networks, as well as the state-of-the-art arbitrary multiplexing analysis TMA. On a set of Internet-like topologies of different sizes and utilizations, we show that LUDB-FF almost always beats the currently best tool-supported competing analysis SFA-FIFO by a margin of up to 75%.

Opportunities for Future Research. With LUDB-FF we have created the foundation to investigate improvements known from the analysis of arbitrary multiplexing networks in networks of FIFO queueing systems as are, for example, defined in modern standards like IEEE Time-Sensitive Networking. We also hope to leverage newer advances such as finite modeling of curve domains [22,25] as well as GPGPU computing [26] and machine learning [21] in the future.

A SFA-FIFO

SFA-FIFO is the simple hop-by-hop analysis where we set every occurring FIFO parameter given some (output) arrival curve α_f and (residual) service curve $\beta_f^{\text{l.o.}}$ to $\underline{\theta}(\beta_f^{\text{l.o.}}, \alpha_f)$ (see Definition 3). Algorithm 3 below shows the details of the approach where $1, ..., n$ represents the server indices on the path p and Flows(i) with $i \in \{1, ..., n\}$ denotes all flows that cross server i.

Algorithm 3. Compute Service Curve Algorithm (SFA-FIFO)

Input $(F, \alpha_F^{\text{Source}(p)}, p)$ Flows for which to compute the (aggregate) residual service on path $p := 1, ..., n$
Output Residual service curve for flow aggregate F

1: **procedure** COMPUTESERVICECURVE($F, \alpha_F^{\text{Source}(p)}, p$)
2: $\quad\beta^{\text{l.o.}} \leftarrow \delta_0$
3: $\quad$**for** $i = 1$ to n **do**
4: $\quad\quad F^i \leftarrow \text{Flows}(i)$
5: $\quad\quad F_x \leftarrow F^i \backslash F$
6: $\quad\quad \alpha^i_{F_x} \leftarrow$ COMPUTEARRIVALCURVE(F_x, i)
7: $\quad\quad \beta_F^{\text{l.o.},i}(t) \leftarrow \left[\beta^i(t) - \alpha_F(t - \underline{\theta}(\beta^i, \alpha^i_{F_x}))\right]^\uparrow \cdot 1_{\{t > \underline{\theta}(\beta^i, \alpha^i_{F_x})\}}$
8: $\quad\quad \beta^{\text{l.o.}} \leftarrow \beta^{\text{l.o.}} \otimes \beta_F^{\text{l.o.},i}$
9: $\quad$**end for**
10: $\quad$return $\beta^{\text{l.o.}}$
11: **end procedure**

References

1. Bisti, L., Lenzini, L., Mingozzi, E., Stea, G.: Estimating the worst-case delay in FIFO tandems using network calculus. In: Proceedings of the ICST ValueTools (2008)
2. Bisti, L., Lenzini, L., Mingozzi, E., Stea, G.: DEBORAH: a tool for worst-case analysis of FIFO tandems. In: Proceedings of the ISoLA (2010)
3. Bisti, L., Lenzini, L., Mingozzi, E., Stea, G.: Numerical analysis of worst-case end-to-end delay bounds in FIFO tandem networks. Real-Time Syst. **48**(5), 527–569 (2012)

4. Bondorf, S., Nikolaus, P., Schmitt, J.B.: Quality and cost of deterministic network calculus - design and evaluation of an accurate and fast analysis. In: Proceedings of the ACM on Measurement and Analysis of Computing Systems (POMACS), vol. 1, no. 1 (2017)
5. Bondorf, S., Nikolaus, P., Schmitt, J.B.: Catching corner cases in network calculus - flow segregation can improve accuracy. In: Proceedings of the GI/ITG MMB (2018)
6. Bondorf, S., Schmitt, J.B.: The DiscoDNC v2 - a comprehensive tool for deterministic network calculus. In: Proceedings of the EAI ValueTools (2014)
7. Bondorf, S., Schmitt, J.B.: Boosting sensor network calculus by thoroughly bounding cross-traffic. In: Proceedings of the IEEE INFOCOM (2015)
8. Bondorf, S., Schmitt, J.B.: Calculating accurate end-to-end delay bounds - you better know your cross-traffic. In: Proceedings of the EAI ValueTools (2015)
9. Le Boudec, J.Y., Thiran, P.: Network Calculus: A Theory of Deterministic Queuing Systems for the Internet. Springer-Verlag, Berlin (2001)
10. Bouillard, A.: Stability and performance bounds in cyclic networks using network calculus. In: Proceedings of the FORMATS (2019)
11. Bouillard, A.: Trade-off between accuracy and tractability of network calculus in FIFO networks (2020). arxiv:2010.09263
12. Bouillard, A., Boyer, M., Le Corronc, E.: Deterministic Network Calculus: From Theory to Practical Implementation. Wiley, Hoboken (2018)
13. Bouillard, A., Stea, G.: Exact worst-case delay for FIFO-multiplexing tandems. In: Proceedings of the EAI ValueTools (2012)
14. Bouillard, A., Stea, G.: Exact worst-case delay in FIFO-multiplexing feed-forward networks. IEEE/ACM Trans. Netw. **23**(5), 1387–1400 (2015)
15. Bouillard, A., Thierry, É.: An algorithmic toolbox for network calculus. Discret. Event Dyn. Syst. **18**(1), 3–49 (2008)
16. Boyer, M., Graillat, A., Dupont de Dinechin, B., Migge, J.: Bounding the delays of the MPPA network-on-chip with network calculus: models and benchmarks. Perform. Eval. **143**, 102124 (2020)
17. Bu, T., Towsley, D.: On distinguishing between internet power law topology generators. In: Proceedings of the IEEE INFOCOM (2002)
18. Cattelan, B., Bondorf, S.: Iterative design space exploration for networks requiring performance guarantees. In: Proceedings of the IEEE/AIAA DASC (2017)
19. Cruz, R.L.: A calculus for network delay, part I: network elements in isolation. IEEE Trans. Inf. Theory **37**(1), 114–131 (1991)
20. Fidler, M.: Extending the network calculus pay bursts only once principle to aggregate scheduling. In: Proceedings of the QoS-IP (2003)
21. Geyer, F., Bondorf, S.: DeepTMA: Predicting effective contention models for network calculus using graph neural networks. In: Proceedings of the IEEE INFOCOM (2019)
22. Guan, N., Yi, W.: Finitary real-time calculus: efficient performance analysis of distributed embedded systems. In: Proceedings of the IEEE RTSS (2013)
23. Jafari, F., Jantsch, A., Lu, Z.: Weighted round robin configuration for worst-case delay optimization in network-on-chip. IEEE Trans. Very Large Scale Integr. (VLSI) Syst. **24**(12), 3387–3400 (2015)
24. Jafari, F., Lu, Z., Jantsch, A.: Least upper delay bound for VBR flows in networks-on-chip with virtual channels. ACM Trans. Des. Autom. Electron. Syst. **20**(3), 1–33 (2015)
25. Lampka, K., Bondorf, S., Schmitt, J.B., Guan, N., Yi, W.: Generalized finitary real-time calculus. In: Proceedings of the IEEE INFOCOM (2017)

26. Luangsomboon, N., Hesse, R., Liebeherr, J.: Fast min-plus convolution and deconvolution on GPUs. In: Proceedings of the EAI ValueTools (2017)
27. Scheffler, A., Fögen, M., Bondorf, S.: The deterministic network calculus analysis: reliability insights and performance improvements. In: Proceedings of the IEEE CAMAD (2018)
28. Schmitt, J.B., Gollan, N., Bondorf, S., Martinovic, I.: Pay bursts only once holds for (some) non-FIFO systems. In: Proceedings of the IEEE INFOCOM (2011)
29. Schmitt, J.B., Zdarsky, F.A., Martinovic, I.: Improving performance bounds in feed-forward networks by paying multiplexing only once. In: Proceedings of the GI/ITG MMB (2008)
30. Schmitt, J.B., Zdarsky, F.A., Thiele, L.: A comprehensive worst-case calculus for wireless sensor networks with in-network processing. In: Proceedings of the IEEE RTSS (2007)
31. She, H., Lu, Z., Jantsch, A., Zhou, D., Zheng, L.R.: Performance analysis of flow-based traffic splitting strategy on cluster-mesh sensor networks. Int. J. Distrib. Sens. Netw. **8**(3), 232937 (2012)
32. Zhao, L., Pop, P., Craciunas, S.S.: Worst-case latency analysis for IEEE 802.1qbv time sensitive networks using network calculus. IEEE Access **6**, 41803–41815 (2018)
33. Zhao, L., Pop, P., Zheng, Z., Daigmorte, H., Boyer, M.: Latency analysis of multiple classes of AVB traffic in TSN with standard credit behavior using network calculus. IEEE Trans. Ind. Electron. **68**(10), 10291–10302 (2021). https://doi.org/10.1109/TIE.2020.3021638
34. Zhao, L., Pop, P., Zheng, Z., Li, Q.: Timing analysis of avb traffic in tsn networks using network calculus. In: IEEE RTAS (2018)
35. Zhou, B., Howenstine, I., Limprapaipong, S., Cheng, L.: A survey on network calculus tools for network infrastructure in real-time systems. IEEE Access **8**, 223588–223605 (2020)

SEH: Size Estimate Hedging for Single-Server Queues

Maryam Akbari-Moghaddam(✉) and Douglas G. Down

Department of Computing and Software, McMaster University, Hamilton, ON, Canada
{akbarimm,downd}@mcmaster.ca

Abstract. For a single server system, Shortest Remaining Processing Time (SRPT) is an optimal size-based policy. In this paper, we discuss scheduling a single-server system when exact information about the jobs' processing times is not available. When the SRPT policy uses estimated processing times, the underestimation of large jobs can significantly degrade performance. We propose a simple heuristic, Size Estimate Hedging (SEH), that only uses estimated processing times for scheduling decisions. A job's priority is increased dynamically according to an SRPT rule until it is determined that it is underestimated, at which time the priority is frozen. Numerical results suggest that SEH has desirable performance for estimation error variance that is consistent with what is seen in practice.

Keywords: Estimated job sizes · M/G/1 · Gittins' index policy · Size estimate hedging

1 Introduction

Over the past decades, there has been significant study on the scheduling of jobs in single-server queues. When preemption is allowed and processing times are known to the scheduler, the Shortest Remaining Processing Time (SRPT) policy is optimal in the sense that, regardless of the processing time distribution, it minimizes the number of jobs in the system at each point in time and hence, minimizes the mean sojourn time (MST) [23,24]. However, scheduling policies such as SRPT are rarely deployed in practical settings. A key disadvantage is that the assumption of knowing the exact job processing times prior to scheduling is not always practical to make. However, it is often possible to estimate the job processing times and use this approximate information for scheduling. The Shortest Estimated Remaining Processing Time (SERPT) policy is a version of SRPT that employs the job processing time estimates as if they were error-free and thus, schedules jobs based on their estimated remaining times. Motivated by the fact that estimates can often be obtained through machine learning techniques, Mitzenmacher [18] studies the potential benefits of using such estimates for simple scheduling policies. For this purpose, a price for misprediction, the

A. Abate and A. Marin (Eds.): QEST 2021, LNCS 12846, pp. 168–185, 2021.
https://doi.org/10.1007/978-3-030-85172-9_9

ratio between a job's expected sojourn time using its estimated processing time and the job's expected sojourn time when the job processing time is known is introduced, and a bound on this price is given. The results in [18] suggest that naïve policies work well, and even a weak predictor can yield significant improvements under policies such as SERPT. However, this insight is only made when the job processing times have relatively low variance. As discussed below, when job processing times have high variance, underestimating even a single very large job can severely affect the smaller jobs' sojourn times.

The work in [18] has the optimistic viewpoint that it is possible to obtain improved performance by utilizing processing time estimates in a simple manner. The more pessimistic view is that when job processing times are estimated, estimation errors naturally arise, and they can degrade a scheduling policy's performance, if the policy was designed to exploit exact knowledge of job processing times [15]. The SERPT policy may have poor performance when the job processing times have high variance and large jobs are underestimated. Consider a situation where a job with a processing time of 1000 enters the system and is underestimated by 10%. The moment the job has been processed for 900 units (its estimated processing time), the server assumes that this job's estimated remaining processing time is zero, and until it completes, the job will block the jobs already in the queue as well as any new arrivals. This situation becomes more severe when both the actual job processing time and the level of underestimation increase. However, when the job processing times are generated from lower variance distributions, the underestimation of large jobs will not cause severe performance degradation [16].

The Shortest Estimated Processing Time (SEPT) policy is a version of the Shortest Processing Time (SPT) policy that skips updating the estimated remaining processing times and prioritizes jobs based only on their estimated processing times. Experimental results show that SEPT has impressive performance in the presence of estimated job processing times, as well as being easier to implement than SERPT [8].

In this paper, we will discuss the problem of single-server scheduling when only estimates of the job processing times are available. In Sect. 2, we discuss the existing literature for scheduling policies that handle inexact job processing time information. Most of the existing literature analyzes and introduces size-based policies when the estimation error is relatively small, restricting applicability of the results. Furthermore, many simulation-based examinations only consider certain workload classes and are not validated over a range of job processing times and estimation error distributions. We propose a scheduling policy that exhibits desirable performance over a wide range of job processing time distributions, estimation error distributions, and workloads.

The Gittins' Index policy [10], a dynamic priority-based policy, is optimal in minimizing the MST in an M/G/1 queue [1]. When there are job processing time estimates, the Gittins' Index policy utilizes information about job estimated processing time, and the job processing time and estimation error distributions to decide which job should be processed next. The assumption of knowing these distributions before scheduling may be problematic in real environments. Furthermore, scheduling jobs using the Gittins' Index policy introduces computational overhead that may be prohibitive. While there are significant barriers to implementing the Gittins' Index policy, our proposed policy is motivated by the form of the Gittins' Index policy.

We make the following contributions: While the SEPT policy performs well in the presence of estimated job processing times [8], we first introduce a heuristic that combines the merits of SERPT and SEPT. Secondly, we specify the Gittins' Index policy given multiplicative estimation errors and restricted to knowing only the estimation error distribution. We show that our proposed policy, which we call the Size Estimate Hedging (SEH) policy, has performance close to the Gittins' Index policy. Similar to SERPT and SEPT, the SEH policy only uses the job processing time estimates to prioritize the jobs. Finally, we provide numerical results obtained by running a wide range of simulations for both synthetic and real workloads. The key observations suggest that SEH outperforms SERPT except in scenarios where the job processing time variance is extremely low. SEH outperforms SEPT whether the variance of the job processing times is high or low. With the presence of better estimated processing times in the system (low variance in the estimation errors), SEH outperforms SEPT and has performance close to the optimal policy (SRPT) if the estimation errors are removed. On the other hand, we observe that when the estimation errors have high variance, there is little value in using the estimated processing times. We also notice that the system load does not significantly affect the relative performance of the policies under evaluation. The SEH policy treats underestimated and overestimated jobs fairly, in contrast with other policies that tend to favor only one class of jobs. When the job processing time variance is high, the SEH and SEPT policies obtain a near-optimal mean slowdown value of 1, indicating that underestimated large jobs do not delay small jobs. In terms of mean slowdown, SEH outperforms SEPT across all levels of job processing time variance.

The rest of the paper is organized as follows. Section 2 presents the existing literature in scheduling single-server queues with estimated job processing times. Section 3 defines our SEH policy and discusses its relationship to a Gittins' Index approach. Our simulation experiments are described in detail in Sect. 4. We provide the results of our simulations in Sect. 5 and conclude and discuss future directions in Sect. 6.

2 Related Work

Scheduling policies and their performance evaluation in a preemptive M/G/1 queue have been a subject of interest for some time. Size-based policies are known to perform better than size-oblivious policies with respect to sojourn times. In fact, the SRPT policy is optimal in minimizing the MST [23]. However, size-based policies have a considerable disadvantage: When the exact processing times are not known to the system before scheduling, which is often the case in practical settings, their performance may significantly degrade. Dell'Amico et al. [9] study the performance of SRPT with estimated job processing times and demonstrate the consequences of job processing time underestimations under different settings. Studies in Harchol-Balter et al. [13] and Chang et al. [4] discuss the effect of inexact processing time information in size-based policies for web servers and MapReduce systems, respectively. Our paper assumes that the processing time is not available to the scheduler until the job is fully processed, but that processing time estimations are available. The related literature for this setting is reviewed in the following paragraph.

Lu et al. [15] were the first to study this setting. They show that size-based policies only benefit the performance when the correlation between a job's real and estimated processing time is high. The results in Wierman and Nuyens [27], Bender et al. [3], and Becchetti et al. [2] are obtained by making assumptions that may be problematic in practice. A strict upper bound on the estimation error is assumed in [27]. On the other hand, [2] and [3] define specific job processing time classes and schedule the jobs based on their processing time class, which can be problematic for very small or very large jobs. This setting is also known as semi-clairvoyant scheduling. In this work, we do not assume any bounds on the estimation error or assign jobs to particular processing time classes. Consistent with this body of work, we do find that SEH is not recommended for systems with large estimation error variance. However, we do find that it performs well for levels of estimation error variance that are typically found in practice.

When the job processing time distribution is available, the Gittins' Index policy [10] assigns a score to each job based on the processing time it has received so far, and the scheduler chooses the job with the highest score to process at each point in time. This policy is proven to be optimal for minimizing the MST in a single-server queue when the job processing time distribution is known [1]. This policy is specified in the next section.

3 Size Estimate Hedging: A Simple Dynamic Priority Scheduling Policy

3.1 Model

Consider an M/G/1 queue where preemption is allowed and we are interested in minimizing the MST. We assume that a job's processing time is not known upon arrival; however, an estimated processing time is provided to the scheduler. We concentrate on a multiplicative error model where the error distribution is independent of the job processing time distribution. The estimated processing time $\hat{S}$ of a job is defined as $\hat{S} = SX$ where S is the job processing time and X is the job processing time estimation error. We assume that the value of $\hat{S}$ is known upon each job's arrival and is denoted by $\hat{s}$. The choice of a multiplicative error model results in having an absolute error proportional to the job processing time S, thus avoiding situations where the estimation errors tend to be worse for small jobs than for large jobs. Furthermore, Dell'Amico et al. [9] and Pastorelli et al. [19] suggest that a multiplicative error model is a better reflection of reality. To define our scheduling policies, we also require the notion of a quantum of service. The job with the highest priority is processed for a quantum of service Δ until either it completes or a new job arrives. At that point, priorities are recomputed.

3.2 Gittins' Index Approach

The Gittins' Index Policy is an appropriate technique for determining scheduling policies when the job processing time and estimation error distributions are known. For a waiting job i, an index $G(a_i)$ is calculated, where a_i is the elapsed processing time. At each time epoch, the Gittins' Index policy processes the job with the highest index $G(a)$ among all of the present waiting jobs [10]. The Gittins' rule takes the job's elapsed processing time into account and calculates the optimal quantum of service $\Delta^*(a)$ that it should receive.

The associated efficiency function $J(a, \Delta), a, \Delta \geq 0$ of a job with processing time S, elapsed processing time a and quantum of service Δ is defined as

$$J(a, \Delta) = \frac{P(S - a \leq \Delta | S > a)}{E[\min\{S - a, \Delta\} | S > a]}. \tag{1}$$

The numerator is the probability that the job will be completed within a quantum of service Δ, and the denominator is the expected remaining processing time a job with elapsed processing time a and quantum of service Δ will require to be completed.

The server (preemptively) processes the job with the highest index at each decision epoch. Decisions are made when (i) a new job arrives to the queue, (ii) the current job under processing completes, or (iii) the current job receives its optimal quantum of service and does not complete. If there are multiple jobs that have the same highest index and all have zero optimal quanta of service, the processor will be shared among them as long as this situation does not change. If there is only one job with the highest index and zero optimal quantum of service, its index should be updated throughout its processing [1].

Although the Gittins' Index policy is optimal in terms of minimizing the mean sojourn time in an $M/G/1$ queue [1], the assumption of knowing the job size and estimation error distributions might not always be practical to make. Furthermore, forming the Gittins' Index policy's efficiency function has significant computational overhead. As a result, this policy may be a problematic choice for real environments where the scheduling speed is important. However, examining the form of optimal policies has helped us in the construction of a simple heuristic. In particular, the notion of defining a policy in terms of an index allows us to make precise our notion of combining the relative merits of SRPT and SEPT.

3.3 Motivation

When a job enters the system under SERPT, there is no basis on which to assume that the estimated processing time, $\hat{s}$, is incorrect. However, when the elapsed processing time reaches $\hat{s}$, we are certain that the job processing time has been underestimated. In addition, Dell'Amico et al. [8] show that SEPT performs well when dealing with estimated processing times and in the presence of estimation errors, in particular severe underestimates. So, we would like to combine these two policies. A convenient way to do this is to introduce a Gittins'-like score function, where a higher score indicates a higher priority. We will be aggressive and use the score function for SERPT until the point that we know a job is underestimated and then freeze the score, which is similar to what SEPT's constant score function does (see (4) below). In this way, instead of switching to SEPT's score function, we would like to give credit for the jobs' cumulative elapsed processing times.

The score functions for SRPT, SERPT, and SEPT are provided in (2), (3), and (4), respectively.

$$G(a, s) = \frac{1}{s - a}, \tag{2}$$

$$G(a, \hat{s}) = \begin{cases} \frac{1}{\hat{s} - a}, & \hat{s} > a, \\ \infty, & \hat{s} \leq a, \end{cases} \tag{3}$$

$$G(a, \hat{s}) = \frac{1}{\hat{s}}. \tag{4}$$

We note that (2) and (3) have an increasing score function, and (4) always assigns a constant score for a particular job.

3.4 The SEH Policy

Combining the score functions for SERPT and SEPT, we now define our policy. As discussed in the previous section, we would like to transition between SERPT when we cannot determine if a job processing time is underestimated to a fixed priority like SEPT when it is determined that underestimation has occurred. One consequence of using this policy is that any underestimated small job can still receive a "high" score and be processed, while underestimated large jobs will have a much lower score and do not interfere, even with underestimated small jobs. Furthermore, not needing to know the job processing time and estimation error distribution, the SEH Policy does not have much overhead. Thus, it can schedule the jobs at a speed comparable to the SEPT policy.

We introduce the score function of our SEH policy as

$$G(a, \hat{s}) = \begin{cases} \frac{1}{\hat{s} - a(1 - \frac{a}{2\hat{s}})}, & 0 \leq a < \hat{s}, \\ \frac{2}{\hat{s}}, & a \geq \hat{s}, \end{cases} \tag{5}$$

where the scheduling decisions are only made at arrivals and departures.

With the score function in (5), a job's score will increase up to the point that it receives processing equal to its estimated processing time and then receives a constant score of $\frac{2}{\hat{s}}$ until it completes. The choice of 2 was made after some experimentation, it would be worthwhile to explore the sensitivity of the performance to this choice.

3.5 Gittins' Index Vs. SEH

In this section, we show that the form of our policy is consistent with the Gittins' index in the setting that we only know the error estimate distribution. In particular, we have no a priori or learned knowledge of the processing time distribution.

With our estimation model in mind, (1) can be rewritten as

$$J(a, \Delta, \hat{s}) = \frac{P(\frac{\hat{s}}{X} - a \leq \Delta | \frac{\hat{s}}{X} > a)}{E[\min\{\frac{\hat{s}}{X} - a, \Delta\} | \frac{\hat{s}}{X} > a]}. \tag{6}$$

The Gittins' index $G(a, \hat{s}), a \geq 0$, is defined by

$$G(a, \hat{s}) = \sup_{\Delta \geq 0} J(a, \Delta, \hat{s}).$$

The optimal quantum of service is denoted as

$$\Delta^*(a, \hat{s}) = \sup\{\Delta \geq 0 | G(a, \hat{s}) = J(a, \Delta, \hat{s})\}.$$

Suppose that the lower and upper limits on the estimation error distribution are l and u, respectively (l may be zero and u may be ∞). After some calculation, the Gittins' index can then be written as

$$G(a, \hat{s}) = \begin{cases} \frac{1}{\hat{s} - aE[X | X \leq \frac{\hat{s}}{a}]}, & \frac{\hat{s}}{a} < u, \\ \frac{1}{\hat{s} - aE[X]}, & otherwise, \end{cases} \tag{7}$$

where $\Delta^* = \frac{\hat{s}}{l} - a$. For instance, the Gittins' index for a $Log - N(\mu, \sigma^2)$ error distribution is

$$G(a, \hat{s}) = \frac{1}{\hat{s} - ae^{\mu + g(a, \hat{s})}}, \tag{8}$$

where

$$g(a, \hat{s}) = \frac{\sigma^2 \phi[\frac{\ln(\frac{\hat{s}}{a}) - \mu - \sigma^2}{\sigma}]}{2\phi[\frac{\ln(\frac{\hat{s}}{a}) - \mu}{\sigma}]},$$

and ϕ is the cumulative distribution function of the $Log - N(0, \sigma^2)$ distribution. Note that for the Log-Normal distribution as the job processing time error distribution, the second case in (7) cannot happen. For the remainder of the paper, we will refer to this policy as the Gittins' Index policy. We recognize that this is a slight abuse of terminology, as we are ignoring the job processing time distribution.

Taking the score in (8) into account, for any job with an estimated processing time $\hat{s}$, the score calculated with the Gittins' Index policy continuously increases until the job completes. Figure 1(a) shows this score for a job with an estimated processing time of 20 and an estimation error generated from a $Log - N(0, \sigma^2)$ distribution as a function of its elapsed processing time. We observe that for larger values of elapsed processing time, the slope of the score is decreasing.

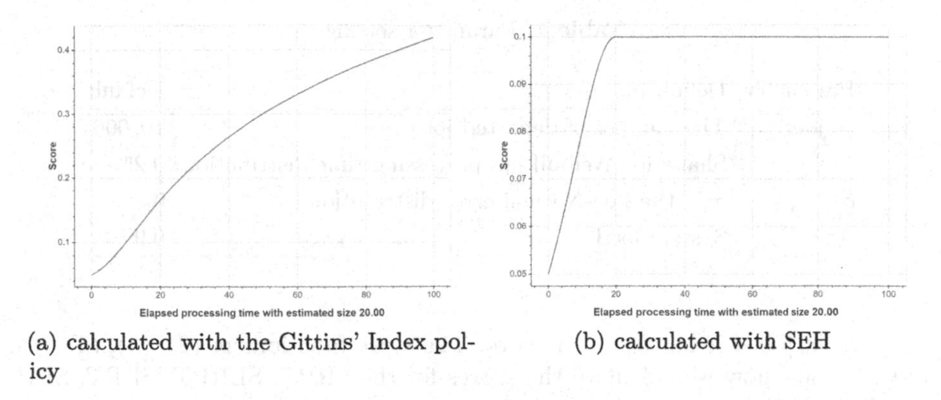

(a) calculated with the Gittins' Index policy

(b) calculated with SEH

Fig. 1. Job score as a function of the elapsed processing time

Figure 1(b) shows the score calculated with the SEH policy for a job with an estimated processing time of 20 as a function of its elapsed processing time. The score shown in Fig. 1(a) is consistent with the score function having decreasing slope at some point beyond the point at which the elapsed processing time reaches the estimated processing time, as in Fig. 1(b). Of course, the change in slope for SEH is more severe, but we will see in our numerical experiments that the performance of the two policies is quite close. SEH has less computational overhead and more importantly, does not require knowledge of the estimation error distribution.

4 Evaluation Methodology

4.1 Policies Under Evaluation

In this section, we introduce the size-based scheduling policies considered for evaluation. As our baseline policy, we consider the SRPT policy when the exact job processing times, given by s, are known before scheduling. The SRPT policy is an "ideal" policy since it assumes that there are no errors in estimating the processing time.

- **SERPT policy**—The SERPT policy is a version of SRPT that uses the estimates of job processing times as if they were the true processing times.
- **SEPT policy**—The SPT policy skips the SRPT policy's updating of remaining processing times and only schedules jobs based on their estimated processing time.
- **SEH and Gittins' Index policy**—Our proposed SEH policy and the Gittins' Index policy are explained in detail in Sect. 3.4 and Sect. 3.5, respectively.

All these policies fit into the "scoring" framework, and they assign scores to each job and process the jobs in the queue in the descending order of their scores. Moreover, preemption is allowed, and a newly-arrived job can preempt

Table 1. Parameter settings

Parameter	Definition	Default
# jobs	The number of departed jobs	$10,000$
k	Shape for Weibull job processing time distribution	0.25
σ	σ in the Log-Normal error distribution	0.5
ρ	System load	0.9

the current job if it has a higher score. The score functions in (2), (3), (4), (5), and (7) show how we calculate the scores for the SRPT, SERPT, SEPT, SEH, and Gittins' Index policy, respectively.

4.2 Performance Metrics

We evaluate the policies defined in Sect. 4.1 with respect to two performance metrics: MST and Mean Slowdown. When the job processing times have large variance, the sojourn times for small jobs and large jobs differ significantly. Thus, we use the per job slowdown, the ratio between a job's sojourn time and its processing time [26].

4.3 Simulation Parameters

We would like to evaluate the policies over a wide range of job processing time and error distributions. To generate this range of distributions, we fix the form of the distribution and vary the parameters. We use the same settings that Dell'Amico et al. [9] use in their work. Table 1 provides the default parameter values that we use in our simulation study. We now provide details of our simulation model. Note that our policy fits into the SOAP framework of Scully et al. [25], however as we are also evaluating mean slowdown, we chose simulation for evaluation.

Job Processing Time Distribution—We consider an $M/G/1$ queue where the processing time is generated according to a Weibull distribution. This allows us to model high variance processing time distributions, which better reflect the reality of computer systems (see [7,12] for example). In general, the choice of a Weibull distribution gives us the flexibility to model a range of scenarios. The shape parameter k in the Weibull distribution allows us to evaluate both high variance (smaller k) and low variance (larger k) processing time distributions.

Considering that the job processing time distribution plays a significant role in the scheduling policies' performance and size-based policies show different behaviors with high variance job processing time distributions, we choose $k = 0.25$ as our default shape for the Weibull job processing time distribution. With this choice for k, the scheduling policies' performance is highly influenced by a few very large jobs that constitute a substantial percentage of the system's overall workload. We vary k between 0.25 and 2, considering specific values of 0.25, 0.375, 0.5, 0.75, 1, and 2. We show that the SEH policy performs best in the presence of high variance job processing time distributions.

Job Processing Time Error Distribution—We have chosen the Log-Normal distribution as our error distribution so that a job has an equal probability of being overestimated or underestimated. The Gittins' index for this estimation error distribution is shown in (8). The σ parameter controls the correlation between the actual and estimated processing time, as well as the estimation error variance. By increasing the σ value, the correlation coefficient becomes smaller, and the estimation error variance increases, resulting in the occurrence of more large underestimations/overestimations (more imprecise processing times). We choose $\sigma = 0.5$ as the default value that corresponds to a median relative error factor of 1.40. We vary σ between 0.25 and 1 with specific values of 0.25, 0.375, 0.5, 0.75, and 1 to better illustrate the effect of σ on the evaluated performance.

System Load—Following Lu et al. [15], we consider $\rho = 0.9$ as the default load value and vary ρ between 0.5 (lightly loaded) and 0.95 (heavily loaded) with increments of 0.05 and an additional system load of 0.99.

Number of Jobs—The number of jobs in each simulation run is 10,000 and a simulation run ends when the first 10,000 jobs that arrived to the system are completed. We fix the confidence level at 95%, and for each simulation setting, we continue to perform simulation runs until the width of the confidence interval is within 5% of the estimated value. For low variance processing time distributions (larger k), 30 simulation runs suffice; however, more simulation runs are required for high variance processing time distributions (smaller k).

5 Simulation Results

In this section, we evaluate the performance of the policies in Sect. 4.1 by running experiments on both synthetic and real workloads. We run different simulations by generating synthetic workloads based on different job processing time and error parameters and we analyze these parameters' effect on the performance of each of the policies.

For evaluating our results in practical environments, we consider a real trace from a Facebook Hadoop cluster in 2010 [5] and show that the policies' performance is consistent with the results we obtained with synthetic workloads. The key observations, validated both on synthetic and real workloads, are highlighted as follows:

- The Gittins' Index policy outperforms SERPT for all the evaluated values of k and σ. We show the same observation with our proposed SEH policy except for values of k that correspond to very low job processing time variance.

- The Gittins' Index and SEH policies outperform SEPT with lower values of σ (better estimated processing times) and have an MST near the optimal MST obtained without any estimation errors.
- SEH performs well in reducing both the MST of overestimated jobs and underestimated jobs.
- The load parameter does not have a significant effect on the relative values of the MST obtained with the evaluated policies.
- The Gittins' Index, SEH and SEPT policies have a near-optimal mean slowdown of 1 when the estimated processing times have high variance.
- The SEH performs best across all values of k in terms of minimizing the mean slowdown.

In what follows, we discuss the numerical results and how they support these key observations.

Synthetic Workloads—We first note that the job processing time k parameter and the estimation error σ parameter have the greatest impact on the policies' performance. Thus, we focus on varying these parameters. We show that the Gittins' Index policy outperforms SERPT across all evaluated values of k and σ and our SEH policy outperforms SERPT except for the values of k and σ that correspond to distributions with extremely low variance. For the scenarios where we do not state the parameter values explicitly, the parameters in Table 1 (see Sect. 4.3) are considered.

Figure 2 captures the impact of job processing time variance and displays the MST of the Gittins' Index, SEH, SERPT, and SEPT policies normalized against the MST obtained with SRPT with σ having the default value of 0.5. We observe that for a high variance job processing time distribution ($k = 0.25$), SERPT performs very poorly compared to the other policies due to the presence of large, underestimated jobs. We note that the SERPT policy performs well if the variance of the processing times is sufficiently low. Based on Fig. 2, we notice that the gap between SEPT and the Gittins' Index policy grows slightly when the job processing time variance is lower. The gap between SEH and the Gittins' Index policy also grows but not to the same degree as SEPT. For $k > 0.75$, the performance of the Gittins' Index policy, SEH, and SERPT are quite close. In fact, we observe that our SEH policy performs very close to the Gittins' Index policy across all values of k. Furthermore, we notice that as the variance in processing times gets smaller, the gap between what is achievable by the policy under evaluation and what is achievable if there were no errors is larger than for the high variance scenarios.

The shape parameter k affects the job processing time variance and the scheduling policies' performance the most, especially when the job processing time distribution has high variance. We can be optimistic about using estimates if the variance is low, but we have to be careful in choosing the scheduling policy if the job processing time variance is high. The literature focuses on high variance workloads, and we will continue evaluating the policies on such workloads. In Fig. 3, we display the normalized MST of the policies against the MST of the

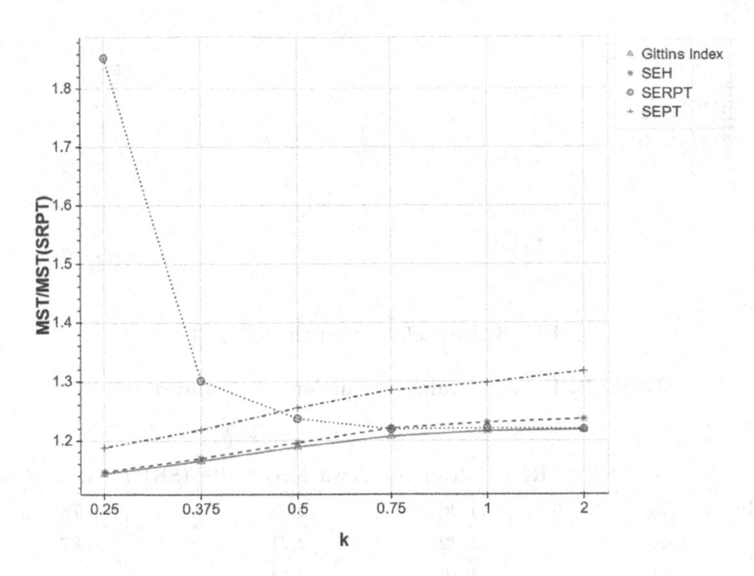

Fig. 2. Impact of k on the MST

SRPT policy under varying σ, $\rho = 0.9$, and the default $k = 0.25$. We notice that the Gittins' Index, SEH, and SEPT policies are relatively insensitive to the σ value, while the gap between these three policies and SERPT increases with increasing σ. In fact, the Gittins' Index and SEH policies outperform SEPT with $\sigma \leq 0.5$ and have an MST near the optimal MST obtained without any estimation errors. We conclude that the impact of the Gittins' Index policy and SEH becomes more prominent when the estimates improve.

In Fig. 3, we observe that while choosing a more aggressive policy like the Gittins' Index and SEH policies is a good choice under lower values of σ, SEPT is preferred when $\sigma = 1$. The reason is that lower values of k (here, $k = 0.25$), cause more large jobs in the system. Furthermore, for values of $\sigma \geq 1$, the estimation errors have high variance and thus the estimated processing times can be very imprecise. We notice that both SEH and the Gittins' Index policy suffer from a slight promotion of severely underestimated jobs that leads to temporary blockage for the other jobs. What has happened in this case is that the estimates of the processing times have degraded to the point that they are not useful. In particular, one should instead base scheduling decisions on the processing time distribution, so for example in scenarios with high variance in both processing times and estimation errors, a policy which ignores the estimates, such as Least Attained Service (LAS) would be warranted. The LAS scheduling policy [21], also known as Shortest Elapsed Time [6] and Foreground-Background [14], preemptively prioritizes the job(s) that have been processed the least. If more than one job has received the least amount of processing time, the jobs will share the processor in a processor-sharing mode. Analytic results in [22,28] show that LAS minimizes MST when the job processing time distribution has a decreasing hazard rate and there are no processing time estimates available.

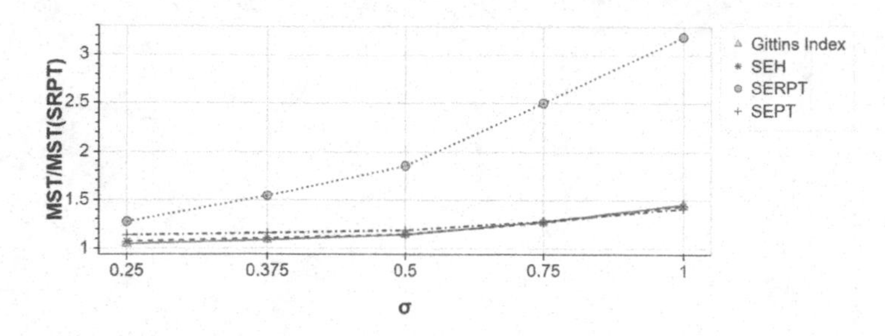

Fig. 3. Impact of σ on the MST

Table 2. Policies evaluation under $\sigma = 1$ and $\sigma = 2$

Policy	$\sigma = 1$		$\sigma = 2$	
	MST/ MST(SRPT)	Mean Slowdown	MST/ MST(SRPT)	Mean Slowdown
Gittins' Index	1.45	1.26	2.68	6.78
SEH	1.44	1.22	2.71	6.87
SEPT	1.41	1.16	2.54	4.71
LAS	1.81	1.27	1.81	1.27
SRPT	1	1.06	1	1.06

These observations are consistent with the results in Table 2 which considers the same settings as in Fig. 3 when $\sigma = 1$ and $\sigma = 2$. SERPT has poor performance compared the other policies under $\sigma \geq 1$ and thus is not included. Pastorelli et al. [19] show that lower values of σ ($\sigma < 1$) are what one sees in practice. It would be interesting to look at the optimal Gittins' Index policy that includes both the job processing time and estimation error distributions, as it would capture this effect. Although doing so can help develop policies that are effective even at high values of σ, deriving the Gittins' index would be quite complicated with this extra condition, but it could give insight into designing simpler policies.

Figure 4(a), Fig. 4(b), and Fig. 4(c) show the result of simulations with the default values in Table 1 and varying the system load between 0.5 and 0.99 for all jobs, only the overestimated jobs, and only the underestimated jobs, respectively. If we concentrate only on one class of jobs (overestimated or underestimated), the policy that minimizes the MST the most can be different. We observe that the Gittins' Index and SEH policies perform best in minimizing the overall MST given different system loads. The Gittins' Index policy performs best in reducing the MST of underestimated jobs and the SEH policy has desirable performance in reducing the MST of all jobs, the overestimated jobs, and the underestimated jobs. Figure 4(a) shows that the load parameter does not have a significant effect on the MST since the ratio between the MST of each policy and the MST of SRPT remains almost unchanged.

The mean slowdown is the other metric we consider to evaluate the performance of the policies. High values of mean slowdown indicate that some jobs

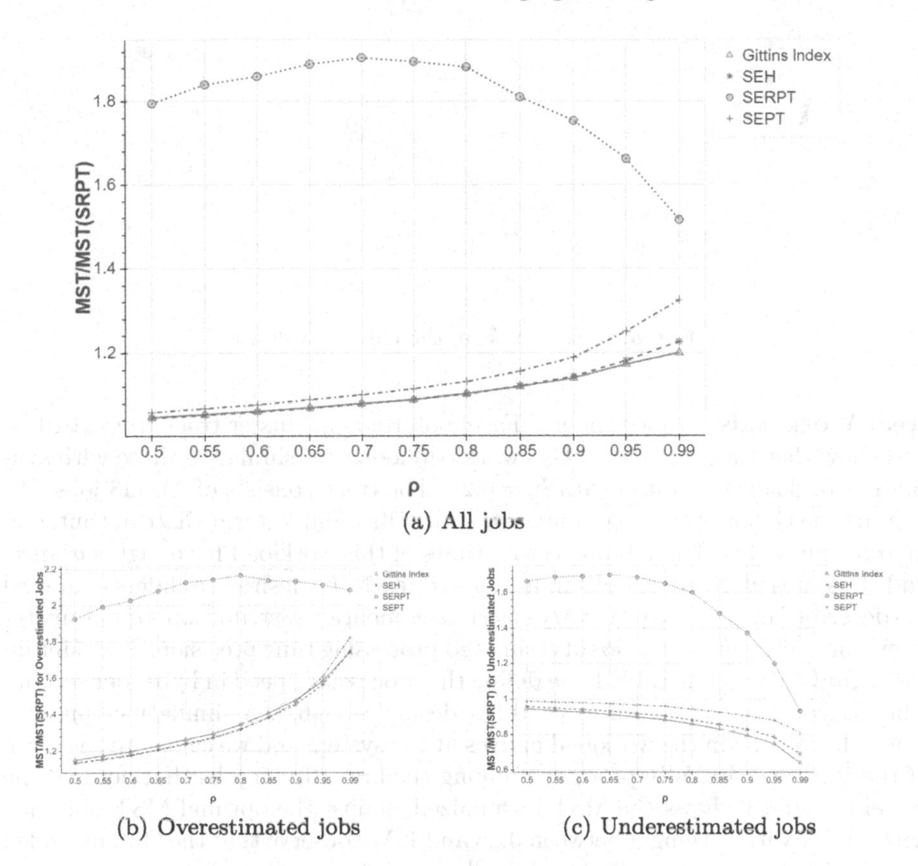

(a) All jobs

(b) Overestimated jobs

(c) Underestimated jobs

Fig. 4. Impact of ρ on the MST

spend a disproportionate amount of time waiting. In Fig. 5, we show the mean slowdown for different values of k with $\rho = 0.9$ and a σ value of 0.5. The mean slowdown of SERPT is not included since it is several orders of magnitude higher for $k \leq 0.5$. We see that the Gittins' Index, SEH, and SEPT policies have similar performance. All policies have a near-optimal mean slowdown of 1 for high variance job processing time distributions (smaller k). The reason is that the very small jobs (that make up the majority of the jobs) are processed the moment they enter the system, and no large job blocks them. We also observe that SEH performs best across all values of k in terms of minimizing the mean slowdown.

We conclude our experiments with synthetic workloads by indicating that the Gittins' Index and SEH policies perform better than SERPT under different parameter settings. The only exception is extreme situations like the low variance job processing time distributions (larger k) where SERPT outperforms SEH and works analogously to the Gittins' Index policy.

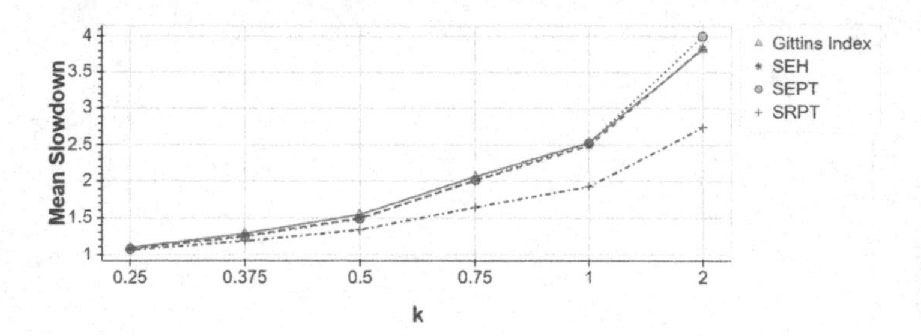

Fig. 5. Impact of k on the mean slowdown

Real Workloads—We consider a Facebook Hadoop cluster trace from 2010 [5] and show that the results with this workload look very similar to those with synthetic workloads generated with $k = 0.25$. The trace consists of 24, 443 jobs. We assume each job's processing time is the sum of its input, intermediate output, and final output bytes. The job processing times of this workload have high variance, and thus, we run hundreds of simulations to reach the desired confidence interval (as described in Sect. 4.3). We vary the error estimation distribution's σ parameter to evaluate different scenarios of estimated processing time precision. To maintain the default settings in Table 1, we define the processing speed in bytes per second. The arrival rate λ is chosen to yield the desired $\rho = 0.9$. A simulation run ends when the last job in the workload arrives at the system and we calculate the MST of the jobs that are fully processed among the first 10, 000 jobs that entered the system. Figure 6 shows the MST normalized against the optimal MST obtained with SRPT with varying σ between 0.25 and 1. We observe that the Gittins' Index and SEH policies perform best across all values of σ.

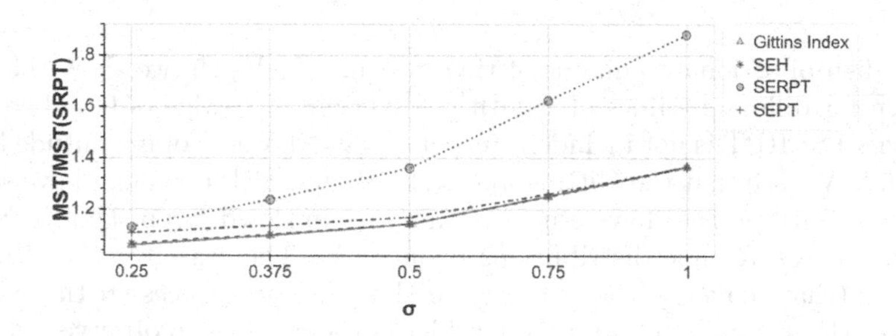

Fig. 6. MST of the Facebook Hadoop workload

In Fig. 7, we display the mean slowdown obtained with the policies under evaluation. Similar to Fig. 5, we have not included the mean slowdown of SERPT since it is several orders of magnitude higher. We observe that for $\sigma \leq 0.5$, where the estimates are better, the SEH policy has lower mean slowdown than the Gittins' Index and SEPT policies, however, SEPT starts to outperform the Gittins' Index and SEH policies when σ increases, consistent with our observations for synthetic workloads.

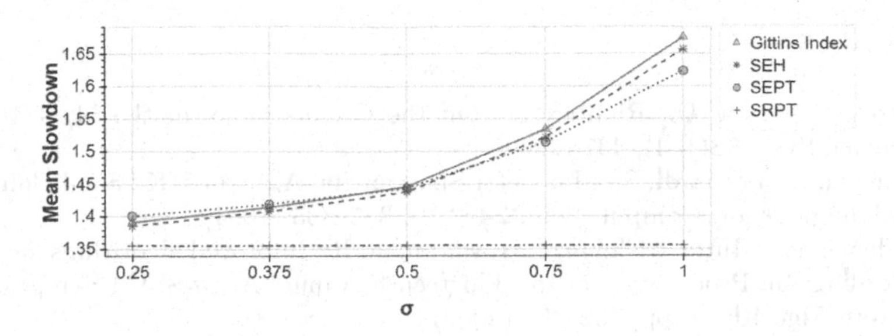

Fig. 7. Mean slowdown of the Facebook Hadoop workload

6 Conclusion and Future Work

The SRPT policy, which is optimal for scheduling in single-server systems, may have problematic performance when job processing times are estimated. This work has considered the problem of scheduling with the presence of job processing time estimates. A multiplicative error model is used to produce estimation errors proportional to the job processing times. We have introduced a novel heuristic that combines the merits of SERPT and SEPT and requires minimal calculation overhead and no information about the job processing time and estimation error distributions. We have shown that this policy is consistent with a Gittins'-like view of the problem. Our numerical results demonstrate that the SEH policy has desirable performance in minimizing both the MST and mean slowdown of the system when there is low variance in the estimation error distribution. It outperforms SERPT except in scenarios where the job processing time variance is extremely low. Examining the SEH policy under other error models as well as analytic bounds as to how far it is from optimal could be investigated in future work. It would also be useful to examine how well policies designed for worst case performance would perform with respect to the performance metrics considered in this paper. The work of Purohit et al. [20] is an intriguing candidate, as it runs two policies in parallel to provide worst case performance guarantees, even when there are large estimation errors.

Not much work has been done in the area of multi-server scheduling in the presence of estimation errors. One major reason is that determining optimal policies for multi-server queues is much more challenging compared to the single-server case. Mailach and Down [17] suggest that when SRPT is used in a multi-server system, the estimation error affects the system's performance to a lesser degree than in a single-server system. Grosof et al. [11] prove that multi-server SRPT is asymptotically optimal when an $M/G/k$ system is heavily loaded. Our work only evaluates the performance of SEH in a single-server framework so we leave the extension and evaluation of this policy in multi-server queues for future investigation.

Acknowledgment. The authors would like to thank Ziv Scully for useful discussions on the limitations of the SEH policy.

References

1. Aalto, S., Ayesta, U., Righter, R.: On the Gittins index in the M/G/1 queue. Queueing Syst. **63**(1–4), 437 (2009)
2. Becchetti, L., Leonardi, S., Marchetti-Spaccamela, A., Pruhs, K.: Semi-clairvoyant scheduling. Theor. Comput. Sci. **324**(2–3), 325–335 (2004)
3. Bender, M.A., Muthukrishnan, S., Rajaraman, R.: Improved algorithms for stretch scheduling. In: Proceedings of the Thirteenth Annual ACM-SIAM Symposium on Discrete Algorithms, pp. 762–771 (2002)
4. Chang, H., Kodialam, M., Kompella, R.R., Lakshman, T., Lee, M., Mukherjee, S.: Scheduling in mapreduce-like systems for fast completion time. In: 2011 Proceedings IEEE INFOCOM, pp. 3074–3082. IEEE (2011)
5. Chen, Y., Alspaugh, S., Katz, R.: Interactive analytical processing in big data systems: A cross-industry study of mapreduce workloads. arXiv preprint arXiv:1208.4174 (2012)
6. Coffman, E.G., Denning, P.J.: Operating Systems Theory, vol. 973. prentice-Hall Englewood Cliffs, Hoboken (1973)
7. Crovella, M.E., Taqqu, M.S., Bestavros, A.: Heavy-tailed probability distributions in the World Wide Web. Pract. Guide Heavy Tils **1**, 3–26 (1998)
8. Dell'Amico, M.: Scheduling with inexact job sizes: The merits of shortest processing time first. arXiv preprint arXiv:1907.04824 (2019)
9. Dell'Amico, M., Carra, D., Michiardi, P.: PSBS: practical size-based scheduling. IEEE Trans. Comput. **65**(7), 2199–2212 (2015)
10. Gittins, J.C.: Bandit processes and dynamic allocation indices. J. Roy. Stat. Soc. Ser. B (Methodol.) **41**(2), 148–164 (1979)
11. Grosof, I., Scully, Z., Harchol-Balter, M.: SRPT for multiserver systems. Perform. Eval. **127**, 154–175 (2018)
12. Harchol-Balter, M.: The effect of heavy-tailed job size distributions on computer system design. In: Proceedings of the ASA-IMS Conference on Applications of Heavy Tailed Distributions in Economics, Engineering and Statistics (1999)
13. Harchol-Balter, M., Schroeder, B., Bansal, N., Agrawal, M.: Size-based scheduling to improve web performance. ACM Trans. Comput. Syst. (TOCS) **21**(2), 207–233 (2003)
14. Kleinrock, L.: Queueing Systems: vol. 1, Theory (1975)
15. Lu, D., Sheng, H., Dinda, P.: Size-based scheduling policies with inaccurate scheduling information. In: The IEEE Computer Society's 12th Annual International Symposium on Modeling, Analysis, and Simulation of Computer and Telecommunications Systems, 2004. (MASCOTS 2004). Proceedings, pp. 31–38. IEEE (2004)
16. Mailach, R.: Robustness to estimation errors for size-aware scheduling. Ph.D. thesis, McMaster University, Department of Computing and Software, Canada (2017)
17. Mailach, R., Down, D.G.: Scheduling jobs with estimation errors for multi-server systems. In: 2017 29th International Teletraffic Congress (ITC 29), vol. 1, pp. 10–18. IEEE (2017)
18. Mitzenmacher, M.: Scheduling with predictions and the price of misprediction. arXiv preprint arXiv:1902.00732 (2019)
19. Pastorelli, M., Barbuzzi, A., Carra, D., Dell'Amico, M., Michiardi, P.: HFSP: size-based scheduling for Hadoop. In: 2013 IEEE International Conference on Big Data, pp. 51–59. IEEE (2013)

20. Purohit, M., Svitkina, Z., Kumar, R.: Improving online algorithms via ML predictions. In: Advances in Neural Information Processing Systems, pp. 9661–9670 (2018)
21. Rai, I.A., Urvoy-Keller, G., Biersack, E.W.: Analysis of LAS scheduling for job size distributions with high variance. In: Proceedings of the 2003 ACM SIGMETRICS international conference on Measurement and modeling of computer systems, pp. 218–228 (2003)
22. Righter, R., Shanthikumar, J.G.: Scheduling multiclass single server queueing systems to stochastically maximize the number of successful departures. Probab. Eng. Inf. Sci. **3**(3), 323–333 (1989)
23. Schrage, L.: Letter to the editor-a proof of the optimality of the shortest remaining processing time discipline. Oper. Res. **16**(3), 687–690 (1968)
24. Schrage, L.E., Miller, L.W.: The queue M/G/1 with the shortest remaining processing time discipline. Oper. Res. **14**(4), 670–684 (1966)
25. Scully, Z., Harchol-Balter, M., Scheller-Wolf, A.: Soap: one clean analysis of all age-based scheduling policies. Proc. ACM Measurement Anal. Comput. Syst. **2**(1), 1–30 (2018)
26. Wierman, A.: Fairness and scheduling in single server queues. Surv. Oper. Res. Manag. Sci. **16**(1), 39–48 (2011)
27. Wierman, A., Nuyens, M.: Scheduling despite inexact job-size information. In: Proceedings of the 2008 ACM SIGMETRICS International Conference on Measurement and Modeling of Computer Systems, pp. 25–36 (2008)
28. Yashkov, S.: Processor-sharing queues: some progress in analysis. Queueing Syst. **2**(1), 1–17 (1987)

An Approximate Bribe Queueing Model for Bid Advising in Cloud Spot Markets

Bogdan Ghiţ[1](✉) and Asser Tantawi[2]

[1] Databricks Inc., Amsterdam, The Netherlands
bogdan.ghit@databricks.com
[2] IBM T.J. Watson Research Center, Yorktown Heights, NY 10598, USA
tantawi@us.ibm.com

Abstract. We consider the scheduling system of a container cloud spot market where the user specifies the requested number of containers and their resource requirements, along with a bid value. Jobs are preemptively ordered based on their bid values as the available capacity, which is excess capacity made available for the spot market, may vary over time. Due to this variation, the number of allocated containers to a job may vary during its lifetime, resulting in users experiencing periods of degraded performance, potentially leading to job slowdown. We want to model and analyze such a scheduling system starting from first principles, inspired by the M/M/1 bribe queue. Thus, we introduce a simple, empirical queueing model which parametrically relates job slowdown to bid values given load and bid distribution. We demonstrate the accuracy of our approximation and parameter estimation through simulation.

1 Introduction

Cloud providers make excess resource capacity available at discounted prices through a so called spot market [5,10,17]. The unit of sale is typically a Virtual Machine (VM) instance, but variations may also include containers or collection of VM instances in case of a container or batch service, respectively. Users submit their bids for such resource units which have a time-varying price per unit that is controlled by the provider. When the price goes above the user bid, the user loses the corresponding resources. Such a market is attractive to users because spot instance have relatively low prices. However, a major drawback of the spot market is that users need to deal with potential unit revocations [1], which are difficult to anticipate. For both the service provider and users, there is a crucial need for a prediction tool to provide (1) revenue estimates as a function of price and (2) quality of service as a function of bid, respectively.

We consider an enhanced management of excess resource capacity through a scheduling system, where a user specifies a bid value, which acts as a priority level. As the available capacity shrinks or higher priority jobs are submitted, the

This work was done while B. Ghit was an intern at the IBM T.J. Watson Research Center.

A. Abate and A. Marin (Eds.): QEST 2021, LNCS 12846, pp. 186–194, 2021.
https://doi.org/10.1007/978-3-030-85172-9_10

scheduler reclaims resources from lower priority running jobs, by preempting them and putting them back in the queue. From a modeling point of view, we consider a preemptive priority scheduler of jobs using containers. At job submission time, the user specifies the requested number of containers and their resource requirements, along with a bid value. The number of allocated containers may vary during the lifetime of a job, anywhere from zero to the requested number. The tasks of a job are managed by a task scheduler and run on the allocated containers. The job continues to execute, with potential degraded performance, as the number of allocated containers varies. The deallocation of a container causes the currently running task(s) on the container to be aborted. If the number allocated goes down to zero, the job is put back in the queue.

We seek to obtain a simple and empirical expression for the job slowdown as a function of bid value. To this end, we propose a parametric approximate expression inspired by the M/M/1 *bribing queue* [9]. We are also concerned with the dynamic estimation of parameters in order to adapt the queueing model and provide accurate performance predictions in the face of time-varying workloads. We achieve this by employing an extended Kalman filter [12] on the slowdown and bid values measured over a period of time. We validate the accuracy of our approximation through simulation experiments.

The main contributions of this paper are: (1) a simple, empirical, parametric closed-form approximate expression for the job slowdown as a function of bid value for scheduling jobs in a container cloud spot market, and (2) a methodology, based on filtering techniques, for dynamically estimating the model parameters at runtime based on measurements.

2 Problem Description

We consider a container cloud spot market, enhanced with a job scheduler, providing differentiated performance based on bids. As the market price fluctuates, a job may wait in the queue and/or be preempted during its life time. Partial preemption is possible if only a fraction of the containers that a job needs are de-allocated. This is in contrast with a typical spot market (without queueing and only for single instances (containers)), where jobs are either rejected or completely aborted.

We assume a parallel job model such as data analytics applications in a real environment with multiple resources. A job is decomposed into multiple tasks that run on units of the available capacity called *compute slots*. Without loss of generality, we refer to a *compute slot* as a container, which is the unit of allocation. Typically, a worker runs in a container and is responsible for the execution of the tasks assigned to it by the task scheduler. We use the terms worker and container interchangeably throughout the paper. In this paper we are *not* concerned with the task graph of the job, nor the scheduling of individual tasks. Rather, we remain at the job level and consider the job, as a whole, and its allocated workers. We assume that running workers are always busy executing tasks. Because the number of tasks is typically much larger than the allocated

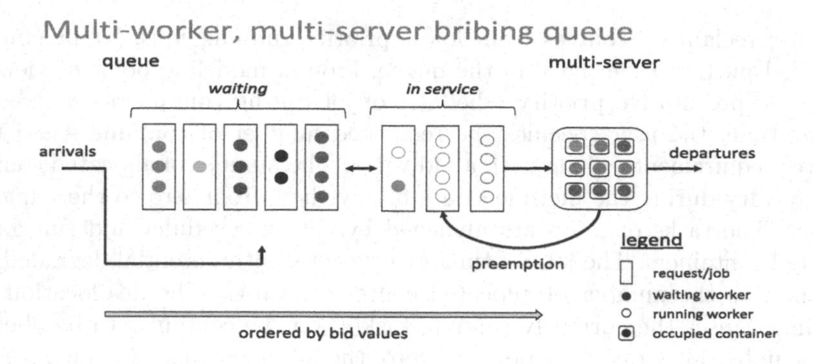

Fig. 1. An overview of our queueing system.

containers, jobs are often multi-waved, thereby running only a fraction of their tasks at a time.

Figure 1 depicts the scheduling of jobs in this environment. Jobs arrive independently according to some stochastic process. A job comes with a bid value and the jobs are ordered based on their bid values, so that lower bid values are in the back of the queue. A job requests some number of containers to run its tasks. Jobs and containers are drawn as rectangles and circles within the rectangles, respectively. The system has a certain container capacity, represented as squares. A circle within a square represents a worker assigned to a container. Jobs wait in the queue until they are allocated at least one container. At such a time they move to the *in service* area. They remain in service until they complete and depart from the system. While *in service*, the number of allocated containers may grow and shrink, depending on container availability and preemption. Allocated and unallocated containers are drawn as hollow and solid circles, respectively. Preempted tasks may need to be re-executed. If all the containers of a job are deallocated, the job goes back to the waiting queue.

A job could be in one of three states, as far as its container allocation is concerned: waiting, executing, or partially executing. A job is in a waiting state if all its requested containers are not allocated yet (all solid circles). Alternatively, a job is in an executing state if all of its requested containers are allocated (all hollow circles). And, a partially executing job is one with a non-zero (and non-one) fraction of its requested containers are allocated. Such a job is progressing with a degraded performance. Thus, jobs in the *in service* area are either executing or partially executing. In general, containers may not be the same size in terms of their allocated resources (CPU and memory). However, in the case of homogeneous containers, we may only have at most one job in the partially executing state.

3 Modeling and Analysis

We seek a functional relationship between job slowdown and bid. In addition to the bid value, the job slowdown depends on many factors such as the current load in the system, the resource requirements of the job, the number of containers allocated to run the job, and the bid values of other jobs. To assess the relationship between the job slowdown and the bid value, given such factors, we need to develop an analytic model which can be used to predict either the slowdown given a bid value, or the bid value which would result in a given slowdown. Both predictions could be valuable to users to set the expectation for job performance and to advise setting a bid value, respectively.

3.1 Definitions and Assumptions

Let job arrivals constitute a Poisson process with rate λ and let X be the random variable representing the job bid value. Without loss of generality, we assume that the bid value is in the set $\mathcal{X} = [0, 1]$. The probability distribution function of X is denoted by $B(x) = Pr[X \leq x], x \in \mathcal{X}$ which is continuous and differentiable.

A job leaves the system after completing all of its work, which we denote by W. Jobs request homogeneous workers that have the same amount of resources. We further assume that it is always possible to divide the remaining work among the running workers, with no running (allocated) worker staying idle. The service time R of a job is the duration over which the running workers execute all the work needed. Thus, if a job is allocated all requested workers the service time is given by $R = W/K$, where K denotes the number of workers. This analysis is for the preemptive-resume case.[1] Let $\mu = 1/\overline{R}$ be the service rate, where $\overline{R}$ is the average job service time. We assume homogeneous slots, i.e. equal amount of resources per slot, and that one worker fits exactly into one slot. Thus, the offered load is given by $\rho = \lambda\overline{K}/N\mu < 1$, where $\overline{K}$ is the average number of workers per job and N is the system capacity.

We denote the average response time of a job with bribe value x by $T(x)$. We further define the job slowdown as the ratio of the average response time $T(x)$ and the average service time $1/\mu$, denoted by $S(x) = T(x)/(1/\mu)$.[2]

3.2 Bribery Queueing Model

The simplest case for our queueing system is when K is fixed at $K = 1$, W is exponentially distributed, and $N = 1$, resulting in the M/M/1 bribing[3] queue [9]. For such a model, the slowdown of a job with bribe value x is given by:

$$S(x) = \frac{1}{(1 - \rho(1 - B(x)))^2}. \tag{1}$$

[1] Note that in the case of exponential service time, the preemptive-repeat and preemptive-resume cases result in similar expressions for the average response time.

[2] Note that we define the slowdown as the ratio of two average values, and not the average of a ratio of two values. The latter alternative definition would have (1) resulted in a more complex derivation and conditional expression on the service time and, more importantly, (2) necessitated *a priori* knowledge of job service time, which may not be available in practice.

[3] We will use the words bribe and bid interchangeably throughout this paper.

Let S be the random variable representing the slowdown across all jobs. The bribe value which yields a given slowdown of $s \in S$ is obtained by inverting Eq. 1, as

$$x(s) = B^{-1}\left(1 - \frac{1 - 1/\sqrt{s}}{\rho}\right). \tag{2}$$

Our queueing system, described in Sect. 2, along with other *modern* job models in data centers and clouds, are quite challenging to analyze [6]. Though simplistic and limiting, the single-server bribing queue has an appealingly concise expression. We seek an approximate expression for our generalized model by introducing a parameter vector, Θ, consisting of two model parameters, $\Theta = [\theta_0, \theta_1]$, which act as scale and shape parameters, respectively, such that $0 \leq \theta_0 < 1$ and $\theta_1 > 0$. First, θ_0 acts as a (virtual) replacement for the server utilization, ρ, which may not be available to an external observer. Second, θ_1 captures the variation in the expression due to the model features described previously. Hence, we extend Eq. 1 and write a closed-form approximate expression for the job slowdown as a function of B, the bid distribution, and Θ, the parameter vector, in addition to x, the bid value, as

$$S(x; B, \Theta) = \frac{1}{[1 - \theta_0(1 - B(x))^{\theta_1}]^2}. \tag{3}$$

The bid which results in a given job slowdown may be obtained by inverting the above equation, similar to Eq. 2.

3.3 Parameter Estimation

This section addresses the issue of dynamically estimating the parameters in our model of the scheduling system. The model has bid values as input and corresponding slowdown values as output. There are two sets of parameters: (1) bidding parameters which characterize the bid distribution, $B(x)$, and (2) model parameters which characterize the queueing model, i.e. the relationship between a bid value x and its corresponding slowdown value $S(x)$. In practice, one is not given such a distribution or parameter values. Thus, we need to have a dynamic estimator which derives them dynamically, based on observations of the job bid sequence $\{x_0, x_1, \cdots, x_i\}$ and corresponding attained slowdown sequence $\{s_0, s_1, \cdots, s_i\}$, $i \gg 0$. Based on such sequences, the estimator builds a model and keeps updating the two sets of parameters dynamically. The model produced by the estimator may be used to predict a slowdown $\tilde{S}(x)$ for a given bid value x, or a bid value $\tilde{x}$ for a desired slowdown value s.

Our design for such a dynamic estimator is depicted in Fig. 2. We separate the estimation process into two independent processes: one for estimating the bidding parameters and another for estimating the model parameters. Firstly, we use the job bid sequence $\{x_0, x_1, \cdots, x_i\}$ to derive a bid distribution. Then, using the latter, along with the attained slowdown sequence $\{s_0, s_1, \cdots, s_i\}$, we use a filter to estimate the model parameters dynamically. Both the bid distribution and parameter values could then be used for prediction.

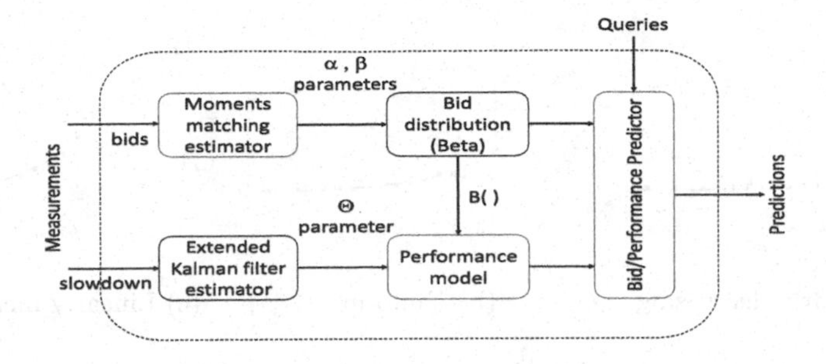

Fig. 2. Our framework for dynamic estimation and prediction.

We select a probability distribution for the bid distribution from the first and second moments of the distribution. In general, given the first few moments of a probability distribution over a finite range, the maximum-entropy distribution may be obtained using Lagrange multipliers [3], which may be approximated by the Beta distribution [13]. We characterize the bid distribution with two parameters: α and β, associated with the first and second moments of X. Let $\tilde{r}$ and $\tilde{s}^2$ be the sample average and sample variance of the observed bid values over a given time interval. Thus, using the method of moments [4], we can estimate the parameters $\tilde{\alpha}$ and $\tilde{\beta}$, as $\tilde{\alpha} = \tilde{r}\left(\frac{\tilde{r}(1-\tilde{r})}{\tilde{s}^2} - 1\right)$ and $\tilde{\beta} = (1-\tilde{r})\left(\frac{\tilde{r}(1-\tilde{r})}{\tilde{s}^2} - 1\right)$, respectively.

As changes occur in the system, such as the nature of workload, load intensity, and cluster configuration, the model parameters change accordingly. Hence, we need a method by which an estimate of the parameter vector $\tilde{\Theta}$ is obtained. In particular, we employ a system where the state vector corresponds to the parameter vector, $\tilde{\Theta}$, the observation corresponds to the measured slowdown, and the system environment includes the bid value. The system transfer function is given by the model functional expression which relates the input to the output, which is non-linear. We employ an extended Kalman filter technique [19,20] to linearize the transfer function by taking first derivatives.

We set the state evolution matrix $\mathbf{F}$ to the identity matrix and the (evolution) covariance matrix $\mathbf{Q}$ to a fraction f of the squared values of the initial state variables. For the (system) covariance matrix $\mathbf{R}$, we use an approximation based on the 95% confidence interval of the t-distribution divided by a factor γ which is a fraction of the actual measurement window in relation to one which yields steady state measurements. We set $f = 5\%$ and $\gamma = 0.5$.

4 Simulations

To validate the job slowdown approximation given by Eq. 3, we simulate a system with 12 slots over 4,000 s in steady state. Jobs arrive as a Poisson process with an average rate of 3.2 jobs/s. The job service time is Gamma distributed with average 1 s and variance 0.5 s^2. The requested number of slots per job is uniformly distributed between 1 and 5 slots. This constitute an offered load of 80%. As

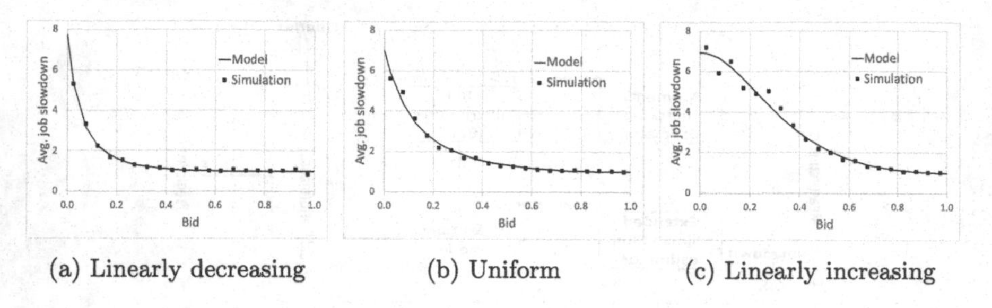

(a) Linearly decreasing (b) Uniform (c) Linearly increasing

Fig. 3. The average job slowdown versus the bid value with the prediction model and simulations for different bid distributions.

(a) Linearly decreasing (b) Uniform (c) Linearly increasing

Fig. 4. Estimates of bid distribution parameters for different bid distributions.

for the bid distribution, we consider three cases: linearly decreasing,[4] uniform, and linearly increasing. The density functions $b(x)$ are $2(1-x)$, 1, and $2x$, respectively, $x \in [0,1]$. And, the distribution functions $B(x)$ are $x(2-x)$, x, and x^2, respectively.

We divide the bid range into 20 bins, each with a width of 0.05, and we calculate the average job slowdown for each such bin. The data points are used in a regression analysis of Eq. 3, solving for $[\tilde{\theta}_0, \tilde{\theta}_1]$ that minimizes the mean squared error (mse) between the model and simulation values of the average slowdown. Figure 3 depicts the average slowdown as a function of bid using the model and simulation for three bid distributions. We observe that our model anticipates with high accuracy the simulation results for the entire range of bid values, irrespective of the shape of the bid distribution. The estimates for the three bid distributions were $\tilde{\theta}_0 = 0.64$ and $\tilde{\theta}_1 = 2.44$, with $mse = 3.47 * 10^{-03}$, $\tilde{\theta}_0 = 0.62$ and $\tilde{\theta}_1 = 2.26$, with $mse = 2.56 * 10^{-02}$ and $\tilde{\theta}_0 = 0.62$ and $\tilde{\theta}_1 = 2.27$, with $mse = 3.21 * 10^{-02}$, respectively.

Figure 4 shows the estimates of the three bid distributions. Because our estimator uses samples of the bid distribution, parameters α and β fluctuate around their values, governed by $\alpha/\beta = 0.5$, 1, and 2, respectively.

[4] In practice, users may favor bidding low.

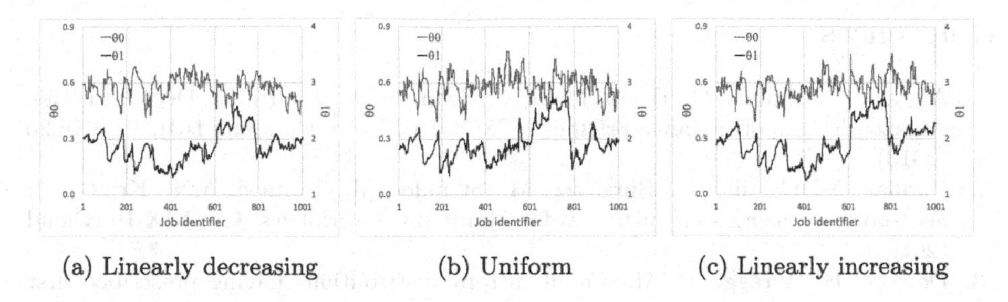

(a) Linearly decreasing (b) Uniform (c) Linearly increasing

Fig. 5. Estimates of model parameters for different bid distributions.

Figure 5 shows the model parameters over time for the three bid distributions. We observe that $\tilde{\theta}_1$ has a catalyzing effect and drops to lower values when $\tilde{\theta}_0$ is overestimated thus adjusting the prediction model.

5 Related Work

Since Amazon EC2 released its spot markets in 2009, a sizable body of research analyzed the operation of such systems in the cloud. The characterization and prediction of spot prices of the AWS spot markets [1] inspired the design of user bidding strategies that optimize cost while also achieving uninterrupted service. Such strategies can be derived either by means of statistical analysis of historical spot prices [8,18] or through more advanced modeling techniques such as Markov chains [2,16].

Modeling and predicting the performance of multi-task MapReduce-based applications has been studied in various settings [7,15]. Common approaches build an estimator by choosing a relationship between an output variable that needs to be predicted and several system properties that can be measured and used for prediction. To provide good predictions, the estimator employs machine learning techniques and needs large amounts of training data based on low-level application performance characteristics [11,14].

6 Conclusion

We presented a simple, empirical, parametric approximate expression for the job slowdown as a function of bid value for job scheduling in container cloud spot markets. The approximate expression extends the M/M/1 *bribing queue* using two parameters. Further, we provided a methodology for the dynamic estimation of the parameters using the method of moments matching and extended Kalman filtering, a control-theoretic approach. We validated our approximation and prediction methodology using simulation experiments. We incorporated our approximate model in a spot advisor that is employed to either (1) set an expectation for the performance of a job given a particular bid value or (2) suggest a minimum bid value required to attain a given service level.

References

1. Agmon Ben-Yehuda, O., Ben-Yehuda, M., Schuster, A., Tsafrir, D.: Deconstructing Amazon EC2 spot instance pricing. ACM Trans. Econ. Comput. **1**(3), 16:1–16:20 (2013)
2. Chohan, N., Castillo, C., Spreitzer, M., Steinder, M., Tantawi, A.N., Krintz, C.: See spot run: using spot instances for MapReduce workflows. USENIX HotCloud (2010)
3. Dowson, D., Wragg, A.: Maximum-entropy distributions having prescribed first and second moments (corresp.). IEEE Trans. Inf. Theor. **19**(5), 689–693 (1973)
4. Forbes, C., Evans, M., Hastings, N., Peacock, B.: Statistical Distributions. Wiley, Hoboken (2011)
5. Ghit, B., Epema, D.: Better safe than sorry: grappling with failures of in-memory data analytics frameworks. In: ACM HPDC (2017)
6. Harchol-Balter, M.: Open problems in queueing theory inspired by datacenter computing. Queueing Syst. **97**(1), 3–37 (2021). https://doi.org/10.1007/s11134-020-09684-6
7. Herodotou, H., et al.: Starfish: a self-tuning system for big data analytics. In: CIDR, vol. 11, no. 2011, pp. 261–272 (2011)
8. Javadi, B., Thulasiramy, R.K., Buyya, R.: Statistical modeling of spot instance prices in public cloud environments. In: 2011 Fourth IEEE International Conference on IEEE Utility and Cloud Computing (UCC), pp. 219–228. IEEE (2011)
9. Kleinrock, L.: Optimum bribing for queue position. Oper. Res. **15**(2), 304–318 (1967)
10. Liu, H.: Cutting MapReduce cost with spot market. HotCloud (2011)
11. Shi, J., Zou, J., Lu, J., Cao, Z., Li, S., Wang, C.: MRTuner: a toolkit to enable holistic optimization for mapreduce jobs. In: VLDB Endowment, vol. 7, no. 13, pp. 1319–1330 (2014)
12. Simon, D.J.: Optimal State Estimation: Kalman, H Infinity, and Nonlinear Approaches. Wiley, Hoboken (2006)
13. Unuvar, M., Doganata, Y., Tantawi, A.: Configuring cloud admission policies under dynamic demand. In: 2013 IEEE 21st International Symposium on Modeling, Analysis Simulation of Computer and Telecommunication Systems (MASCOTS), pp. 313–317, August 2013
14. Venkataraman, S., Yang, Z., Franklin, M.J., Recht, B., Stoica, I.: Ernest: efficient performance prediction for large-scale advanced analytics. In: USENIX NSDI (2016)
15. Verma, A., Cherkasova, L., Campbell, R.H.: ARIA: automatic resource inference and allocation for MapReduce environments. In: ACM ICAC (2011)
16. Zafer, M., Song, Y., Lee, K.-W.: Optimal bids for spot VMs in a cloud for deadline constrained jobs. In: IEEE CLOUD (2012)
17. Zaharia, M., Konwinski, A., Joseph, A.D., Katz, R.H., Stoica, I.: Improving MapReduce performance in heterogeneous environments. In: USENIX OSDI (2008)
18. Zheng, L., Joe-Wong, C., Tan, C.W., Chiang, M., Wang, X.: How to bid the cloud. ACM SIGCOMM Comput. Commun. Rev. **45**(4), 71–84 (2015)
19. Zheng, T., Woodside, M., Litoiu, M.: Performance model estimation and tracking using optimal filters. IEEE Trans. Softw. Eng. **34**(3), 391–406 (2008)
20. Zheng, T., Yang, J., Woodside, M., Litoiu, M., Iszlai, G.: Tracking time-varying parameters in software systems with extended Kalman filters. In: IBM Press Centre for Advanced Studies on Collaborative Research (2005)

Learning and Verification

DSMC Evaluation Stages: Fostering Robust and Safe Behavior in Deep Reinforcement Learning

Timo P. Gros(✉), Daniel Höller, Jörg Hoffmann, Michaela Klauck, Hendrik Meerkamp, and Verena Wolf

Saarland University, Saarland Informatics Campus, Saarbrücken, Germany
{timopgros,hoeller,hoffmann,klauck,meerkamp,wolf}@cs.uni-saarland.de

Abstract. Neural networks (NN) are gaining importance in sequential decision-making. Deep reinforcement learning (DRL), in particular, is extremely successful in learning action policies in complex and dynamic environments. Despite this success however, DRL technology is not without its failures, especially in safety-critical applications: (i) the training objective maximizes *average* rewards, which may disregard rare but critical situations and hence lack local robustness; (ii) optimization objectives targeting safety typically yield degenerated reward structures which for DRL to work must be replaced with proxy objectives. Here we introduce methodology that can help to address both deficiencies. We incorporate *evaluation stages* (ES) into DRL, leveraging recent work on deep statistical model checking (DSMC) which verifies NN policies in MDPs. Our ES apply DSMC at regular intervals to determine state space regions with weak performance. We adapt the subsequent DRL training priorities based on the outcome, (i) focusing DRL on critical situations, and (ii) allowing to foster arbitrary objectives. We run case studies in Racetrack, an abstraction of autonomous driving that requires navigating a map without crashing into a wall. Our results show that DSMC-based ES can significantly improve both (i) and (ii).

1 Introduction

In recent years, neural networks (NN), especially deep neural networks, have accomplished major successes across many computer science domains, like image classification [25], natural language processing [21], and game-playing [41]. The latter was especially accomplished by combining reinforcement learning (RL) and deep neural networks, so called *deep reinforcement learning* (DRL). DRL was used successfully for sequential decision-making, e.g., mastering Atari games [28,29], playing the games Go and Chess [40–42], or solving the Rubik's cube [1],

---

Authors are listed alphabetically. This work was partially supported by the German Research Foundation (DFG) under grant No. 389792660, as part of TRR 248, see https://perspicuous-computing.science, and by the European Regional Development Fund (ERDF).

The original version of this chapter was revised: an error in the algorithm on page 206 were corrected. The correction to this chapter is available at https://doi.org/10.1007/978-3-030-85172-9_25

A. Abate and A. Marin (Eds.): QEST 2021, LNCS 12846, pp. 197–216, 2021.
https://doi.org/10.1007/978-3-030-85172-9_11

and is beginning to be used in real-world (motivated) examples, such as vehicle routing [30], robotics [17], and autonomous driving [36].

Despite this success however, DRL technology is not without fail, especially in safety-critical applications. While neural network action policies achieve good performance in many sequential decision-making processes, that performance pertains to *average* rewards as optimized by DRL training. That objective however may average out poor local behavior, and thus disregard rare but critical situations (e.g. a child running in front of a car). In other words, we do not get system-level guarantees, even in the ideal case where the learned policy is near-optimal with respect to its training objective. We refer to this deficiency as a lack of *local robustness*. Dedicated exploration strategies have been developed to ensure inclusion of rare experiences during training [11,12]. These focus on reducing the variance of the accumulated reward, e.g. by importance sampling, but they are not flexible enough to enforce desirable behavior robustly across the whole state space.

This problem is exacerbated by the fact that optimization objectives specifically targeting safety typically yield degenerated reward structures. This is true in particular for the natural objective to maximize the probability of reaching a goal condition – without getting stuck in an unsafe (terminal) state. That objective yields reward 1 in goal states and 0 elsewhere, an extremely sparse reward structure not suited for (D)RL training in large state spaces (a widely known fact, see e.g. [3,18,24,35,38]). Hence, for (D)RL training to be able to identify a useful policy, proxy objectives are used, such as discounted cumulative reward giving positive feedback for goal states and (highly) negative feedback for unsafe states.[1]

In summary, two deficiencies of current DRL methods in safety-critical systems are that (i) training for average reward lacks local robustness, and (ii) safety objectives like goal probability cannot be used for effective training. Let us illustrate these points in an example, taken from the Racetrack benchmark which we will also use in our case studies later on. Racetrack is a commonly used benchmark for Markov decision process (MDP) solution algorithms in AI [6,8,33,44]. The challenge is to navigate to a goal line on a discrete map without crashing into a wall, where actions accelerate/decelerate the car. Racetrack is thus a simple (but highly extensible [5]) abstraction of autonomous driving. Consider Fig. 1, which measures the performance of a NN policy trained with DRL, using a discounted reward structure with +100 reward for goal states and −50 reward for crashes.

Figure 1 (a) evaluates policy performance according to the reward structure it is trained on; whereas Fig. 1 (b) evaluates goal probability, which is the objective we ideally want to optimize. Both heat maps visualize performance when

[1] One can combine such a proxy with the goal probability objective, though multiple objectives are difficult to achieve with a one-dimensional reward signal and standard backpropagation algorithms for neural nets [26]; anyway, training objective vs. ideal objective are still not identical here. Reward shaping is an alternative option that can in principle preserve the optimal policy [31], but this is not always possible, and manual work is needed for individual learning tasks (substantial work sometimes, see e.g. [46]).

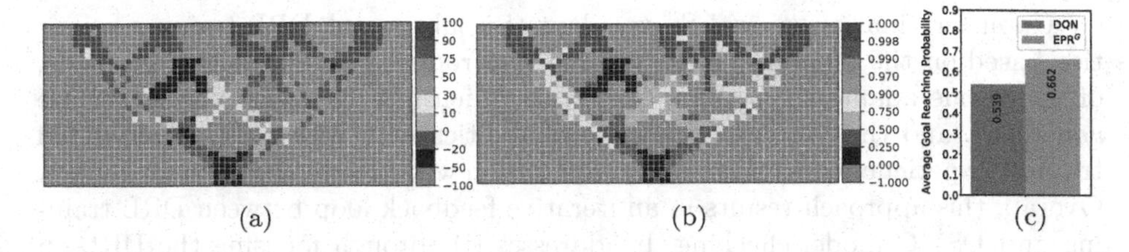

Fig. 1. Example performance measures of a DRL policy on a Racetrack example map. (Color figure online)

starting the policy from each map cell. We clearly see deficiency (i) from the high variance in colors, in particular black and red areas with (very) low expected reward (a) /goal probability (b). Regarding deficiency (ii), while expected reward correlates with goal probability, crashes are "more tolerable" in the reward structure than for goal probability (if we set high negative rewards for crashes then the policy learns to drive in circles). This is difficult to see in the heat maps as the reward scale in (a) cannot be directly compared to the probability scale in (b). Figure 1 (c) hence complements this picture by average goal probability in the critical areas of the map, as achieved by the standard DRL method *deep Q-network*, vs. EPR^G which is one of the new methods we introduce here. EPR^G takes goal probability into account directly, which clearly pays off.

We address deficiencies (i) and (ii) through incorporating *evaluation stages (ES)* into DRL, conducted at regular intervals during training (i.e., periodically after a given number of training episodes) to determine state space regions with weak performance. The "performance" evaluation here is flexible, and can be done either (i) with respect to the training objective, or (ii) with respect to the true objective (for example: goal probability in EPR^G above) in case a proxy objective is used for training.

To design such flexible ES, we leverage recent work on *deep statistical model checking* (DSMC) [15], an approach to explicitly examine properties of NN action policies. The approach assumes that the NN policy resolves the nondeterminism in an MDP, resulting in a Markov chain which is analyzed by statistical model checking [10]. This provides flexible methodology for evaluating policy performance starting from individual states.

The target of an evaluation stage being to identify "weak regions", the question arises which individual states to apply DSMC to. Our answer to this question, at present, is based on the assumption that the possible initial states for the problem at hand (the states from which policy execution may start) can be partitioned into a feasibly small set of state-space regions. In Racetrack, regions are identified by the location of the car. The approach we propose is to sample a single representative state s from each region, and evaluate s through DSMC.[2]

[2] The benefit of our proposed ES thus hinges, in particular, on how meaningful these representative states are for policy performance. While this is a limitation, partitioning by physical location like in Racetrack could be a canonical candidate in many scenarios.

Upon termination of an ES, we adapt the subsequent DRL training priorities based on the outcome. Specifically, we introduce two alternative methods, of which one adapts the probabilities with which new training experiences are generated, and the other adapts the probabilities with which the accumulated training experiences are taken into consideration within individual learning steps. Overall, this approach results in an iterative feedback loop between DRL training and DSMC model checking. It addresses (i) through focusing the DRL on critical situations, and addresses (ii) as DSMC can evaluate arbitrary temporal properties.

We implement this approach on top of deep Q-learning [6,29], and we run experiments in case studies from Racetrack. The results show that DSMC-based ES can indeed (i) make policy reward more robust across complex maps, and (ii) improve goal probability when using a discounted-reward proxy for DRL training.

In summary, our contributions are as follows:

- We introduce evaluation stages as an idea to improve local robustness and goal-probability training in DRL.
- We design and implement two variants of this approach, adapting a state-of-the-art DRL algorithm.
- We evaluate the approach in Racetrack and show that it can indeed have beneficial effects regarding deficiencies (i) and (ii).

Related Work: Recent work of Hasenbeig et al. [20] proposes a method to include a property encoded as an LTL formula and to synthesize policies that maximize the probability of that LTL property. However, while this allows to specify complex tasks, it addresses neither of our two deficiencies.

Further, our work relates to the area of *safe reinforcement learning.* Several works investigate the usage of *shields* [2,4,22] or *permissive schedulers* [23] to restrict the agent from entering unsafe states, even during training. However, these approaches can only be applied if a shield/permissive scheduler was computed beforehand, which is a model-based task. In contrast, our approach is model-free; it does not need to compute a shield or permissive scheduler beforehand, and does not restrict the action (and thus also state-) space. Instead, the task is learned entirely through self-play and Monte Carlo-based evaluation runs. Moreover, our approach is also applicable in more general scenarios, when there are not just safe and unsafe states but more fine-grained state distinctions.

2 Background

We briefly introduce the necessary background on Markov decision processes, deep Q-learning, and deep statistical model checking.

2.1 Markov Decision Processes

The underlying model of both DSMC and DRL is that of a (state-discrete) Markov decision process in discrete time. Let $\mathcal{D}(S)$ denote the set of probability distributions over S for any non-empty set S.

Definition 1 (Markov Decision Process). *A Markov decision process (MDP) is a tuple $\mathcal{M} = \langle \mathcal{S}, \mathcal{A}, \mathcal{T}, \mu \rangle$ consisting of a finite set of* states $\mathcal{S}$, *a finite set of* actions $\mathcal{A}$, *a partial* transition probability function $\mathcal{T} : \mathcal{S} \times \mathcal{A} \rightharpoonup \mathcal{D}(\mathcal{S})$, *and an* initial distribution $\mu \in \mathcal{D}(\mathcal{S})$. *We say that an action $a \in \mathcal{A}$ is* applicable *in state $s \in \mathcal{S}$ if $\mathcal{T}(s, a)$ is defined. We denote by $\mathcal{A}(s) \subseteq \mathcal{A}$ the set of actions applicable in s.*

MDPs are typically associated with a *reward* structure r, specifying numerical rewards that are obtained when following a transition, i.e., $r : \mathcal{S} \times A \times \mathcal{S} \rightarrow \mathbb{R}$. In the following, we call the support of μ the set of initial states I, i.e., $I = \{s \in \mathcal{S} \mid \mu(s) > 0\}$.

Usually, an MDP's behavior is considered jointly with an entity resolving the otherwise non-deterministic choices in a state. Given a state, a so-called *action policy* (or scheduler, or adversary) determines which of the applicable actions to apply.

Definition 2 (Action Policy). *A (history-independent)* action policy *is a function $\pi : \mathcal{S} \times \mathcal{A} \rightarrow [0, 1]$ such that $\pi(s, \cdot)$ is a probability distribution on $\mathcal{A}$ and, for all $s \in \mathcal{S}$, $\pi(s, a) > 0$ implies that $a \in \mathcal{A}(s)$.*

We remark that history-independent action policies are often also called *memoryless* because their decisions depend only on the given state and not on the history of formerly visited states. We call an action policy *deterministic* if in each state s, π selects an action with probability one. We then simply write $\pi(s)$ for the corresponding action.

In the sequel, for a given MDP $\mathcal{M}$ and action policy π, we will write $S_0, S_1, S_2, \ldots$ for the states visited at times $t = 0, 1, 2, \ldots$. Let A_t be the action selected by policy π in state S_t and $R_{t+1} = r(S_t, A_t, S_{t+1})$ the reward obtained when transitioning from S_t to S_{t+1} with action A_t. Note that – as we are dealing with finite-state MDPs – the probability measure associated with these random variables is well defined and $\{S_t\}_{t \in \mathbb{N}_0}$ is a Markov chain with state space $\mathcal{S}$ induced by policy π. For further details we refer to Puterman [34].

The induced Markov chain can be analyzed using statistical model checking [39,47]. For statistical model checking of MDPs, different approaches have been proposed to handle nondeterminism [7,10].

2.2 Deep Q-learning

In the following, let

$$G_t = \sum_{k=t+1}^{T} \gamma^{k-t-1} R_k \tag{1}$$

denote the discounted, accumulated reward, also called *return*, from time t on, where $\gamma \in [0, 1]$ is a discount factor, and T is the final time step [44]. The discount factor determines the importance between short and long term rewards; if $\gamma = 0$, the discounted return will be equal to the reward accumulated in one step only, if $\gamma = 1$ all future rewards will be worth the same, and if $\gamma \in (0, 1)$ the long term rewards will be less important than the short term ones.

Q-learning is a well known algorithm to approximate action policies that maximize said accumulated reward [6]. For a fixed policy π, the so-called *action-value* or *q-value* $q_\pi(s, a)$ at time t is defined as the expected return G_t that is achieved by taking an action $a \in A(s)$ in state s and following the policy π afterwards, i.e.,

$$q_\pi(s,a) = \mathbb{E}_\pi\left[G_t \middle| S_t = s, A_t = a\right] = \mathbb{E}_\pi\left[\sum_{k=0}^{\infty} \gamma^k R_{t+k+1} \middle| S_t = s, A_t = a\right]. \quad (2)$$

Policy π is optimal, if it maximizes the expected return. We write $q_*(s, a)$ for the corresponding *optimal action-value*. Intuitively, the optimal action-value $q_*(s, a)$ is equal to the expected sum of the reward that we receive when taking action a from state s, and the (discounted) highest optimal action-value that we receive afterwards. For optimal π, the Bellman optimality equation [44] gives

$$q_*(s,a) = \mathbb{E}_\pi\left[R_{t+1} + \gamma \cdot \max_{a'} q_*\left(S_{t+1}, a'\right) \middle| S_t = s, A_t = a\right]. \quad (3)$$

Vice versa, one can evidently obtain the optimal policy if the optimal action values are known by selecting $\pi(s) = \text{argmax}_{a \in A(s)}\, q_*(s, a)$.

By estimating the optimal q-values, one can obtain (an approximation of) an optimal policy. During tabular *Q-learning*, the action values are approximated separately for each state-action pair [6]. In the case of large state spaces, *deep Q-learning* can be used to replace the Q-table by a *neural network* (NN) as a function approximator [29]. NNs can learn low-dimensional feature representations and express complex non-linear relationships. Deep reinforcement learning is based on training deep neural networks to approximate optimal policies. Here, we consider a neural network with weights θ estimating the Q-value function as a *deep Q-network (DQN)* [28]. We denote this Q-value approximation by $Q_\theta(s, a)$ and optimize the network w.r.t. the target

$$y_\theta(s,a) = \mathbb{E}_\theta\left[R_{t+1} + \gamma \cdot \max_{a'} Q_\theta(S_{t+1}, a') \mid S_t = s, A_t = a\right], \quad (4)$$

where the expectation is taken over trajectories induced by the policy represented by the parameters θ. The corresponding loss function in iteration i of the learning process is

$$L(\theta_i) = \mathbb{E}_{\theta_i}\left[\left(y_{\theta'}(S_t, A_t) - Q_{\theta_i}(S_t, A_t)\right)^2\right]. \quad (5)$$

Here, the so-called *fixed target* means that in Eq. (5) θ' does not depend on the current iteration's weights of the (so called *local*) neural network θ_i but on weights that were stored in earlier iterations (so called *target* network), to avoid an unstable training procedure [29]. We approximate $\nabla L(\theta_i)$ and optimize the loss function by stochastic gradient descent.

In contrast to Mnih et al. [29], we do not update the target network after a fixed number of learning stochastic gradient descend update steps, but perform a soft update instead, i.e., whenever we update the local network in iteration i, the weights of the target network are given by $\theta' = (1 - \tau) \cdot \theta_i + \tau \cdot \theta'$ with $\tau \in (0, 1)$ [13,43].

Stochastic gradient descent assumes independent and identically distributed samples. However, when directly learning from self-play, this assumption is disrupted as the next state depends on the current decision. To mitigate this problem, we do not directly learn from observations, but store them in an *experience replay buffer* [29]. Whenever a learning step is performed, we uniformly sample from this replay buffer to consider (approximately) uncorrelated tuples. Thus, the loss is given by

$$L(\theta_i) = \mathbb{E}_{(s,a,r,s') \sim U(D)} \left[\left(r + \gamma \cdot \max_{a'} Q_{\theta'}(s', a') - Q_{\theta_i}(s, a) \right)^2 \right]. \quad (6)$$

We generate our experience tuples by exploring the state space epsilon-greedily, i.e., during the Monte Carlo simulation we follow the policy that is implied by the current network weights with a chance of $(1 - \epsilon)$ and otherwise choose a random action. We start with a high exploration coefficient $\epsilon = \epsilon_{start}$ and exponentially decay it, i.e., for every iteration set $\epsilon = \epsilon \cdot \epsilon_{decay}$ with $\epsilon_{decay} < 1$, until a certain threshold ϵ_{end} is met. Afterwards, we constantly use $\epsilon = \epsilon_{end}$. Common termination criteria for the learning process are fixing the number of episodes or using a threshold on the expected return achieved by the current policy. The overall algorithm is displayed in Sect. 3 (Algorithm 1), together with the extensions and changes we will introduce.

A common improvement to the DQN algorithm sketched above, which we will also consider in this paper, is the so-called *prioritized replay buffer* [37]. Not all samples are equally useful to improve the policy. In particular, those samples with a relatively small individual loss do not contribute to the learning process as much as those with a high loss. Thus, the idea of prioritized experience replay is to sample from the aforementioned replay buffer with a probability that reflects the loss. Specifically, the priority δ of a sample (s, a, s', r) in iteration i is given by

$$\delta = \left(\left(Q_{\theta'}(s, a) - \left(r + \gamma \cdot \max_{a'} Q_{\theta_i}(s', a') \right) \right) + \epsilon_p \right)^{\alpha}, \quad (7)$$

where ϵ_p is a hyperparameter to ensure that all samples have non-zero probability, and α is used to control the amount of prioritization. $\alpha = 0$ means that there is no prioritization, $\alpha = 1$ means full prioritization, $\alpha \in (0, 1)$ defines a balance. In Eq. (6), instead of sampling uniformly, the probability at which a sample is picked from the buffer is then proportional to its priority, i.e. we divide the samples' priority by the sum of all priorities. In the following, we will abbreviate DQN with such prioritized experience replay as *DQNPR*.

2.3 Deep Statistical Model Checking

Deep Statistical Model Checking. [15] is a method to analyze a NN-represented policy π taking action decisions (resolving the nondeterminism) in an MDP $\mathcal{M}$.

Namely, the induced Markov chain $\mathcal{C}$ is examined by statistical model checking. Given an MDP $\mathcal{M}$, DSMC assumes that the policy π has been trained based on $\mathcal{M}$ completely prior to the analysis without influencing the training process at all. This approach is promising in terms of scalability as the analysis of $\mathcal{C}$ merely requires to evaluate the NN on input states: there is no need for other deeper and more complex NN analyses. Gros et al. [15] implemented this approach for the statistical model checker MODES [10] in the MODEST TOOLSET [19].

3 RL with Evaluation Stages

We now introduce our approach of RL with evaluation stages, addressing the DRL deficiencies discussed in the introduction: (i) training for average reward lacks local robustness; (ii) safety objectives like goal probability cannot be used for effective training. We next discuss a basic design decision, then describe our two alternative methods, and then specify how they are realized on top of deep Q-learning.

3.1 Initial State Partitioning and Notations

Recall that I denotes the initial states of the MDP, i.e., the support of the initial distribution μ. As already mentioned, an important premise of our work is that I can be partitioned into a manageable number of *regions*. We denote that partition by $\mathbb{P} = \{J_1, J_2, \ldots, J_k\}$ where the regions are non-empty $J_i \neq \emptyset$, cover the set of all initial states $\bigcup_{i \in 1,2,\ldots,k} J_i = I$, and are disjoint $J_i \cap J_j = \emptyset$ for $i \neq j$. During the evaluation stages, we consider one representative $s_i \in J_i$ from each region. The underlying assumption is that the representatives are sufficiently meaningful to identify important deficiencies in policy behavior.[3]

The evaluation stages may consider arbitrary optimization objectives in principle, and use arbitrary methods to measure the objective values of the states s_i. Here we compute E using DSMC, measuring expected reward or goal probability. We denote the outcome of evaluation as an *evaluation function*, a function $E : \mathbb{P} \rightarrow [0, 1]$ mapping each region J_i to the evaluation value of its representative state s_i. For optimization objectives that are not probabilities, we assume here a normalization step into the interval $[0, 1]$, with 0 being the worst value and 1 the best. In particular, for expected rewards, the natural method we use in our experiments is to set $E(r^{\min}) = 0$ and $E(r^{\max}) = 1$ and interpolate linearly in between.

We also use the representative states to define an initial probability distribution over the regions J_i:

$$\beta(P_i) = \mu(s_i) / \sum_{j=1}^{k} \mu(s_j) \tag{8}$$

[3] In our Racetrack case studies, we use the map cells as the basis of $\mathbb{P}$ – i.e., states sharing the same physical location. We believe that this partitioning method may work for many application scenarios involving physical space. Alternatively, one may, for example, partition state-variable ranges into intervals.

3.2 Evaluation-Based Initial Distribution (EID)

Given the initial distribution μ of the MDP, with the insights gained through the DSMC evaluation stages, we can adapt the initial distribution to guide the training process after an evaluation stage. Recall that β is the initial distribution of a region in the original MDP. The probability to start in a region J_i for the EID method is then given by

$$p(J_i) = \frac{(1 - E(J_i)) \cdot \beta(J_i)}{\sum_j (1 - E(J_j)) \cdot \beta(J_j)}, \tag{9}$$

i.e., we shift the initial distribution for the regions such that we start with a higher probability in areas with low quality and vice versa. Once region J_i is selected, we uniformly sample a starting state from J_i.

The idea of EID is that by generating experiences from regions with poor behavior, we improve the robustness of the policy as the NN will learn to select the most appropriate actions in these regions.

3.3 Evaluation-Based Prioritized Replay (EPR)

As discussed, the principle of prioritized experience replay buffers is to sample states according to their loss, i.e., we more often sample states where the loss is high and less often where the loss is low (see Eq. (7)). Here, our idea is to base the priorities on the outcome of the evaluation instead.

The samples (s, a, r, s') in the replay buffer may be arbitrary and, in particular, may not contain possible initial states. Yet the evaluation is done for initial states only. To be able to judge individual transition samples, we evaluate each sample in terms of the initial state $s_0 \in I$ from which it was generated, i.e., from which the respective training episode started. This arrangement is meaningful as improving the policy for s_0 necessarily involves further training on its successor states. For each transition sample, we store the partition J_i of the initial state s_0 in the replay buffer. The replay priority δ is then set to

$$\delta = (1 - E(J_i) + \epsilon_p)^\alpha, \tag{10}$$

where $s_0 \in J_i$ is the initial state of the training episode, and ϵ_p and α have the same functionality as in Eq. (7). After every evaluation stage, we update the priorities of the replay buffer according to Eq. (10). The probability of picking experience (s, a, r, s') during training from the buffer is then proportional to the above replay priority.

3.4 Deep Q-learning with Evaluation Stages

EID is applicable to any (deep) reinforcement learning algorithm, and EPR to any such algorithm using a replay buffer. Here, we implement both methods on top of deep Q-learning [29]. Algorithm 1 shows pseudocode for deep Q-learning

Algorithm 1. Deep Q-learning with Evaluation Stages

1: **for** episodes $i = 0$ **to** $M - 1$ **do**
2: sample $s_0 \in I$ from μ // [DQN, DQNPR, EPR]
3: sample $s_0 \in I$ according to Equation (9) // [EID]
4: **for** steps $t = 0$ **to** $T - 1$ **do**
5: with probability ϵ select random action $a_t \in A(s_t)$
6: otherwise with probability $1 - \epsilon$ select $a_t = \text{argmax}_{a \in A(s_t)} Q_\theta(s, a)$
7: execute a_t; observe r_{t+1} and s_{t+1}
8: compute $\delta = \begin{cases} \text{constant} & \text{// [DQN, EID]} \\ \text{Equation (7)} & \text{// [DQNPR]} \\ \text{Equation (10)} & \text{// [EPR]} \end{cases}$
9: store $(s_t, a_t, r_{t+1}, s_{t+1}, \delta)$ in replay buffer D
10: **every** C steps **do**
11: sample a minibatch of samples $(s_j, a_j, r_{j+1}, s_{j+1}, \delta)$ from D w.r.t. δ
12: set target $y_j = \begin{cases} r_{j+1} & s_{j+1} \text{ is terminal state} \\ r_{j+1} + \gamma \cdot \max_{a'} Q_{\theta'}(s_{j+1}, a') & \text{else} \end{cases}$
13: perform a gradient descent step on loss $(y_j - Q_\theta(s_j, a_j))^2$
14: soft-update the network weights $\theta' = (1 - \tau) \cdot \theta + \tau \cdot \theta'$
15: **end every**
16: **end for**
17: **if** $i > P$ **then** // [EID, EPR]
18: **every** L episodes **do** // [EID, EPR]
19: compute and store $E(J_i)$ for all $J_i \in \mathbb{P}$ // [EID, EPR]
20: **end every** // [EID, EPR]
21: **end if** // [EID, EPR]
22: **end for**

with soft updates (denoted DQN) and its previously discussed variant DQNPR, as well as the extensions for EID and EPR.

The unmarked lines in Algorithm 1 are inherited from the original algorithm and are applied in all versions. Lines that are marked differently are only applied in the versions they are marked with, e.g., line 2 is part of DQN, DQNPR and EPR but not of EID. The colored lines mark the extensions of EID (line 3, blue) and the extensions of both EID and EPR (lines 17–21, green). The DSMC-based evaluation stages are inserted after a threshold P of pre-training episodes was met (line 17), and then are repeated every L episodes (line 18). Thus, the total number of training episodes M is given by $M = P + N \cdot L$ where N is the number of performed evaluation stages. The priority δ (marked in orange, line 8) depends on the algorithm:

- Both original deep Q-learning DQN and EID sample uniformly from the replay buffer, so δ is set to a constant value.
- For DQNPR [37], δ is initialized with the maximal temporal difference loss observed throughout the training procedure, and updated in every learning step according to Eq. (7).
- EPR sets the priority to a constant prior to the first ES, and afterwards according to Eq. (10).

4 Case Studies

We next describe the Racetrack benchmark, which we use to evaluate our approach.

4.1 Racetrack

Racetrack originally is a pen and paper game, adopted as a benchmark in the AI community [6,9,27,32,33], particularly for reinforcement learning [5,14,16]. The task is to steer a car on a map towards the goal line without crashing into walls. The map is given by a two-dimensional grid, where each map cell either is free, part of the goal line, or a wall. We assume that initially the car may start on any free map cell with velocity 0 with equal probability (i.e., μ is uniform and I is the set of all non-wall positions with zero velocity).

Figure 2 shows the three maps that we consider in the following. Barto-big (Fig. 2a) was originally introduced by Barto et al. [6]. We designed the other two maps, Maze (Fig. 2b) and River (Fig. 2c), as examples with a more localized structure highlighting the problem of local robustness.

The position and velocity of the car each is a pair of integers, for the x- and y-dimension. In each step, the agent can accelerate the car by at most one unit in each dimension, i.e., the agent can add an element of $\{-1, 0, 1\}$ to each of x and y, resulting in nine different actions. The ground is slippery, meaning that the action might fail, in which case the acceleration/deceleration does not happen and the car's velocity remains unchanged. Each action application fails with a fixed probability that we will refer to as *noise*.

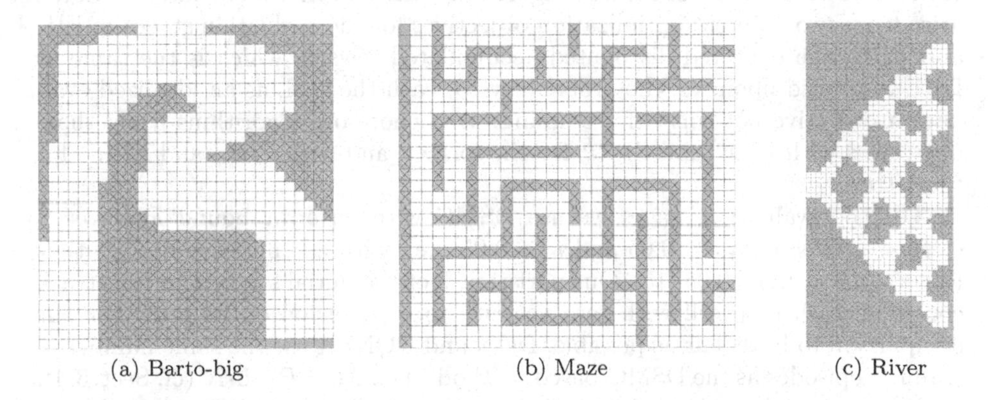

(a) Barto-big (b) Maze (c) River

Fig. 2. Three Racetrack maps, where the goal line is marked in green and wall cells are colored gray. (Color figure online)

The velocity after applying an action defines the car's new position. The car then moves in a straight line from the old position to the new position. If that line intersects with a wall cell, the car crashes and the game is lost. If that line intersects with a goal cell, the game is won. In both cases, the game terminates. We use the following simple reward function:

$$r\left(s \xrightarrow{(ax,ay)} s'\right) = \begin{cases} 100 & \text{if } s' = \top \\ -50 & \text{if } s' = \bot \\ -5 & \text{if } s' = s \\ 0 & \text{otherwise} \end{cases} \tag{11}$$

This reward function is positive if the game was won ($\top$), negative if the game was lost ($\bot$), and slightly negative if the state did not change (incentivizing the agent to not stand still); otherwise no reward signal is given. The incentive to reach the goal as quickly as possible is given through the discount factor γ that is chosen to be smaller than 1, making short-term rewards more important than long-term ones (see Eq. (2)). This reward function encodes the objective to reach the goal as quickly as possible and to not crash into a wall; the concrete values were found experimentally optimizing the performance of the vanilla DQN algorithm.

We remark that one can view the above reward structure as a proxy for the probability to reach the goal. We will consider both perspectives in our experiments, as described next.

4.2 Experiments Setup

The policies (also: *agents*) in our experiments are trained using the different variants of Algorithm 1. Specifically we run DQN and DQNPR, as well as two variants each of our DSMC-based algorithms EID and EPR. The latter variants arise from two different optimization objectives for the evaluation stages in EID and EPR: the expected discounted accumulated reward, which is the same as DRL is trained upon; vs. the probability to reach the goal, as an idealized evaluation objective not suited for training. We denote our algorithms using these objectives with EID^R and EPR^R for the former, and with EID^G and EPR^G for the latter.

For the evaluation stages we use DSMC with an error bound $P(error > \epsilon_{err}) < \kappa$, where $\epsilon_{err} = 0.05$ and $\kappa = 0.05$, i.e., with a confidence of 95% that the error is at most 0.05 [15]. Our partition $\mathbb{P}$ of the initial states I in Racetrack considers each map cell with zero velocity to be a region on its own. For our comparison to be as fair as possible, DQN and DQNPR use the same number of training episodes as the DSMC-based methods, i.e., $M = P + L \cdot N$ (cf. Sect. 3.4).

We use a high noise level, namely 50%, for the Barto-big (Fig. 2a) and river maps (Fig. 2b), to make the decision-making problems challenging. The maze map (Fig. 2c), with its long and narrow paths, is already challenging with much less uncertainty, so we set the noise to 10% there.

All compared approaches use the same neural network structure. We consider multilayer perceptrons (MLPs), aka. feed-forward networks, with a ReLu activation function for every single neuron. We specifically consider the same

NN structure as [15], with input and output layers fixed by Racetrack, and two hidden layers with 64 neurons each.

As deep reinforcement learning is known to be sensitive to different random seeds (affecting the exploration of the state space), we perform multiple trainings and report about the average result. Moreover, we fix the random seeds across algorithms in individual runs, so that the first P episodes are equal. The detailed hyperparameter settings can be found in Appendix A.

5 Results

We now analyze whether the inclusion of evaluation stages in the EID and EPR algorithms can improve (i) local robustness and (ii) goal probability performance, compared to the standard algorithms DQN and DQNPR. We first set the evaluation objective to be identical to the expected-reward training objective (EID^R and EPR^R) and analyze whether local robustness is improved; second, we set the evaluation objective to be the goal probability instead (EID^G and EPR^G) and analyze whether the policy's performance for that objective (both on average and local) improves by applying DSMC analysis after training.

5.1 Local Robustness (Deficiency (i))

Consider the heat maps in Fig. 3. For each cell on the map, we plot the expected cumulative discounted reward – the return – when starting from that map cell with zero velocity. In other words, the heat maps have one colored entry for every initial state $s_0 \in I$. We compute the return value for each s_0 using DSMC, with $\epsilon_{err} = 0.01$ and $\kappa = 0.01$, i.e., with a confidence of 99% that the error is at most 0.01.

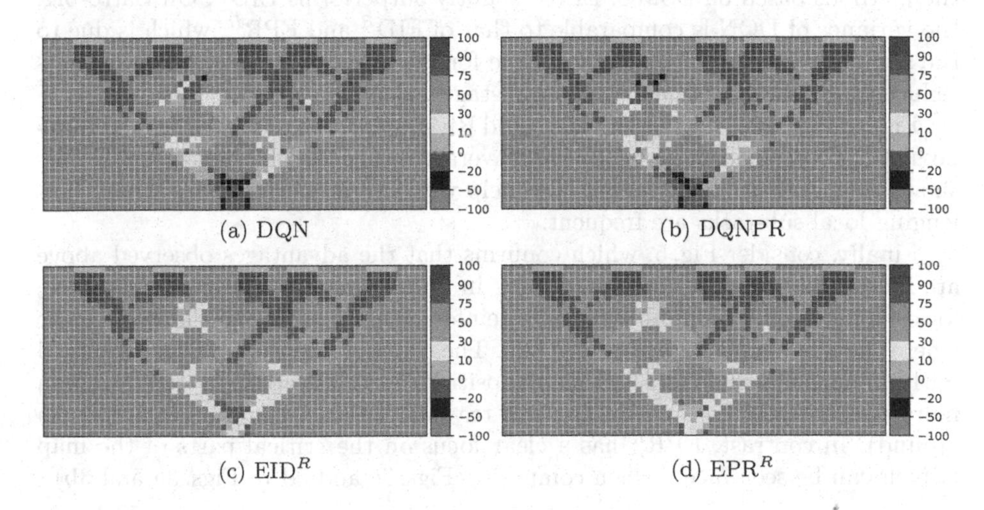

Fig. 3. Return per map cell on the River map. (Color figure online)

Clearly, the intended improvement of local robustness is achieved by EID^R and EPR^R compared to DQN and DQNPR: the return of the algorithms with evaluation stages is much better in specific areas of the map. This pertains foremost to the bottom end of the map, far away from the goal at the top; and to the "dead-end street" colored red in (a) and (b), where there is no direct connection to the nearest goal and the agents have to temporarily increase the distance to the goal. While the return of EID^R and EPR^R may also seem low in these critical parts, recall that the noise level here is 50% so it is not possible to navigate through this map without a high crash risk.

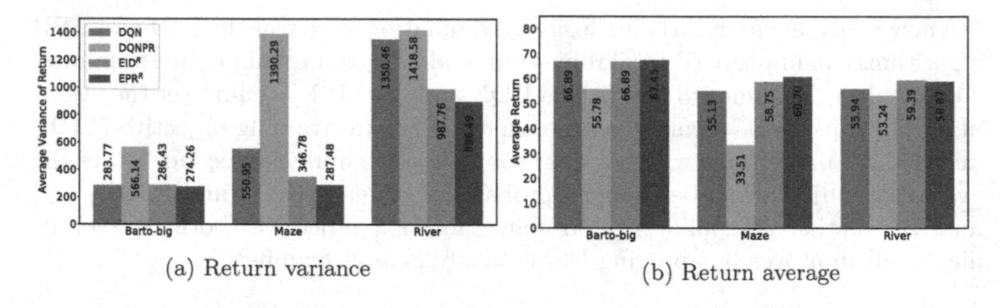

(a) Return variance

(b) Return average

Fig. 4. Variance and average of return on all maps.

Figure 4a summarizes these findings, for all maps, in terms of the variance of the return across the map (the variance of return per map cell).

On the Maze and River maps, the variance of EID^R and EPR^R is much smaller than that of DQN and DQNPR, confirming their improved local robustness. The variance reduction reaches up to about 50% compared to DQN/DQNPR. Among the methods based on DSMC, EPR^R slightly outperforms EID^R. On Barto-big, the variance of DQN is comparable to that of EID^R and EPR^R, which is due to the simpler structure of that map, while for the other two maps the variance is reduced by including DSMC evaluation stages.

Figure 4b shows that, on the Maze and River maps, the improved local robustness also results in somewhat improved average return for EID^R and EPR^R. This shows that evaluation stages can also help with overall performance when challenging local sub-tasks are frequent.

Finally, consider Fig. 5, which confirms that the advantages observed above are indeed due to more intense training in critical parts of the map. We show, for each map cell, the number of times reinforcement learning considered a state where the car was positioned in that cell. The training intensity of DQN is spread fairly homogenously across the map (positions in the dead-end street are seen more often merely because, in any run traversing those, the car needs to turn around). In contrast, EPR^R has a clear focus on the critical parts of the map (which can be seen nicely when comparing Figs. 5c and 5d to Figs. 3a and 3b).

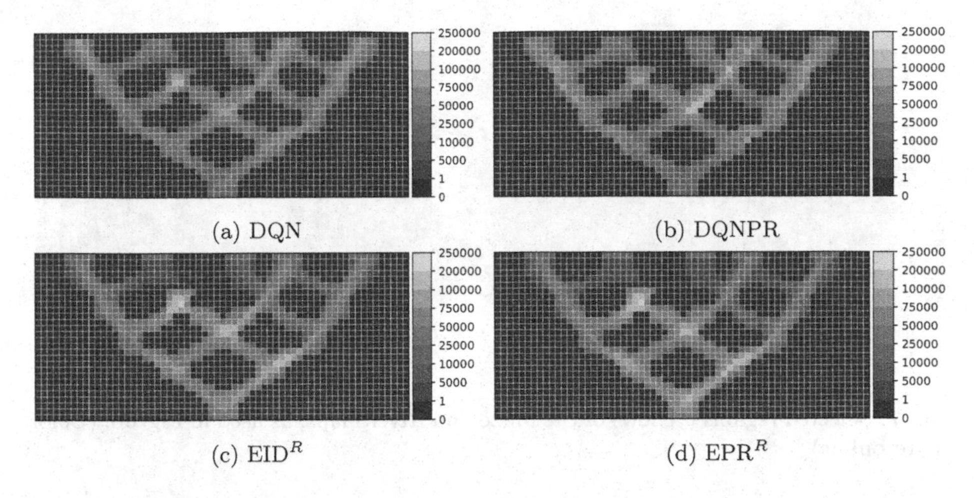

Fig. 5. Number of times each cell was encountered during training on the River map.

5.2 Fostering Goal Probability (Deficiency (ii))

We now turn to deficiency (ii), goal probability performance when training on expected reward. As discussed above, the reward structure is such that goal-reaching is rewarded, but also punishes crashes into the wall. We now show that, indeed, goal-reaching performance can be improved by introducing evaluation stages. EID^G and EPR^G improve the learning signal w.r.t. this objective. In what follows, we compute the goal probability for each map cell – for each initial state $s_0 \in I$ – using DSMC, again with $\epsilon_{err} = 0.01$ and $\kappa = 0.01$, i.e., with a confidence of 99% that the error is at most 0.01.

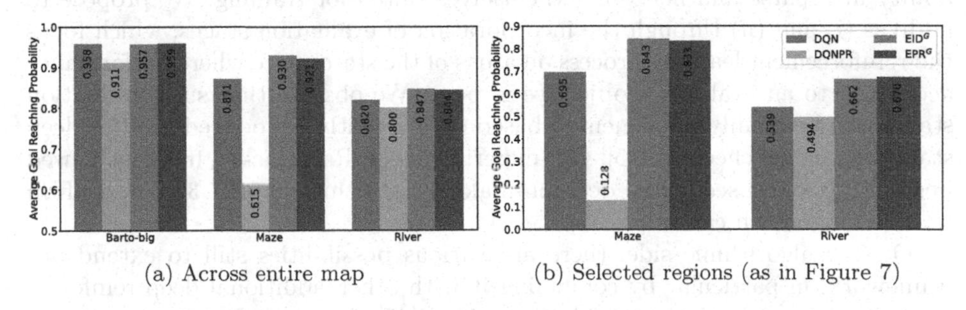

Fig. 6. Average goal probability when training on expected reward, without (DQN and DQNPR) vs. with (EID^G and EPR^G) goal-probability evaluation stages.

Figure 6 shows the corresponding results, (a) for all maps across the entire map, and (b) exemplarily for the Maze map, only for the critical regions. In Fig. 6a, we see that, again, on the Maze and River maps our proposed methods significantly increase the average goal probability. In Barto-big, this does not happen due to the simpler structure of that map.

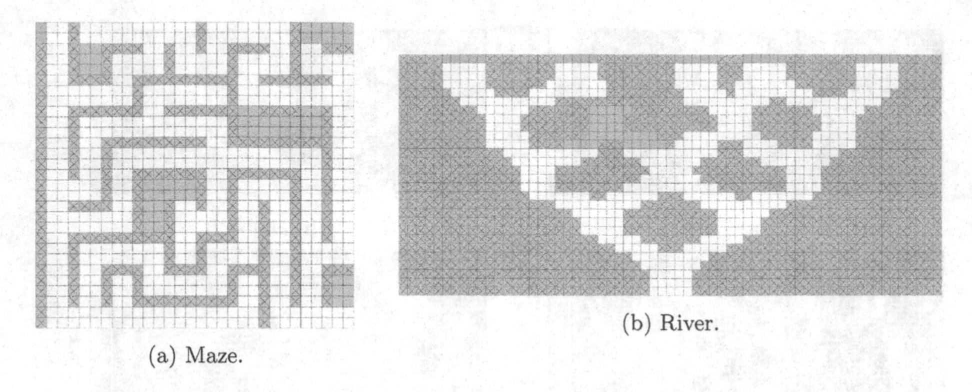

(a) Maze.

(b) River.

Fig. 7. Selected regions (yellow) of the Maze and River maps, as used in Fig. 6b. (Color figure online)

As one would expect, the improvement is higher in critical areas of the maps. To illustrate this, Fig. 6b shows average goal probability for selected regions of the Maze and River maps, as shown in Fig. 7. These regions are the "dead-end streets" which the policy will need to back out from.

6 Conclusion and Future Work

Despite its enormous successes, deep reinforcement learning suffers from important deficiencies in safety-critical systems. Apart from the general inscrutability of neural networks, these include that (i) training on average performance measures lacks local robustness, and that (ii) safety-related objectives like goal probability are sparse and hence not themselves suited for training. We propose to address (i) and (ii) through the incorporation of evaluation stages, which focus the reinforcement learning process on areas of the state space where performance according to an evaluation objective is poor. We observe that such evaluation stages can be readily implemented based on a recently introduced tool for deep statistical model checking [15]. Our experiments on Racetrack, a frequently used benchmark for AI sequential decision-making algorithms [6,9,27,32,33], confirm that this approach can work.

On the algorithmic side, there are various possibilities still to extend our framework, in particular by combining it with other/additional deep reinforcement learning algorithms. Double Q-learning [45], for example, may be promising given the lackluster performance of DQNPR in our experiments. Further, the implementation of our framework on top of policy-based approaches is of interest.

Apart from that, an important direction for future work is the broader empirical exploration of our approach. A straightforward possibility are extensions of Racetrack to include obstacles, traffic, fuel, etc. on a roadmap towards more realistic abstractions of autonomous driving as outlined by [5]. But our approach is of course not limited to Racetrack, and may in principle be applicable in arbitrary contexts where deep reinforcement learning is used. We believe that safety-critical cyber-physical systems should be the prime target, seeing as (i) and (ii) are key in that context, and seeing as the initial state partition required by our approach can be naturally obtained by (coarse discretizations of) physical location. In this context, a particular question to address will be the partition granularity trade-off, between the amount of information available during evaluation stages, and the overhead for conducting them.

A Hyperparameters

Parameter	Description	Value
DQN:		
ϵ_{start}	exploration coefficient in the beginning of the training	0.99
ϵ_{end}	exploration coefficient in the end of the training	0.05
ϵ_{decay}	exponential decay factor of the training coefficient that is applied every episode until ϵ_{end} is reached	0.999
γ	discount factor	0.99
M	number of used training episodes	110000
T	maximal length episodes	100
	learning rate in gradient descend optimization	0.0005
τ	soft update coefficient	0.001
	size of replay buffer	10^8
DQNPR:		
α	prioritization coefficient	1
ϵ_p	minimal priority	10^{-6}
ES-based:		
P	number of pre-training episodes	10000
N	number of evaluation stages	100
L	number of episodes between the evaluation stages	1000
κ	error rate/ half-width parameter during training	0.05
κ	error rate/ half-width parameter during evaluation	0.01
ϵ_{err}	error probability/ confidence during training	0.05
ϵ_{err}	error probability/ confidence during evaluation	0.01

References

1. Agostinelli, F., McAleer, S., Shmakov, A., Baldi, P.: Solving the Rubik's cube with deep reinforcement learning and search. Nat. Mach. Intell. **1**, 356–363 (2019)
2. Alshiekh, M., Bloem, R., Ehlers, R., Könighofer, B., Niekum, S., Topcu, U.: Safe reinforcement learning via shielding. In: Thirty-Second AAAI Conference on Artificial Intelligence (2018)
3. Amit, R., Meir, R., Ciosek, K.: Discount factor as a regularizer in reinforcement learning. In: International Conference on Machine Learning, pp. 269–278. PMLR (2020)

4. Avni, G., Bloem, R., Chatterjee, K., Henzinger, T.A., Könighofer, B., Pranger, S.: Run-time optimization for learned controllers through quantitative games. In: Dillig, I., Tasiran, S. (eds.) CAV 2019. LNCS, vol. 11561, pp. 630–649. Springer, Cham (2019). https://doi.org/10.1007/978-3-030-25540-4_36
5. Baier, C., et al.: Lab conditions for research on explainable automated decisions. In: Heintz, F., Milano, M., O'Sullivan, B. (eds.) TAILOR 2020. LNCS (LNAI), vol. 12641, pp. 83–90. Springer, Cham (2021). https://doi.org/10.1007/978-3-030-73959-1_8
6. Barto, A.G., Bradtke, S.J., Singh, S.P.: Learning to act using real-time dynamic programming. Artif. Intell. **72**(1–2), 81–138 (1995)
7. Bogdoll, J., Hartmanns, A., Hermanns, H.: Simulation and statistical model checking for modestly nondeterministic models. In: Schmitt, J.B. (ed.) MMB&DFT 2012. LNCS, vol. 7201, pp. 249–252. Springer, Heidelberg (2012). https://doi.org/10.1007/978-3-642-28540-0_20
8. Bonet, B., Geffner, H.: GPT: a tool for planning with uncertainty and partial information. In: Proceedings of the IJCAI Workshop on Planning with Uncertainty and Incomplete Information, pp. 82–87 (2001)
9. Bonet, B., Geffner, H.: Labeled RTDP: improving the convergence of real-time dynamic programming. In: Proceedings of the International Conference on Automated Planning and Scheduling, pp. 12–21 (2003)
10. Budde, C.E., D'Argenio, P.R., Hartmanns, A., Sedwards, S.: A statistical model checker for nondeterminism and rare events. In: Beyer, D., Huisman, M. (eds.) TACAS 2018. LNCS, vol. 10806, pp. 340–358. Springer, Cham (2018). https://doi.org/10.1007/978-3-319-89963-3_20
11. Ciosek, K., Whiteson, S.: Offer: off-environment reinforcement learning. In: Proceedings of the AAAI Conference on Artificial Intelligence, vol. 31 (2017)
12. Frank, J., Mannor, S., Precup, D.: Reinforcement learning in the presence of rare events. In: Proceedings of the 25th International Conference on Machine Learning, pp. 336–343 (2008)
13. Fujita, Y., Nagarajan, P., Kataoka, T., Ishikawa, T.: ChainerRL: a deep reinforcement learning library. J. Mach. Learn. Res. **22**(77), 1–14 (2021)
14. Gros, T.P., Groß, D., Gumhold, S., Hoffmann, J., Klauck, M., Steinmetz, M.: TraceVis: towards visualization for deep statistical model checking. In: Proceedings of the 9th International Symposium On Leveraging Applications of Formal Methods, Verification and Validation. From Verification to Explanation (2020)
15. Gros, T.P., Hermanns, H., Hoffmann, J., Klauck, M., Steinmetz, M.: Deep statistical model checking. In: Proceedings of the 40th International Conference on Formal Techniques for Distributed Objects, Components, and Systems (FORTE 2020) (2020). https://doi.org/10.1007/978-3-030-50086-3_6
16. Gros, T.P., Höller, D., Hoffmann, J., Wolf, V.: Tracking the race between deep reinforcement learning and imitation learning. In: Gribaudo, M., Jansen, D.N., Remke, A. (eds.) QEST 2020. LNCS, vol. 12289, pp. 11–17. Springer, Cham (2020). https://doi.org/10.1007/978-3-030-59854-9_2
17. Gu, S., Holly, E., Lillicrap, T., Levine, S.: Deep reinforcement learning for robotic manipulation with asynchronous off-policy updates. In: 2017 IEEE International Conference on Robotics and Automation (ICRA), pp. 3389–3396. IEEE (2017)
18. Hare, J.: Dealing with sparse rewards in reinforcement learning. arXiv preprint arXiv:1910.09281 (2019)

19. Hartmanns, A., Hermanns, H.: The modest toolset: an integrated environment for quantitative modelling and verification. In: Ábrahám, E., Havelund, K. (eds.) TACAS 2014. LNCS, vol. 8413, pp. 593–598. Springer, Heidelberg (2014). https://doi.org/10.1007/978-3-642-54862-8_51
20. Hasanbeig, M., Abate, A., Kroening, D.: Logically-constrained reinforcement learning. arXiv preprint arXiv:1801.08099 (2018)
21. Hinton, G., et al.: Deep neural networks for acoustic modeling in speech recognition: the shared views of four research groups. IEEE Signal Process. Mag. **29**(6), 82–97 (2012)
22. Jansen, N., Könighofer, B., Junges, S., Serban, A., Bloem, R.: Safe Reinforcement Learning Using Probabilistic Shields (2020)
23. Junges, S., Jansen, N., Dehnert, C., Topcu, U., Katoen, J.-P.: Safety-constrained reinforcement learning for MDPs. In: Chechik, M., Raskin, J.-F. (eds.) TACAS 2016. LNCS, vol. 9636, pp. 130–146. Springer, Heidelberg (2016). https://doi.org/10.1007/978-3-662-49674-9_8
24. Knox, W.B., Stone, P.: Reinforcement learning from human reward: discounting in episodic tasks. In: 2012 IEEE RO-MAN: The 21st IEEE International Symposium on Robot and Human Interactive Communication, pp. 878–885 (2012). https://doi.org/10.1109/ROMAN.2012.6343862
25. Krizhevsky, A., Sutskever, I., Hinton, G.E.: ImageNet classification with deep convolutional neural networks. In: NIPS, pp. 1097–1105 (2012)
26. Liu, C., Xu, X., Hu, D.: Multiobjective reinforcement learning: a comprehensive overview. IEEE Trans. Syst. Man Cybern. Syst. **45**(3), 385–398 (2014)
27. McMahan, H.B., Gordon, G.J.: Fast exact planning in Markov decision processes. In: Proceedings of the International Conference on Automated Planning and Scheduling, pp. 151–160 (2005)
28. Mnih, V., et al.: Playing atari with deep reinforcement learning. arXiv preprint arXiv:1312.5602 (2013). Accessed 15 Sept 2020
29. Mnih, V., et al.: Human-level control through deep reinforcement learning. Nature **518**, 529–533 (2015)
30. Nazari, M., Oroojlooy, A., Snyder, L., Takac, M.: Reinforcement learning for solving the vehicle routing problem. In: Bengio, S., Wallach, H., Larochelle, H., Grauman, K., Cesa-Bianchi, N., Garnett, R. (eds.) Advances in Neural Information Processing Systems, vol. 31, pp. 9839–9849. Curran Associates, Inc. (2018)
31. Ng, A.Y., Harada, D., Russell, S.J.: Policy invariance under reward transformations: theory and application to reward shaping. In: Proceedings of the 16th International Conference on Machine Learning (ICML 1999), pp. 278–287 (1999)
32. Pineda, L.E., Lu, Y., Zilberstein, S., Goldman, C.V.: Fault-tolerant planning under uncertainty. In: Twenty-Third International Joint Conference on Artificial Intelligence, pp. 2350–2356 (2013)
33. Pineda, L.E., Zilberstein, S.: Planning under uncertainty using reduced models: revisiting determinization. In: Proceedings of the International Conference on Automated Planning and Scheduling, vol. 24 (2014)
34. Puterman, M.L.: Markov Decision Processes: Discrete Stochastic Dynamic Programming, 1st edn. Wiley, New York (1994)
35. Riedmiller, M., et al.: Learning by playing solving sparse reward tasks from scratch. In: International Conference on Machine Learning, pp. 4344–4353. PMLR (2018)
36. Sallab, A.E., Abdou, M., Perot, E., Yogamani, S.: Deep reinforcement learning framework for autonomous driving. Electron. Imaging **2017**(19), 70–76 (2017)

37. Schaul, T., Quan, J., Antonoglou, I., Silver, D.: Prioritized experience replay. In: Bengio, Y., LeCun, Y. (eds.) 4th International Conference on Learning Representations, ICLR (2016)
38. Schwartz, A.: A reinforcement learning method for maximizing undiscounted rewards. In: Proceedings of the Tenth International Conference on Machine Learning, vol. 298, pp. 298–305 (1993)
39. Sen, K., Viswanathan, M., Agha, G.: On statistical model checking of stochastic systems. In: International Conference on Computer Aided Verification, pp. 266–280 (2005)
40. Silver, D., et al.: Mastering the game of go with deep neural networks and tree search. Nature **529**(7587), 484–489 (2016)
41. Silver, D., et al.: A general reinforcement learning algorithm that masters chess, shogi, and go through self-play. Science **362**(6419), 1140–1144 (2018)
42. Silver, D., et al.: Mastering the game of go without human knowledge. Nature **550**(7676), 354–359 (2017)
43. Stooke, A., Abbeel, P.: rlpyt: a research code base for deep reinforcement learning in Pytorch. arXiv preprint arXiv:1909.01500 (2019)
44. Sutton, R.S., Barto, A.G.: Reinforcement Learning: An Introduction, Adaptive Computation and Machine Learning, 2nd edn. The MIT Press, Cambridge (2018)
45. Van Hasselt, H., Guez, A., Silver, D.: Deep reinforcement learning with double q-learning. In: Proceedings of the AAAI Conference on Artificial Intelligence, vol. 30 (2016)
46. Vinyals, O., et al.: Grandmaster level in StarCraft II using multi-agent reinforcement learning. Nature **575**, 350–354 (2019)
47. Younes, H.L.S., Kwiatkowska, M., Norman, G., Parker, D.: Numerical vs. statistical probabilistic model checking: an empirical study. In: Jensen, K., Podelski, A. (eds.) TACAS 2004. LNCS, vol. 2988, pp. 46–60. Springer, Heidelberg (2004). https://doi.org/10.1007/978-3-540-24730-2_4

Active and Sparse Methods in Smoothed Model Checking

Paul Piho(✉) and Jane Hillston

University of Edinburgh, Edinburgh, UK
paul.piho@ed.ac.uk

Abstract. Smoothed model checking based on Gaussian process classification provides a powerful approach for statistical model checking of parametric continuous time Markov chain models. The method constructs a model for the functional dependence of satisfaction probability on the Markov chain parameters. This is done via Gaussian process inference methods from a limited number of observations for different parameter combinations. In this work we incorporate sparse variational methods and active learning into the smoothed model checking setting. We use these methods to improve the scalability of smoothed model checking. In particular, we see that active learning-based ideas for iteratively querying the simulation model for observations can be used to steer the model-checking to more informative areas of the parameter space and thus improve sample efficiency. We demonstrate that online extensions of sparse variational Gaussian process inference algorithms provide a scalable method for implementing active learning approaches for smoothed model checking.

1 Introduction

Stochastic modelling coupled with verification of logical properties via model checking has provided useful insights into the behaviour of the stochastic models from epidemiology, systems biology and networked computer systems. A large number of interesting models in these fields are too complex for the application of exact model checking methods [13]. To improve the scalability of model checking there has been significant work on statistical model checking that aims to estimate the satisfaction probability of logical properties based on independently sampled trajectories of a stochastic model [3].

This paper considers statistical model checking in the context of parametrised continuous time Markov chain models. Statistical model checking methods have generally considered single parametrisations of a model. Based on a large number of independent sample trajectories, one can estimate the probability of the model satisfying a specified logical property defined over individual sample trajectories. In order to gain insight across the entire parameter space associated with a model, it can be necessary to repeat the estimation procedures with different parametrisations to cover the whole space, which leads to poor scalability.

A. Abate and A. Marin (Eds.): QEST 2021, LNCS 12846, pp. 217–234, 2021.
https://doi.org/10.1007/978-3-030-85172-9_12

As an alternative, a model checking approach based on Gaussian process classification, named smoothed model checking, was proposed in [6]. The main result of that paper was to show that under mild conditions the function which maps parameter values to satisfaction probabilities is smooth. Thus the problem can be solved as a Gaussian process classification problem where the aim is to estimate the function describing the satisfaction probability over the parameter space. Model checking results, returning a label true or false, of individual simulation trajectories are used as the training data to infer how the satisfaction probability depends on the model parameters. This method can be used to greatly reduce the number of simulation trajectories needed to estimate the satisfaction probability in exchange for some accuracy.

There are two aspects that limit the speed of such model checking procedures. Firstly, the computational cost of gathering the individual trajectories and secondly, the cost of (approximate) Gaussian process inference itself. Both benefit from keeping the number of gathered trajectories as low as possible while minimising the impact a smaller set of training data has on the accuracy of the methods. In order to keep the gathered sample size small we propose a method based on active learning. In particular, we make the observation that the parameter space of models is usually constrained to physically reasonable ranges. However, even when constrained to such ranges there can be large parts of the parameter space where the probability of satisfying a formula undergoes little change. Adaptively identifying parts of the parameter space where the satisfaction probability changes in order to decide where to concentrate the computational effort leads to improved algorithms for smoothed model checking.

The main contribution of this paper is considering smoothed model checking in a sparse online setting and proposing active learning strategies for querying the parameter space of a model. We make use of state of the art sparse variational Gaussian process inference methods and streaming variational inference with inducing points [8]. This combination of sparse inference methods and active learning improves the scalability of smoothed model checking in two ways. The use of sparse inference methods reduces the complexity of the underlying Gaussian process inference, while active learning query methods improve the sample efficiency by concentrating the model checking efforts to the parts of the parameter space where the satisfaction probability undergoes most change.

2 Related Work

A wealth of literature exists on statistical model checking of stochastic systems. The use of statistical methods in the domain of formal verification is motivated by the fact that in order to perform statistical model checking it is only necessary to be able to simulate the model. Thus these methods can be used for systems where exact verification methods are infeasible including black-box systems [14]. In its classical formulation, this involves hypothesis testing [20] with respect to the desired (or undesired) property based on independent trials, or in this case, stochastic simulations.

In addition to the frequentist approaches based on hypothesis testing, there have been Bayesian approaches [12] to estimate the satisfaction probability of a given logical formula. Our work follows the approach presented in [6] where the dependence of the satisfaction probability on model parameters is modelled as a Gaussian process classification problem.

The problem of deciding where to concentrate the model checking efforts is closely related to optimal experimental design. Experimental design problems are commonly treated as optimisation problems where the goal is to allocate resources in a way that allows the experimental goals to be reached more rapidly and thus with smaller costs [19]. This idea is also known in the machine learning literature as active learning [22]. The idea is to design learning algorithms that interactively query an oracle to label new data points.

In the context of model checking, active learning was used in [7] to solve a threshold synthesis problem which is closely related to the model checking problem considered in this paper. That approach used a base grid on the parameter space for initial estimation. The estimates were then refined around values where the satisfaction probability was close to a defined threshold. However, the threshold for synthesis has to be defined a priori making the introduced active step not applicable when we are interested in the satisfaction probability. We further address the scalability of the ideas presented by the authors of [7] by considering sparse approximation results for Gaussian process based model checking.

Finally, Bayesian optimisation is another example of active learning. Bayesian optimisation methods refine the posterior distribution over the black-box objective function based on function evaluations. An example relevant to model checking was given in [4] where Bayesian optimisation was used to optimise parameters of stochastic models to maximise robustness of the given logical specification.

3 Background

3.1 Continuous Time Markov Chains

Stochastic models are widely used to model a variety of phenomena in natural and engineered systems. We focus on a type of stochastic model commonly used in biological modelling, epidemiology and performance evaluation domains. Specifically, we consider continuous time Markov chain models (CTMCs). To define a CTMC we start by noting that it is a continuous-time stochastic process and thus defined as an indexed collection of random variables $\{\mathbf{X}\}_{t\in\mathbb{R}_{\geq 0}}$. We consider CTMCs defined over a finite state space S with an $|S| \times |S|$ matrix Q whose entries $q(i,j)$ satisfy

1. $0 \leq -q(i,i) < \infty$, 2. $0 \leq q(i,j)$ for $i \neq j$, 3. $\sum_j q(i,j) = 0$.

A CTMC is then defined by the following: for time indices $t_1 < t_2 \ldots < t_{n+1}$ and states $i_1, i_2, \ldots, i_{n+1}$ we have

$$\mathbb{P}(\mathbf{X}_{t_{n+1}} = i_{n+1} | \mathbf{X}_{t_n} = i_n, \ldots, \mathbf{X}_{t_1} = i_1) = p(i_{n+1}; t_{n+1} | i_n; t_n)$$

where $p(j;t|i;s)$ is the solution to the following Kolmogorov forward equation

$$\frac{\partial}{\partial t}p(j;t|i;s) = \sum_k p(k;t|i;s)q(k,j), \qquad \text{on } (s,\infty) \text{ with } p(j;s|i;s) = \delta_{ij}$$

with δ_{ij} being the Kronecker delta taking the value 1 if i and j are equal and the value 0 otherwise. By convention the sample trajectories of CTMCs are taken to be right-continuous.

In the rest of the paper we consider parametrised models $\mathcal{M}_{\mathbf{x}}$ and assume that the model $\mathcal{M}$ for a fixed parametrisation $\mathbf{x} \in \mathbb{R}^k$ defines a CTMC. Thus, the model $\mathcal{M}$ specifies a function mapping parameters $\mathbf{x}$ to generator matrix Q of the underlying CTMC. A commonly studied special class of CTMC models are population CTMCs where each state of the CTMC corresponds to a vector of counts. These counts are used to model the aggregate counts of groups of indistinguishable agents in a system. In biological modelling and epidemiology such models are often defined as chemical reaction networks (CRN).

Example 1. Let us consider the following susceptible-infected-recovered (SIR) model defined as a CRN

$$S + I \xrightarrow{k_I} I + I \qquad I \xrightarrow{k_R} R$$

where S gives the number of susceptible, I the infected and R the recovered individuals in the system. The first type of transition corresponds to infected and susceptible individuals interacting, resulting in the number of infected individuals increasing and the number of susceptible decreasing. The second type of transition corresponds to recovery of an infected individual and results in the number of infected decreasing and the number of recovered increasing. The states of the underlying CTMC keep track of the counts of different individuals in the system. For the example let us set the initial conditions to $(95, 5, 0)$—at time 0 there are 95 susceptible, 5 infected and 0 recovered individuals in the system. The parameters k_I and k_R give the infection and recovery rates respectively. We revisit this example throughout the paper to illustrate the presented concepts.

3.2 Smoothed Model Checking

Smoothed model checking was introduced in [6] as a scalable method for statistical model checking where Gaussian process classification methods were used to infer the functional dependence between a parametrisation of a model and the satisfaction probability given a logical specification.

As described in Sect. 3.1, suppose we have a model $\mathcal{M}_{\mathbf{x}}$ parametrised by vector of values $\mathbf{x} \in \mathbb{R}^d$ such that the model $\mathcal{M}$ for a fixed parametrisation $\mathbf{x}$ defines a CTMC. Additionally assume we have a logical property φ we want to check against. The logical properties we consider here are defined as a mapping from the time trajectories over the states of $\mathcal{M}_{\mathbf{x}}$ to $\{0, 1\}$ corresponding to whether the property holds for a given sample trajectory of $\mathcal{M}_{\mathbf{x}}$ or not. One way to define such mappings would be, for example, to specify the properties in metric interval temporal logic (MiTL) [15] or signal temporal logic (STL) [10] and map the paths satisfying the properties to 1 and those not satisfying the properties to 0. Through sampling multiple trajectories for the same parametrisation we gain an estimate of the satisfaction probability corresponding to the parametrisation.

With that in mind, a logical property φ with respect to $\mathcal{M}_{\mathbf{x}}$ can be seen to give rise to a Bernoulli random variable. The binary outcomes of the random variable correspond to whether or not a randomly sampled trajectory of $\mathcal{M}_{\mathbf{x}}$ satisfies the property φ. We introduce the notation $f_{\varphi}(\mathbf{x})$ for the parameter of the said Bernoulli random variable given the model parameters $\mathbf{x}$. In particular, samples from the distribution $Bernoulli(f_{\varphi}(\mathbf{x}))$ model whether a randomly sampled trajectory of $\mathcal{M}_{\mathbf{x}}$ satisfies φ—for a parameter value $\mathbf{x}$ the logical property is said to be satisfied with probability $f_{\varphi}(\mathbf{x})$.

A naive approach for estimating $f_{\varphi}(\mathbf{x})$ at a given parametrisation $\mathbf{x}$ is to gather a large number N of sample trajectories and give simple Monte Carlo estimates for the $f_{\varphi}(\mathbf{x})$ by dividing the number of trajectories where the property holds by the total number of sampled trajectories N. An accurate estimate requires a large number of samples. However, having such an estimate at a set of given parametrisations does not provide us with a rigorous way to estimate the satisfaction function at a nearby point.

In [6] the authors considered population CTMCs. It was shown that the introduced satisfaction probability $f_{\varphi}(\mathbf{x})$ is a smooth function of $\mathbf{x}$ under the following conditions: the transition rates of the CTMC $\mathcal{M}_{\mathbf{x}}$ depend smoothly on the parameters $\mathbf{x}$; and the transition rates depend polynomially on the state vector $\mathbf{X}$ of the CTMC.

The result was exploited by treating the estimation of the satisfaction function $f_{\varphi}(\mathbf{x})$ as a Gaussian process classification problem. The main benefit of this approach is that, based on sampled model checking results, we can reconstruct an approximation for the functional dependence between the parameters and satisfaction probability. This makes it easy to make predictions about the satisfaction probability at previously unseen parametrisations.

Simulating $\mathcal{M}_{\mathbf{x}}$ we gather a finite set of observations $\mathcal{D} = \{(\mathbf{x}_i, y_i) | i = 1, \cdots, n\}$ where $\mathbf{x}_i$ are the parametrisations of the model and y_i correspond to model checking output over single trajectories. For classification problems, a Gaussian process prior with mean m and kernel k is placed over a latent function

$$g_{\varphi}(\mathbf{x}) \sim GP(m(\mathbf{x}), k(\mathbf{x}, \mathbf{x}')) .$$

Here, let us consider the standard squared exponential kernel defined by

$$k(\mathbf{x}, \mathbf{x}') = a^2 \exp\left(-\frac{|\mathbf{x} - \mathbf{x}'|^2}{2l}\right)$$

where a^2 is the amplitude and ℓ is the length scale parameter governing how far two distinct points have to be in order to be considered uncorrelated.

The function g_{φ} is then squashed through the standard logistic or probit transformation σ so that the composition $\sigma(g_{\varphi}(\mathbf{x}))$ takes values between 0 and 1. The quantity $\sigma(g_{\varphi}(\mathbf{x}))$ is interpreted as the probability that φ holds given model

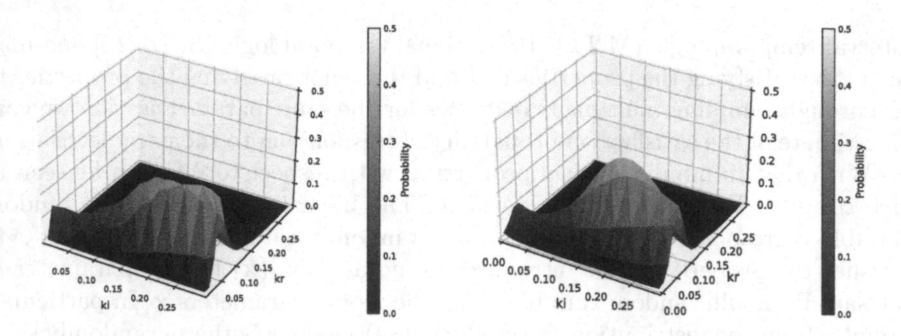

Fig. 1. Left is the baseline satisfaction probability surface. Satisfaction probability estimated for parameters on a regular 20×20. At each parameter the estimate is based on 3000 SSA trajectories. Right is the smoothed model checking satisfaction probability surface. Training data is constructed from a set of 10 SSA trajectories at parameter points on a regular 15×15 grid.

parametrisation $\mathbf{x}$ and thus estimates the probability $f_\varphi(\mathbf{x})$ that a simulation trajectory for parameters $\mathbf{x}$ satisfies the property φ.

The general aim of Gaussian process inference is to find the distribution $p(g_\varphi(\mathbf{x}^*)|\mathcal{D})$ over the values g_φ at some test point $\mathbf{x}^*$ given the set of training observations $\mathcal{D}$. This distribution is then used to produce a probabilistic prediction at parameter $\mathbf{x}^*$ of $\sigma(g_\varphi(\mathbf{x}^*)) \approx f_\varphi(\mathbf{x}^*)$. We present details of inference in the next section. This section is ended by returning to the running SIR example.

Example 2. The property we consider is the following: there always exists an infected agent in the population in the time interval $(0.0, 100.0)$ and in the time interval $(100.0, 120.0)$ the number of infected becomes 0. Constraining the parameters to the ranges $k_I \in [0.005, 0.3]$ and $k_R \in [0.005, 0.3]$ gives satisfaction probabilities as depicted in Fig. 1. There each estimate on the 20×20 grid is calculated based on 3000 stochastic simulation algorithm (SSA) sample runs of the model. For comparison, Fig. 1 also gives the results of the smoothed model checking where 10 sample trajectories are drawn for each parameter on the 12×12 grid. The smoothed model checking approximation for the model checking problem shows good agreement with the baseline surface and is much faster to perform.

3.3 Variational Inference with Inducing Points

In order to infer the latent Gaussian process g_φ based on training data $\mathcal{D}$ we have to deal with two problems. Firstly the inference is analytically intractable due to the non-Gaussian likelihood model provided by Bernoulli observations. To counter this there exists a wealth of approximate inference schemes like Laplace approximation, expectation propagation [11,16], and variational inference methods [23]. Here we consider variational inference. The second problem is that the methods for inference in Gaussian process models have cubic complexity in the number of training cases. To address that, there exist sparse approximations based on inducing variables. Sparse variational methods [9,23] are popular methods for reducing the complexity of Gaussian process inference by constructing an approximation based on a small set of inducing points that are typically selected from training data. In this section we detail the inference procedure.

Variational inference methods choose a parametric class of variational distributions for the posterior and minimising the KL-divergence between the real posterior and the approximate posterior. To accommodate large training data sets we work with sparse variational methods. We start by defining the prior distribution

$$p(\mathbf{g}_\varphi, \mathbf{u}) = \mathcal{N}\left(\begin{bmatrix}\mathbf{g}_\varphi \\ \mathbf{u}\end{bmatrix} \middle| \mathbf{0}, \begin{bmatrix}\mathbf{K}_{nn} & \mathbf{K}_{nm} \\ \mathbf{K}_{nm}^T & \mathbf{K}_{mm}\end{bmatrix}\right)$$

where $\mathbf{g}_\varphi$ is a vector of n latent function values $[g_\varphi(\mathbf{x}_1), \cdots, g_\varphi(\mathbf{x}_n)]$. Similarly, $\mathbf{u}$ is a vector of m latent function values $[g_\varphi(\mathbf{z}_1), \cdots, g_\varphi(\mathbf{z}_m)]$ evaluated at chosen *inducing points* $\mathbf{z}_i$. The matrices $\mathbf{K}_{nn}$, $\mathbf{K}_{nm}$ and $\mathbf{K}_{mm}$ are defined by the kernel function. In particular, the (i, j)-th element of the matrix $\mathbf{K}_{nn}$ is given by $k(\mathbf{x}_i, \mathbf{x}_j)$. Similarly, $\mathbf{K}_{nm}$ gives the kernel matrix between the training points $\mathbf{x}$ and the inducing points $\mathbf{z}$ and $\mathbf{K}_{mm}$ gives the kernel matrix between the locations of inducing points. We then fit the variational posterior at those points rather than the whole set of training data points. The assumption we are making is that $p(g_\varphi(\mathbf{x}_*)|\mathbf{g}_\varphi, \mathbf{u}) = p(g_\varphi(\mathbf{x}_*)|\mathbf{u})$. That is, the inducing values $\mathbf{u}$ are a sufficient statistic for a function value at a test point $\mathbf{x}_*$.

Under this assumption we make predictions at a test point $\mathbf{x}^*$ as follows:

$$p(g_\varphi(\mathbf{x}_*)|\mathbf{y}) = \int p(g_\varphi(\mathbf{x}_*), \mathbf{u}|\mathbf{y}) d\mathbf{u} = \int p(g_\varphi(\mathbf{x}_*)|\mathbf{u}) p(\mathbf{u}|\mathbf{y}) d\mathbf{u}\,.$$

Thus, we need posterior distribution $p(\mathbf{u}|\mathbf{y})$ at the inducing points. Here, as mentioned, we consider variational approximations where $p(\mathbf{u}|\mathbf{y})$ is approximated by a multivariate Gaussian $q(\mathbf{u})$ making the expression for $p(g_\varphi(\mathbf{x}_*)|\mathbf{y})$ tractable. Finding the parameters of $q(\mathbf{u})$ is done by minimising the KL divergence between the approximate posterior $q(\mathbf{u})$ and true posterior $p(\mathbf{u}|\mathbf{y})$. In particular, we have

$$\mathcal{D}_{KL}(q(\mathbf{u}), p(\mathbf{u}|\mathbf{y})) = \int q(\mathbf{u}) \log \frac{q(\mathbf{u})}{p(\mathbf{u}|\mathbf{y})} du = -\left\langle \log \frac{p(\mathbf{y}, \mathbf{u})}{q(\mathbf{u})} \right\rangle_{q(\mathbf{u})} + \log p(\mathbf{y}) \quad (1)$$

where $\langle \cdot \rangle_{q(\mathbf{u})}$ denotes the expectation with respect to distribution $q(\mathbf{u})$. The term $\log p(\mathbf{y})$ is known as the log marginal likelihood. In the following we use the well-known Jensen's inequality[1] to derive a lower bound for the log marginal likelihood. As log function is concave we get the following:

$$\begin{aligned}\log p(\mathbf{y}) &= \log \int p(\mathbf{y}, \mathbf{u}) d\mathbf{u} = \log \left\langle \frac{p(\mathbf{y}, \mathbf{u})}{q(\mathbf{u})} \right\rangle_{q(\mathbf{u})} \geq \left\langle \log \frac{p(\mathbf{y}, \mathbf{u})}{q(\mathbf{u})} \right\rangle_{q(\mathbf{u})} \\ &= \langle \log p(\mathbf{y}|\mathbf{u}) \rangle_{q(\mathbf{u})} - \mathcal{D}_{KL}(q(\mathbf{u}), p(\mathbf{u}))\,. \end{aligned} \quad (2)$$

[1] For a concave function f and a random variable X we have the following well-known inequality: $f\langle X \rangle \geq \langle f(X) \rangle$.

The right-hand side of the inequality 2 is known as the evidence-based lower bound or ELBO. Now note that the first term in the expression for KL-divergence in Eq. 1 is exactly the derived ELBO. As the KL-divergence is always non-negative and ELBO serves as a lower bound for the log marginal likelihood $p(\mathbf{y})$, then maximising the ELBO minimises the KL-divergence between the approximate and true posteriors $q(\mathbf{u})$ and $p(\mathbf{u}|\mathbf{y})$.

Choosing our approximating family of variational distribution to be multivariate Gaussian makes the KL term in ELBO easy to evaluate. The integral in the expectation term $\langle \log p(\mathbf{y}|\mathbf{u})\rangle_{q(\mathbf{u})}$ can be computed via numerical approximation schemes making it possible to use ELBO as a utility function for optimising the parameters of the approximate posterior $q(\mathbf{u})$ via gradient ascent. When $q(\mathbf{u})$ is chosen to be a multivariate Gaussian these parameters are the mean $\boldsymbol{\mu}$ and covariance matrix $\boldsymbol{\Sigma}$. With the approximation $p(\mathbf{u}|\mathbf{y}) \approx q(\mathbf{u})$ the predictions are given by the integral

$$p(g_\varphi(\mathbf{x}_*)|\mathbf{y}) \approx \int p(g_\varphi(\mathbf{x}_*)|\mathbf{u})q(\mathbf{u})d\mathbf{u}\,.$$

This can be shown [18] to be a probability density function of a Normal distribution with the following mean μ and variance σ^2

$$\begin{aligned}
\mu_* &= k(\mathbf{x}_*, \mathbf{u})K_{mm}^{-1}\boldsymbol{\mu} \\
\sigma_*^2 &= k(\mathbf{x}_*, \mathbf{x}_*) - k(\mathbf{x}_*, \mathbf{u})K_{mm}^{-1}\left[K_{mm} - \boldsymbol{\Sigma}\right]^{-1}\left[k(\mathbf{x}_*, \mathbf{u})K_{mm}^{-1}\right]^T .
\end{aligned}$$

Note that the terms in ELBO depend on the chosen kernel and in particular the kernel hyperparameters. As mentioned, the ELBO is maximised directly via gradient ascent with respect to the parameters of the variational distribution. The kernel hyperparameter can be tuned in the same fashion. A common approximate technique we use in this paper is to interleave the optimisation steps in the variational distribution parameters with optimisation steps in the hyperparameters.

We have made the assumption that the posterior distribution is fitted at a selection of inducing points $\mathbf{z}$ such that number of inducing points is much smaller than the whole training data set. There exists a variety of possible methods to select the inducing points. For simplicity in this paper we use a regular grid over the parameter space as our inducing points.

3.4 Active Learning

Active learning methods in machine learning are a family of methods which may query data instances to be labelled for training by an *oracle* [21]. The fundamental question asked by active learning research is whether or not these methods can achieve higher accuracy than passive methods with fewer labelled examples. This is closely related to the established area of optimal experimental design, where the goal is to allocate experimental resources in a way that reduces uncertainty about a quantity or function of interest [19,22].

In the case of Gaussian process classification problems like smoothed model checking, an active learning procedure can be set up as follows. An *active learner* consists of a classifier learning algorithm $\mathcal{A}$ and a *query function* $\mathcal{Q}$. The query function is used to select an unlabelled sample u from the pool of unlabelled samples $\mathcal{U}$. This sample is then labelled by an oracle. In the case of stochastic model checking, the pool of unlabelled samples $\mathcal{U}$ corresponds to a subset of the possible parametrisations for the model. An oracle is implemented by running the stochastic simulation for the selected parametrisation and then model checking the resulting trajectory.

The above describes a pool-based active learner. Common formulations of such pool-based learners select a single unlabelled sample at each iteration to be sampled. However, in many applications it is more natural to acquire labels for multiple training instances at once. In particular, the query function $\mathcal{Q}$ selects a subset $U \subset \mathcal{U}$. We see in the next section that the sparse inference methods can be extended to a setting where batches of training data become available over time making it natural to decide on a query function that selects batches of queries. The main difficulty of selecting a batch of queries instead of a single query is that the instances in the subset U need to be both informative and diverse in order to make the best use of the available labelling resources.

4 Active Model Checking

The shape and properties of the functional dependence of satisfaction for a logical specification with respect to parameters are generally not known a priori and can exhibit a variety of properties. For example, in the running example much of the sampling was performed in completely flat regions of the parameter space. Thus the key challenge addressed in this section is where to sample to make the posterior estimates as informative as possible about the underlying mechanics. We aim to decide on the regions where the satisfaction probability surface is not flat and concentrate most of our model checking effort there. To that end we introduce the main contribution of this paper—active sparse model checking.

The general outline of the procedure is given by Algorithm 1. The first step, given by the procedure *generate_initial_data*, is to simulate the initial data set $\mathcal{D}_{old}$ via stochastic simulation of the CTMC model $\mathcal{M}$ for a sample of the parameter space $\mathcal{X}$ and checking whether or not the individual trajectories satisfy the property φ or not. The initial set of parameter samples can, for example, be a regular grid or sampled uniformly from the parameter space.

In general the inducing points are then chosen based on the results $\mathcal{D}_{old}$ and adjusted as new data is seen. However, for simplicity we are going to set the inducing points so that they form a regular grid over the parameter space and keep them fixed throughout the active iterations. The posterior at inducing points $\mathbf{z_v}$ is then initialised as a multivariate Gaussian $q(\mathbf{v}) = \mathcal{N}(\mathbf{0}, \mathbf{I})$) with 0 mean and identity covariance matrix. Each iteration of the model checking loop will first update the variational posterior $q(\mathbf{v})$ via *update_variational*. The details of how this is done in a sparse streaming setting are given in Sect. 4.1. Secondly,

Algorithm 1. Active smoothed model checking

```
1: procedure MODEL CHECKING(model M, property φ, parameter space X)
2:   D_old ← generate_initial_data(M, φ, X)
3:   D_new ← D_old
4:   z_u ← inducing_points(D_old)
5:   z_v ← z_u
6:   q(v) ← initialise_posterior(v)
7:   while true do
8:     q(v) ← update_variational(q(u), v, D_new, D_old)
9:     D_old ← D_old ∪ D_new
10:    D_new ← query_new(q(v), M, φ, X)
11:  end while
12:  return q(v)
13: end procedure
```

each iteration uses the fitted approximate posterior to query new points in the parameter space to perform model checking through the *query_new* procedure. The proposed query functions are discussed in Sect. 4.2.

There are two issues to be resolved before the procedure can be implemented. First is that the direct use of ELBO as introduced in Sect. 3.3 does not suffice in the online setting where new data becomes available in batches. Second is the challenge of choosing an appropriate query function that is going to suggest more points in the parameter space at which to gather more model checking data. These will be addressed in the following sections.

4.1 Streaming Setting

In order to incorporate active learning ideas into the Gaussian process based model checking approach we need to address the problem that not all of the training data is available a priori. For our purposes it is important to be able to conduct inference in a streaming setting where data is gradually added to the model. A naive approach would refit a Gaussian process from scratch every time a new batch of data arrives. However, with potentially large data sets this becomes infeasible. To perform sparse variational inference in a scalable way the method needs to avoid revisiting previously considered data points. In particular, we consider the method proposed in [8] that derives a correction to ELBO that allows us to incorporate streaming data incrementally into the posterior estimate.

The main question is how to update the variational approximation to the posterior at time step n, denoted $q_{old}(\mathbf{u})$, to form an approximation at the time step $n+1$, denoted $q_{new}(\mathbf{v})$. In the following we note the variational posteriors q_{old} and q_{new} at $\mathbf{g}_\varphi$ and inducing values $\mathbf{u}$ and $\mathbf{v}$, respectively, are approximations to the true posteriors given observations $\mathbf{y}_{old}$ and $\mathbf{y}_{new}$. It was shown in [8] that the lower bound of $\log p(\mathbf{y}_{new}|\mathbf{y}_{old})$ becomes

$$\int q_{new}(\mathbf{g}_\varphi, \mathbf{v}) \log p(\mathbf{y}_{new}|\mathbf{g}_\varphi, \mathbf{v}) d(\mathbf{g}_\varphi, \mathbf{v}) - \mathcal{D}_{KL}(q_{new}(\mathbf{v}), p(\mathbf{v})) \\ -\mathcal{D}_{KL}(q_{new}(\mathbf{u}), q_{old}(\mathbf{u})) + \mathcal{D}_{KL}(q_{new}(\mathbf{u}), p(\mathbf{u})) .$$

The above can be interpreted as follows: the first two terms give the ELBO under the assumption that the new data seen at iteration $n+1$ is the whole data set; the final two terms take into account the old likelihood through the approximate posteriors at old inducing points and the prior $p(\mathbf{u})$. This allows us to implement an online version of the smoothed model checking where the observation data arrives in batches.

4.2 Query Strategies

As discussed in Sect. 3.4, in order to implement an active learning method for model checking we need to decide which new parameters are tested based on the existing information. In the following we consider two query strategies for active model checking.

Predictive Variance. The first approach is a commonly used experimental design strategy which aims to minimise the predictive variance. Recall that in smoothed model checking for a property φ we fit a latent Gaussian process g_φ. The posterior satisfaction probability for parameter $\mathbf{x}_*$ given the GP g_φ is then calculated via

$$p(y_* = 1|\mathcal{D}, \mathbf{x}_*) = \int \sigma(g_\varphi(\mathbf{x}_*)) p(g_\varphi(\mathbf{x}_*)|\mathcal{D}) dg_\varphi(\mathbf{x}_*) .$$

The above can also be seen as the expectation of $\sigma(g_\varphi(\mathbf{x}_*)$ with respect to the distribution $g_\varphi(\mathbf{x}_*)$, denoted $\mathbb{E}\left[\sigma(g_\varphi(\mathbf{x}_*)\right]$. Similarly, we can consider the variance of this estimate

$$\mathbb{E}\left[\sigma(g_\varphi(\mathbf{x}_*)^2\right] - \mathbb{E}\left[\sigma(g_\varphi(\mathbf{x}_*)\right]^2 .$$

Our aim is then to iteratively train the Gaussian process model so that predictive variance over the parameter space is minimised.

Before giving the outline of the proposed procedure we address the issue of redundancy in the query points. As pointed out in Sect. 3.4, simply taking a set of points with the highest predictive variance leads to querying parameters that are clustered together. We can overcome this problem by clustering the pool of unlabelled samples $\mathcal{U}$ from which the query choice is made. In particular, the points with the highest predictive variance are chosen from a pool of samples where the redundancy is already reduced. Informally, this leads to the following basic outline of the procedure:

1. Sample an initial set of training points or parametrisations $\mathbf{x}$ of the model (via uniform or Latin hypercube sampling or taking points on a regular grid) and conduct model checking based on sampled trajectories. These points are used to fit the first iteration of the Gaussian process model.

2. For the next iteration we randomly sample another set of points U and cluster them via regular kmeans clustering algorithm. From the set of cluster centres U_k the query function $\mathcal{Q}$ selects a set of points for model checking. The query function is simply defined by taking the subset U^* of cluster centres U_k where the predictive variance, as defined above, is the highest. This concentrates the sampling to points where the model is most uncertain about its prediction.
3. The points in U^* are labelled by simulating the model for the parametrisations in U^* and checking the resulting trajectories against the logic specification φ. The results are incorporated into the Gaussian process model via the streaming method discussed in Sect. 4.1.
4. Repeat points 2 and 3 until a set computational budget is exhausted.

Predictive Gradient. The second strategy we consider is based on the predictive mean $\bar{g}_\varphi(\mathbf{x})$ of the Gaussian process. Our aim is to concentrate the sampling at the locations where the predictive mean undergoes the most rapid change. This requires gradients of the predictive mean.

We recall from Sect. 3.3 that for a variational posterior $q(\mathbf{u})$ with mean $\boldsymbol{\mu}$ and covariance $\boldsymbol{\Sigma}$, the posterior mean at a point $\mathbf{x}$ is given by

$$\bar{g}_\varphi(\mathbf{x}) = k(\mathbf{x}, \mathbf{z_u}) K_{mm}^{-1} \boldsymbol{\mu} \stackrel{\text{def}}{=} k(\mathbf{x}, \mathbf{z_u}) \boldsymbol{\alpha} \,.$$

Only the first part, the kernel function, depends on $\mathbf{x}$. Thus, in order to get the derivative of the predictive mean we need to differentiate $k(\mathbf{x}, \mathbf{z_u})$. Recall that in this paper we chose to work with the squared exponential kernel given by

$$k(\mathbf{x}, \mathbf{z}_{\mathbf{u}_i}) = \exp\left(-\frac{|\mathbf{x} - \mathbf{z}_{\mathbf{u}_i}|^2}{2\ell}\right) .$$

We have used $\mathbf{z}_{\mathbf{u}_i}$ to denote a single inducing point in the set of inducing points $\mathbf{z_u}$. Thus, the derivative of $k(\mathbf{x}, \mathbf{z}_{\mathbf{u}_i})$ with respect to $\mathbf{x}$ is given by

$$\frac{dk(\mathbf{x}, \mathbf{z}_{\mathbf{u}_i})}{d\mathbf{x}} = -\frac{\mathbf{x} - \mathbf{z}_{\mathbf{u}_i}}{\ell} \exp\left(-\frac{|\mathbf{x} - \mathbf{z}_{\mathbf{u}_i}|^2}{2\ell}\right) = -\frac{\mathbf{x} - \mathbf{z}_{\mathbf{u}_i}}{\ell} k(\mathbf{x}, \mathbf{z}_{\mathbf{u}_i}) . \tag{3}$$

Equation 3 is given for a single inducing point $\mathbf{z}_{\mathbf{u}_i}$. In order to compute the derivative of the posterior mean we need to concatenate this derivative for all m inducing points. Thus, we get

$$\frac{d\bar{g}_\varphi(\mathbf{x})}{d\mathbf{x}} = -\ell^{-1} \begin{bmatrix} \mathbf{x} - \mathbf{z}_{\mathbf{u}_1} \\ \cdots \\ \mathbf{x} - \mathbf{z}_{\mathbf{u}_m} \end{bmatrix} (k(\mathbf{x}, \mathbf{z_u}) \odot \boldsymbol{\alpha})$$

where $\odot$ denotes element-wise multiplication. Given this we can proceed as in the case of the predictive variance. The only change is that instead of considering the predictive variance for each sampled set of parameters we calculate the norm of $\frac{d\bar{g}_\varphi(\mathbf{x})}{d\mathbf{x}}$ and define the query function to choose a subset U^* of cluster centres U_k with the highest norms.

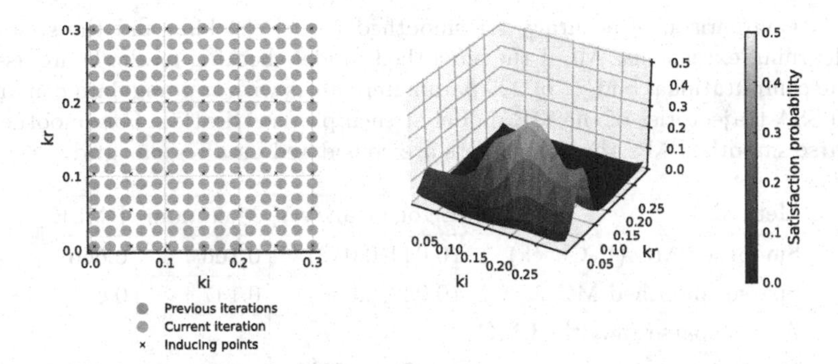

Fig. 2. Mean satisfaction probability surface for sparse smoothed model checking. Inducing points are set as a regular 7×7 grid. Training data is constructed from a set of 10 SSA trajectories at parameter points on a 15×15 grid.

4.3 Implementation

The prototype implementation in written in the Julia programming language and makes use of the tools provided as part of Julia Gaussian Processes repositories [1] to set up the Gaussian process models. The variational inference-based fitting which maximises the ELBO with respect to parameters of the posterior distributions as well as the kernel hyperparameters was implemented for all of the sparse methods. For the standard smoothed model checking we use the U-Check tool [5] available on GitHub [2]. The CTMC models are defined as CRNs with tools provided as part of the SciML ecosystem for scientific simulations [17]. The simulations were carried out on a laptop with Intel i7-10750H CPU.

4.4 Results

In this section we evaluate the proposed active learning methods for model checking on the running SIR example. The methods are compared to the baseline naive stochastic simulation-based model checking and smoothed model checking without the sparse approximation and the active step. We present several metrics for comparing the smoothed model checking results with the empirical mean based on stochastic simulation. The first is the mean and standard deviation of the difference between the mean probability predicted by the fitted Gaussian processes and the empirical mean from the stochastic simulation results at each of the points on the 20×20 grid. Secondly, we consider the maximum difference between the predicted mean probability and the naive empirical mean. Finally, we give the root-mean-square error (RMSE) $\sqrt{\frac{1}{N}\sum\left(\bar{g}_{\varphi}(\mathbf{x}_i) - \bar{f}_{\varphi}(\mathbf{x}_i)\right)^2}$ where $\bar{g}_{\varphi}(\mathbf{x}_i)$ is the predicted mean satisfaction probability for parametrisation $\mathbf{x}_i$. We denote by $\bar{f}_{\varphi}(\mathbf{x}_i)$ the empirical estimate of the satisfaction probability at $\mathbf{x}_i$ given 3000 sample trajectories.

In the active learning experiments we start with a 12×12 grid followed by an active iteration where an additional 81 parameter points are chosen to refine the

Table 1. Comparison of accuracy for smoothed model checking and the sparse and active learning extensions. All of the smoothed model checking methods are assigned the same computational budget of 225 parameter values with training data constructed from 10 SSA trajectories of the SIR model at each parametrisation. For smoothed MC and sparse smoothed MC the parameters are considered on a regular grid.

Method	Error mean/var	Maximum	RMSE
Smoothed MC (U-Check)	(0.044, 0.042)	0.166	0.666
Sparse smoothed MC	(0.042, 0.036)	0.147	0.6
Active sparse smoothed MC			
Predictive variance	(0.033, 0.029)	**0.131**	0.479
Predictive gradient	**(0.03, 0.026)**	0.14	**0.436**
Random sampling	(0.049, 0.039)	0.149	0.681

Table 2. Comparison of computation times (in seconds) for smoothed model checking and the sparse and active learning extensions. The hyperparameter tuning time is included in the inference column.

Method	SSA	Inference	Active query	Total
Naive statistical MC	191.5	N/A	N/A	191.5
Smoothed MC (U-Check tool)	0.21	16.8	N/A	17.0
Sparse smoothed MC	0.48	4.1	N/A	4.5
Active sparse smoothed MC				
Predictive variance	0.57	5.7	0.003	6.2
Predictive gradient	0.55	3.1	0.2	3.9
Random sampling	0.60	2.9	0.00	3.5

approximation for a total of 225 training points. At each parameter point the model checking is conducted for 10 sample trajectories. The inducing points are initialised by choosing a regular 7×7 grid and kept constant for the remainder of the fitting procedure. Similarly we present the results for sparse smoothed model checking for a 15×15 grid with the 7×7 grid of inducing points, as well as smoothed model checking where inducing points are not chosen.

Figure 2 gives the mean satisfaction probability surface based on the fitted sparse Gaussian process. Figure 3 presents the evolution of the predictive mean surface through two active learning iterations. All figures are accompanied by the scatter plots showing where the samples were drawn.

The results for accuracy are summarised in Tables 1 and 2. Table 1 gives the comparisons for each point on the 20×20 grid where the naive model checking was conducted and satisfaction probability estimates exceeding 0.02. This is done to concentrate the analysis to the parts of the parameter space where the surface is not completely flat. As expected, the main benefit of the sparse methods comes from significant reductions in computation costs. Surprisingly however, the stan-

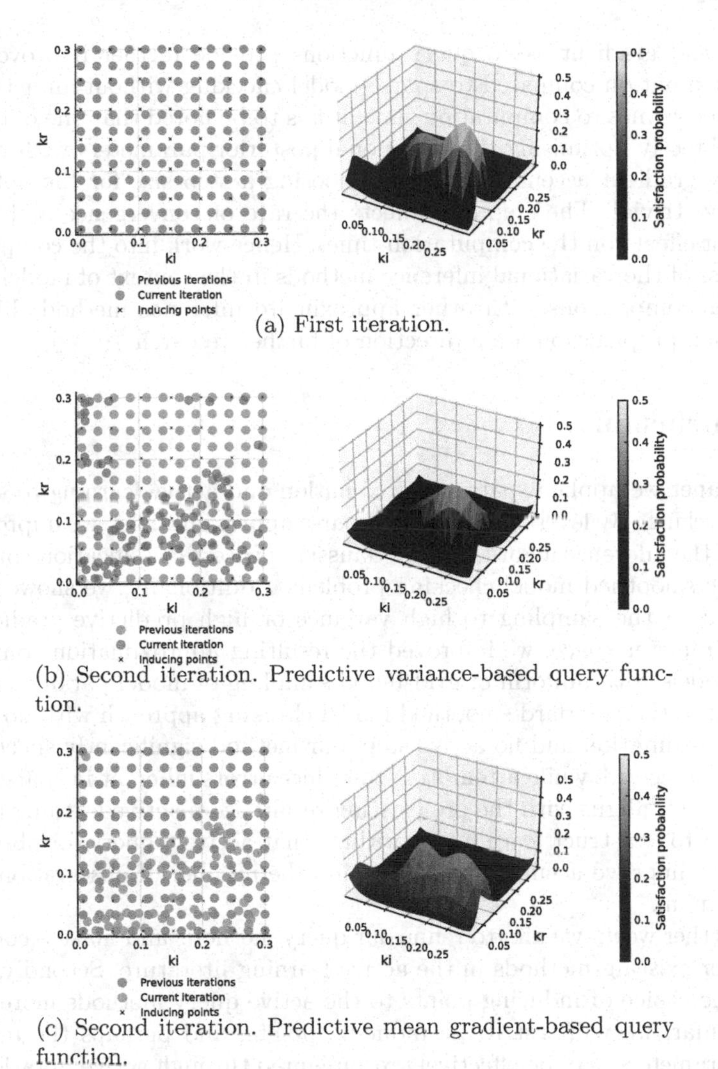

(a) First iteration.

(b) Second iteration. Predictive variance-based query function.

(c) Second iteration. Predictive mean gradient-based query function.

Fig. 3. Mean satisfaction probability surfaces for the active and sparse smoothed model checking methods with 2 iterations. The first iteration of the active learning-based methods fits the Gaussian process inference model based on 10 model checking results for each parameter on a regular 12×12 grid. The active step is then used to exhaust the total computational budget of 225 parameters and refine the approximation.

dard sparse method without active step performs better than the method provided by the U-Check tool with respect to the error metrics considered. The reasons for this may be the differences in the recovered hyperparameters—notably, amplitude ≈2.7 for U-Check versus ≈4.5 for the sparse methods. These differences can be attributed to different methods for tuning the hyperparameters as well as the sparsity assumption. Note that the active methods with predictive variance and gradient-based query functions provide further improvements in the approximation compared to sparse model checking without an active step.

When it comes to computation time, it has to be noted that one of the downsides of directly optimising the variational posterior parameters with respect to ELBO by gradient ascent means that choosing a step size for the optimiser is not always trivial. The step size affects the rate of convergence and can have significant effects on the computation times. Hence work into the computational robustness of the variational inference methods in the context of model checking as well as comparisons with other approximate inference methods like sparse expectation propagation are a direction of further research.

5 Conclusions

In this paper we applied sparse approximation and active learning to smoothed model checking. By leveraging existing sparse approximations, we improved scalability of the inference algorithms for Gaussian process classification corresponding to the smoothed model checking problem. Additionally, we showed that by concentrating the sampling to high variance or high predictive gradient areas of the parameter space, we improved the resulting approximation compared to sparse models with uniform or grid-based sampling of model parameters. When compared to the standard smoothed model checking approach with no inducing point approximation and no active step, our method significantly speeds up the inference procedure while attempting to reduce errors inherent in sparse approximations. This aligns with the pre-existing results from active learning literature which aim to construct learning algorithms that actively query for observations in order to improve accuracy while keeping the number of observations needed to a minimum.

As further work, we aim to refine our query methods and make a comparison with other existing methods in the active learning literature. Secondly, we plan to link the choice of inducing points to the active query methods more directly. In particular, we will test if the inducing points, and perhaps the underlying kernel parameters, can be effectively reconfigured through active iterations. This would further improve the approximation to the satisfaction probability surface at parts of the parameter space where satisfaction probability undergoes change. Finally we will consider alternative kernel functions. The kernel function chosen in this paper is a standard first approach in many settings but is best suited for modelling very smooth functions—not necessarily the case with satisfaction probability surfaces for parametric CTMCs.

References

1. Gaussian processes for machine learning in Julia. https://github.com/JuliaGaussianProcesses. Accessed 07 May 2021
2. U-check tool. https://github.com/dmilios/U-check. Accessed 05 July 2021

3. Agha, G., Palmskog, K.: A survey of statistical model checking. ACM Trans. Model. Comput. Simul. **28**(1), 6:1-6:39 (2018)
4. Bartocci, E., Bortolussi, L., Nenzi, L., Sanguinetti, G.: System design of stochastic models using robustness of temporal properties. Theor. Comput. Sci. **587**, 3–25 (2015)
5. Bortolussi, L., Milios, D., Sanguinetti, G.: U-check: model checking and parameter synthesis under uncertainty. In: Campos, J., Haverkort, B.R. (eds.) QEST 2015. LNCS, vol. 9259, pp. 89–104. Springer, Cham (2015). https://doi.org/10.1007/978-3-319-22264-6_6
6. Bortolussi, L., Milios, D., Sanguinetti, G.: Smoothed model checking for uncertain continuous-time Markov chains. Inf. Comput. **247**, 235–253 (2016)
7. Bortolussi, L., Silvetti, S.: Bayesian statistical parameter synthesis for linear temporal properties of stochastic models. In: Beyer, D., Huisman, M. (eds.) TACAS 2018. LNCS, vol. 10806, pp. 396–413. Springer, Cham (2018). https://doi.org/10.1007/978-3-319-89963-3_23
8. Bui, T.D., Nguyen, C.V., Turner, R.E.: Streaming sparse Gaussian process approximations. In: Advances in Neural Information Processing Systems 30. Annual Conference on Neural Information Processing Systems 2017, pp. 3299–3307 (2017)
9. Csato, L., Opper, M.: Sparse online Gaussian processes. Neural Comput. **14**, 641–668 (2002)
10. Donzé, A., Maler, O.: Robust satisfaction of temporal logic over real-valued signals. In: Chatterjee, K., Henzinger, T.A. (eds.) FORMATS 2010. LNCS, vol. 6246, pp. 92–106. Springer, Heidelberg (2010). https://doi.org/10.1007/978-3-642-15297-9_9
11. Hernandez-Lobato, D., Hernandez-Lobato, J.M.: Scalable Gaussian process classification via expectation propagation. In: Gretton, A., Robert, C.C. (eds.) Proceedings of the 19th International Conference on Artificial Intelligence and Statistics (2016). Proceedings of Machine Learning Research, vol. 51, pp. 168–176. PMLR
12. Jha, S.K., Clarke, E.M., Langmead, C.J., Legay, A., Platzer, A., Zuliani, P.: A Bayesian approach to model checking biological systems. In: Degano, P., Gorrieri, R. (eds.) CMSB 2009. LNCS, vol. 5688, pp. 218–234. Springer, Heidelberg (2009). https://doi.org/10.1007/978-3-642-03845-7_15
13. Kwiatkowska, M., Norman, G., Parker, D.: Stochastic model checking. In: Bernardo, M., Hillston, J. (eds.) SFM 2007. LNCS, vol. 4486, pp. 220–270. Springer, Heidelberg (2007). https://doi.org/10.1007/978-3-540-72522-0_6
14. Legay, A., Lukina, A., Traonouez, L.M., Yang, J., Smolka, S.A., Grosu, R.: Statistical model checking. In: Steffen, B., Woeginger, G. (eds.) Computing and Software Science. LNCS, vol. 10000, pp. 478–504. Springer, Cham (2019). https://doi.org/10.1007/978-3-319-91908-9_23
15. Maler, O., Nickovic, D.: Monitoring temporal properties of continuous signals. In: Lakhnech, Y., Yovine, S. (eds.) FORMATS/FTRTFT -2004. LNCS, vol. 3253, pp. 152–166. Springer, Heidelberg (2004). https://doi.org/10.1007/978-3-540-30206-3_12
16. Minka, T.P.: Expectation propagation for approximate Bayesian inference. In: Breese, J.S., Koller, D. (eds.) UAI: Proceedings of the 17th Conference in Uncertainty in Artificial Intelligence, pp. 362–369. Morgan Kaufmann (2001)
17. Rackauckas, C., Nie, Q.: Differential equations.jl-a performant and feature-rich ecosystem for solving differential equations in Julia. J. Open Res. Softw. **5**(1) (2017)
18. Rasmussen, C.E., Williams, C.K.I.: Gaussian Processes for Machine Learning. Adaptive Computation and Machine Learning. MIT Press, Cambridge (2006)

19. Santner, T.J., Williams, B.J., Notz, W.I.: The Design and Analysis of Computer Experiments. Springer Series in Statistics, Springer, New York (2003). https://doi.org/10.1007/978-1-4757-3799-8
20. Sen, K., Viswanathan, M., Agha, G.: Statistical model checking of black-box probabilistic systems. In: Alur, R., Peled, D.A. (eds.) CAV 2004. LNCS, vol. 3114, pp. 202–215. Springer, Heidelberg (2004). https://doi.org/10.1007/978-3-540-27813-9_16
21. Settles, B.: From theories to queries. In: Active Learning and Experimental Design workshop, In conjunction with AISTATS. JMLR Proceedings, vol. 16, pp. 1–18. JMLR.org (2011). http://proceedings.mlr.press/v16/settles11a/settles11a.pdf
22. Settles, B.: Active Learning. Synthesis Lectures on Artificial Intelligence and Machine Learning, Morgan & Claypool Publishers (2012). https://doi.org/10.2200/S00429ED1V01Y201207AIM018
23. Titsias, M.K.: Variational learning of inducing variables in sparse gaussian processes. In: Dyk, D.A.V., Welling, M. (eds.) Proceedings of the Twelfth International Conference on Artificial Intelligence and Statistics, AISTATS. JMLR Proceedings, vol. 5, pp. 567–574 (2009)

Safe Learning for Near-Optimal Scheduling

Damien Busatto-Gaston[1], Debraj Chakraborty[1], Shibashis Guha[2](✉), Guillermo A. Pérez[3], and Jean-François Raskin[1]

[1] Université libre de Bruxelles, Brussels, Belgium
[2] Tata Institute of Fundamental Research, Mumbai, India
shibashis@tifr.res.in
[3] University of Antwerp – Flanders Make, Antwerp, Belgium

Abstract. In this paper, we investigate the combination of synthesis, model-based learning, and online sampling techniques to obtain safe and near-optimal schedulers for a preemptible task scheduling problem. Our algorithms can handle Markov decision processes (MDPs) that have 10^{20} states and beyond which cannot be handled with state-of-the art probabilistic model-checkers. We provide probably approximately correct (PAC) guarantees for learning the model. Additionally, we extend Monte-Carlo tree search with advice, computed using safety games or obtained using the earliest-deadline-first scheduler, to safely explore the learned model online. Finally, we implemented and compared our algorithms empirically against shielded deep Q-learning on large task systems.

Keywords: Model-based learning · Monte-Carlo tree search · Task scheduling

1 Introduction

In this paper, we show how to combine synthesis, model-based learning, and online sampling techniques to solve a scheduling problem featuring both hard and soft constraints. We investigate solutions to this problem both from a theoretical and from a more pragmatic point of view. On the theoretical side, we show how safety guarantees (as understood in formal verification) can be combined with guarantees offered by the probably approximately correct (PAC) learning framework [23]. On the pragmatic side, we show how safety guarantees obtained from automatic synthesis can be combined with Monte-Carlo tree search (MCTS) [20] to offer a scalable and practical solution to solve the scheduling problem at hand.

The scheduling problem that we consider is defined as follows. A task system is composed of a set of n preemptible tasks $(\tau_i)_{i \in [n]}$ partitioned into a set F of

This work was supported by the ARC "Non-Zero Sum Game Graphs" project (Fédération Wallonie-Bruxelles), the EOS "Verilearn" project (F.R.S.-FNRS & FWO), and the FWO "SAILor" project (G030020N).

A. Abate and A. Marin (Eds.): QEST 2021, LNCS 12846, pp. 235–254, 2021.
https://doi.org/10.1007/978-3-030-85172-9_13

soft tasks and a set H of hard tasks. Time is assumed to be discrete and measured e.g. in CPU ticks. Each task τ_i generates an infinite number of instances $\tau_{i,j}$, called *jobs*, with $j = 1, 2, \ldots$ Jobs generated by both hard and soft tasks are equipped with deadlines, which are relative to the respective arrival times of the jobs in the system. The computation time requirements of the jobs follow a discrete probability distribution, and are unknown to the scheduler but upper bounded by their relative deadline. Jobs generated by hard tasks must complete before their respective deadlines. For jobs generated by soft tasks, deadline misses result in a penalty/cost. The tasks are assumed to be independent and generated stochastically: the occurrence of a new job of one task does not depend on the occurrences of jobs of other tasks, and both the inter-arrival and computation times of jobs are independent random variables. The scheduling problem consists in finding a *scheduler*, i.e. a function that associates, to all CPU ticks, a task that must run at that moment; in order to: (i) avoid deadline misses by hard tasks; and (ii) minimise the mean cost of deadline misses by soft tasks.

In [13], we modelled the semantics of the task system using a Markov decision process (MDP) and posed the problem of computing an optimal and safe scheduler. However, that work assumes that the distribution of all tasks is known a priori which may be unrealistic. Here, we investigate learning techniques to build algorithms that can schedule safely and optimally a set of hard and soft tasks if only the deadlines and the domains of the distributions describing the tasks of the system are known a priori and not the exact distributions. This is a more realistic assumption. Our motivation was also to investigate the joint application of both synthesis techniques coming from the field of formal verification and learning techniques on an understandable yet challenging setting.

Contributions. First, we show the distributions underlying a task system with only soft tasks are *efficiently* PAC learnable: by executing the task system for a polynomial number of steps, enough samples can be collected to infer ϵ-accurate approximations of the distributions with high probability (Theorem 1).

Then, we consider the general case of systems with both hard and soft tasks. Here, safe PAC learning is *not* always possible, and we identify two algorithmically-checkable sufficient conditions for task systems to be safely learnable (Theorems 2 and 3). These crucially depend on the underlying MDP being a single maximal end-component, as is the case in our setting (Lemma 2). Subsequently, we can use *robustness* results on MDPs to compute or learn near-optimal safe strategies from the learnt models (Theorem 4).

Third, in order to evaluate the relevance of our algorithms, we present experiments of a prototype implementation. These empirically validate the efficient PAC guarantees. Unfortunately, the learnt models are often too large for the probabilistic model-checking tools. In contrast, the MCTS-based algorithm scales to larger examples: e.g. we learn safe scheduling strategies for systems with more than 10^{20} states. Our experiments also show that a strategy obtained using deep Q-learning [2,18] by assigning high costs to missing deadlines of hard tasks does not respect safety, even if one learns for a long period of time and the deadline-miss costs of hard tasks are very high (cf. [1]).

Related Works. In [13], we introduced the scheduling problem considered here but made the assumption that the underlying distributions of the tasks are known. We drop this assumption here and provide learning algorithms. In [1], the framework to combine safety via shielding and model-free reinforcement learning is introduced and applied to several examples using table-based Q-learning as well as deep RL. In [3], shield synthesis is studied for long-run objective guarantees instead of safety requirements. Unlike our work, the transition probabilities on MDPs in both [1] and [3] are assumed to be known. We observe that [1] and [3] do not provide model-based learning and PAC guarantees. While some pre-shielding literature does consider unknown MDPs (see, e.g. [12]), we are not aware of PAC-learning works that focus on scheduling problems.

In [16], we studied a framework to mix reactive synthesis and model-based reinforcement learning for mean-payoff with PAC guarantees. There, the learning algorithm estimates the probabilities on the transitions of the MDP. In our approach, we do not estimate these probabilities directly from the MDP, but learn probabilities for the individual tasks in the task system. The efficient PAC guarantees that we have obtained for the model-based part cannot be obtained from that framework. Finally, in [8] we introduced a first combination of shielding with model-predictive control using MCTS, but did not consider learning.

2 Preliminaries

We denote by $\mathbb{N}$ the set of natural numbers; by $\mathbb{Q}$, the set of rational numbers; and by $\mathbb{Q}_{\geq 0}$ the set $\{q \in \mathbb{Q} \mid q \geq 0\}$ of all non-negative rational numbers. Given $n \in \mathbb{N}$, we denote by $[n]$ the set $\{1, \ldots, n\}$. Given a finite set A, a (rational) *probability distribution* over A is a function $p \colon A \to [0,1] \cap \mathbb{Q}$ such that $\sum_{a \in A} p(a) = 1$. We call A the *domain of* p, and denote it by $\mathsf{Dom}(p)$. We denote the set of probability distributions on A by $\mathcal{D}(A)$. The *support* of the probability distribution p on A is $\mathsf{Supp}(p) = \{a \in A \mid p(a) > 0\}$. A distribution is called *Dirac* if $|\mathsf{Supp}(p)| = 1$. For a probability distribution p, the minimum probability assigned by p to the elements in $\mathsf{Supp}(p)$ is $\pi^p_{\min} = \min_{a \in \mathsf{Supp}(p)} (p(a))$. We say two distributions p and p' are *structurally identical* if $\mathsf{Supp}(p) = \mathsf{Supp}(p')$. Given two structurally identical distributions p and p', for $0 < \epsilon < 1$, we say that p is ϵ-close to p', denoted $p \sim^{\epsilon} p'$, if $\mathsf{Supp}(p) = \mathsf{Supp}(p')$, and for all $a \in \mathsf{Supp}(p)$, we have that $|p(a) - p'(a)| \leq \epsilon$.

Scheduling Problem. An instance of the scheduling problem studied in [13] consists of a task system $\Upsilon = ((\tau_i)_{i \in [n]}, F, H)$, where $(\tau_i)_{i \in [n]}$ are n preemptible tasks partitioned into hard and soft tasks H and F respectively. The latter need to be scheduled on a *single processor*. Formally, the work of [13] relies on a probabilistic model for the computation times of the jobs and for the delay between the arrival of two successive jobs of the same task. For all $i \in [n]$, task τ_i is defined as a tuple $\langle \mathcal{C}_i, D_i, \mathcal{A}_i \rangle$, where: (i) $\mathcal{C}_i$ is a discrete probability distribution on the (finitely many) possible computation times of the jobs generated by τ_i; (ii) $D_i \in \mathbb{N}$ is the deadline of all jobs generated by τ_i which is relative to their arrival time; and (iii) $\mathcal{A}_i$ is a discrete probability distribution on the (finitely many) possible inter-arrival times of the jobs generated by τ_i. We denote by $\pi^{\Upsilon}_{\max}$ the maximum probability appearing in the definition of Υ, that is, across all the distributions $\mathcal{C}_i$ and $\mathcal{A}_i$, for all $i \in [n]$. It is assumed that $\max(\mathsf{Dom}(\mathcal{C}_i)) \leq D_i \leq \min(\mathsf{Dom}(\mathcal{A}_i))$ for all $i \in [n]$; hence, at any point in time, there is at most one job per task in the system. Also note that when a new job of some task arrives at the system, the deadline for the previous job of this task is already over. Finally, we assume that the task system is *schedulable for the hard tasks*, meaning that it is possible to guarantee that jobs associated to hard tasks never miss their deadlines. On the other hand, the full set of tasks may not be schedulable, so that jobs associated with soft tasks may be allowed to miss their deadlines. The potential degradation in the quality when a soft task misses its deadline is modelled by a cost function $cost : F \to \mathbb{Q}_{\geq 0}$ that associates to each soft task τ_j a cost $c(j)$ that is incurred every time a job of τ_j misses its deadline. As a final observation, we recall the *earliest deadline first* (EDF) algorithm that always gives execution time to the job closest to its deadline. EDF is an optimal scheduling algorithm in the following sense: if a task system is schedulable (without any misses at all) then EDF will yield such a feasible schedule [6]. In general, applying EDF on both the hard and soft tasks may cause hard tasks to miss deadlines, as the entire task system may not be schedulable. However, one may apply EDF on hard tasks only, and allow for soft tasks whenever no hard task is available. This version of EDF ensures that all jobs of hard tasks are scheduled in time, but does not guarantee optimality with respect to cost.

Given a task system $\Upsilon = ((\tau_i)_{i\in[n]}, F, H)$ with n tasks, the structure of Υ is $((\mathsf{struct}(\tau_i))_{i\in[n]}, F, H)$ where $\mathsf{struct}(\langle \mathcal{C}, D, \mathcal{A}\rangle) = (\langle \mathsf{Dom}(\mathcal{C}), D, \mathsf{Dom}(\mathcal{A})\rangle)$. We denote by $\mathcal{C}_{\max}$ and $\mathcal{A}_{\max}$ resp. the maximum computation time, and the maximum inter-arrival time of a task in Υ. Formally, $\mathcal{C}_{\max} = \max(\bigcup_{i\in[n]} \mathsf{Dom}(\mathcal{C}_i))$, and $\mathcal{A}_{\max} = \max(\bigcup_{i\in[n]} \mathsf{Dom}(\mathcal{A}_i))$. Note that $\mathcal{A}_{\max} \geq \mathcal{C}_{\max}$. We also let $\mathbb{D} = \max_{i\in[n]}(|\mathsf{Dom}(\mathcal{A}_i)|)$. We denote by $|\Upsilon|$ the number of tasks in the task system Υ. Consider two task systems $\Upsilon_1 = ((\tau^1_i)_{i\in[n]}, F, H)$, and $\Upsilon_2 = ((\tau^2_i)_{i\in[n]}, F, H)$, with $|\Upsilon_1| = |\Upsilon_2|$, $\tau^j_i = \langle \mathcal{C}^j_i, D^j_i, \mathcal{A}^j_i\rangle$ for all $i \in [n]$ and $j \in [2]$. The two task systems Υ_1 and Υ_2 are said to be *ϵ-close*, denoted $\Upsilon_1 \approx^{\epsilon} \Upsilon_2$, if (i) $\mathsf{struct}(\Upsilon^1) = \mathsf{struct}(\Upsilon^2)$, (ii) for all $i \in [n]$, we have $\mathcal{A}^1_i \sim^{\epsilon} \mathcal{A}^2_i$, and (iii) for all $i \in [n]$, we have $\mathcal{C}^1_i \sim^{\epsilon} \mathcal{C}^2_i$.

Markov Decision Processes. Let us now introduce *Markov Decision Process* (MDP) as they form the basis of the formal model of [13], which we recall later. A finite *Markov decision process* is a tuple $\Gamma = \langle V, E, L, (V_{\Box}, V_{\bigcirc}), A, \delta, \mathsf{cost}\rangle$, where: (i) A is a finite set of actions; (ii) $\langle V, E\rangle$ is a finite directed graph and L is an edge-labelling function (we denote by $E(v)$ the set of outgoing edges from vertex v); (iii) the set of vertices V is partitioned into $V_{\Box}$ and $V_{\bigcirc}$; (iv) the graph is bipartite i.e. $E \subseteq (V_{\Box} \times V_{\bigcirc}) \cup (V_{\bigcirc} \times V_{\Box})$, and the labelling function is s.t. $L(v, v') \in A$ if $v \in V_{\Box}$, and $L(v, v') \in \mathbb{Q}$ if $v \in V_{\bigcirc}$; and (v) δ assigns to each vertex $v \in V_{\bigcirc}$ a rational probability distribution on $E(v)$. For all edges e, we let $\mathsf{cost}(e) = L(e)$ if $L(e) \in \mathbb{Q}$, and $\mathsf{cost}(e) = 0$ otherwise. We further assume that, for all $v \in V_{\Box}$, for all e, e' in $E(v)$: $L(e) = L(e')$ implies $e = e'$, i.e. an action identifies uniquely an outgoing edge. Given $v \in V_{\Box}$, and $a \in A$, we define $\mathsf{Post}(v, a) = \{v' \in V_{\bigcirc} \mid (v, v') \in E \text{ and } L(v, v') = a\} \cup \{v'' \in V_{\Box} \mid \exists v' :$

$(v, v') \in E, L(v, v') = a$ and $\delta(v', v'') > 0\}$. For all vertices $v \in V_\Box$, we denote by $A(v)$, the set of actions $\{a \in A \mid \mathsf{Post}(v, a) \cap V_\Box \neq \emptyset\}$. The size of an MDP Γ, denoted $|\Gamma|$, is the sum of the number of vertices and the number of edges, that is, $|V| + |E|$. An MDP $\Gamma = \langle V, E, L, (V_\Box, V_\bigcirc), A, \delta, \mathsf{cost}\rangle$ is said to *structurally identical* to another MDP $\Gamma' = \langle V, E, L', (V_\Box, V_\bigcirc), A, \delta', \mathsf{cost}\rangle$ if for all $v \in V_\bigcirc$, we have that $\mathsf{Supp}(\delta(v)) = \mathsf{Supp}(\delta'(v))$. For two structurally identical MDPs Γ and Γ' with distribution assignment functions δ and δ' respectively, we say that Γ is ϵ-approximate to Γ', denoted $\Gamma \approx^\epsilon \Gamma'$, if for all $v \in V_\bigcirc$: $\delta(v) \sim^\epsilon \delta'(v)$.

An MDP Γ can be interpreted as a game $\mathcal{G}_\Gamma$ between two players: $\Box$ and $\bigcirc$, who own the vertices in $V_\Box$ and $V_\bigcirc$ respectively. A play in an MDP is a path in its underlying graph $\langle V, E, A \cup \mathbb{Q}\rangle$. We say that a prefix $\pi(n)$ of a play π belongs to player $i \in \{\Box, \bigcirc\}$, iff its last vertex $\mathsf{Last}(\pi(n))$ is in V_i. The set of prefixes that belong to player i is denoted by $\mathsf{Prefs}_i(\mathcal{G}_\Gamma)$. A play is obtained by the interaction of the players: if the current play prefix $\pi(n)$ belongs to $\Box$, she plays by picking an edge $e \in E(\mathsf{Last}(\pi(n)))$ (or, equivalently, an action that labels a necessarily unique edge from $\mathsf{Last}(\pi(n))$). Otherwise, when $\pi(n)$ belongs to $\bigcirc$, the next edge $e \in E(\mathsf{Last}(\pi(n)))$ is chosen randomly according to $\delta(\mathsf{Last}(\pi(n)))$. In both cases, the plays prefix is extended by e and the game goes *ad infinitum.*

A (deterministic) *strategy* of $\Box$ is a function $\sigma_\Box : \mathsf{Prefs}_\Box(\mathcal{G}) \to E$, such that $\sigma_\Box(\rho) \in E(\mathsf{Last}(\rho))$ for all prefixes. A strategy $\sigma_\Box$ is *memoryless* if for all finite prefixes ρ_1 and $\rho_2 \in \mathsf{Prefs}(\mathcal{G})$: $\mathsf{Last}(\rho_1) = \mathsf{Last}(\rho_2)$ implies $\sigma_\Box(\rho_1) = \sigma_\Box(\rho_2)$. For memoryless strategies, we will abuse notations and assume that such strategies σ are of the form $\sigma : V_\Box \to E$ (i.e., the strategy associates the edge to play to the current vertex and not to the full prefix played so far). From now on, we will consider memoryless deterministic strategies unless otherwise stated. Let $\Gamma = \langle V, E, L, (V_\Box, V_\bigcirc), A, \delta, \mathsf{cost}\rangle$ be an MDP, and let $\sigma_\Box$ be a *memoryless* strategy. Then, assuming that $\Box$ plays according to $\sigma_\Box$, we can express the behaviour of Γ as a Markov chain $\Gamma[\sigma_\Box]$, where the probability distributions reflect the stochastic choices of $\bigcirc$ (see [13] for the details).

End Components. An *end-component* (EC) $M = (T, A')$, with $T \subseteq V$ and $A' : T \cap V_\Box \to 2^A$, is a *sub-MDP* of Γ such that: for all $v \in T \cap V_\Box$, $A'(v)$ is a subset of the actions available to $\Box$ from v; for all $a \in A'(v)$, $\mathsf{Post}(v, a) \subseteq T$; and, it's underlying graph is strongly connected. A *maximal end-component* (MEC) is an EC that is not included in any other EC.

MDP for the Scheduling Problem. Given a system $\Upsilon = \{\tau_1, \tau_2, \ldots, \tau_n\}$ of tasks, we describe below the modelling of the scheduling problem by an MDP $\Gamma_\Upsilon = \langle V, E, L, (V_\Box, V_\bigcirc), A, \delta, \mathsf{cost}\rangle$ as it appears in [13]. The two players $\Box$ and $\bigcirc$ correspond respectively to the *Scheduler* and the task generator (*TaskGen*) respectively. Since there is at most one job per task that is active at all times, vertices encode the following information about each task τ_i: (i) a *distribution* c_i over the job's possible remaining computation times (rct); (ii) the time d_i up to its deadline; and (iii) a distribution a_i over the possible times up to the next arrival of a new job. We also tag vertices with either $\Box$ or $\bigcirc$ to remember their respective owners and we have a vertex $\bot$ that is reached when a hard task misses a deadline. For a vertex $v = ((c_1, d_1, a_1) \ldots (c_n, d_n, a_n), \Delta)$, for $\Delta \in \{\Box, \bigcirc\}$, let

$\mathsf{active}(v) = \{i \mid c_i(0) \neq 1 \text{ and } d_i > 0\}$ be the tasks that have an active job in v; $\mathsf{dlmiss}(v) = \{i \mid c_i(0) = 0 \text{ and } d_i = 0\}$, those that have missed a deadline in v.

Possible Moves. The possible actions of Scheduler are to schedule an active task or to idle the CPU. We model this by having, from all vertices $v \in V_\square$ one transition labelled by some element from $\mathsf{active}(v)$, or by ε. The moves of TaskGen consist in selecting, for each task one possible *action* out of four: either (i) nothing (ε); or (ii) to finish the current job without submitting a new one (*fin*); or (iii) to submit a new job while the previous one is already finished (*sub*); or (iv) to submit a new job and kill the previous one, in the case of a soft task (*killANDsub*), which will incur a cost.

We consider the following example from [13].

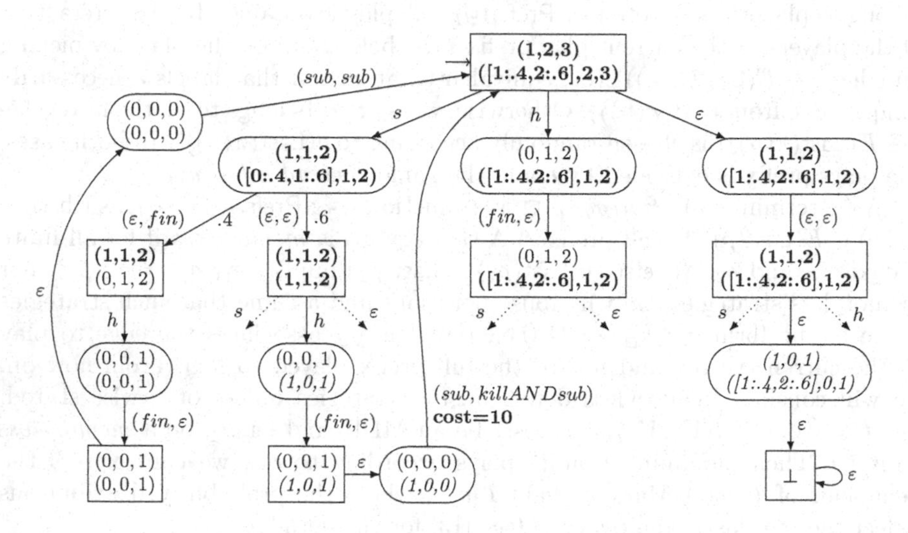

Fig. 1. MDP excerpt for Example 1. **Bold** tasks are active, those in *italics* have missed a deadline.

Example 1. Consider a system with one hard task $\tau_h = \langle \mathcal{C}_h, 2, \mathcal{A}_h \rangle$ s.t. $\mathcal{C}_h(1) = 1$ and $\mathcal{A}_h(3) = 1$; one soft task $\tau_s = \langle \mathcal{C}_s, 2, \mathcal{A}_s \rangle$ s.t. $\mathcal{C}_s(1) = 0.4$, $\mathcal{C}_s(2) = 0.6$, and $\mathcal{A}_s(3) = 1$; and the cost function c s.t. $c(\tau_s) = 10$. Figure 1 presents an excerpt of the MDP Γ_Υ built from the set of tasks $\tau = \{\tau_h, \tau_s\}$ of Example 1. A distribution p with support $\{x_1, x_2, \ldots, x_n\}$ is denoted by $[x_1 : p(x_1), x_2 : p(x_2), \ldots; x_n : p(x_n)]$. When p is s.t. $p(x) = 1$ for some x, we simply denote p by x. Vertices from $V_\square$ and $V_\bigcirc$ are depicted by rectangles and rounded rectangles respectively. Each vertex is labelled by (c_h, d_h, a_h) on the top, and (c_s, d_s, a_s) below.

A strategy to avoid missing a deadline of τ_h consists in first scheduling τ_s, then τ_h. One then reaches the left-hand part of the graph from which $\square$ can avoid $\bot$ whatever $\bigcirc$ does. Other safe strategies are possible: the first step of the algorithm in [13] is to compute all the *safe* nodes (i.e. those from which $\square$ can ensure to avoid $\bot$), and then find an optimal one w.r.t to missed-deadline costs.

There are two optimal memoryless strategies, one in which Scheduler first chooses to execute τ_h, then τ_s; and another where τ_s is scheduled for 1 time unit, and then preempted to let τ_h execute. Since the time difference between the arrival of two consecutive jobs of the soft task τ_s is 3 and the cost of missing a deadline is 10, for both of these optimal strategies, the soft task's deadline is missed with probability 0.6 over this time duration of 3, and hence the mean-cost is 2. There is another safe schedule that is not optimal which only grants τ_h is CPU access, and never schedules τ_s, thus giving a mean-cost of $\frac{10}{3}$. $\square$

Expected Mean-Cost. Let us first associate a value, called the *mean-cost* $\mathsf{MC}(\pi)$ to all plays π in an MDP $\Gamma = \langle V, E, L, (V_\square, V_\bigcirc), A, \delta, \mathsf{cost}\rangle$. First, for a prefix $\rho = e_0 e_1 \ldots e_{n-1}$, we define $\mathsf{MC}(\rho) = \frac{1}{n}\sum_{i=0}^{i=n-1} \mathsf{cost}(e_i)$ (recall that $\mathsf{cost}(e) = 0$ when $L(e)$ is an action). Then, for a play $\pi = e_0 e_1 \ldots$, we have $\mathsf{MC}(\pi) = \limsup_{n\to\infty} \mathsf{MC}(\pi(n))$. Observe that MC is a measurable function. A strategy $\sigma_\square$ is *optimal* for the mean cost from some initial vertex $v_{\mathsf{init}} \in V_\square$ if $\mathbb{E}^{\Gamma[\sigma_\square]}_{v_{\mathsf{init}}}(\mathsf{MC}) = \inf_{\sigma'_\square} \mathbb{E}^{\Gamma[\sigma'_\square]}_{v_{\mathsf{init}}}(\mathsf{MC})$. Such *optimal* strategy always exists, and it is well-known that there is always one which is *memoryless*. Moreover, this problem can be solved in polynomial time through linear programming [11] or in practice using value iteration (as implemented, for example, in the tool Storm [9]). We denote by $\mathbb{E}^{\Gamma}_{v_{\mathsf{init}}}(\mathsf{MC})$ the optimal value $\inf_{\sigma_\square} \mathbb{E}^{\Gamma[\sigma_\square]}_{v_{\mathsf{init}}}(\mathsf{MC})$.

Safety Synthesis. Given an MDP $\Gamma = \langle V, E, L, (V_\square, V_\bigcirc), A, \delta, \mathsf{cost}\rangle$, an initial vertex $v_{\mathsf{init}} \in V$, and a strategy $\sigma_\square$, we define the set of possible *outcomes* in the Markov chain $\Gamma[\sigma_\square]$ as the set of paths $v_{\mathsf{init}} = v_0 v_1 v_2 \ldots$ in $\Gamma[\sigma_\square]$ s.t., for all $i \geq 0$, there is non-null probability to go from v_i to v_{i+1} in $\Gamma[\sigma_\square]$. Let $V_{\mathsf{Outs}_{\Gamma[\sigma_\square]}(v_{\mathsf{init}})} \subseteq V$ denote the set of vertices visited in the set of possible outcomes $\mathsf{Outs}_{\Gamma[\sigma_\square]}(v_{\mathsf{init}})$.

Given Γ with vertices V, initial vertex $v_{\mathsf{init}} \in V$, and a set $V_{\mathsf{bad}} \subseteq V$ of *bad vertices*, the *safety synthesis problem* is to decide whether $\square$ has a strategy $\sigma_\square$ ensuring to visit the safe vertices only, i.e.: $V_{\mathsf{Outs}_{\Gamma[\sigma_\square]}(v_{\mathsf{init}})} \cap V_{\mathsf{bad}} = \emptyset$. If this is the case, we call such a strategy *safe*. The safety synthesis problem is decidable in polynomial time for MDPs (see, e.g., safety games in [22]). Moreover, if a safe strategy exists, then there is a *memoryless* safe strategy. Henceforth, we will consider safe strategies that are memoryless only. We say that a vertex v is safe iff $\square$ has a safe strategy from v, and that an edge $e = (v, v') \in E \cap (V_\square \times V_\bigcirc)$ is safe iff there is a safe strategy $\sigma_\square$ s.t. $\sigma_\square(v) = v'$. So, the *safe edges* $\mathsf{safe}(v)$ from some node v correspond to the choices that $\square$ can safely make from v. The set of safe edges exactly correspond to the set of safe actions that $\square$ can make from v. Then, we let the *safe region* of Γ be the MDP Γ^{safe} obtained from Γ by applying the following transformations: (i) remove from Γ all *unsafe edges*; (ii) remove from Γ all vertices and edges that are not reachable from v_{init}.

Most General Safe Scheduler. Consider a task system Υ that is schedulable for the hard tasks. Then, Scheduler has a winning strategy to avoid $\bot$ in Γ_Υ. We say a non-deterministic strategy in Γ_{Υ} is the *most general safe scheduler* (MGS) for the hard tasks if from any vertex of Scheduler it allows all safe edges[1].

3 Model-Based Learning

We now investigate the case of *model-based learning* of task systems. First, we consider the simpler case of task systems with only soft tasks. We show that those systems are always efficiently PAC learnable. Second, we consider learning task systems with both hard and soft tasks. In that case, we study two conditions for learnability. The first condition allows us to identify task systems that are safely PAC learnable, i.e. learnable while enforcing safety for the hard tasks. The second condition is stronger and allows us to identify task systems that are safely and *efficiently* PAC learnable.

Learning Setting. We consider a setting in which we are given the structure of a task system $\Upsilon = ((\tau_i)_{i\in I}, F, H)$ to schedule. While the structure is known, the actual distributions that describe the behaviour of the tasks are unknown and need to be learnt to behave optimally or near optimally. The learning must be done only by observing the jobs that arrive along time. When the task system contains some hard tasks ($H \neq \emptyset$), all deadlines of such tasks must be enforced.

For learning the inter-arrival time distribution of a task, a *sample* corresponds to observing the time difference between the arrivals of two consecutive jobs of that task. For learning the computation time distribution, a sample corresponds to observing the CPU time a job of the task has been assigned up to completion. Thus if a job does not finish execution before its deadline, we do not obtain a valid sample for the computation time. Given a class of task systems, we say:

- the class is *probably approximately correct (PAC) learnable* if there is an algorithm $\mathbb{L}$ such that for all task systems Υ in this class, for all $\epsilon, \gamma \in (0, 1)$: given $\mathsf{struct}(\Upsilon)$, the algorithm $\mathbb{L}$ can execute the task system Υ, and can compute Υ^M such that $\Upsilon \approx^{\epsilon} \Upsilon^M$, with probability at least $1-\gamma$.
- the class is *safely PAC learnable* if it is PAC learnable, and $\mathbb{L}$ can ensure safety for the hard tasks while computing Υ^M.
- the class is (safely) *efficiently PAC learnable* if it is (safely) PAC learnable, and there is a polynomial q in the size of the task system, in $1/\epsilon$, and in $1/\gamma$, s.t. $\mathbb{L}$ obtains enough samples to compute Υ^M in a time bounded by q.

Note that our notion of efficient PAC learning is stronger than the definition used in classical PAC learning terminology [23] since we take into account the time that is needed to get samples and not only the number of samples needed.

Learning Discrete Finite Distributions. To learn an unknown discrete distribution p defined on a finite domain $\mathsf{Dom}(p)$, we collect i.i.d. samples from that distribution and infer a model of it. Formally, given a sequence $\mathbb{S} = (s_j)_{j\in J}$ of

[1] The existence of a most general safe scheduler follows from the existence of a unique most general (a.k.a. maximally permissive) strategy for safety objectives [19].

samples drawn i.i.d. from the distribution p, we denote by $p(\mathbb{S}) : \mathsf{Dom}(p) \to [0,1]$, the function that maps every element $a \in \mathsf{Dom}(p)$ to its relative frequency in $\mathbb{S}$. Using Hoeffding's inequality, it is easy to prove the following.

Lemma 1. *For all finite discrete distributions p with $|\mathsf{Dom}(p)| = r$, for all $\epsilon, \gamma \in (0,1)$ such that $\pi^p_{\min} > \epsilon$, if $\mathbb{S}$ is a sequence of at least $r \cdot \lceil \frac{1}{2\epsilon^2}(\ln 2r - \ln \gamma) \rceil$ i.i.d. samples drawn from p, then $p \sim^{\epsilon} p(\mathbb{S})$ with probability at least $1 - \gamma$.*

We say that we "PAC learn" a distribution p if for all $\epsilon, \gamma \in (0,1)$ such that $\pi^p_{\min} > \epsilon$, by drawing a sequence $\mathbb{S}$ of i.i.d. samples from p, we have $p \sim^{\epsilon} p(\mathbb{S})$ with probability at least $1 - \gamma$. Given a task system Υ, if we can learn the distributions corresponding to all the tasks in Υ, and hence a model Υ^M, such that each learnt distribution in Υ^M is structurally identical to its corresponding distribution in Υ, the corresponding MDP are structurally identical.

Efficient PAC Learning. Let $\Upsilon = ((\tau_i)_{i \in I}, F, \emptyset)$ be a task system with soft tasks only, and let $\epsilon, \gamma \in (0,1)$. We assume that for all distributions p occurring in the models of the tasks in Υ: $\pi^p_{\min} > \epsilon$. To learn a model Υ^M which is ϵ-close to Υ with probability at least $1 - \gamma$, we apply Lemma 1 in the following algorithm:

1. for all tasks $i = 1, 2, \cdots \in F$, repeat the following learning phase:
   Always schedule task τ_i when a job of this task is active. Collect the samples $\mathbb{S}(\mathcal{A}_i)$ of $\mathcal{A}_i$ and $\mathbb{S}(\mathcal{C}_i)$ of $\mathcal{C}_i$ as observed. Collect enough samples to apply Lemma 1 and obtain the desired accuracy as fixed by ϵ and γ.
2. the models of inter-arrival time distribution and computation time distribution for task τ_i are $p(\mathbb{S}(\mathcal{A}_i))$ and $p(\mathbb{S}(\mathcal{C}_i))$ respectively.

Theorem 1. *There is an algorithm s.t. for all task systems $\Upsilon = ((\tau_i)_{i \in I}, F, H)$ with $H = \emptyset$, for all $\epsilon, \gamma \in (0,1)$, it learns Υ^M s.t. $\Upsilon^M \approx^{\epsilon} \Upsilon$ with probability at least $1 - \gamma$ after executing Υ for $|F| \cdot \mathcal{A}_{\max} \cdot \mathbb{D} \cdot \lceil \frac{1}{2\epsilon^2}(\ln 4\mathbb{D}|F| - \ln \gamma) \rceil$ steps.*

Safe Learning with Hard Tasks. We turn to task systems $\Upsilon = ((\tau_i)_{i \in I}, F, H)$ with both hard and soft tasks. The learning algorithm must ensure that all the jobs of hard tasks meet their deadlines while learning the task distributions. The soft-task-only algorithm is clearly not valid for that more general case. Recall we have assumed schedulability of the task system for the hard tasks[2]. This is a necessary condition for safe learning but it is not a sufficient condition. Indeed, to apply Lemma 1, we need enough samples for all tasks $i \in H \cup F$.

First, we note that when executing any safe schedule for the hard tasks, we will observe enough samples for the hard tasks. Indeed, under a safe schedule for the hard tasks, any job of a hard task that enters the system will be executed to completion before its deadline. We then observe the value of the inter-arrival and computation times for all the jobs of hard tasks that enter the system. Unfortunately, this is not necessarily the case for soft tasks when they execute in the presence of hard tasks. Indeed, it is in general not possible to schedule all the jobs of soft tasks up to completion. We thus need stronger conditions in order to be able to learn the distributions of the soft tasks while ensuring safety.

[2] Note that safety synthesis already identifies task systems that violate this condition.

PAC Guarantees for Safe Learning. Our condition to ensure safe PAC learnability relies on properties of the safe region $\Gamma_\Upsilon^{\mathsf{safe}}$ in the MDP Γ_Υ associated to the task system Υ. First, note that $\Gamma_\Upsilon^{\mathsf{safe}}$ is guaranteed to be non-empty as the task system Υ is guaranteed to be schedulable for its hard tasks by hypothesis. Our condition will exploit the following property of its structure:

Lemma 2. *Let $\Upsilon = ((\tau_i)_{i \in I}, F, H)$ be a task system and let Γ_Υ^{safe} be the safe region of its MDP. Then Γ_Υ^{safe} is a single maximal end-component (MEC).*

Good for Sampling. The safe region $\Gamma_\Upsilon^{\mathsf{safe}}$ of the task system $\Upsilon = ((\tau_i)_{i \in I}, F, H)$ is *good for sampling* if for all soft tasks $i \in F$, there exists a vertex $v_i \in \Gamma_\Upsilon^{\mathsf{safe}}$ such that: (i) a new job of task i enters the system in v_i; and (ii) there exists a strategy σ_i of Scheduler that is compatible with the set of safe schedules for the hard tasks so that from v_i, under schedule σ_i, the new job associated to task τ_i is guaranteed to reach completion before its deadline.

There is an algorithm that executes in polynomial time in the size of $\Gamma_\Upsilon^{\mathsf{safe}}$ and which decides if $\Gamma_\Upsilon^{\mathsf{safe}}$ is good for sampling. Also, remember that only the knowledge of the structure of the task system is needed to compute $\Gamma_\Upsilon^{\mathsf{safe}}$.

Given a task system $\Gamma_\Upsilon^{\mathsf{safe}}$ that is *good for sampling*, given any $\epsilon, \gamma \in (0, 1)$, we safely learn a model Υ^M which is ϵ-close to Υ with probability at least $1 - \gamma$ (PAC guarantees) by applying the following algorithm:

1. Choose any safe strategy σ_H for the hard tasks, and apply it until enough samples $(\mathbb{S}(\mathcal{A}_i), \mathbb{S}(\mathcal{C}_i))$ for each $i \in H$ have been collected according to Lemma 1. The models for tasks $i \in H$ are $p(\mathbb{S}(\mathcal{A}_i))$ and $p(\mathbb{S}(\mathcal{C}_i))$.
2. Then for each $i \in F$, apply the following phases:
   (a) from the current vertex v, schedule some task uniformly at random among the set of tasks that correspond to the safe edges in $\mathsf{safe}(v)$ up to reaching some v_i (while choosing tasks that do not violate safety uniformly at random, we reach some v_i with probability 1.[3] The existence of a v_i is guaranteed by the hypothesis that $\Gamma_\Upsilon^{\mathsf{safe}}$ is good for sampling).
   (b) from v_i, apply the schedule σ_i as defined by the second condition in the *good for sampling condition.* This way we are guaranteed to observe the computation time requested by the new job of task i that entered the system in vertex v_i, no matter how TaskGen behaves. At the completion of this job of task i, we have collected a valid sample of task i.
   (c) go back to (a) until enough samples $(\mathbb{S}(\mathcal{A}_i), \mathbb{S}(\mathcal{C}_i))$ have been collected for soft task i according to Lemma 1.

Theorem 2. *There is an algorithm s.t. for all task systems $\Upsilon = ((\tau_i)_{i \in I}, F, H)$ with a safe region Γ_Υ^{safe} that is good for sampling, for all $\epsilon, \gamma \in (0, 1)$, the algorithm learns a model Υ^M such that $\Upsilon^M \approx^\epsilon \Upsilon$ with probability at least $1 - \gamma$.*

In the algorithm above, to obtain one sample of a soft task, we need to reach a particular vertex v_i from which we can safely schedule a new job for the task

[3] This follows from the fact that there is a single MEC in the MDP by Lemma 2.

i up to completion. As the underlying MDP $\Gamma_{\Upsilon}^{\mathsf{safe}}$ can be large (exponential in the description of the task system), we cannot bound by a polynomial the time needed to get the next sample in the learning algorithm. So, this algorithm does not guarantee efficient PAC learning. We develop in the next paragraph a stronger condition to guarantee efficient PAC learning.

Good for Efficient Sampling. The safe region $\Gamma_{\Upsilon}^{\mathsf{safe}}$ of the task system $\Upsilon = ((\tau_i)_{i\in I}, F, H)$ is *good for efficient sampling* if there exists $K \in \mathbb{N}$ which is bounded polynomially in the size of $\Upsilon = ((\tau_i)_{i\in I}, F, H)$, and if, for all soft tasks $i \in F$ the two following conditions hold:

1. let $V_{\Box}^{\mathsf{safe}}$ be the set of Scheduler vertices in $\Gamma_{\Upsilon}^{\mathsf{safe}}$. There is a non-empty subset $\mathsf{Safe}_i \subseteq V_{\Box}^{\mathsf{safe}}$ of vertices from which there is a strategy σ_i for Scheduler to schedule safely the tasks $H \cup \{i\}$ (i.e. all hard tasks *and* the task i); and
2. for all $v \in V_{\Box}^{\mathsf{safe}}, i \in F$, there is a uniform memoryless strategy $\sigma_{\Diamond\mathsf{Safe}_i}$ s.t.:
   (a) $\sigma_{\Diamond\mathsf{Safe}_i}$ is compatible with the safe strategies (for the hard tasks) of $\Gamma_{\Upsilon}^{\mathsf{safe}}$;
   (b) when $\sigma_{\Diamond\mathsf{Safe}_i}$ is executed from any $v \in V_{\Box}^{\mathsf{safe}}$, then the set Safe_i is reached within K steps. By Lemma 2, since $\Gamma_{\Upsilon}^{\mathsf{safe}}$ has a single MEC, we have that Safe_i is reachable from every $v \in V_{\Box}^{\mathsf{safe}}$.

Here again, the condition can be efficiently decided: there is a polynomial-time algorithm in the size of $\Gamma_{\Upsilon}^{\mathsf{safe}}$ that decides if $\Gamma_{\Upsilon}^{\mathsf{safe}}$ is good for efficient sampling.

Given a task system $\Gamma_{\Upsilon}^{\mathsf{safe}}$ that is *good for efficient sampling*, given $\epsilon, \gamma \in (0,1)$, we safely and efficiently learn a model Υ^M which is ϵ-close of Υ with probability at least than $1-\gamma$ (efficient PAC guarantees) by applying:

1. Choose any safe strategy σ_H for the hard tasks, and apply this strategy until enough samples $(\mathbb{S}(\mathcal{A}_i), \mathbb{S}(\mathcal{C}_i))$ for each $i \in H$ have been collected according to Lemma 1. The models for tasks $i \in H$ are $p(\mathbb{S}(\mathcal{A}_i))$ and $p(\mathbb{S}(\mathcal{C}_i))$.
2. Then for each $i \in F$, apply the following phase:
   (a) from the current vertex v, play $\sigma_{\Diamond\mathsf{Safe}_i}$ to reach the set Safe_i.
   (b) from the current vertex in Safe_i, apply the schedule σ_i as defined above. This way we are guaranteed to observe the computation time requested by all the jobs of task i that enter the system.
   (c) go to (*b*) until enough samples $(\mathbb{S}(\mathcal{A}_i), \mathbb{S}(\mathcal{C}_i))$ are collected for task i as per Lemma 1. The models for task i are given by $p(\mathbb{S}(\mathcal{A}_i))$ and $p(\mathbb{S}(\mathcal{C}_i))$.

For a task system Υ, let $T = \mathcal{A}_{\max} \cdot \mathbb{D} \cdot \lceil \frac{1}{2\epsilon^2}(\ln 4\mathbb{D}|\Upsilon| - \ln \gamma) \rceil$. The properties of the learning algorithm above are used to prove the following theorem:

Theorem 3. *There is an algorithm s.t. for all systems $\Upsilon = ((\tau_i)_{i\in I}, F, H)$ with safe region Γ_{Υ}^{safe} that is good for efficient sampling, for all $\epsilon, \gamma \in (0,1)$, it learns Υ^M s.t. $\Upsilon^M \approx^{\epsilon} \Upsilon$ with probability at least than $1-\gamma$ after scheduling Υ for $T + |F| \cdot (T+K)$ steps.*

Using the Learnt Model. Given a system Υ of tasks, and parameters $\epsilon, \gamma \in (0,1)$, once we have learnt a model Υ^M such that $\Upsilon^M \approx^{\epsilon} \Upsilon$, we construct the MDP $\Gamma_{\Upsilon^M}^{\mathsf{safe}}$. From $\Gamma_{\Upsilon^M}^{\mathsf{safe}}$, we can compute an optimal scheduling strategy that minimises the expected mean-cost of missing deadlines of soft tasks. Such an algorithm is given in [13]. Then, we execute the actual task system Υ under schedule σ. However, since σ has been computed using the model Υ^M, it might not be optimal in the original, unknown taks system Υ. Nevertheless, we can bound the difference between the optimal values obtained in $\Gamma^{\text{safe}}_{\Upsilon^M}$ and $\Gamma^{\text{safe}}_{\Upsilon}$.

The following lemma relates the model that is learnt with the approximate distribution that we have in the MDP corresponding to the learnt model. Given $\epsilon \in (0,1)$, let $s = \min\{1, \pi^{\Upsilon}_{\max} + \epsilon\}$ and $\eta = s^{2n} - (s-\epsilon)^{2n}$, where $n = |\Upsilon|$.

Lemma 3 (From [16]**).** *Let Υ be a task system, let $\epsilon, \gamma \in (0,1)$, let Υ^M be the learnt model such that $\Upsilon^M \approx^{\epsilon} \Upsilon$ with probability at least $1-\gamma$. Then we have that $\Gamma_{\Upsilon^M} \approx^{\eta} \Gamma_{\Upsilon}$ with probability at least $1-\gamma$.*

A strategy σ is said to be *(uniformly) expectation-optimal* if for all $v \in V_{\Box}$, we have $\mathbb{E}_v^{\Gamma[\sigma]}(\mathsf{MC}) = \inf_{\tau} \mathbb{E}_v^{\Gamma[\tau]}(\mathsf{MC})$. The following Lemma captures the idea that some expectation-optimal strategies for MDPs whose transition functions have the same support as that of Γ are 'robust'.

Lemma 4 (From [7, **Theorem 5**]**).** *Consider $\beta \in (0,1)$, and MDPs Γ and Γ' such that $\Gamma \approx^{\eta_\beta} \Gamma'$ with $\eta_\beta \leq \frac{\beta \cdot \pi_{\min}}{8|V_{\Box}|}$, where $\pi_{\min}$ is the minimum probability appearing in Γ. For all memoryless deterministic expectation-optimal strategies σ in Γ', for all $v \in V_{\Box}$, it holds that $\left|\mathbb{E}_v^{\Gamma[\sigma]}(\mathsf{MC}) - \inf_{\tau} \mathbb{E}_v^{\Gamma[\tau]}(\mathsf{MC})\right| \leq \beta$.*

The proof of the above lemma uses Thm. 6 in [21] and Thm. 5 in [7]. Using both Lemma 3 and Lemma 4, we obtain the following guarantees on the quality of the scheduler that our model-based learning algorithm outputs:

Theorem 4. *Given a task system Υ (with min probability $\pi_{\min}$) and $\beta \in (0,1)$. Let $\gamma, \epsilon \in (0,1)$ be s.t. $\epsilon \leq \frac{\beta\pi_{\min}}{8|V_{\Box}| + \beta\pi_{\min}}$. Let Υ^M be s.t. $\Upsilon^M \approx^{\epsilon} \Upsilon$ with probability at least $1-\gamma$, and let σ be a memoryless deterministic expectation-optimal strategy of Γ_{Υ^M}. Then, with probability at least $1-\gamma$, the expected mean-cost of playing σ in Γ_{Υ} is s.t. for all $v \in V_{\Box}$: $\left|\mathbb{E}_v^{\Gamma_{\Upsilon}[\sigma]}(\mathsf{MC}) - \inf_{\tau} \mathbb{E}_v^{\Gamma_{\Upsilon}[\tau]}(\mathsf{MC})\right| \leq \beta$.*

4 Monte Carlo Tree Search with Advice

When the model of the task system is known, or once it has been learned using techniques developed in Sect. 3, our goal is to compute a (near) optimal strategy while ensuring safe scheduling of hard-tasks with certainty.

The challenge is the sizes of the MDPs that are too large for exact model-checking techniques (see Sect. 5). To overcome this problem, we resort to a *receding horizon* framework [14], that bases its decisions on a finite-depth unfolding of the MDP from the current state. In particular, we advocate the use of *Monte Carlo Tree Search* (MCTS) algorithms [4], that are a popular method for sampling the finite-depth unfolding while avoiding an exponential dependency on the horizon. MCTS algorithms aim at discovering and exploring the "most relevant" parts of the unfolding, and they approximate the value of actions in intermediary nodes using a fixed number of trajectories obtained by simulations. The MCTS algorithm builds an exploration tree incrementally. At every step of the algorithm, the *selection phase* selects a path in the current tree, possibly extending it by adding a new node. It is followed by a *simulation phase*, that extends this trajectory further, until the fixed horizon is reached. Finally, a *back-propagation* phase updates the exploration tree based on this new trajectory. A reader looking for a more detailed introduction to MCTS is referred to [5].

MCTS has been successfully applied to large state-spaces. For example, it is an important building block of the AlphaGo algorithm [20] that has obtained super-human performances in the game of Go. Such level of performances cannot be obtained with the plain MCTS algorithm. In Go, the simulation and selection phases are guided by a board scoring function that has been learned using neural-networks techniques and self-play. For our scheduling problem, we also need a solution to this guidance problem and, equally importantly, we must augment the MCTS algorithm in a way that *ensures* safe scheduling of hard tasks.

Symbolic Advice. In a recent previous work [5], we have introduced the notion of (symbolic) advice that provides a generic and formal solution to systematically incorporate domain knowledge in the MCTS algorithm. For our scheduling problem, we use selection advice that prunes parts of the MDP on-the-fly in order to ensure that only safe schedulers are explored. We have considered two possibilities. First, we consider the most general safe scheduler (MGS scheduler) as defined in page 7 to restrict the selection phase to safe scheduling decisions only. Second, we consider the earliest deadline first (EDF) scheduling strategy for hard tasks defined in page 4, that only allows soft tasks when there are no available hard tasks, and restricts to the hard tasks with the earliest deadline otherwise. EDF is guaranteed safe as the set of hard tasks is assumed schedulable. The MGS advice allows for maximal exploration as it leaves open all possible safe scheduling solutions, while the EDF advice can be applied on larger task systems as it does not require any precomputations. These advice are also applicable during the simulation phases.

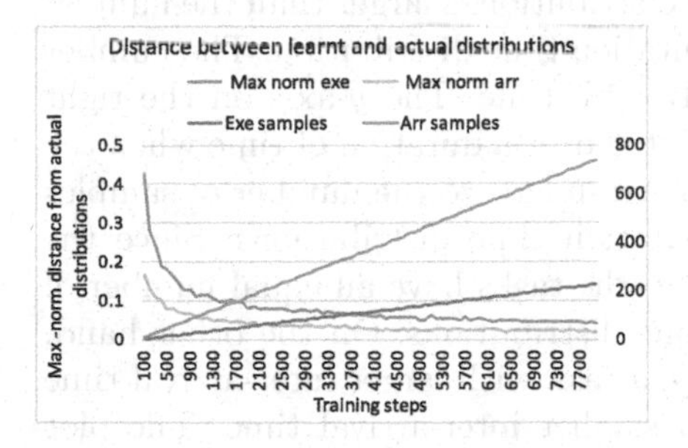

Fig. 2. Learning distributions for a system with 6 soft tasks.

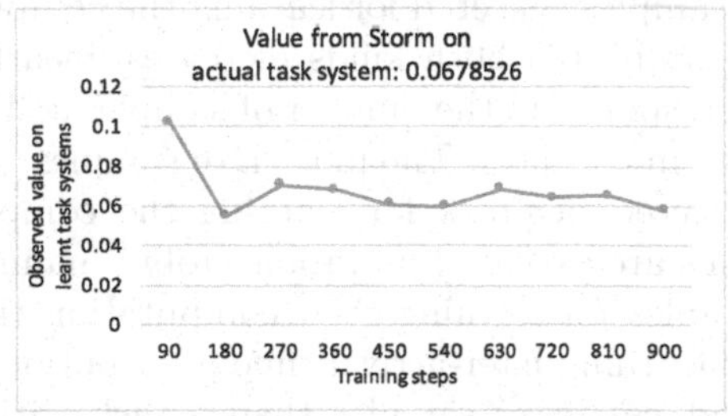

Fig. 3. Model-based learning for 1 hard, 2 soft tasks

5 Experimental Results

In this section, we first report experimental results on model-based learning and observe that the models are learnt efficiently with only a small number of samples. Our MCTS based algorithms can then be applied on the learnt models that are very close to the original ones.[4] We compare the performance of our MCTS-based algorithms with a state-of-the-art deep Q-learning implementation from OpenAI [10] on a set of benchmarks of task systems of various sizes. The experimental results show that our MCTS-based algorithms perform better in practice than safe reinforcement learning (RL)[3].

Models With Only Soft Tasks. In Fig. 2, we show that the distributions of a task system with soft tasks can be learnt efficiently with a small number of samples, corroborating our theory in Sect. 3. This is not the case in general for arbitrary MDPs where in order to collect samples, one may need to reach some specific states of the MDP, and it may take a considerable amount of time to reach such states. However, in this case of systems with only soft tasks, the number of samples increases linearly with time. As a representative task system, we display the learning curve for a system with six soft tasks in Fig. 2. Here "exe" and "arr" refer to the distributions of the computation times and the inter-arrival times respectively. The left y-axis is the max-norm distance between the probabilities in the actual distributions and the learnt distributions across all soft tasks. The x-axis is the number of time steps over which the system is executed. For learning the computation time distribution, the soft tasks are scheduled in a round robin manner. Once a job of a soft task is scheduled, it is executed until completion without being preempted. A sample for learning the computation time distribution of a soft task thus corresponds to a job of the task that is scheduled to execute until completion. Since the system has only soft tasks, a job can always be executed to finish its execution without safety being violated. On the other hand, the samples for learning the inter-arrival time distribution for each task correspond to all the jobs of the task that arrive in the system. Thus over a time duration, for each task, the number of samples collected for learning the inter-arrival time distribution is larger than the number of samples collected for learning the computation time distribution. The number of samples of both kinds increases linearly with time. The y-axis on the right corresponds to the number of samples collected over a duration of time when the system executes. The plot "Exe samples" corresponds to the number of samples collected per task for learning the computation time distributions. Since the tasks are executed in a round robin manner, the tasks have an equal number of samples for learning their computation time distributions. On the other hand, for learning inter-arrival time distributions, a task with larger inter-arrival time produces fewer samples than a task with smaller inter-arrival time. The plot "Arr samples" corresponds to the minimum of the number of jobs, over all the

[4] Here we do not learn to the point where our PAC guarantees hold. Rather, we are interested in how fast the learnt model converges to the real model in practice.

tasks, that arrived in the system. Each point in the graphs is obtained as a result of averaging over 50 simulations.

Safe Model-Based Learning. For safe model-based learning of systems with both hard and soft tasks, first, we verify that the task system satisfies the *good for efficient sampling* condition, and hence admits safe efficient PAC learning. We consider a small representative task system, and report the value of the optimal expected mean-cost strategy as computed by STORM on the learnt model as a function of the number of steps for which the system is executed (training steps). This converges quickly to the optimal expected value of the actual task system, roughly equal to 0.06 (see Fig. 3). We also note that the expected value computed by STORM is not necessarily monotonic as it is computed on the learnt model and this model changes over time with the samples that it receives, and the expected value may also sometimes be smaller than the value on the actual model. The results show that this approach is effective in terms of the quality of learning and the number of samples required.

MCTS. In the above approach, the main bottleneck towards scalability is the extraction of an optimal strategy from the learnt model using probabilistic model-checkers like STORM. This is because the underlying MDP grows exponentially with the number of tasks. Therefore we advocate the use of receding horizon techniques instead, that optimize the cost based on the next h steps for some horizon h. In our examples, the unfoldings have approximately 2^h states, so we use MCTS to explore them in a scalable way.

Deep Q-Learning. One of the most successful model-free learning algorithm is the *Q-learning* algorithm, due to Watkins and Dayan [24]. It aims at learning (near) optimal strategies in a (partially unknown) MDP for the *discounted sum* objective. In our scheduling problem, we search for (near) optimal strategies for the mean-cost and *not* for the discounted sum, as we want to minimise the limit average of the cost of missing deadlines of soft tasks. However, if the discount factor is close to 1, both values coincide [17,21]. In our experiments, we use an implementation of deep Q-learning available in the OPENAI repository [10]. We make use of shielding [1,3,8], a technique that restricts actions in the learning process so that only those actions that are safe for the hard tasks can be used.

Experimental Setup for MCTS and Deep Q-Learning. We compare some variants of model-based learning augmented with MCTS and some variants of deep Q-learning in the context of scheduling. The first option is to set a very high penalty on missing the deadline of a hard task, and then to apply either MCTS or deep Q-learning. However, safety is not guaranteed in this case, and we report on whether a violation was observed or not. We call this variant unsafe MCTS and unsafe deep Q-learning respectively as a consequence. The second option is to enforce safety in MCTS and deep Q-learning by computing the most general safe scheduler for hard tasks, and then using the MGS advice for MCTS or the MGS shield for deep Q-learning. The third option is to use the earliest-deadline-first (EDF) scheme on hard tasks instead of MGS as an advice or a shield. Note that the second and the third options are required to ensure safety, and thus are applicable to systems that have at least one hard task, and hence are not applicable (NA) to systems with only soft tasks.

Table 1. Comparison of MCTS and reinforcement learning.

Task	size	Storm output	MCTS unsafe	MCTS MGS	MCTS EDF	Deep-Q unsafe	Deep-Q MGS	Deep-Q EDF
4S	10^5	0.38	0.52	NA	NA	0.56	NA	NA
5S	10^6	T.O	0	NA	NA	0.13	NA	NA
10S	10^{18}	T.O	0	NA	NA	0.96	NA	NA
simple	10^2	0	0.72	0	0	1.08	0.1	0
1H, 2S	10^4	0.07	0.67	0.14	0.28	0.24	0.11	0.22
1H, 3S	10^5	0.28	1.13	0.45	0.49	∞	0.47	0.47
2H, 1S	10^4	0	0.92	0	0.2	∞	0.02	0.3
2H, 5S	10^{10}	T.O.	3.44	1.93	2.14	∞	2.39	2.48
3H, 6S	10^{14}	T.O.	4.17	2.88	2.97	∞	3.42	3.47
2H, 10S	10^{22}	T.O.	0.3	0.03	0.03	∞	1.42	1.6
4H, 12S	10^{30}	T.O.	2.1	1.2	1.3	∞	2.68	2.87

Experimental Results. In the first column of Table 1, we describe the task systems that we consider. A description 2H, 5S refers to a task system with two hard tasks and five soft tasks, while 4S refers to a task system with four soft tasks and no hard tasks. The simple system refers to a 1H, 2S task system where all the arrival time distributions are Dirac. The output of STORM for the smaller task systems is given in the third column. We report sizes of the MDPs, computed with STORM whenever possible. Otherwise we report an approximation of the size of the state space obtained by taking the product of $(c_i + 1)(a_i + 1)$ over the set of tasks, where c_i and a_i are the greatest elements in the support of the distributions $\mathcal{C}_i$ and $\mathcal{A}_i$. Recall that the size of the state space is exponential in the number of tasks in the system. In the columns where safety is not guaranteed, ∞ denotes an observed violation (a missed deadline for a hard task).

For MCTS, at every step we explore 500 nodes of the unfolding of horizon 30, and the value of each node is initialized using 100 uniform simulations. This computation takes 1–4 min in our Python implementation for different benchmarks, running on a standard laptop. It is reasonable to believe that a substantial speedup could be obtained with well-optimised code and parallelism. For deep Q-learning, we train each task system for 10000 steps. The implementation of deep-Q learning in the OEPNAI respository uses the Adam optimizer [15]. The size of the replay buffer is set to 2000. The learning rate used is 10^{-3}. The probability ϵ of taking a random action is initially set to 1. This parameter reduces over the training steps, and becomes equal to 0.02 at the end of the training. The network used is a multi-layer perceptron which, by default, uses two fully connected hidden layers, each with 64 nodes. Since we are interested in mean-cost objective, the discount factor γ is set to 1. We observed that reducing the value of γ leads to poorer results. The values reported for both MCTS and deep Q-learning are obtained as an average cost over 600 steps.

Conclusions. While deep Q-learning provides good results for small task systems with 3–4 tasks with several thousands of states, this method does not perform well for the benchmarks with large number of tasks. We trained the task system with 10 soft tasks with deep Q-learning for several million steps, but the state space was found to be too large to learn a good strategy, and the resulting output produced a cost that is much higher than that observed with MCTS.

Overall, our experimental results show that MCTS consistently provides better results, in particular when the task systems are large, with huge state spaces. This can be explained by the fact that MCTS optimizes locally using information about multiple possible "futures" while deep Q-learning rather optimizes globally using information about the uniquely observed trace. We observe that the performance of MCTS with EDF advice is only slightly worse than MCTS with MGS advice. EDF guarantees safety and does not require computing the most general safe strategy, therefore it forms a good heuristic for systems with many hard tasks, where MGS computation becomes too expensive.

In future work, we consider using Deep-Q learning in either a selection advice for MCTS or as a complement to simulations when evaluating new states.

Appendix

A Proof of Theorem 1

There is a learning algorithm such that for all task systems $\Upsilon = ((\tau_i)_{i\in I}, F, H)$ *with* $H = \emptyset$*, for all* $\epsilon, \gamma \in (0,1)$*, the algorithm learns a model* Υ^M *such that* $\Upsilon^M \approx^\epsilon \Upsilon$ *with probability at least* $1-\gamma$ *after executing* Υ for $|F| \cdot \mathcal{A}_{\max} \cdot \mathbb{D} \cdot \lceil \frac{1}{2\epsilon^2}(\ln 4\mathbb{D}|F| - \ln\gamma)\rceil$ *steps.*

Proof. Using Lemma 1, given $\epsilon, \gamma' \in (0,1)$, for every distribution p of the task system, a sequence $\mathbb{S}$ of $\mathbb{D} \cdot \lceil \frac{1}{2\epsilon^2}(\ln 2\mathbb{D} - \ln\gamma')\rceil$ i.i.d. samples suffices to have $p(\mathbb{S}) \sim^\epsilon p$ with probability at least $1-\gamma'$. Since in the task system Υ, there are $2|F|$ distributions, with probability at least $1 - 2|F|\gamma'$, we have that the learnt model $\Upsilon^M \approx^\epsilon \Upsilon$. Thus for $\gamma' = \frac{\gamma}{2|F|}$, and using $2\exp(-2m\epsilon^2) \le \frac{\gamma}{2|F|\mathbb{D}}$, we have that for each distribution, a sequence of $\mathbb{D}\cdot\lceil\frac{1}{2\epsilon^2}(\ln 4\mathbb{D}|F| - \ln\gamma)\rceil$ samples suffices so that $\Upsilon^M \approx^\epsilon \Upsilon$ with probability at least $1-\gamma$.

Since samples for computation time distribution and inter-arrival time distribution for each soft task can be collected simultaneously, and observing each sample takes a maximum of $\mathcal{A}_{\max}$ time steps, and we collect samples for each soft task by scheduling one soft task after another, the result follows. □

B Proof of Lemma 2

Let $\Upsilon = ((\tau_i)_{i\in I}, F, H)$ *be a task system and let* $\Gamma_\Upsilon^{\mathrm{safe}}$ *be the safe region of its MDP. Then* $\Gamma_\Upsilon^{\mathrm{safe}}$ *is a single MEC.*

Proof. We first assume that the task system $\Upsilon = ((\tau_i)_{i \in I}, F, H)$ is schedulable. Otherwise, $\Gamma_\Upsilon^{\mathsf{safe}}$ is empty and the Lemma is trivially true. Let V and E be the set of vertices and the set of edges of $\Gamma_\Upsilon^{\mathsf{safe}}$ respectively. First, observe that, since we want to prove that the whole MDP $\Gamma_\Upsilon^{\mathsf{safe}}$ corresponds to an MEC, we only need to show that its underlying graph (V, E) is strongly connected. Indeed, since (V, E) contains all vertices and edges from $\Gamma_\Upsilon^{\mathsf{safe}}$, it is necessarily maximal, and all choices of actions from any vertex will always lead to a vertex in V.

In order to show the strongly connected property, we fix a vertex $v \in V$, and show that there exists a path in $\Gamma_\Upsilon^{\mathsf{safe}}$ from v to v_{init}. Since all vertices in V are, by construction of $\Gamma_\Upsilon^{\mathsf{safe}}$, reachable from the initial vertex v_{init}, this entails that all vertices v' are also reachable from v, hence, the graph is strongly connected.

Let us first assume that $v \in V_\square$, i.e., v is a vertex where Scheduler has to take a decision. Let $v_{\mathsf{init}} = v_0, v_0', v_1, v_1', \cdots, v_{n-1}', v_n = v$ be the path π leading to v, where all vertices v_j belong to Scheduler, and all v_j' are are vertices that belong to TaskGen.

Then, from path π, we extract, for all tasks τ_i the sequence of *actual inter-arrival times* $\sigma_i = t^i(1), t^i(2), \ldots, t^i(k_i)$ defined as follows: for all $1 \leq j \leq k_i$, $t^i(j) \in \mathsf{Supp}(\mathcal{A}_i)$ is the time elapsed (in CPU ticks) between the arrival of the $j-1$th job the jth job of task i along π (assuming the initial release occurring in the initial state v_{init} is the 0-th release). In other words, letting $T^i(j) = \sum_{k=1}^{j} t^i(k)$, the jth job of τ_i is released along π on the transition between $v'_{T^i(j-1)}$ and $v_{T^i(j)}$. Observe thus that all tasks $i \in [n]$ are in the same state in vertex v_{init} and in vertex $v_{T^i(j)}$, i.e. the time to the deadline, and the probability distributions on the next arrival and computation times are the same in v_{init} and $v_{T^i(j)}$. However, the vertices $v_{T^i(j)}$ can be different for all the different tasks, since they depend on the sequence of job releases of τ_i along π. Nevertheless, we claim that π can be extended, by repeating the sequence of arrivals of all the tasks along π, in order to reach a vertex where all tasks have just submitted a job (i.e. v_{init}). To this aim, we first extend, for all tasks $i \in [i]$, σ_i into $\sigma_i' = \sigma_i, t^i(k_i+1)$, where $t^i(k_i+1) \in \mathsf{Supp}(\mathcal{A}_i)$ ensures that the k_i+1 arrival of a τ_i occurs *after* v.

For all $i \in [n]$, let Δ_i denote $\sum_{j=1}^{k_i+1} t^i(j)$, i.e. Δ_i is the total number of CPU ticks needed to reach the first state after v where task i has just submitted a job (following the sequence of arrival σ_i' defined above). Further, let $\Delta = \mathrm{lcm}(\Delta_i)_{i \in [n]}$. Now, let π' be a path in $\Gamma_\Upsilon^{\mathsf{safe}}$ that respects the following properties:

1. π is a prefix of π';
2. π' has a length of Δ CPU ticks;
3. π' ends in a $\square$ vertex v'; and
4. for all tasks $i \in [n]$: τ_i submits a job at time t along π' iff it submits a job at time $t \mod \Delta_i$ along π.

Observe that, in the definition of π', we do not constrain the decisions of Scheduler after the prefix π. First, let us explain why such a path exists. Observe that the sequence of task arrival times is legal, since it consists, for all tasks i, in repeating Δ/Δ_i times the sequence σ_i' of inter-arrival times which is legal since it is extracted from path π (remember that nothing that Scheduler player does can restrict the times at which TaskGen introduces new jobs in the system). Then, since Υ is schedulable, we have the guarantee that all $\Box$ vertices in $\Gamma_\Upsilon^{\mathsf{safe}}$ have at least one outgoing edge. This is sufficient to ensure that π' indeed exists. Finally, we observe π' visits v (since π is a prefix of π'), and that the last vertex v' of π' is a $\Box$ vertex obtained just after *all tasks* have submitted a job, by construction. Thus $v' = v_{\mathsf{init}}$, and we conclude that, from all $v \in V_\Box$ which is reachable from v_{init}, one can find a path in $\Gamma_\Upsilon^{\mathsf{safe}}$ that leads back to v_{init}.

This reasoning can be extended to account for the nodes $v \in V_\bigcirc$: one can simply select any successor $\overline{v} \in V_\Box$ of v, and apply the above reasoning from $\overline{v}$ to find a path going back to v_{init}. $\Box$

C Proof of Theorem 2

There is a learning algorithm such that for all task systems $\Upsilon = ((\tau_i)_{i \in I}, F, H)$ with a safe region $\Gamma_\Upsilon^{\mathsf{safe}}$ that is good for sampling, for all $\epsilon, \gamma \in (0, 1)$, the algorithm learns a model Υ^M such that $\Upsilon^M \approx^\epsilon \Upsilon$ with probability at least $1 - \gamma$.

Proof. For the hard tasks, as mentioned above, we can learn the distributions by applying the safe strategy σ_H to collect enough samples $(\mathbb{S}(\mathcal{A}_i), \mathbb{S}(\mathcal{C}_i))$ for each $i \in H$.

We assume an order on the set of soft tasks. First for all τ_i for $i \in F$, since $\Gamma_\Upsilon^{\mathsf{safe}}$ is good for sampling, we note that the set V_i of vertices v_i (as defined in the definition of good for sampling condition) is non-empty. Recall from Lemma 2 that $\Gamma_\Upsilon^{\mathsf{safe}}$ has a single MEC. Thus from every vertex of $\Gamma_\Upsilon^{\mathsf{safe}}$, Scheduler by playing uniformly at random reaches some $v_i \in V_i$ with probability 1, and hence can visit the vertices of V_i infinitely often with probability 1. Now given ϵ and γ, using Theorem 1, we can compute an m, the number of samples corresponding to each distribution required for safe PAC learning of the task system. Since by playing uniformly at random, Scheduler has a strategy to visit the vertices of V_i infinitely often with probability 1, it is thus possible to visit these vertices at least m times with arbitrarily high probability.

Also after we safely PAC learn the distributions for task τ_i, since there is a single MEC in $\Gamma_\Upsilon^{\mathsf{safe}}$, there exists a uniform memoryless strategy to visit a vertex v_{i+1} corresponding to task τ_{i+1} with probability 1. Hence the result. $\Box$

References

1. Alshiekh, M., Bloem, R., Ehlers, R., Könighofer, B., Niekum, S., Topcu, U.: Safe reinforcement learning via shielding. In: AAAI, pp. 2669–2678. AAAI Press (2018)
2. Arulkumaran, K., Deisenroth, M.P., Brundage, M., Bharath, A.A.: Deep reinforcement learning: a brief survey. IEEE Signal Process. Mag. **34**(6), 26–38 (2017)
3. Avni, G., Bloem, R., Chatterjee, K., Henzinger, T.A., Könighofer, B., Pranger, S.: Run-time optimization for learned controllers through quantitative games. In: CAV, pp. 630–649 (2019)

4. Browne, C., et al.: A survey of Monte Carlo tree search methods. IEEE Trans. Comput. Intell. AI Games **4**(1), 1–43 (2012). https://doi.org/10.1109/TCIAIG.2012.2186810
5. Busatto-Gaston, D., Chakraborty, D., Raskin, J.: Monte carlo tree search guided by symbolic advice for MDPs. In: CONCUR, pp. 40:1–40:24 (2020). https://doi.org/10.4230/LIPIcs.CONCUR.2020.40
6. Buttazzo, G.C.: Hard Real-Time Computing Systems: Predictable Scheduling Algorithms and Applications, vol. 24. Springer, Boston (2011). https://doi.org/10.1007/978-1-4614-0676-1
7. Chatterjee, K.: Robustness of structurally equivalent concurrent parity games. In: FOSSACS, pp. 270–285 (2012)
8. Chatterjee, K., Novotný, P., Pérez, G.A., Raskin, J.F., Zikelic, D.: Optimizing expectation with guarantees in pomdps. In: AAAI, pp. 3725–3732 (2017)
9. Dehnert, C., Junges, S., Katoen, J., Volk, M.: A storm is coming: a modern probabilistic model checker. In: CAV (2017)
10. Dhariwal, P., et al.: Openai baselines (2017). https://github.com/openai/baselines
11. Filar, J., Vrieze, K.: Competitive Markov Decision Processes. Springer, New York (1997). https://doi.org/10.1007/978-1-4612-4054-9
12. Fu, J., Topcu, U.: Probably approximately correct MDP learning and control with temporal logic constraints. In: Fox, D., Kavraki, L.E., Kurniawati, H. (eds.) Robotics: Science and Systems X, University of California, Berkeley, USA, July 12–16, 2014 (2014). https://doi.org/10.15607/RSS.2014.X.039, http://www.roboticsproceedings.org/rss10/p39.html
13. Geeraerts, G., Guha, S., Raskin, J.F.: Safe and optimal scheduling for hard and soft tasks. In: FSTTCS. LIPIcs, vol. 122, pp. 36:1–36:22 (2018)
14. Kearns, M.J., Mansour, Y., Ng, A.Y.: A sparse sampling algorithm for near-optimal planning in large Markov decision processes. Mach. Learn. **49**(2–3), 193–208 (2002). https://doi.org/10.1023/A:1017932429737
15. Kingma, D.P., Ba, J.: Adam: a method for stochastic optimization. In: ICLR (2015)
16. Kretínský, J., Pérez, G.A., Raskin, J.F.: Learning-based mean-payoff optimization in an unknown MDP under omega-regular constraints. In: CONCUR. LIPIcs (2018)
17. Mertens, J.F., Neyman, A.: Stochastic games. Int. J. Game Theory **10**(2), 53–66 (1981)
18. Mnih, V., et al.: Human-level control through deep reinforcement learning. Nature **518**(7540), 529–533 (2015)
19. Ramadge, P.J., Wonham, W.M.: Supervisory control of a class of discrete event processes. SIAM J. Control Opt. **25**(1), 206–230 (1987)
20. Silver, D., et al.: Mastering the game of go with deep neural networks and tree search. Nature **529**(7587), 484–489 (2016). https://doi.org/10.1038/nature16961
21. Solan, E.: Continuity of the value of competitive Markov decision processes. J. Theoret. Prob. **16**, 831–845 (2003)
22. Thomas, W.: On the synthesis of strategies in infinite games. In: STACS, pp. 1–13 (1995)
23. Valiant, L.G.: A theory of the learnable. Commun. ACM **27**(11), 1134–1142 (1984)
24. Watkins, C.J.C.H., Dayan, P.: Technical note Q-learning. Mach. Learn. **8**, 279–292 (1992)

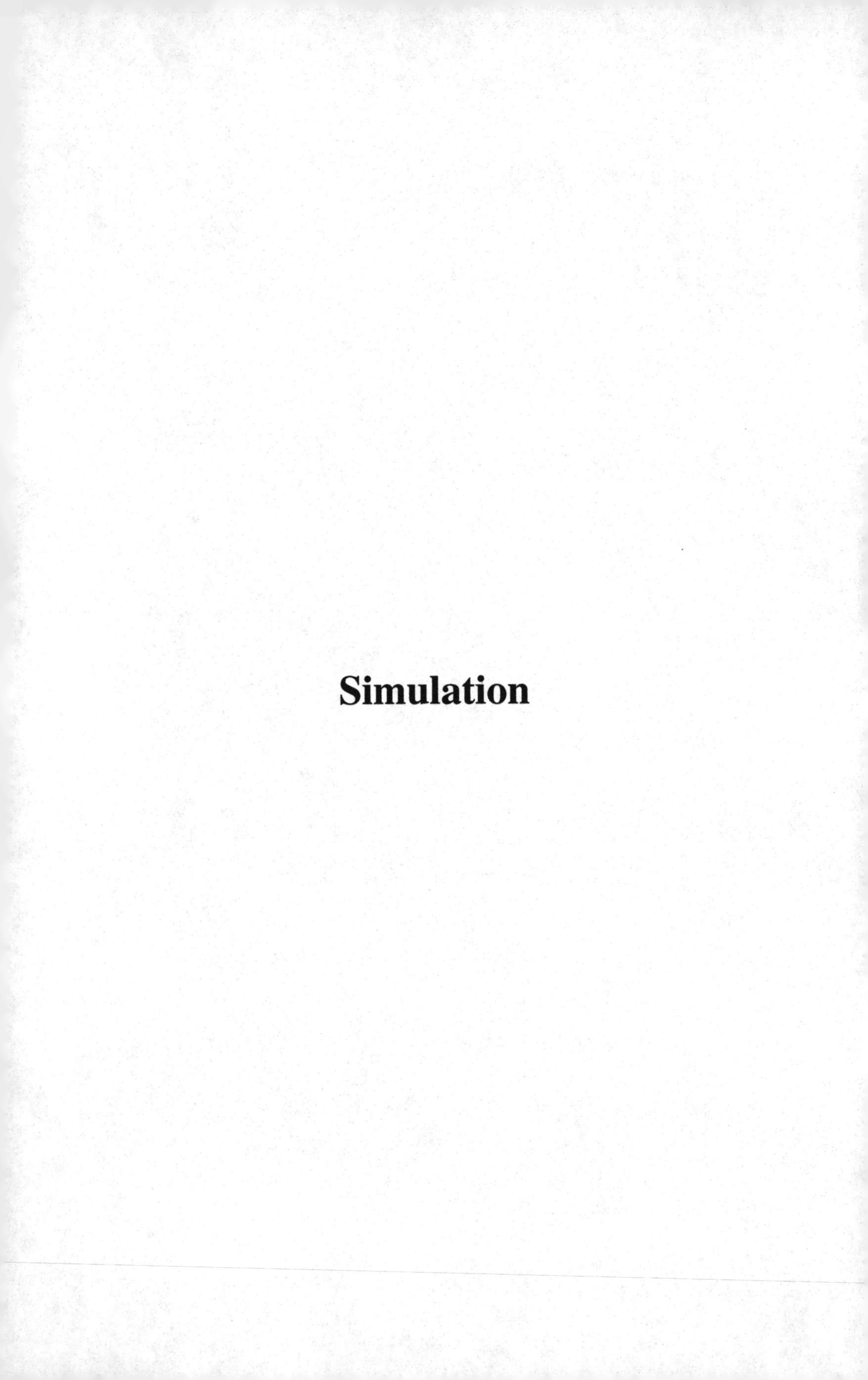

Simulation

Symbolic Simulation of Railway Timetables Under Consideration of Stochastic Dependencies

Rebecca Haehn(✉), Erika Ábrahám, and Nils Nießen

RWTH Aachen University, Aachen, Germany
{haehn,abraham}@cs.rwth-aachen.de, niessen@via.rwth-aachen.de

Abstract. In this paper we propose an *exact* symbolic simulation method to compute the impact of delays in railway systems. We use macroscopic railway infrastructure models and model primary delays of trains in a timetable by discrete probability distributions. Our method is capable of computing exact probabilistic quantities like delay probability distributions and expected delays for timetable trains, or expected capacity usage of infrastructure elements within a given finite time window. In turn, these quantities allow us to examine timetable robustness and to identify problematic infrastructure elements. We evaluate our approach on realistic case studies and discuss possible further improvements.

Keywords: Simulation · Railway timetables · Delay propagation

1 Introduction

To manage increasing railway traffic, besides infrastructure extensions, it is highly important to optimally exploit the existing infrastructures' utilization. On the one hand, a high number of passenger and freight trains should be able to use the infrastructure but, on the other hand, a high level of service quality should be maintained. From a certain view, these two objectives are negatively correlated: the level of customer satisfaction strongly depends on the trains' punctuality, but increasing traffic comes with increasing delays.

Simulation can be used to estimate delay propagation. The systems RailSys [4,15] and OpenTrack [3,12] use detailed *microscopic* models [14] and *synchronous* simulation, meaning that they simulate all train rides simultaneously in a single run. In contrast, the system MOSES/WiZug [16] is based on *macroscopic* models with fewer details [14]. This system operates in an *asynchronous* way, simulating the train rides sequentially one after the other, starting with the trains that have the highest priority. It is applied specifically for rail freight transportation. Meanwhile, the system LUKS [1,10] uses microscopic models and a combination of asynchronous and synchronous train ride simulations.

This research is funded by the German Research Council (DFG) – Research Training Group UnRAVeL (RTG 2236).

A. Abate and A. Marin (Eds.): QEST 2021, LNCS 12846, pp. 257–275, 2021.
https://doi.org/10.1007/978-3-030-85172-9_14

All the above systems implement *Monte Carlo* simulation. They model primary delays (i.e. delays that are not caused by other delays) as random variables, compute a large number of simulations for different random primary delay values and compute probability estimations statistically from the simulation outcomes. Instead of a Monte Carlo simulation, the approach presented in [5] and implemented in the system OnTime [2,7] uses an *analytical* procedure to compute delay propagations based on mesoscopic models and activity graphs. In [8], we used macroscopic models to compute probability distributions for delays, but completely neglecting stochastic dependencies. All these works give useful estimations, but none of them can compute *exact probability distributions for the train delays under consideration of stochastic dependencies.*

In this paper, we propose the (to our best knowledge) first method to solve this problem. We use macroscopic railway models and model primary delays as discrete random variables. Instead of the sequential simulation of individual delay scenarios as in Monte Carlo approaches, we simulate all possible scenarios simultaneously. Thereby we do not represent each delay scenario separately, but we take up the *symbolic* simulation idea from [8] and define local, partial scenarios to cover a set of global scenarios symbolically. However, in contrast to [8] where we neglected stochastic dependencies, in this work, we add a mechanism to store all relevant information about the stochastic dependencies and replace our approximative answers from [8] by exact answers in the presented approach. Based on the analysis outcome, we are able to examine timetables' robustness by evaluating performance indicators like expected values for train arrival times and infrastructure capacity usage, but also to identify problematic infrastructure elements with high expected delay increment. In the presented approach we abstract from the actual delay distributions and compute the results parametric of the distributions. This makes it possible to consider different delay distributions without having to execute the algorithm repeatedly.

Theoretically, we could formalize our system as a discrete-time Markov chain (DTMC) [13] and use a probabilistic model checker like PRISM [11] or STORM [9] for the analysis. However, the DTMC model size would be exponential in the number of trains. For example, for a typical number of 100 trains in a timetable there would be 2^{100} reachable states even without considering delay propagation. Thus the state space would be too large for any meaningful input. The hardness of the problem is reflected by the fact that besides our previous work [8], currently there exists only one further formal approach in the OnTime tool [7] to solve such problems symbolically, but also these methods can provide only approximative answers.

Outline. We describe our models in Sect. 2 and our symbolic simulation approach in Sect. 3. An experimental evaluation can be found in Sect. 4. We conclude the paper in Sect. 5.

2 Railway Systems

A railway system consists of an infrastructure network and trains moving through it according to a timetable. Similar to [8], in this work, we model railway systems on a *macroscopic* level, i.e. we neglect certain details like signals and exact routes inside stations.

We use $\mathbb{R}$ to denote the set of all real numbers, $\mathbb{N}$ ($\mathbb{N}_{>0}$) for the non-negative (positive) integers, $[x, y]$ for intervals in $\mathbb{R}$, and $[x..y]$ for intervals in $\mathbb{N}$.

2.1 Modeling Railway Infrastructure Networks

We use directed graphs to model railway infrastructure networks, where the vertices model the infrastructure's operation control points and the (directed) edges model unidirectional tracks connecting them. In addition, a capacity function c assigns to each infrastructure element its number of parallel tracks available. We assume that all tracks of an infrastructure element are equivalent in the sense that they could replace each other. There are also bidirectional tracks that can be used in both directions. We model bidirectional connections by two edges, one in each direction, both with the total bidirectional capacity, and assure that at each time point and each track, only one of the directions is used. To do so, we remember these edges in $B \subseteq E$ and "couple" them by a function $b : B \to B$.

Definition 1. *An* infrastructure network *is a tuple* $I = (V, E, c, B, b)$ *with a finite set* V *of* vertices, *a set* $E \subseteq V \times V$ *of (directed)* edges, *a* capacity function $c : (V \cup E) \to \mathbb{N}_{>0}$, *a* bidirectional edge *set* $B \subseteq E$ *and a* coupling function $b : B \to B$ *such that for all* $e \in B$ *and* $b(e) = e'$ *we have* $b(e') = e$ *and* $c(e) = c(e')$. *An* infrastructure element $x \in V \cup E$ *is either a vertex or an edge.*

Let in the following $I = (V, E, c, B, b)$ be an infrastructure network and $\mathbb{T} = [t_{min}, t_{max}] \subseteq \mathbb{R}$ a time window with $t_{min}, t_{max} \in \mathbb{N}$, $t_{min} < t_{max}$.

Definition 2. *A* timed path *is a finite sequence* $\pi = (v^1(a^1 \mapsto d^1), \ldots, v^k(a^k \mapsto d^k))$ *such that for all* $i, i' \in [1..k]$: *(i)* $v^i \in V$ *and if* $i < k$ *then* $(v^i, v^{i+1}) \in E$, *(ii)* $a^i \le d^i$ *and if* $i < k$ *then* $d^i \le a^{i+1}$ *for* arrival *and* departure *times* $a^i, d^i \in \mathbb{T}$, *and (iii)* $i \ne i'$ *implies* $v^i \ne v^{i'}$ *(loop-free).*

A train (ride) $z = (type, \pi)$ *specifies a train type (e.g. freight train) from a finite ordered domain and a timed path* π. *A railway* timetable $T = \{z_1, \ldots, z_n\}$ *is a finite ordered set of trains. We call* i *the* identity *of train* z_i.

Assume in the following a timetable $T = \{z_1, \ldots, z_n\}$ with $z_i = (type_i, \pi_i)$ and $\pi_i = (v_i^1(a_i^1 \mapsto d_i^1), \ldots, v_i^{k_i}(a_i^{k_i} \mapsto d_i^{k_i}))$ for $i \in [1..n]$. T is *executable* if it doesn't exceed the available capacities in the absence of delays.

Definition 3. *Let* $occ_T : (\mathbb{T} \times (V \cup E)) \to 2^{[1..n]}$ *with*

$$occ_T(t, v) = \{i \in [1..n] \mid \exists j \in [1..k_i].v_i^j = v \wedge a_i^j \le t \le d_i^j + \delta\} \text{ and}$$
$$occ_T(t, (v, v')) = \{i \in [1..n] \mid \exists j \in [1..k_i - 1].v_i^j = v \wedge v_i^{j+1} = v' \wedge d_i^j \le t \le a_i^{j+1} + \delta\}$$

for each $t \in \mathbb{T}$, $v \in V$ *and* $(v, v') \in E$, *where* $\delta \in \mathbb{N}$ *is a* safety distance *(number of time units each track is blocked for safety reasons after a train has left it).* T *is* executable *iff for all* $t \in \mathbb{T}$: *(i)* $\forall x \in V \cup (E \backslash B)$. $|occ_T(t, x)| \le c(x)$ *and (ii)* $\forall e \in B$. $|occ_T(t, e) \cup occ_T(t, b(e))| \le c(e)$.

In the following we assume that the timetables we consider are executable. Despite this assumption, in reality delays occur. We differentiate between primary and secondary delays. *Primary* delays are directly caused by varying and uncertain factors like signal failures, defects on the train or many passengers who take longer than planned to board a train. *Secondary* delays of a train arise when the train needs to wait because the capacity of the next needed infrastructure element is fully occupied by other trains. We assume that trains can stop at all infrastructure elements when the next infrastructure element is fully occupied.

2.2 Modeling Primary Delays

We model the duration of primary delays by discrete random variables, one for each train and each infrastructure element used by the train.

Definition 4. *For each $i \in [1..n]$, $j \in [1..k_i]$ and $j' \in [1..k_i-1]$, we define stochastically independent discrete* random variables $p_i^{v_j}$ *and* $p_i^{(v_i^{j'}, v_i^{j'+1})}$ *with sample space $\mathbb{N}$. Let P be the set of all these random variables. For each $p_i^x \in P$ we denote the* probability *that p_i^x has the value $\Delta \in \mathbb{N}$ as $\mathbb{P}(p_i^x = \Delta) \in [0,1]$ and its finite* support *as $\mathcal{D}(p_i^x) = \{\Delta \in \mathbb{N} | \mathbb{P}(p_i^x = \Delta) > 0\}$. It holds that $\sum_{\Delta \in \mathcal{D}(p_i^x)} \mathbb{P}(p_i^x = \Delta) = 1$ for each $p_i^x \in P$.*

Each random variable $p_i^x \in P$ represents the number of time units by which the delay of train z_i increases at infrastructure element x without the cause being another train. Technically, we need additional random variables p_i^{entry} to represent the system entry delay of each train z_i for $i \in [1..n]$, i.e. the time by which the trains' arrival at its first vertex is delayed. We leave this out of the definition to simplify the presentation of our approach in Sect. 3. In the implementation in Sect. 4 we consider these random variables.

The probability distribution of each random variable may depend for example on the types of the trains, the specific infrastructure element and the time of the day. Note that $\mathbb{P}(p_i^x = 0) = 1$ excludes primary delay of z_i at x. These random variables are stochastically independent. Our algorithm can be used to examine the impact of different probability distributions, e.g. based on past measurements [18,19].

A partial scenario puts restrictions on random values but does not necessarily fix their values, in contrast to a complete scenario that fixes all (stochastically independent) primary delays. In our symbolic simulation in Sect. 3, we will use partial scenarios as symbolic representations for sets of complete scenarios.

Definition 5. *A* random inclusion *s for $p_i^x \in P$ has the form $p_i^x \lhd D$ for some $D \subseteq \mathcal{D}(p_i^x)$, $D \neq \emptyset$; we define $\mathbb{P}(s) = \sum_{\Delta \in D} \mathbb{P}(p_i^x = \Delta)$. A* scenario *$S$ is a set that contains exactly one random inclusion for each random variable. Let $\mathbb{S}$ be the set of all scenarios. For $S \in \mathbb{S}$ and $(p_i^x \lhd D) \in S$ we define $S(p_i^x) = D$, and set $\mathbb{P}(S) = \Pi_{s \in S} \mathbb{P}(s)$. We call S* complete *iff $|S(p_i^x)| = 1$ for each $p_i^x \in P$.*

We say that $S \in \mathbb{S}$ refines *$S' \in \mathbb{S}$ (written $S \preceq S'$) iff $S(p_i^x) \subseteq S'(p_i^x)$ for all $p_i^x \in P$; we also say that S'* contains *S. We call S and S'* compatible *iff*

$S(p_i^x) \cap S'(p_i^x) \neq \emptyset$ for all $p_i^x \in P$. For two compatible scenarios S and S' we define $S \triangle S'$ as the scenario $\{p_i^x \lhd (S(p_i^x) \cap S'(p_i^x)) \mid p_i^x \in P\}$.

A scenario S is a symbolic representation of all complete scenarios contained in it; it is easy to see that $\mathbb{P}(S)$ is the sum of the probabilities of all those represented complete scenarios. We sometimes also write $p_i^x = \Delta$ for $p_i^x \lhd \{\Delta\}$, silently skip trivial random inclusions $p_i^x \lhd \mathcal{D}(p_i^x)$ and write $\emptyset$ for the maximal scenario with only trivial constraints.

Example 1. Let $P = \{p_1, p_2\}$, with $\mathcal{D}(p_1) = \{0, 1\}$ and $\mathcal{D}(p_2) = \{0, 1, 2, 3\}$. The scenario $\{p_2 = 0\}$ contains the complete scenario $\{p_1 = 0, p_2 = 0\}$ and $\{p_1 = 1, p_2 = 0\}$; these are also contained in the scenario $\{p_2 \lhd \{0, 3\}\}$. The set $\{p_1 \lhd \emptyset\}$ is not a scenario, because it excludes all random values for p_1. An example for incompatible scenarios are $\{p_2 = 0\}$ and $\{p_2 \lhd \{1, 2\}\}$.

2.3 Timetable Execution

Having fixed all probabilistic decisions by a complete scenario, the secondary delays are uniquely determined. Considering different scenarios and their probabilities, the timetable can be thoroughly analysed. For example, infrastructure elements can be identified where delays strongly increase with high probability. One could also try to identify trains that are especially problematic, i.e. expected to cause high secondary delays. Note that it is practically intractable to execute all $\prod_{p \in P} |\mathcal{D}(p)|$ different scenarios as this would take too long. To solve this problem, we will introduce *symbolic* simulation in the next section.

In the following we describe how to compute the secondary delays for a given scenario; we say that we *execute* the timetable under the given scenario. We assume that trains are not cancelled and re-routing is not applied to deal with delays. Therefore, the number of trains is fixed, as well as their physical paths. However, the arrival and departure times may change. Note that we neglect when different trains in the timetable are physically identical. This could be taken into account by modeling the train rides of each physical train during the considered time window as one consecutive train. For easier presentation of the algorithms we impose loop-freeness, but it would be straightforward to store the index of an infrastructure element instead of just the element itself in train rides and relax Definition 2 to allow paths with loops and thus enabling the sequential connection of train rides.

Assume a complete scenario S for the timetable $T = \{z_1, \ldots, z_n\}$ with $z_i = (type_i, \pi_i)$ and $\pi_i = (v_i^1(a_i^1 \mapsto d_i^1), \ldots, v_i^{k_i}(a_i^{k_i} \mapsto d_i^{k_i}))$ for $i \in [1..n]$. The execution of T under S is basically a timetable as well, but with potentially postponed arrival and departure times. The time distances between these arrival and departure times have to not only account for the driving and halting times in T but also for the delay values Δ_i^x with $S(p_i^x) = \{\Delta_i^x\}$ for all $p_i^x \in P$. We expect that a train drives as soon as it can. That is never earlier than planned in the original timetable. Additionally, the planned driving, respectively halting times and the delay defined in S have to be considered. Apart from that a train drives as soon as the next infrastructure element on its path is available.

Even if T is executable, due to delays it might happen at some time point that more trains want to use an infrastructure element than its capacity allows. In such cases we need a conflict resolution mechanism to determine unambiguously when an infrastructure element is available for a certain train. In our formal definitions we use an abstract mechanism that can be instantiated by any concrete priority ordering.

Remark 1. In practice, the priority of a train depends on various factors. In our implementation, we prioritize trains according to the train type, the planned arrival time and the train identity. More precisely, assume two trains z_i and z_j that want to enter an infrastructure element x at the same time, with planned arrival times a_i resp. a_j at x (where the arrival time at an edge is interpreted as the departure time at its source vertex). Then z_i has a higher priority than z_j to arrive at x if:

- $type_i < type_j$ (e.g. in Germany ICEs are prioritized over freight trains), or
- $type_i = type_j$ and $a_i < a_j$, or
- $type_i = type_j$ and $a_i = a_j$ and $i < j$.

This ordering could be exchanged by any other locally computable criterion, however, it is currently not implemented to perform global delay resolution, e.g. not letting a train drive because it would further delay a train with a higher priority that is scheduled to arrive at a later time point.

We note that timetables often already account for possible delays and contain some longer than necessary halting times such that these could be reduced to a certain degree if needed to decrease a train's delay. There might also be the option for trains to drive slightly faster than planned in the timetable to make up for delays. To simplify the presentation, our approach presented in the following section does not support these optimizations. However, their inclusion is straightforward and we did integrate them in our implementation. There we assume that halting times can be reduced to three minutes for passenger trains and ten minutes for freight trains, while driving times can be reduced by 5%, i.e. multiplied with 0.95.

3 Symbolic Simulation

For a better understanding of the concept, in the following we also neglect bidirectional edges. Their inclusion is really easy and in the implementation we considered them to safely avoid capacity over-approximation.

Executions of complete scenarios can be computed by simulation. Computing executions for all complete scenarios would allow exact probability computations for delays. However, due to the large number of complete scenarios, this approach is not practical. Instead, we propose a *symbolic* approach, based on possibly partial scenarios.

Firstly, we reduce the analysis over a continuous time horizon to the analysis over finitely many discrete time points. We do so by considering only *relevant* time points, which are collected in the course of the analysis. These are the time points in the timetable extended by the time points where some delayed train is supposed to move to its next infrastructure element. By considering only relevant time points we avoid visiting time points at which no train can change its infrastructure element.

We store for all trains and all relevant time points where they might reside together with the scenario under which it happens. That means, we compute the train's position under all possible stochastic dependencies, but try to keep the corresponding (symbolic) scenarios as coarse as possible to cover as many cases as we can as long as they all lead to the same train status, but as fine as needed to keep the computations exact also for future movements.

In the following, we call a tuple (i, S, t) of a train identity i, a scenario S and the time point t of the train's next planned movement (i.e. infrastructure element change) a *train instance.* Note that, since we do not cancel trains, each train must be represented for every complete scenario, so for every complete scenario each train has to be represented by exactly one of its train instances. Initially, each train has a single instance with the coarsest scenario $S = \emptyset$ (when skipping trivial inclusions); we split train instances into refined parts if the train's movement is affected by other train instances with different scenarios or its own primary delay. As we will see later, some additional splits might also be necessary due to the restrictions of the scenario representation (Definition 5) we chose.

3.1 Initialization

Our algorithm receives the following input:

- $I = (V, E, c)$: infrastructure network (neglecting bidirectional track information);
- $\delta \in \mathbb{N}$: safety distance;
- $T = \{z_1, \ldots, z_n\}$: timetable for the time window $\mathbb{T} = [t_{min}, t_{max}]$;
- $P \subseteq \{p_i^x \mid i \in [1..n],\ x \in V \cup E\}$: set of random variables according to Definition 4 (implicitly carrying also their probability distributions).

As global variables, we use the following sets:

- *times*: ascendingly ordered set of the time points $t \in \mathbb{T}$ at which some trains want to change infrastructure element; this sequence is naturally partial and gets extended during computations, but we assure that initially and after each processed time instance, no train can change infrastructure element before the smallest time point in *times*;
- for each infrastructure element $x \in V \cup E$:
  - *occupy*$[x]$: set of train instances occupying x at the current time point;
  - *block*$[x]$: set of train instances that left x but are still blocking it at the current time point due to the safety distance, here the time point encodes the end time of blocking[1];

[1] Actually, in *block*$[x]$ we do not need all details stored in the train instances; all what we need is a unique representation of a track-blocking until a certain time point. We store the train instances here to have a unique data type for the global sets *occupy*$[x]$, *block*$[x]$ and *req*$[x]$.

Algorithm 1. Initialization

```
1: procedure INITIALIZE( )
2:     V ← V ∪ {source, target}; c(source) ← ∞; c(target) ← ∞; times ← ∅;
3:     for each v ∈ V do
4:         E ← E ∪ {(source, v), (v, target)}; c((source, v)) ← ∞; c((v, target)) ← ∞;
5:     for each x ∈ (V ∪ E) do occupy[x] ← ∅; block[x] ← ∅; req[x] ← ∅; cap[x] ← ∅;
6:     for each i ∈ {1, ..., n} do
7:         times ← times ∪ {a_i^1};
8:         occupy[(source, v_i^1)] ← occupy[(source, v_i^1)] ∪ {(i, ∅, a_i^1)};
```

- *req*$[x]$: set of train instances that want to move to x at the current time point;
- *cap*$[x]$: set of scenarios with the corresponding number of trains occupying or blocking x in these scenarios at the current time point.

These variables are initialized in Algorithm 1. Initially there are no trains on the actual infrastructure, also no trains block the infrastructure yet. Each train resides at a virtual edge with infinite capacity that leads from a virtual vertex *source* to the train's initial vertex. This is necessary, because it might not be possible for each train to start at the planned time, if its initial vertex is fully occupied. In reality, physical trains do not leave the infrastructure, but will move on to new rides. To model trains moving on, we also connect each vertex to a virtual node *target* and let trains with completed rides move onto these virtual edges with infinite capacity.

3.2 Algorithm

We can now describe our symbolic algorithm to compute exact probabilistic information about delays. As mentioned before, we compute the timetable executions iteratively over the time, considering only time values at which some train wants to move. For each such time value t we iterate first over the vertices, then over the edges[2] and collect for each infrastructure element the trains that want to arrive at the respective infrastructure element at t. Since we store train instances (in contrast to global configurations), in order to determine which train instances are allowed to move within the current free capacities, we need to check compatibilities between the scenarios of train instances. If two train instances have different but compatible scenarios then they "exist" together in the commonly contained stochastic cases, but not in the remaining ones contained only in one of the scenarios. In order to maintain the representation of each complete scenario exactly once for each train, we "split" train instances when needed. We first demonstrate this symbolic concept on a small example.

[2] In contrast to edges, arrival and departure times might be equal for vertices, i.e. the train might not want to stop at the given vertex but move on directly to the next infrastructure element. Processing vertices first allows us to implement entering the vertex first and entering the outgoing edge afterwards for the same time point.

Example 2. Consider the following part of a network with two vertices v_1 and v_2 having capacities $c(v_1) = c(v_2) = 3$ and a bidirectional edge $e = (v_1, v_2)$ with capacity $c(e) = 1$ between them. Assume two train rides

$$z_1 = (\quad v_1(0 \mapsto 1), \quad v_2(2 \mapsto 3) \quad)$$
$$z_2 = (\quad v_2(0 \mapsto 0), \quad v_1(1 \mapsto 2) \quad)$$

where z_1 has higher priority. We assume primary delays only upon start with random values 0 or 1, with probabilities $\mathbb{P}(p_1 = 0) = 0.9$, $\mathbb{P}(p_1 = 1) = 0.1$ for z_1 in vertex v_1, and $\mathbb{P}(p_2 = 0) = 0.8$, $\mathbb{P}(p_2 = 1) = 0.2$ for z_2 in v_2. Below we illustrate the working of our algorithm for the first two iterations, where for each relevant time point, we first specify the state after updating movement into the vertices, then the state after updating movements onto the edges.

v_1 ⟷ v_2

$(source, v_1)$	v_1	(v_1, v_2)	v_2	$(source, v_2)$
		initially		
$(1, \{p_1 = 0\}, 0)$ $(1, \{p_1 = 1\}, 1)$				$(2, \{p_2 = 0\}, 0)$ $(2, \{p_2 = 1\}, 1)$
		$t = 0$ after entering vertices		
$(1, \{p_1 = 1\}, 1)$	$(1, \{p_1 = 0\}, 1)$		$(2, \{p_2 = 0\}, 0)$	$(2, \{p_2 = 1\}, 1)$
		$t = 0$ after entering edges		
$(1, \{p_1 = 1\}, 1)$	$(1, \{p_1 = 0\}, 1)$	$(2, \{p_2 = 0\}, 1)$		$(2, \{p_2 = 1\}, 1)$
		$t = 1$ after entering vertices		
	$(1, \{p_1 = 0\}, 1)$ $(1, \{p_1 = 1\}, 2)$ $(2, \{p_2 = 0\}, 2)$		$(2, \{p_2 = 1\}, 1)$	
		$t = 1$ after entering edges		
	$(1, \{p_1 = 1\}, 2)$ $(2, \{p_2 = 0\}, 2)$	$(1, \{p_1 = 0\}, 2)$ $\left(2, \left\{\begin{matrix} p_1 = 1, \\ p_2 = 1 \end{matrix}\right\}, 2\right)$	$\left(2, \left\{\begin{matrix} p_1 = 0 \\ p_2 = 1 \end{matrix}\right\}, 1\right)$	

The trains can move undisturbed through the network, as planned in the timetable, if there are sufficient capacities. When a train is delayed, it might affect other trains though. This can be seen for time $t = 1$, where two trains need to use the bidirectional edge between v_1 and v_2. As z_1 has the higher priority and the edge is initially available, z_1 can use the edge. Since the edge has only capacity 1, the train z_2 can now only use the edge, if z_1 does not, i.e. in the scenarios, in which the instance of z_1 currently at the edge does not exist. This requires to split the train instance of z_2 at v_2 into two instances, one that can use the edge and one that has to wait for z_1 to leave the edge. For both instances we store the scenarios in which the respective instance exists.

In contrast to this approach the simulation algorithm in [8] considers only probabilities and not scenarios. When executing the same example with the algorithm in [8], we would only store the probabilities of the train instances, but not

Algorithm 2. Symbolic simulation

```
1: procedure SIMULATE( )
2:     INITIALIZE(); let time point t ← t_min;
3:     while t ≤ t_max ∧ times ≠ ∅ do
4:         t ← times.GETSMALLEST(); times ← times \ {t};
5:         for each x ∈ V ∪ E do                      // first vertices then edges
6:             REQUESTS(t, x);                                  // update req[x]
7:             OCCUPATION(t, x);                                // update cap[x]
8:             while req[x] ≠ ∅ do // requests have to be sorted (highest priority first)
9:                 r ← req[x].POP(); UPDATE(t, x, r);    // update occupy and block
```

Algorithm 3. Collecting requests

```
1: procedure REQUESTS(t ∈ 𝕋, x ∈ V ∪ E)
2:     req[x] ← ∅;
3:     for each y ∈ pre(x) do            // either incoming edges or source vertex of x
4:         for each (i, S, t') ∈ occupy[y] do
5:             if t' ≤ t then req[x] ← req[x] ∪ {(i, S, t')};
```

their scenarios. For example, instead of storing the train instance $(1, \{p_1 = 0\}, 0)$, in [8] we only store $(1, 0.9, 0)$ for the initial train instance on the edge from *source* to v_1. Without knowing the stochastic context, the capacities are computed in [8] assuming that all train instances potentially staying in an infrastructure element are indeed there with their given probabilities, i.e., without considering the compatibilities of their scenarios.

The main SIMULATE method is presented in Algorithm 2. After calling the initialization method, we declare a local variable t to store the next time point to be processed in line 2. The loop in lines 3–9 processes all relevant time points from the set *times* in ascending order. We need to consider only these time points, because for all other time values either Algorithm 3 would return no requests for all infrastructure elements or all collected requests could not be scheduled anyway, as they could not be scheduled previously and since then no capacity became available.

After getting the next (smallest) relevant time point in line 4, we process first all vertices and then all edges in the loop in lines 5–9. It is necessary to process the vertices first, as it is a common occurrence that a train does not halt at a vertex and thus has the same arrival and departure time there. This means that the train is supposed to arrive at its next infrastructure element at the same time if it is available. By processing the vertices first, it is possible for train instances to move to a vertex and the following edge on their path in the same time step. In line 6 we first collect for each infrastructure element $x \in V \cup E$ all train instances that want to move to this element at the current time point; the result is stored in the sets *req*[x]. This update is implemented in the REQUESTS method (Algorithm 3). Next we call the OCCUPATION method (Algorithm 4) to compute

Algorithm 4. Computing the occupation of an infrastructure element

1: **procedure** OCCUPATION($t \in \mathbb{T}$, $x \in V \cup E$)
2: $block[x] \leftarrow \{(\cdot,\cdot,t') \in block[x] \mid t' \geq t\}$;
3: $trains \leftarrow occupy[x] \cup block[x]$;
4: $cap[x] \leftarrow$ SPLIT($\emptyset, 0, trains, x$);
5: **procedure** SPLIT($S \in \mathbb{S}, \#z \in \mathbb{N}, trains \subseteq [1..n] \times \mathbb{S} \times \mathbb{T}, x \in V \cup E$)
6: **if** $trains = \emptyset$ or $\#z = c(x)$ **then return** $\{(S, \#z)\}$;
7: choose $(\cdot, S', \cdot) \in trains$ and remove it from $trains$;
8: let set $cap' \leftarrow \emptyset$;
9: **if** S and S' are compatible **then**
10: $cap' \leftarrow \{(S \triangle S', \#z + 1)\}$;
11: $cap' \leftarrow cap' \cup \{(S'', \#z) \mid S'' \in$ SCENARIODIFF($S, \{S'\}$)$\}$;
12: **else** $cap' \leftarrow \{(S, \#z)\}$;
13: **return** $\bigcup_{(S'',\#z'') \in cap'}$ SPLIT($S'', \#z'', trains, x$);

Algorithm 5. Updating a train instance's position

1: **procedure** UPDATE($t \in \mathbb{T}$, $x \in V \cup E$, $r = (i, S, t^*) \in [1..n] \times \mathbb{S} \times \mathbb{T}$)
2: let set $\mathcal{S} \leftarrow \emptyset$;
3: **if** $c(x) = \infty \vee |\{j | (j, \cdot, \cdot) \in occupy[x] \cup block[x] \cup req[x]\}| < c(x)$ **then** $\mathcal{S} \leftarrow \{S\}$
4: **else** $\mathcal{S} \leftarrow$ AVAILABLE(x, r);
5: **for each** $S' \in \mathcal{S}$ **do**
6: **for each** $t' \in \mathcal{D}(p_i^x)$ **do**
7: $occupy[x] \leftarrow occupy[x] \cup \{(i, S' \triangle \{p_i^x = t'\}, t + t' + t'')\}$;
8: $times \leftarrow times \cup \{t + t' + t''\}$; // t'' is waiting/driving time
9: $block[pre(i,x)] \leftarrow block[pre(i,x)] \cup \{(i, S', t + \delta)\}$; $times \leftarrow times \cup \{t + \delta\}$;
10: $occupy[pre(i,x)] \leftarrow occupy[pre(i,x)] \setminus \{(i, S, t^*)\}$;
11: **for each** $S' \in$ SCENARIODIFF($S, \mathcal{S}$) **do**
12: $occupy[pre(i,x)] \leftarrow occupy[pre(i,x)] \cup \{(i, S', t^*)\}$;

for each infrastructure element its currently free capacities (or rather the number of trains occupying or blocking it) under all relevant scenarios. Note that it is not always necessary to call OCCUPATION here, however, for simplicity we omit the corresponding case distinctions. Finally, we iterate over the train instances that want to move to a new infrastructure element in descending priority order, and compute whether there is sufficient capacity left in the relevant scenario by calling the UPDATE method (Algorithm 5). Below we describe all sub-algorithms in more detail.

After SIMULATE terminates, we can collect all instances of a train to extract stochastic information of interest like e.g. expected arrival time at its last station or expected delay. Now we profit from the computationally expensive work to maintain exact computations and represent each train in each complete scenario by exactly one train instance, such that the probabilities of the trains instances' scenarios give us precise analysis results. Furthermore, we can derive information

Algorithm 6. Computing the availability of an infrastructure element

1: **procedure** AVAILABLE($x \in V \cup E$, $r = (i, S, t) \in [1..n] \times \mathbb{S} \times \mathbb{T}$)
2: let set of sets $\mathcal{S} \leftarrow \emptyset$; let set $cap' \leftarrow \emptyset$;
3: **for each** $(S', \#z) \in cap[x]$ **do**
4: **if** $\#z = c(x)$ or S' is not compatible with S **then**
5: $cap' \leftarrow cap' \cup (S', \#z)$; CONTINUE;
6: $\mathcal{S} \leftarrow \mathcal{S} \cup \{(S' \triangle S)\}$; // r moves to x
7: **if** $S' \preceq S$ **then** $cap' \leftarrow cap' \cup (S', \#z + 1)$; // update $cap[x]$
8: **else** $cap' \leftarrow cap' \cup \{(S' \triangle S, \#z + 1)\}$;
9: $cap' \leftarrow cap' \cup \{(S'', \#z) \mid S'' \in$ SCENARIODIFF$(S', \{S\})\}$;
10: $cap[x] \leftarrow cap'$; **return** $\mathcal{S}$

Algorithm 7. Computing the difference of a scenario and a set of scenari

1: **procedure** SCENARIODIFF($S \in \mathbb{S}$, $\mathcal{S}' \in 2^{\mathbb{S}}$)
2: **for each** $S' \in \mathcal{S}'$ **do**
3: **if** $S \preceq S'$ **then return** $\emptyset$;
4: **else if** S is not compatible with S' **then** $\mathcal{S}' \leftarrow \mathcal{S}' \setminus \{S'\}$;
5: **if** $\mathcal{S} = \emptyset$ **then return** $\{S\}$;
6: choose $S' \in \mathcal{S}'$ and $p \in P$ with $S(p) \not\subseteq S'(p)$; // note: $S(p) \cap S'(p) \neq \emptyset$
7: $S_+ \leftarrow S \triangle \{p \triangleleft (S(p) \cap S'(p))\}$;
8: $S_- \leftarrow S \triangle \{p \triangleleft (S(p) \setminus S'(p))\}$;
9: **return** SCENARIODIFF$(S_-, \mathcal{S} \setminus \{S'\}) \cup$ SCENARIODIFF$(S_+, \mathcal{S})$;

for infrastructure elements like e.g. expected capacity usage or delay increment of trains. We plan as future work to also visualize these data in a GUI, even though it is not yet supported currently by our tool.

REQUESTS (Algorithm 3). This method is called for each time point t and infrastructure element x. First we initialize the request set $req[x]$ for x to the empty set in line 2. Then we iterate over all predecessors y of x, which are all incoming edges for a vertex resp. the source vertex for an edge, from where train instances can move to x. The current train instances on the predecessor element y are collected in the set $occupy[y]$. These train instances carry the train identity, their respective scenario and the time point at which they want to move to the next element. We recognize those train instances that currently want to move to x on the fact that the latter time instance is not in the future; exactly these instances are collected in $req[x]$ in line 5.

OCCUPATION (Algorithm 4). This method takes as input a time point t and an infrastructure element x, and computes the number of trains occupying or blocking x in all relevant scenarios. It starts in line 2 with updating the blocking set by removing all blocking entries that are "timed out", i.e. where the end of blocking lies in the past. To simplify notation, note that by "·" we refer to data components that are currently not needed and thus not named. Next we store those train instances that are occupying or blocking x at time t in the local set *trains* in line 3. As these train instances determine the available capacities in x at t, we pass on this set to the SPLIT method in line 4 to compute the number of all occupying/blocking trains in the different scenari, and store this information in $cap[x]$.

SPLIT (Algorithm 4). This method takes as input a scenario S, a non-negative integer $\#z$, a set *trains* of train instances, and an infrastructure element x. The method recursively splits S into incompatible sub-scenarios such that in all stochastic cases covered by a sub-scenario the same train instances from *trains* "exist"; the number of these train instances is counted for each sub-scenario, and the set of all these scenario-integer pairs is returned. Thus calling SPLIT from OCCUPATION with the scenario $\emptyset$ and *trains* containing all occupiers/blockers of x, it will split the scenario space into sub-spaces in which the same train instances from *trains* "exist" and occupy/block x and thus the used capacity can be uniquely determined.

At the beginning, we start with the information that $\#z$ train instances occupy/block an infrastructure element x at a time point t in scenario S, and additional train instances in *trains* might increase the capacity usage when their scenarios are compatible with S. If *trains* is empty or the capacity is already fully used (line 6) then the current result $\{(S, \#z)\}$ is final and returned. Otherwise, in line 7 we process a train instance with scenario S' from *trains*. The splits will be stored in the local set cap' (line 8). In line 9 we check whether our chosen train instance "exists" in S; if not then we recursively call SPLIT to examine the remaining other train instances. Otherwise, if the considered scenario S is compatible with the train instance's scenario S' then we split S into a part in which our selected train instance with scenario S' exists (line 10) and other sub-scenarios in which the train does not exist. Note that the first split is easy to compute as $S \triangle S'$, but the latter might need multiple splits and is computed with the help of the SCENARIODIFF method.

UPDATE (Algorithm 5). This method takes as input a time t, an infrastructure element x and a train instance r that wants to move to x at t, determines whether capacities allow this and if yes models the movement. In case of infinite capacity or if the number of train identities that might want to use x is below the capacity, we know that the train can move (line 3). Otherwise, with the help of the AVAILABLE method we compute a set of scenarios in which the movement is possible (line 4). For each of these scenarios (line 5) and each possible delay of r in x (line 6), we update the occupier set (line 7) and the relevant time points (line 8) and model the blocking time on the previous infrastructure element (line 9) to represent r's move to x. Finally, for those sub-scenarios for which a movement is not possible, we add back corresponding train instances to the occupier set of the previous infrastructure element (lines 11–12). Note that if x is a vertex then in order to identify the respective predecessor we need to store this information somewhere; to ease notation, we did not do so in this method description but our implementation of course stores this information in the corresponding data types.

AVAILABLE (Algorithm 6). This method computes and returns all scenarios in which the request of a train instance r to move to the infrastructure element x at the current time point can be scheduled based on the remaining capacities. The computation uses the information in $cap[x]$, which contains the number of occupiers/blockers of x at t for all relevant scenarios. It iterates over these scenarios (line 3), checks whether a movement is possible in the examined scenario S' and if yes it stores the scenario in which the movement is possible (line 6) and updates $cap[x]$ to increase the stored capacity usage (lines 7–9); if the movement is possible in the whole scenario S' then the number of users is increased (line 8), otherwise we need to split S' into a part where the train instance can move (line 8) and where it is not possible (line 9). Note that the remaining bookkeeping for *occupy*, *times* and *block* is done in the UPDATE method.

SCENARIODIFF (Algorithm 7). This method computes the "difference" of a scenario S and a set $\mathcal{S}$ of scenarios. To do so, we "split" S into a set $\mathcal{S}'$ of scenarios that together represent all complete scenarios from S that are not represented by any scenario from $\mathcal{S}$. If S is contained in one of the "subtracted" scenarios then the result is empty (line 4), and if one of the "subtracted" scenarios is not compatible with S then we can omit it (line 4). Otherwise, we choose one "subtracted" scenario S' and a suitable random variable and split S into an S'-compatible part S_+ (line 7) and a S'-disjoint part S_- (line 8) and make a recursive call to accommodate further refinements.

In summary, we made three simplifying assumptions, all of which could be relaxed without any major changes in the algorithms:

- a train ride is loop-free;
- a physical train path is modeled as several train rides;
- safety distance is a constant value.

An assumption, which is not easy to relax is the modeling of delays by *discrete* probability distributions. In future work we will also consider continuous distributions to allow more precise modeling.

4 Experimental Results

We implemented the algorithm presented in Sect. 3 in C++. To represent probabilities we use high-precision floating point values since computations with exact representations of real numbers are much slower, but floating points could be easily replaced by exact arithmetic computations. All experiments were run on a computer with a 1.80 GHz × 8 Intel Core i7 CPU and 16 GB of RAM. As input we used some real-world railway infrastructure networks that have been generated from confidential infrastructure data in XML form, provided by DB Netz AG (German Railways). The second, respectively third, column in Table 1 lists the number of vertices, respectively edges in the infrastructure networks. I_1 is a sub-network of I_2.

Table 1. Railway systems - infrastructure network and timetable properties

Input	$\|V\|$	$\|E\|$	$\mathbb{T}$	$\|T\|$	Runtime [s]	
					Symb.	[8]
1: I_1, T_1	1195	2532	$[240, 300]$	90	66.2	5.4
2: I_1, T_2	1195	2532	$[480, 300]$	172	149.7	5.9
3: I_2, T_3	2646	5622	$[240, 300]$	201	161.2	12.1

Time was modeled in the unit of minutes using high-precision floating point values. A day is modeled by $[0; 1440]$ with 0 representing 12:00 am. Additionally, unlike in the algorithms above, for the evaluation we do not simulate each time value where some change may occur, but group them to consecutive time values that are at least one second apart. This reduces the amount of computations, in particular of unnecessary checks whether something changes, significantly. The execution of the algorithm is sped up by about factor 10 by this measure. The imprecisions caused by this are not expected to be significant as we are concerned with long planning horizons and not real-time dispatching. This is also not a restriction in the implementation, it would also be possible to simulate every time value where something might happen. We consider different time intervals, specified in Table 1, column four.

For the infrastructure networks and the respective time window, different executable timetables were considered. The number of trains in those timetables is listed in Table 1, column five. The timetables are based on the DB data, but had to be slightly modified to in order to match the network's level of detail.

Due to the currently limited scalability, we consider primary delays only at the start of train rides. The discretized probability distribution for the system entry delay p_i^{entry} of each train $i \in [1..n]$ is based on the delay distribution

$$\mathbb{P}(X \leq t) = F(t) = \begin{cases} 0, & \text{if } t < 0 \\ 1 - p \cdot e^{-\lambda t}, & \text{if } t \geq 0 \end{cases}$$

suggested in [17], with $p \in [0; 1]$ the share of delayed trains and $\lambda \in \mathbb{R}$ the reciprocal of the average delay of delayed trains, for random variable X. For p and λ we use the values suggested by the DB guidelines [6], depending on the train's type. We then select three to four values representing $[0; \infty)$, split into varyingly wide intervals and assign those the probability corresponding to the interval they represent. Compared to these system entry delays the primary delays during the train rides are lower and also four to ten times less likely [6]. Also the number of train instances would grow exponentially in the number of infrastructure elements on the train paths, even without considering trains influencing each other, as for each train its number of instances would at least double at every infrastructure element. Thus we decided to consider only system entry delays in our experiments. Technically, the implementation can be easily extended to consider primary delays during the train ride. We set the safety distance δ between two consecutive trains to 2.0 min.

The only relevant difference between the algorithm we presented above and the implementation is that in the implementation we consider primary delays only at initial vertices. We argued above that this is a reasonable restriction. Additionally, the implementation allows in contrast to the algorithm to consider only time values that are at least a second apart. This exchange between computation speed and precision is optional though.

We compare the algorithm we presented in this paper, in the following called SYMBOLIC, with the one presented in [8]. The time window extends the timetable with 20 min, to accommodate delays. The running times for SYMBOLIC are given in the second to last column of Table 1, respectively the last column for Algorithm [8], which needed less than 10% of those times. The difference in computational effort is due to the more complex computations caused by using scenarios instead of just probabilities for each train instance. In the following we visualize some results only for selected inputs for more clarity.

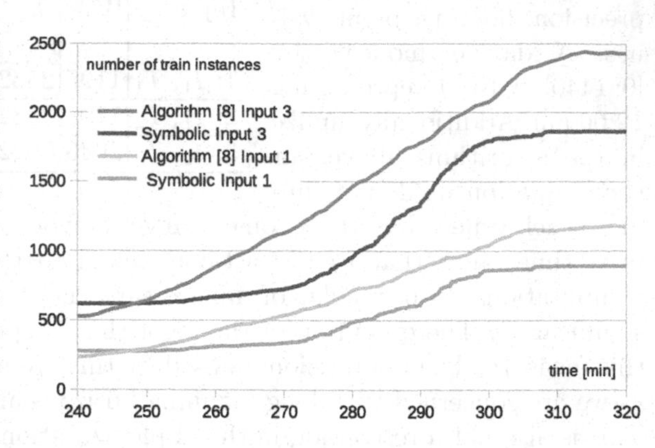

Fig. 1. Number of train instances over time for SYMBOLIC in blue and for Algorithm [8] in green, both for the first (I_1, T_1) and third input (I_2, T_3) (Color figure online)

The number of train instances the simulation computes, depicted exemplary for two of the inputs in Fig. 1, is for all three considered examples lower for SYMBOLIC. This is somewhat surprising as we expected the exactness of SYMBOLIC to come at the cost of more train instances, however, seemingly the approximation of [8] causes an even stronger increment of the number of train instances, at least for smaller time windows.

In general neglecting the stochastic dependencies in the simulation had a larger effect than we expected. This is clearly visible in Fig. 3, where we show for the second input the expected number of trains that reached their target and the trains that should reach their target according to the timetable T_2. The approach presented in this paper is much closer to the actual timetable, despite both approaches using the same delay distributions.

Another indication for this are the numbers of trains that arrive at their target with a certain delay, shown in Fig. 2. There we can see that, as one would expect, most trains arrive in time or only marginally late for the approach presented in this paper. The exact number is even cut off to properly see the remaining values and would be at 140 for 1 min delay. For Algorithm [8] this peak is much lower and there are lots of trains being delayed for 30 min and more, a peak of about 50 that is again cut off. A possible explanation for this is that interactions between trains repeatedly cause the simulation of spurious delays when neglecting stochastic dependencies. By spurious delays we mean delays that only occur in the simulation, due to the simplifying inaccuracy to ignore stochastic dependencies. This holds especially for freight trains, because they have low priorities.

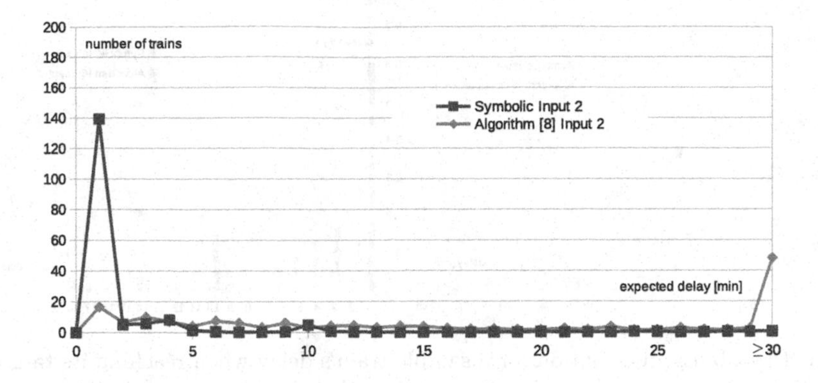

Fig. 2. Number of trains that arrive at their target with a certain expected delay for SYMBOLIC in blue and for Algorithm [8] in green, for the second input (I_1, T_2) (Color figure online)

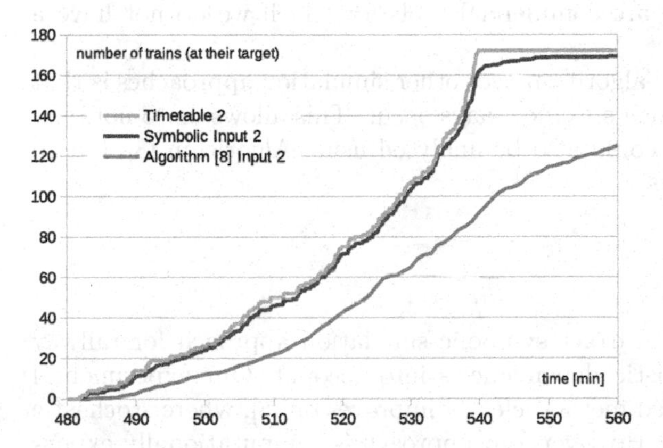

Fig. 3. Number of trains expected to have reached their target over time for SYMBOLIC in blue and for Algorithm [8] in green, for the second input (I_1, T_2) (Color figure online)

This would explain that, for example, for the first input we compute an average delay of 3 min and 42 s when reaching the target, but without considering stochastic dependencies the computed average delay is over 16 min, for the same initial delay distributions. This is visualized for one specific train in Fig. 4, there we show the delay distribution with which this train reaches its target. In the left picture, we can see that SYMBOLIC computes different train instances that arrive with approximately the same delay. This does not happen in Algorithm [8]. These train instances have different scenario representations, which makes it possible to examine which trains caused a certain delay.

In contrast to Monte Carlo simulations, our algorithm is exact (up to floating-point computations) for the given model. While Monte Carlo simulations usually compute about 50 to 100 timetable executions with random values for the primary delays, our algorithm considers all possible scenarios. With more than $3^{|T|}$ complete scenarios, a Monte Carlo simulation may give some idea about the timetables properties, but the result depends on the random scenarios considered. Most of the software tools using Monte Carlo simulation that were mentioned in the introduction though use a more detailed model than we do

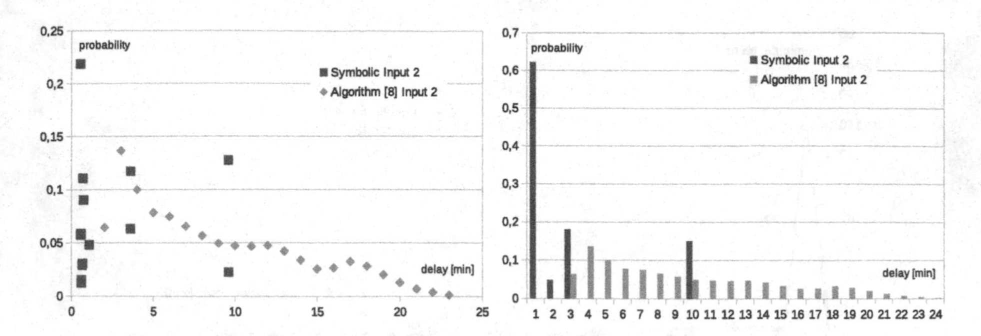

Fig. 4. Probability distribution for a sample train's delay when reaching its target (in T_2), left for the computed train instances, right discretized to minutes

in this paper, which increases their precision compared to this approach. Unfortunately, we can not compare this implementation with the systems mentioned in the introduction as those are commercial tools for which we do not have a license.

Another advantage of this algorithm over other simulation approaches is that we store the scenarios for which specific delays occur. This allows us to not only examine the properties that could also be analysed using Algorithm [8], but to examine the causes for delays.

5 Conclusion

In this paper we presented an exact symbolic simulation approach for railway timetables that takes stochastic dependencies into account. Our experimental results show that the proposed method clearly improves on [8], where stochastic dependencies were neglected. However, our approach is computationally expensive and needs improvement for better scalability. Besides parallelization, ongoing work aims at a major reduction of the number of train instances by merging suitable cases. We target reduction while maintaining exact results, as well as over-approximative reductions embedded in a CEGAR (counterexample-guided abstraction refinement) framework. We are also working on a formal proof of correctness for the presented approach. Another work thread aims at rigorous result evaluation and visualization to help to identify problematic issues in timetables.

References

1. LUKS (2021). https://www.via-con.de/en/development/luks/. Accessed 28 Apr 2021
2. OnTime (2021). https://www.trafit.ch/en/ontime. Accessed 28 Apr 2021
3. OpenTrack Railway Technology (2021). http://www.opentrack.ch/opentrack/opentrack_e/opentrack_e.html. Accessed 28 Apr 2021
4. RailSys (2021). https://www.rmcon-int.de/railsys-en/. Accessed 28 April 2021

5. Büker, T., Seybold, B.: Stochastic modelling of delay propagation in large networks. J. Rail Transp. Plann. Manag. **2**(1), 34–50 (2012). https://doi.org/10.1016/j.jrtpm.2012.10.001
6. DB Netz AG: Fahrwegkapazität. In: Richtlinie 405 / DB Netz, Deutsche Bahn Gruppe. DB Netz, Deutsche Bahn Gruppe, [Frankfurt am Main] (2008)
7. Franke, B., Seybold, B., Büker, T., Graffagnino, T., Labermeier, H.: Ontime – network-wide analysis of timetable stability. In: 5th International Seminar on Railway Operations Modelling and Analysis (2013)
8. Haehn, R., Ábrahám, E., Nießen, N.: Probabilistic simulation of a railway timetable. In: Huisman, D., Zaroliagis, C.D. (eds.) 20th Symposium on Algorithmic Approaches for Transportation Modelling, Optimization, and Systems (ATMOS 2020). Schloss Dagstuhl-Leibniz-Zentrum für Informatik (2020). https://doi.org/10.4230/OASIcs.ATMOS.2020.16. https://drops.dagstuhl.de/opus/volltexte/2020/13152
9. Hensel, C., Junges, S., Katoen, J., Quatmann, T., Volk, M.: The probabilistic model checker storm. CoRR abs/2002.07080 (2020). https://arxiv.org/abs/2002.07080
10. Janecek, D., Weymann, F.: Luks - analysis of lines and junctions. In: Proceedings of the 12th World Conference on Transport Research (WCTR 2010), Lisbon, Portugal (2010)
11. Kwiatkowska, M., Norman, G., Parker, D.: PRISM 4.0: verification of probabilistic real-time systems. In: Gopalakrishnan, G., Qadeer, S. (eds.) CAV 2011. LNCS, vol. 6806, pp. 585–591. Springer, Heidelberg (2011). https://doi.org/10.1007/978-3-642-22110-1_47
12. Nash, A., Huerlimann, D.: Railroad simulation using OpenTrack. Comput. Rail. IX, 45–54 (2004). https://doi.org/10.2495/CR040051
13. Norris, J.R.: Markov Chains. Cambridge Series in Statistical and Probabilistic Mathematics, Cambridge University Press, Cambridge (1997). https://doi.org/10.1017/CBO9780511810633
14. Radtke, A.: Infrastructure Modelling. Eurailpress, Hamburg (2014)
15. Radtke, A., Bendfeldt, J.: Handling of railway operation problems with RailSys. In: Proceedings of the 5th World Congress on Rail Research (WCRR 2001), Cologne, Germany (2001)
16. Schneider, W., Nießen, N., Oetting, A.: MOSES/WiZug: Strategic modelling and simulation tool for rail freight transportation. In: Proceedings of the European Transport Conference, Straßbourg (2003)
17. Schwanhäusser, W.: Die Bemessung der Pufferzeiten im Fahrplangefüge der Eisenbahn. Verkehrswissenschaftliches Institut Aachen: Veröffentlichungen, Verkehrswiss. Inst. d. Rhein.-Westfäl. Techn. Hochsch. (1974)
18. Yuan, J.: Stochastic modelling of train delays and delay propagation in stations, vol. 2006. Eburon Uitgeverij BV (2006)
19. Yuan, J., Medeossi, G.: Statistical analysis of Train Delays and Movements. Eurailpress, Hamburg (2014)

Simulation of N-Dimensional Second-Order Fluid Models with Different Absorbing, Reflecting and Mixed Barriers

Marco Gribaudo[1], Mauro Iacono[2(✉)], and Daniele Manini[3]

[1] Dipartimento di Elettronica, Informatica e Bioingegneria, Politecnico di Milano, via Ponzio 5, 20133 Milan, Italy

[2] Dipartimento di Matematica e Fisica, Università degli Studi della Campania, "L. Vanvitelli", viale Lincoln 5, 81100 Caserta, Italy
mauro.iacono@unicampania.it

[3] Dipartimento di Informatica, Università di Torino, corso Svizzera 185, 10123 Turin, Italy

Abstract. Simulation of second-order fluid models requires specific techniques due to the continuous randomness of the considered processes. Things become particularly difficult when considering several dimensions, where correlation occurs, and classical concepts like absorption and reflection require specific extensions. In this work, we will focus on three different types of behaviors, with two correlations structures: either independence or total correlation. For the considered scenario, we will describe how to produce suitable traces of the underlying continuous stochastic process.

Keywords: Second-order fluid models · Simulation · Absorbing and reflecting barrier

1 Introduction

Second order fluid models provide an interesting tool to describe systems with continuous components that vary in a random way. While in first order fluid models continuous components of the system evolve in a deterministic way, characterized by a given rate, second order models add variance to include randomness. Second order models have been used in performance evaluation several times, but rarely considering correlation among different continuous components. This has limited the effectiveness and applicability of the models, and they have rarely been fully exploited. In particular, fluid models can be used to describe systems with a random continuous stream of objects, such as cyber-physical systems and peer-to-peer file transfers. Such streams, in most of the cases, are characterized at least by two continuous variables: the "source" and the "destination". These variables are strongly correlated, and this correlation poses challenges for both analytical and simulation purposes.

A. Abate and A. Marin (Eds.): QEST 2021, LNCS 12846, pp. 276–292, 2021.
https://doi.org/10.1007/978-3-030-85172-9_15

In first order fluid models correlation between inputs and outputs is governed by rate adaption: flux speeds are varied to make sure that inputs and outputs match when buffers are either empty or full. In second order models this is achieved using boundary conditions which correlate the various streams. Such conditions become more complex when considering several continuous variables.

Although second order fluid models have been mainly studied analytically in literature, the underlying equations become very easily impossible to solve from a numerical point of view. For this reason Monte-Carlo simulation can be an interesting alternative to exploit the modeling power of such tools, while maintaining the ability to compute results with reasonable amounts of time and resources. Simulation of models with a single reflecting barrier has been studied in [1]. Here we extend those results to simulate processes with different types of barriers in n-dimensions.

To summarize, in this work, we focus on three different types of boundary behaviors, with two correlations structures: either independence or total correlation. The main contribution of this work is a technique to produce suitable traces of the underlying continuous stochastic process.

The paper is organized as follows: related work is discussed in Sect. 2, then we formalize the considered class of models and focus on their analysis when there is a single continuous variable in Sect. 3. We next extend the results to consider a higher number of fluid variables in Sect. 4, and we conclude the paper in Sect. 5.

2 Related Work

The differential equations that describe a fluid model are hard to solve and the symbolic solution of the equations can be obtained only for trivial cases. In the case of transient analysis the system has an initial state which can be exploited as considered in [2–5], to mention a few, while in the case of stationary analysis the equations that describe a fluid model are Ordinary Differential Equations (ODEs) without initial condition. Indeed this problem has been solved for first order models by the analysis of first passage time probabilities, see for instance [6–11] and the references therein. The key of these solutions lies in the matrix characterisation of the distribution of the phase visited at the end of a busy period of the fluid queue.

The problem remains open for second order models (also known as modulated diffusion processes), where the solution is obtained from a set of boundary equations, ODEs and a normalising condition. For example, in case of fluid level independent transition and fluid drift, the solution of the ODE is obtained by computing eigenvalues and eigenvectors of a matrix [12]. Usually those approaches are very sensitive to the computation of the eigenvalues and may lead to severe numerical errors. An alternative approach using modal decomposition is proposed in [13], where authors develop a numerically stable algorithm for a general fluid-flow model to examine the impact of variance in the case of homogeneous on-off sources, in the infinite and finite buffer case. Second order models are introduced in [12,14]. In these works the authors consider a *white noise* factor which represents the variability of the traffic during the transmission periods. The fluid level is described by a reflected Brownian motion modulated by a continuous time Markov chain (CTMC). When the CTMC is in state i, the fluid level is modelled by a reflected Brownian motion with drift r_i and variance parameter σ_i^2. The authors of [15] provide a stability analysis of such models when the modulating process is general stationary and ergodic (not necessarily Markovian). Another notable example is reported in [16] where the authors present a simple approximate compositional method for analysing a network of fluid queues with Markov-modulated input processes at equilibrium. Fluid models have been successfully applied to several interesting practical examples, such as in [17], where the effect of a power-save mode on the battery life of a device subject to stochastically determined charging and discharging periods is studied.

Second order models can have two different types of boundaries: absorbing or reflecting [18]. A special approximation method is proposed in [19] for approximating the absorbing boundary based on the solution of the system with reflecting boundary.

In [20] authors address the problem of performing steady state solution of modulating diffusion processes using neither discretisation nor singular value decomposition. The approach is similar to the one used in [21,22] for first order models and is focused on the boundary behaviours. Moreover, the authors applied second order fluid models to multimedia systems performance evaluation in [23].

In this paper the focus is on the application of a simulation-based approach to models in which the upper and lower boundary of each state can be either absorbing or reflecting. An important impulse towards the research work described in this paper is due to the work of J. Michael Harrison, which extensively studied Brownian motion based analysis of stochastic flow systems problems [24]. The author uses the notion of barrier in solving tandem queues in heavy traffic with diffusion approximation [25]; in this contribution we study the implementation of a suitable simulation oriented approach encompassing barriers based on those methods, aiming at second order fluid models.

3 Simulation of First and Second Order Fluid Models with Reflecting and Absorbing Barriers

From a simulation point of view, the type of second order fluid model we are interested in can be defined by the following tuple:

$$\mathcal{M} = (S, s_0, X, \mathbf{x}_0, \mathbf{Q}, R, V, B)\,, \tag{1}$$

where $S = \{s_1, \ldots, s_N\}$ is a set of N states in which the system can operate, and $s_0 \in S$ is the one in which the system begins its evolution. $X = \{x_1, \ldots, x_C\}$ is a set of C continuous variables, whose initial value is defined by vector $\mathbf{x}_0 = |x_{01}, \ldots, x_{0C}|$. $\mathbf{Q} = |\ldots q_{ij} \ldots|$ is an infinitesimal generator, with $q_{ij} \geq 0$ for $i \neq j$, and $q_{ii} = -\sum_{j \neq i} q_{ij}$. The system stays in a state s_i for an exponentially distributed time, with rate $-q_{ii}$, then it jumps to state s_j with probability $\frac{q_{ij}}{-q_{ii}}$. $R = \{\mathbf{r}_1, \ldots, \mathbf{r}_N\}$ is a set of N vectors of C components each, and $V = \{\Sigma_1, \dots, \Sigma_N\}$ is a set of N matrices of size $C \times C$. When the system is in a state $s_i \in S$, the fluid components evolve according to the average defined by $\mathbf{r}_i$, with covariance matrix Σ_i, as it will be better clarified in Sect. 4. $B = \{b_1, \dots, b_K\}$ is a set of K boundary conditions/action specifications that are used either to perform rate adaption (for first order models) or to define boundary behavior (for second order models). Before entering in more details with the description of the elements of B, we need to define the state of the model. At a given time $t > 0$, the state $\mathcal{Z}(t)$ of the model is defined by the following tuple:

$$\mathcal{Z}(t) = (s(t), \mathbf{x}(t)) \ . \tag{2}$$

In particular, $s(t) \in S$ is the *discrete component* of the state of the model, and describes the current operation mode of the system. We call $s(t)$ discrete, because it remains constant for finite time instants, and changes to different values with jumps, according to both the infinitesimal generator Q and the boundary actions B. $\mathbf{x}(t)$ is the continuous component of the state, and it is a C component vector. This part is called continuous component of the state, since it continuously varies with time. The initial state of the system is thus defined as:

$$\mathcal{Z}(0) = (s_0, \mathbf{x}_0) \ . \tag{3}$$

With the state definition in Eq. 3, we could describe the elements of the boundary action set B. In particular, we define:

$$b_k = (s_k, \mathbf{u}_k, d_k, a_k, p_k) \ , \tag{4}$$

where $\mathbf{u}_k$ is a C component vector, and d_k is a real constant. The boundary action b_k is triggered at time t, if:

$$s(t) = s_k \wedge \mathbf{x}(t) \cdot \mathbf{u}_k = d_k \ , \tag{5}$$

where $\mathbf{x}(t) \cdot \mathbf{u}_k$ is the conventional dot product of two vectors (i.e. $\mathbf{x}(t) \cdot \mathbf{u}_k = \sum_i \mathbf{x}_i(t) \cdot \mathbf{u}_{ki}$). Component $a_k = \{$**jump**,**reflect**$\}$ defines the type of action that must be performed when the boundary action is triggered, and p_k is a parameter required by the corresponding behavior. With the **jump** action, the system changes state, moving from $s(t^-) = s_k$ to $s(t^+) = p_k$. In this case, $p_k \in S$ describes the destination of the jump, and t^- and t^+ denote time just before and after the state change. Note that, without loss of generality, **jump** actions can be used to model both rate adaption in first order models and absorbing behaviors in second order models. The **reflect** action, specific for second order models, defines the reflecting barrier specific to second order fluid models. In this case, p_k is a C component vector that defines the *reflection direction*, as it will be clarified in Sect. 4.

In this work we focus on producing a trace of the considered process. The key difference with conventional discrete event systems is that the continuous component of the state, $\mathbf{x}(t)$, evolves with time. In general, we are interested in simulating the transient evolution of the model, producing a trace $\mathcal{Z}(t_l)$, sampled in L different time points $t_l \in \{0, t_1, \dots, t_L\}$ with $0 < t_1 < \dots < t_L$. Considering the changes of state due to the infinitesimal generator Q is straightforward and can be done with conventional techniques: for this reason, we will not focus on this aspect of the simulation, and, without loss of generality, we will restrict our analysis to a time frame T, such that for any $t_l \leq T$ we have $s(t) = s_i$.

To perform the simulation of the continuous part of the model and compute $\mathbf{x}(t_l), \forall t_l \leq T$, we need essentially to be able to perform two tasks:

1. starting from $\mathbf{x}(t_l)$, determine $\mathbf{x}(t_{l+1})$ for any $t_l < t_{l+1} \leq T$;
2. determine if a barrier k exists that was hit by the continuous component of the model during any considered time interval $t_l < t_{l+1}$, that is, if $\exists t, t_l \leq t < t_{l+1}$ such that $\mathbf{x}(t) \cdot \mathbf{u}_k = d_k$.

In the rest of this section, we will first consider this problem for first order models, then for second order models, revisiting known results in the proposed framework. To also simplify the discussion, we will focus on a single barrier, that is on $K = 1$. Extension to $K > 1$ is straightforward and already analyzed in literature (see, for example, [26]) for first order models, while it is a research topic for second order models, and will be postponed to future work.

3.1 Simulation of First Order Fluid Models

In first order models the fluid level grows linearly according to the deterministic rate r_i. Since we focus on a single continuous variable, if $x(t)$ denotes the fluid level at time t and t' is a time instant such that $t' > t$, then:

$$x(t') = x(t) + r_i \cdot (t' - t) , \tag{6}$$

provided that the continuous variable does not reach a boundary in the (t, t') interval. The evolution of the process is thus deterministic in time, once the initial state and the time intervals have been set. We can then move from time t_l to time t_{l+1} with a very simple recurrence equation:

$$x(t_{l+1}) = x(t_l) + r_i \cdot (t_{l+1} - t_l) . \tag{7}$$

Conversely, we can determine that the boundary is crossed if we have that:

$$(u_k \cdot x(t_l) - d_l)) \cdot (u_k \cdot x(t_{l+1}) - d_l)) < 0 , \tag{8}$$

which essentially checks whether the two values of the continuous variable are on the same side (same sign of the differences with respect to d_l), or on the opposite one (different signs) of the barrier.

3.2 Simulation of Brownian Motion for Second Order Models

In case of second order models, it has been proven (see, for example, [1]) that the model evolves following a Brownian motion characterized by a given drift r_i and a given variance σ_i. In these models, we have that:

$$x(t') = x(t) + N\left(r_i \cdot (t' - t), \sigma_i^2 \cdot (t' - t)\right) , \tag{9}$$

where $N(\mu, \sigma)$ is a sample from a normal distribution characterized by average μ and variance σ^2. In other words, the fluid level change in the (t, t') interval is normally distributed with mean $r_i \cdot (t' - t)$ and variance $\sigma_i^2 \cdot (t' - t)$. Note that Eq. (9) is valid if the continuous variable does not reach a boundary in the (t, t') interval. In other words, the value of the continuous variable $x(t_{l+1})$, starting from $x(t_l)$ can be computed as:

$$x(t_{l+1}) = x(t_l) + r_i \cdot (t_{l+1} - t_l) + \sigma_i \cdot N(0,1) \cdot \sqrt{(t_{l+1} - t_l)}, \quad (10)$$

with $N(0,1)$ being a sample from the standard normal distribution, characterized by zero mean and unitary variance. Determining if the continuous variable crosses a boundary in a second order model requires a little bit more of care, and the way in which it can be performed is tightly connected to the type of barrier (i.e., absorbing or reflecting) being considered. To approach the problem, we start considering the meaning of a barrier in a first order system, and its interpretation in second order models.

3.3 Considering Boundaries

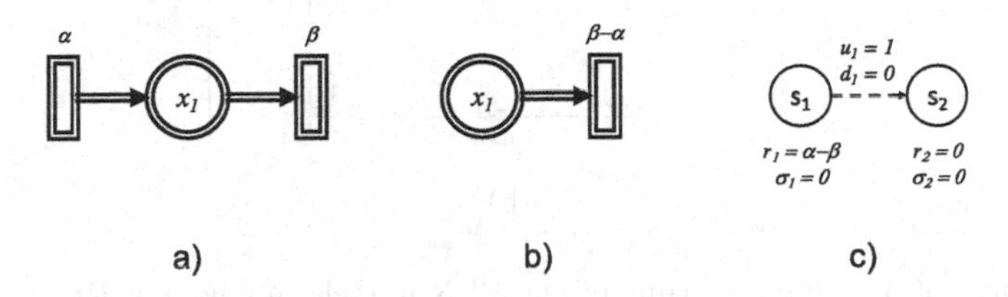

Fig. 1. Boundary in a simple first order fluid model, shown as a Fluid Petri Net (with rate $\beta > \alpha$): a) and b) two versions of the same process; c) the corresponding simulation model.

To better explain the importance of boundary conditions, let us focus on the two examples shown in Fig. 1a) and 1b) by means of a Fluid Petri Net [27]. Case a) shows a process in which we have a continuous input at rate α, and a continuous output at rate β (e.g., a stable producer/consumer system). Case b) presents instead a case in which we only have a continuous output at rate $\beta - \alpha$ (e.g., a pool depletion system). Both models are initialized to state s_1, with an initial fluid level $x_1 = x_0$; they have the same underlying first order fluid model process, depicted in Fig. 1c): at time $t = \frac{x_0}{\beta - \alpha}$, the process jumps to state s_2 to perform rate adaption. In case a), the output of the transition on the right is reduced to α, while in case b it is reduced to 0. The final result in both cases is a fluid rate $r_2 = 0$, which ensures that the process is absorbed at the boundary.

When considering second order models, the two scenarios presented in Fig. 1a) and 1b) are modeled in a different way. The first case is modeled with a single state s_1, with $\sigma_1 > 0$, with a reflecting barrier placed at $x_1 = 0$. Reflection makes sure that the randomness in both input and output processes are correctly modeled, since there could be time instants in which the random rate of the input transition is greater than the one of the output, making thus the fluid increase in the considered continuous place. This is the key of diffusion approximation of queuing systems using second order fluid models. The system shown instead in Fig. 1b) is modeled with an absorbing boundary, which causes the system to jump to state s_2 when $x_1 = 0$ is reached, resulting in a stop of the flow. The main difference with respect to first order models ($\sigma_1 > 0$) is that the time at which the system reaches state s_2 is no longer deterministic, but it rather is a random variable $\mathcal{T}$, whose average is $E[\mathcal{T}] = \frac{x_0}{\beta-\alpha}$. Absorption is needed, since this case corresponds to the time required to empty an initially full container: once all its content is gone, since there is no input, the continuous level will never grow anymore.

Figure 2 shows a few traces of the first order model, identical for cases a) and b), and of the differences between first and second order models.

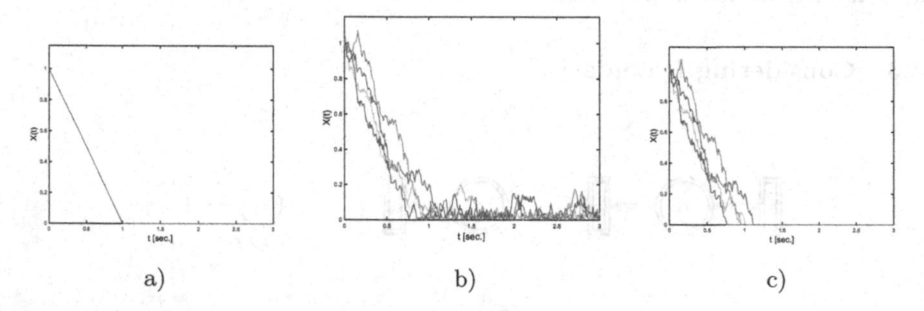

Fig. 2. Traces of 4 simulation runs of the FPN models in Fig. 1 with $\alpha = 1$, $\beta = 2$, $x_0 = 1$ and $\sigma_1 = 0.25$ (in the second order case): a) Trace of the first order models; b) Traces of the second order model corresponding to Fig. 1a); c) Traces of the second order model corresponding to Fig. 1b).

3.4 Reflecting Barrier in One Dimension

According to [1], Brownian motion with a reflecting barrier located at $x = 0$ can be defined as the difference of two processes: the unconstrained Brownian motion and the barrier process. In particular, let us call $w(t)$ the unconstrained process and $y(t)$ the barrier process: then the current fluid level defined by variable $x(t)$ is:

$$x(t) = w(t) - y(t), \tag{11}$$

where $w(t)$ evolves as a conventional Brownian motion as defined in Eq. 10, and $y(t)$ evolves whenever the process crosses the barrier:

$$y(t) = \inf_{\tau \leq t} w(\tau). \tag{12}$$

Figure 3 shows an example of the evolution of the unconstrained process $w(t)$, of the barrier $y(t)$ and of the resulting reflected process $x(t)$. The challenge in

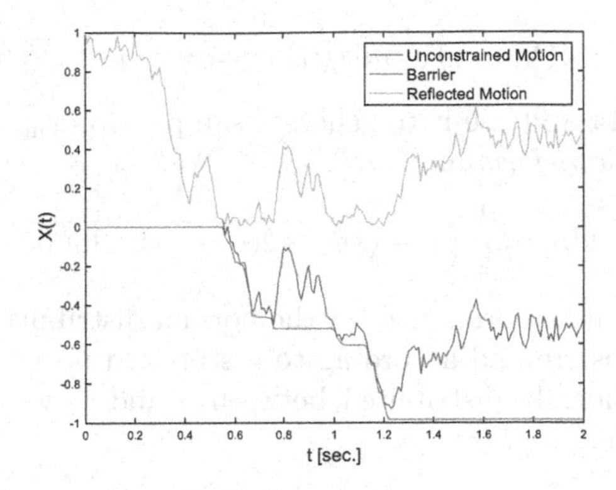

Fig. 3. The components of a reflected Brownian motion process with $r_i = -0.5$, $\sigma_i = 0.5$ and $x_0 = 1$: the unconstrained process, the barrier, and the reflected process.

simulating this type of reflected process is that during each time interval $[t_l, t_{l+1})$ the process might reach values that are below the ones assumed at beginning and at the end of the sampling period, and it might cross the barrier in some unseen point in the middle. This is, for instance, shown in Fig. 4, where the process starts at $t_l = 0$ with $x(t_l) = 1$, and ends at $t_{l+1} = 0.1$ with $x(t_{l+1}) = 1.63$, and the minimum is reached at around $t_{\min} = 0.001$, with $x(t_{\min}) = 0.85$. The process can then be correctly simulated in the following way.

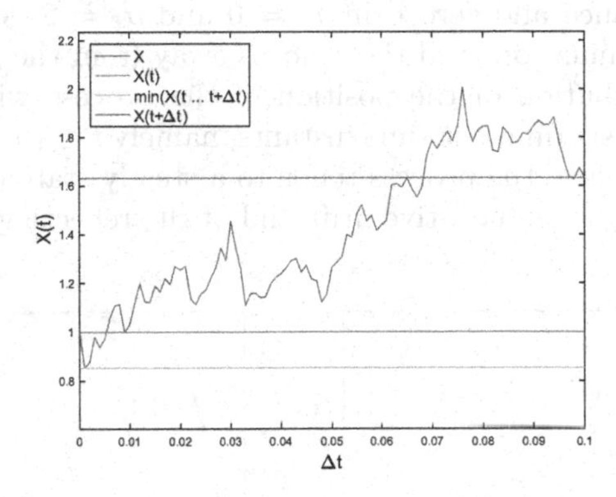

Fig. 4. A detail of the evolution of Brownian motion with respect to a barrier at 0 within a discretization step $\Delta t = 0.1$ s, with $r_i = 0$ and $\sigma_i = 2$: the process, the starting and ending point, and the minimum value reached during the evolution.

Let us call $x_{\max}(t_l, t_{l+1})$ the maximum reached by a process in the time interval $[t_l, t_{l+1})$, i.e.:

$$x_{\max}(t_l, t_{l+1}) = \max_{t_l \leq t < t_{l+1}} x(t). \tag{13}$$

If we consider a process characterised by zero mean and unitary variance ($r_i = 0$ and $\sigma_i = 1$), which starts at $x(t_l) = 0$, it can be proven (see [1]) that:

$$F(m|y) = \text{Prob}\left(x_{\max}(t_l, t_{l+1}) \leq m | x(t_{l+1}) = y\right) = 1 - e^{\frac{-2m(m-y)}{t_{l+1}-t_l}}, m \geq y. \quad (14)$$

We can invert Eq. 14 and use it to generate samples for $x_{\max}(t_l, t_{l+1})$ using the *Inverse Sampling Transform* method:

$$F^{-1}(u|y) = \frac{1}{2}\left(y + \sqrt{y^2 - 2(t_{l+1} - t_l).\ln(1-u)}\right) \quad (15)$$

Exploiting the symmetry properties of the normal distribution, given two samples $\nu = N(0,1)$, distributed according to a standard normal distribution, and $\eta = \text{Unif}(0,1)$, uniformly distributed between 0 and 1, we can compute both $x(t_{l+1})$ and $\gamma = \min_{t_l, t_{l+1}} x(t)$:

$$w(t_{l+1}) = w(t_l) - \nu \cdot \sigma_i \cdot \sqrt{t_{l+1} - t_l} + r_i \cdot (t_{l+1} - t_l) \quad (16)$$

$$\gamma = \min_{t_l, t_{l+1}} w(t) = w(t_l) - F^{-1}(\eta|\nu) \cdot \sigma_i \cdot \sqrt{t_{l+1} - t_l} + r_i \cdot (t_{l+1} - t_l) \quad (17)$$

$$y(t_{l+1}) = \min(y(t_l), \gamma) \quad (18)$$

$$x(t_{l+1}) = w(t_{l+1}) - y(t_{l+1}). \quad (19)$$

Figure 5a) shows four Brownian motion traces, with a reflecting barrier at $x = 0$, starting from $x_0 = 1$, and for a time interval $t \in [0, 2]$, with different drifts ($r_i = \{-0.5, 0, 0, 0.25\}$) and standard deviations ($\sigma_i = \{0.5, 0.5, 2, 0.25\}$). Note how the trace with the negative drift $r_i = -0.5$ clearly bounces repeatedly against the barrier at $X(t) = 0$, apparently never touching it, while the trace with a higher variance and zero drift $r_i = 0$ and $\sigma_i = 2$ gets reflected at the beginning of the simulation, and then moves away from the boundary. Figure 5 b) shows the distribution of the position of the process with $r_i = -0.5$ and $\sigma_i = 0.5, x_0 = 1$, at six different time instants, namely $t \in \{0.5, 1, 1.5, 2, 5, 10\}$. It is interesting to see how the process tends to a steady state distribution caused by the combination of the negative drift and of the reflecting barrier.

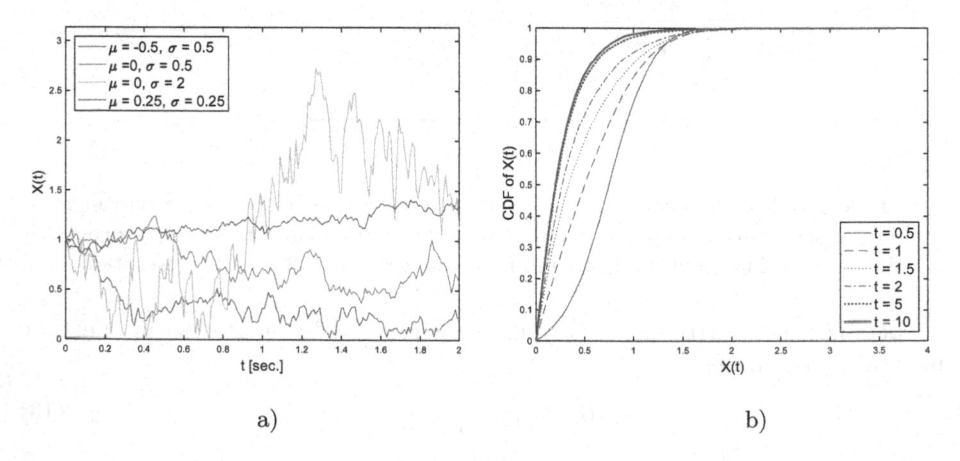

Fig. 5. a) Traces of a Brownian motion with a reflecting barrier and parameters shown in the legend, all starting at $x_0 = 1$. b) Distribution of the position at $t \in \{0.5, 1, 1.5, 2, 5, 10\}$ for the case with $r_i = -0.5, \sigma_i = 0.5, x_0 = 1$.

3.5 Absorbing Barrier in One Dimension

An absorbing barrier can be easily implemented by replacing Eqs. 18 and 19 with the following:

$$x(t_{l+1}) = \max(0, w(t_{l+1})). \tag{20}$$

Barrier crossing can be simply identified, starting from Eq. 17, by testing if $\gamma < 0$. Absorption is then implemented, as for the first order case, by moving the system to a second state s_2, where both the drift and variance are zero, i.e. $r_2 = \sigma_2 = 0$.

Figure 6a) shows four Brownian motion traces, with an absorbing barrier at $x = 0$, starting from $x_0 = 1$, and for a time interval $t \in [0, 2]$, with different drifts ($r_i = \{-0.5, 0, 0, 0.25\}$) and standard deviations ($\sigma_i = \{0.5, 0.5, 2, 0.25\}$). In particular, the same random number generator used for the traces shown in Fig. 5 has been used, to better show the effect of the different barrier. In this case both the traces with negative drift and the one with zero drift and high variance get absorbed, while the other two continues to evolve in the considered time frame. Figure 6 b) shows the distribution of the position of the process with an absorbing barrier and $r_i = -0.5$ and $\sigma_i = 0.5, x_0 = 1$, at six different time instants, namely $t \in \{0.5, 1, 1.5, 2, 5, 10\}$. It is interesting to see how in this case, as the time increases, the distribution tends to be deterministic, with the probability mass centered on the barrier.

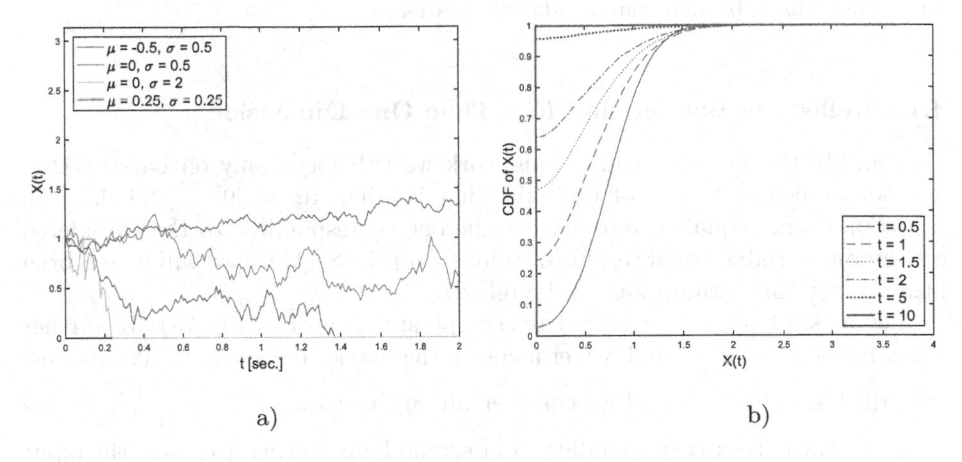

Fig. 6. a) Traces of a Brownian motion with an absorbing barrier and different parameters, all starting $x_0 = 1$. b) Distribution of the position at $t \in \{0.5, 1, 1.5, 2, 5, 10\}$ for the case with $r_i = -0.5, \sigma_i = 0.5, x_0 = 1$.

Extensions to barriers implementing an upper threshold or a different lower boundary can be applied by simply translating or reflecting the process, and will not be further investigated.

4 Extending to More Than One Dimension

In case of more than one dimension, the definition of reflecting barrier becomes more complex, and a new aspect must be considered: the *reflection direction*. To motivate this extra requirement, we start considering the type of processes that describe the continuous evolution of two simple examples: a battery operated device, with energy harvesting (Fig. 7a), and a simple streaming application that continuously fills a finite buffer of packets, and decodes the ones at the head of the buffer (Fig. 7b).

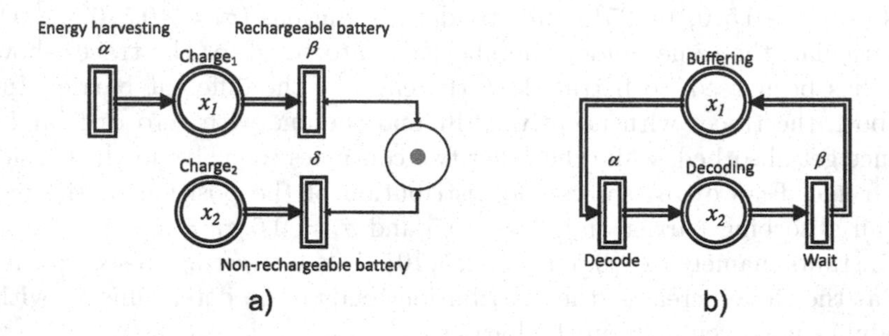

Fig. 7. Two simple FPN models with two second order fluid variables: a) an energy harvesting sensor, b) a streaming video application.

4.1 Reflecting Barriers in More Than One Dimension

To simplify the presentation, in this work we will focus only on barriers that are perpendicular to one of the axis, that is when $\mathbf{u}_k = |0, \ldots, 0, 1, 0, \ldots, 0|$ (all components equal to zero, except the one corresponding to the considered continuous variable equal to plus or minus one). In Sect. 4.3 we will give a brief idea on how this assumption can be relaxed.

Let us start focusing on the battery operated sensor of Fig. 7a). To further simplify the discussion, and better focus on the barrier behavior, we also assume that drift is $\mathbf{r}_1 = |0, 0|$ and we consider unitary covariance $\Sigma_1 = \begin{vmatrix} 1 & 0 \\ 0 & 1 \end{vmatrix}$. In this example, the barrier corresponding to x_1 should be reflecting, to model the input due to energy harvesting, and the one corresponding to x_2 should be absorbing, since the non-rechargeable battery will become useless once completely depleted. In this case, extension to more than one continuous variable is straightforward: it is sufficient to consider $\mathbf{w}(t)$ as a two-dimensional Brownian motion process, and $\mathbf{y}(t)$ as a two components vector that encodes the position of the barrier in both dimensions. Due to the independence of the two continuous variables, barriers can be considered separately: if, initially, we consider the process being unconstrained along x_2, a simulation step can be executed in this way:

$$\mathbf{w}(t_{l+1}) = \mathbf{w}(t_l) - |\nu_1, \nu_2| \cdot \Sigma_1 \cdot \sqrt{t_{l+1} - t_l} + \mathbf{r}_1 \cdot (t_{l+1} - t_l) \quad (21)$$
$$\gamma = w_1(t_l) - F^{-1}(\eta|\nu_1) \cdot \sigma_{i:11} \cdot \sqrt{t_{l+1} - t_l} + r_{i:1} \cdot (t_{l+1} - t_l) \quad (22)$$
$$\mathbf{y}(t_{l+1}) = |\min(y_1(t_l), \gamma), y_2| \quad (23)$$
$$\mathbf{x}(t_{l+1}) = \mathbf{w}(t_{l+1}) - \mathbf{y}(t_{l+1}). \quad (24)$$

An instance generated with the following procedure is represented in Figs. 8a) and 8b), where a trace and a point cloud distribution of the result are shown. It is interesting to see in Fig. 8a) that when the process tries to cross the barrier, the reflecting boundary lowers it to maintain the fluid component positive, as it is clearly visible in the point cloud distribution of Fig. 8b) (which does not display the time component, and uses the content of the two continuous variables to determine the position of the point on the plane).

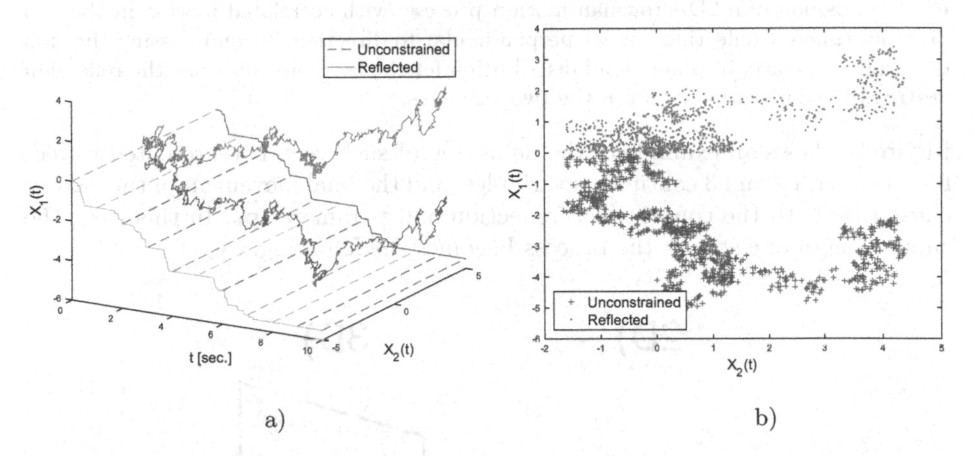

Fig. 8. Position of a 2D Brownian motion process, with independent drift and a reflecting barrier perpendicular to direction x_1 and passing through the origin: a) trace; b) point cloud distribution for the trace.

We might be tempted to use this approach also to deal with the streaming video application shown in Fig. 7b), but this would lead to results that are not correct, as shown in Figs. 9a) and 9b). In this case, the values of the two continuous variables are completely correlated, producing samples always aligned over a line, as shown in the unconstrained curve of Fig. 9b). When the barrier is moved along direction x_1 according to Eq. 23, as shown in Fig. 9a), it also shifts the process along x_2, destroying the invariant that this system should maintain (i.e. $x_1(t) + x_2(t)$ constant $\forall t \geq 0$).

To preserve the correlation between variables, the barrier must be moved not only in the direction perpendicular to the component crossing the boundary, but also in the other directions. In particular, each reflecting barrier must be defined with a *skew* vector $\mathbf{p}_k$, laying on the barrier itself, such that $\mathbf{p}_k \cdot \mathbf{u}_k = 0$.

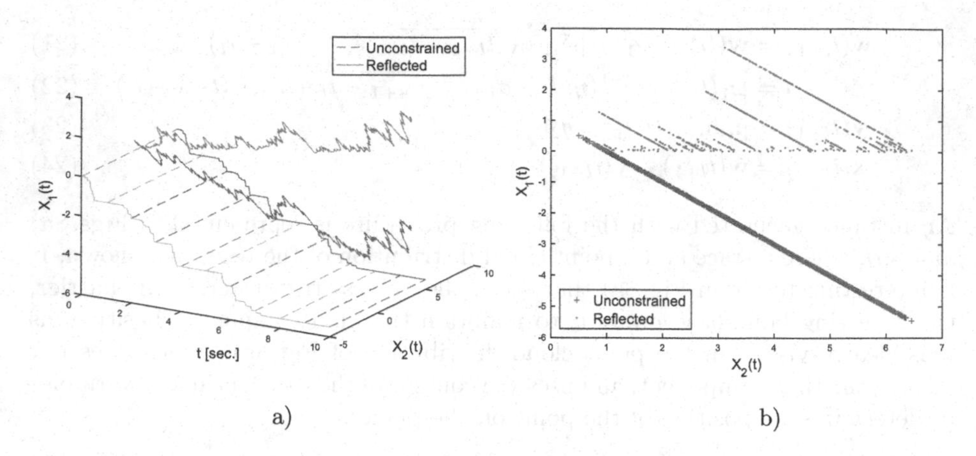

Fig. 9. Position of a 2D Brownian motion process, with correlated motion in the two directions, and a reflecting barrier perpendicular to direction x_1 and passing through the origin: a) trace; b) point cloud distribution for the trace. In this case, the reflection destroys the correlation between the two variables.

Figure 10 shows an example of the definition of such vector, represented in red, for cases with 2 and 3 continuous variables, and the final movement of the barrier caused by both the conventional reflection and parameter $\mathbf{p}_k$. In this case, the simulation of one step of the process becomes the following:

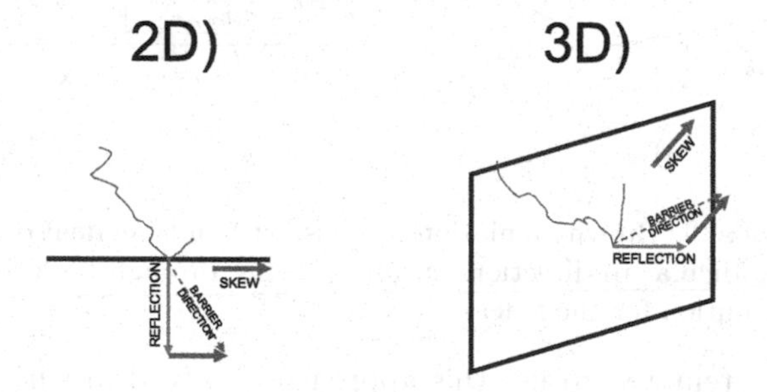

Fig. 10. Skew of the barrier during reflection in two and three dimensions.

$$\mathbf{w}(t_{l+1}) = \mathbf{w}(t_l) - |\nu_1, \nu_2| \cdot \Sigma_1 \cdot \sqrt{t_{l+1} - t_l} + \mathbf{r}_i \cdot (t_{l+1} - t_l) \tag{25}$$

$$\gamma = w_1(t_l) - F^{-1}(\eta|\nu_1) \cdot \sigma_{i:11} \cdot \sqrt{t_{l+1} - t_l} + r_{i:1} \cdot (t_{l+1} - t_l) \tag{26}$$

$$\mathbf{y}(t_{l+1}) = |\min(y_1(t_l), \gamma), y_2| + \mathbf{p}_k \cdot (y_1(t_l) - \min(y_1(t_l), \gamma)) \tag{27}$$

$$\mathbf{x}(t_{l+1}) = \mathbf{w}(t_{l+1}) - \mathbf{y}(t_{l+1}). \tag{28}$$

Factor $(y_1(t_l) - \min(y_1(t_l), \gamma))$ in the second term of the right-hand-side of Eq. 27 accounts for the displacement of the barrier perpendicular to the considered fluid component. This is used to compute the corresponding displacement in the other directions due to the skew vector $\mathbf{p}_k$. Figure 11 shows the effect of having a skew vector $\mathbf{p}_k = |0, 1|$ for the reflection along direction x_1 for the total correlation case, as in the scenario of the streaming application of Fig. 7b). In particular, Fig. 11a), shows that the barrier moves also along axis x_2 to compensate the motion along x_1, and Fig. 11b) confirms that this has the effect of preserving the fluid invariant, having the reflected process perfectly aligned with the unconstrained one.

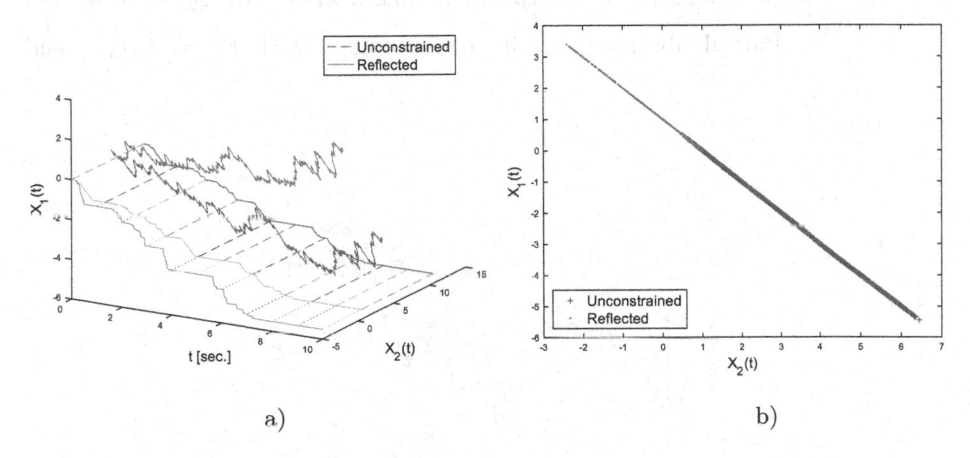

Fig. 11. Position of a 2D Brownian motion process with correlated motion and skew of the barrier: a) trace b) point cloud distribution for the trace.

4.2 Absorbing Barrier in More Than One Dimension

Absorption in more than one dimension can be implemented in the same way as rate adaption in first order models, or absorption in second order models with a single continuous variable. In particular, when the boundary is crossed in a state, the system jumps to a different state that handles the absorption by stopping the in and out flows. However, when more dimensions are involved, we can also have *mixed behaviors*: fluid might be absorbed in one dimension, but continue to evolve in the other (*partial absorption*). In other cases, the system might fully stop fluid motion when the boundary is reached: this can create probability masses over the component perpendicular to the boundary, but leaves a continuous distribution in the other directions. Figure 12 shows the two possible behaviors in a bi-dimensional case: when the boundary is reached, the full absorption stops the flow, creating a straight line. Partial absorption instead constrains the process on the boundary plane, still allowing it to move along direction x_2. Partial absorption is particularly useful to model scenarios such as the battery operated sensor shown in Fig. 7a), where the system might continue to work (even if with a reduced performance and reliability) when the non-rechargeable battery gets depleted. Total absorption can instead be used to model systems where the lack of one resource out of many can stop it from working: for example, a cloud application might stop working when it has consumed any of its network transmission or CPU utilization budgets.

Detection of boundary crossing can be implemented by checking that $\gamma < 0$ in Eq. 26, and movement to the barrier can be obtained by replacing Eqs. 27 and 28 with:

$$\mathbf{x}(t_{l+1}) = |\max(0, w_1(t_{l+1})), w_2(t_{l+1})|. \tag{29}$$

The state s_2 modeling total absorption is characterized by $\mathbf{r}_2 = |0, 0|$ and $\Sigma_2 = \begin{vmatrix} 0 & 0 \\ 0 & 0 \end{vmatrix}$. Partial absorption is instead obtained with $\mathbf{r}_2 = |0, r_{1:2}|$ and $\Sigma_2 = \begin{vmatrix} 0 & 0 \\ 0 & \sigma_{1:2,2} \end{vmatrix}$

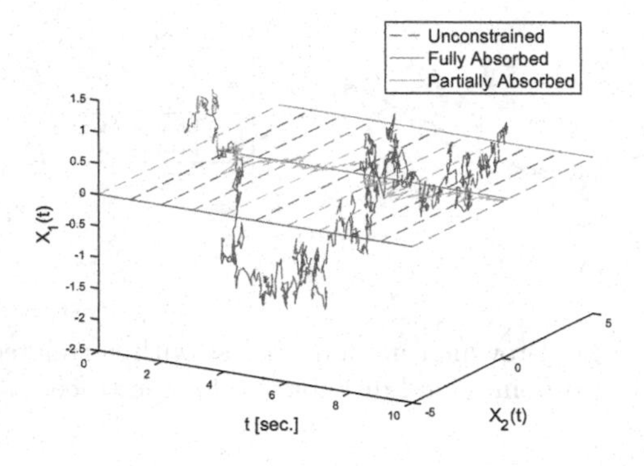

Fig. 12. Different types of absorption in a 2D second order model.

4.3 Extensions

Although we have considered only boundaries that are perpendicular to a given direction, processes with zero mean and with variables that are not correlated, the proposed procedures can be extended to support general cases. Correlation between components can be addressed using the *Cholesky decomposition* of the covariance matrix [1]. Arbitrarily placed boundaries can instead be considered by applying an affine transformation to the base process, so to align the given barrier to be perpendicular to one axis, and passing through the origin. Although such extensions are conceptually simple, they hide a large number of technical details, and will be covered in future work.

5 Conclusions

In this work we have addressed the simulation of second order fluid models with more than one continuous variable, correlated flows, and arbitrarily placed boundaries. The considered topic is quite wide, and cannot be fully explored in single work. However, this paper addresses the most complex parts from a theoretical point of view, leaving the other steps required to fully simulate a model described using the definition proposed in Eq. 1 to future work.

Acknowledgments. We would like to thank Prof. J. Michael Harrison for his kind help and discussions that made this work possible.

References

1. Kroese, D., Taimre, T., Botev, Z.: Handbook of Monte Carlo Methods, Wiley Series in Probability and Statistics. Wiley (2011)
2. Horton, G., Kulkarni, V.G., Nicol, D.M., Trivedi, K.S.: Fluid stochastic Petri Nets: theory, application, and solution techniques. Eur. J. Oper. Res. **105**(1), 184–201 (1998)
3. Wolter, K.: Second order fluid stochastic petri nets: an extension of GSPNs for approximate and continuous modelling. In: Proceedings of World Congress on System Simulation, Singapore, pp. 328–332 (1997)
4. Chen, D.-Y., Hong, Y., Trivedi, K.S.: Second order stochastic fluid flow models with fluid dependent flow rates. Perform. Eval. **49**(1–4), 341–358 (2002)
5. Sericola, B.: Transient analysis of stochastic fluid models. Perform. Eval. **32**(4) (1997)
6. Ramaswami, V.: Matrix analytic methods for stochastic fluid flows. In: Smith, D., Hey, P. (eds.) Proceedings ITC, vol. 16, pp. 1019–1030. Elsevier, Edinburgh (1996)
7. da Silva Soares, A., Latouche, G.: Further results on the similarity between fluid queues and QBDs. In: Latouche, G., Taylor, P. (eds.) Proceedings of the 4th International Conference on Matrix-Analytic Methods, pp. 89–106. World Scientific, Adelaide (2002)
8. Ahn, S., Ramaswami, V.: Fluid flow models and queues - a connection by stochastic coupling. Comm. Statist. Stochastic Models **19**(3), 325–348 (2003)
9. Bean, N.G., O'Reilly, M.M., Taylor, P.G.: Hitting probabilities and hitting times for stochastic fluid flow s. Stoch. Processes Their Appl. **115**, 1530–1556 (2005)
10. Bean, N.G., O'Reilly, M.M. Taylor, P.G.: Algorithms for the first return probabilities for stochastic fluid flows. Stoch. Models **21**(1)
11. da Silva Soares, A., Latouche, G.: Matrix-analytic methods for fluid queues with finite buffers. Perform. Eval. **63**(4), 295–314 (2006)
12. Karandikar, R.L., Kulkarni, V.: Second-order fluid flow models: reflected Brownian motion in a random environment. Oper. Res. **43**, 77–88 (1995)
13. Agapie, M., Sohraby, K.: Algorithmic solution to second order fluid flow. In: Proceedings of IEEE Infocom, Anchorage, Alaska, USA (2001)
14. Asmussen, S.: Stationary distributions for fluid flow models with or without Brownian noise. Stoch. Model. **11**, 1–20 (1995)
15. Rabehasaina, L., Sericola, B.: Stability analysis of second order fluid flow models in a stationary ergodic environment. Ann. Appl. Probab. **13**(4) (2003)
16. Field, T., Harrison, P.: Approximate analysis of a network of fluid queues. ACM (2007)
17. Jones, G.L., Harrison, P.G., Harder, U., Field, T.: Fluid queue models of battery life (2011)

18. Cox, D.R., Miller, H.D.: The Theory of Stochastic Processes. Chapman and Hall Ltd. (1972)
19. Ang, E.-J., Barria, J.: The Markov modulated regulated Brownian motion: a second-order fluid flow model of a finite buffer. Queueing Syst. **35**, 263–287 (2000)
20. Gribaudo, M., Manini, D., Sericola, B., Telek, M.: Second order fluid models with general boundary behaviour. Ann. Oper. Res. **160**(1), 69–82 (2008). uT: 000253211100006
21. Gribaudo, M., German, R.: Numerical solution of bounded fluid models using matrix exponentiation. In: Proceedings 11th GI/ITG Conference on Measuring, Modelling and Evaluation of Computer and Communication Systems (MMB). VDE Verlag, Aachen (2001)
22. German, R., Gribaudo, M., Horváth, G., Telek, M.: Stationary analysis of FSPNs with mutually dependent discrete and continuous parts. In: International Conference on Petri Net Performance Models - PNPM 2003. IEEE CS Press, Urbana (2003)
23. Barbierato, E., Gribaudo, M., Iacono, M., Piazzolla, P.: Second order fluid performance evaluation models for interactive 3D multimedia streaming. In: Bakhshi, R., Ballarini, P., Barbot, B., Castel-Taleb, H., Remke, A. (eds.) EPEW 2018. LNCS, vol. 11178, pp. 205–218. Springer, Cham (2018). https://doi.org/10.1007/978-3-030-02227-3_14
24. Harrison, J.: Brownian Motion and Stochastic Flow Systems. Wiley Series in Probability and Statistics. Wiley (1985)
25. Harrison, J.M.: The diffusion approximation for tandem queues in heavy traffic. Adv. Appl. Probab. **10**(4), 886–905 (1978)
26. Gribaudo, M., Remke, A.: Hybrid Petri nets with general one-shot transitions. Perform. Eval. **105**, 22–50 (2016). https://doi.org/10.1016/j.peva.2016.09.002
27. Gribaudo, M., Sereno, M., Horváth, A., Bobbio, A.: Fluid stochastic petri nets augmented with flush-out arcs: Modelling and analysis. Discret. Event Dyn. Syst. **11**(1–2), 97–117 (2001). https://doi.org/10.1023/A:1008339216603

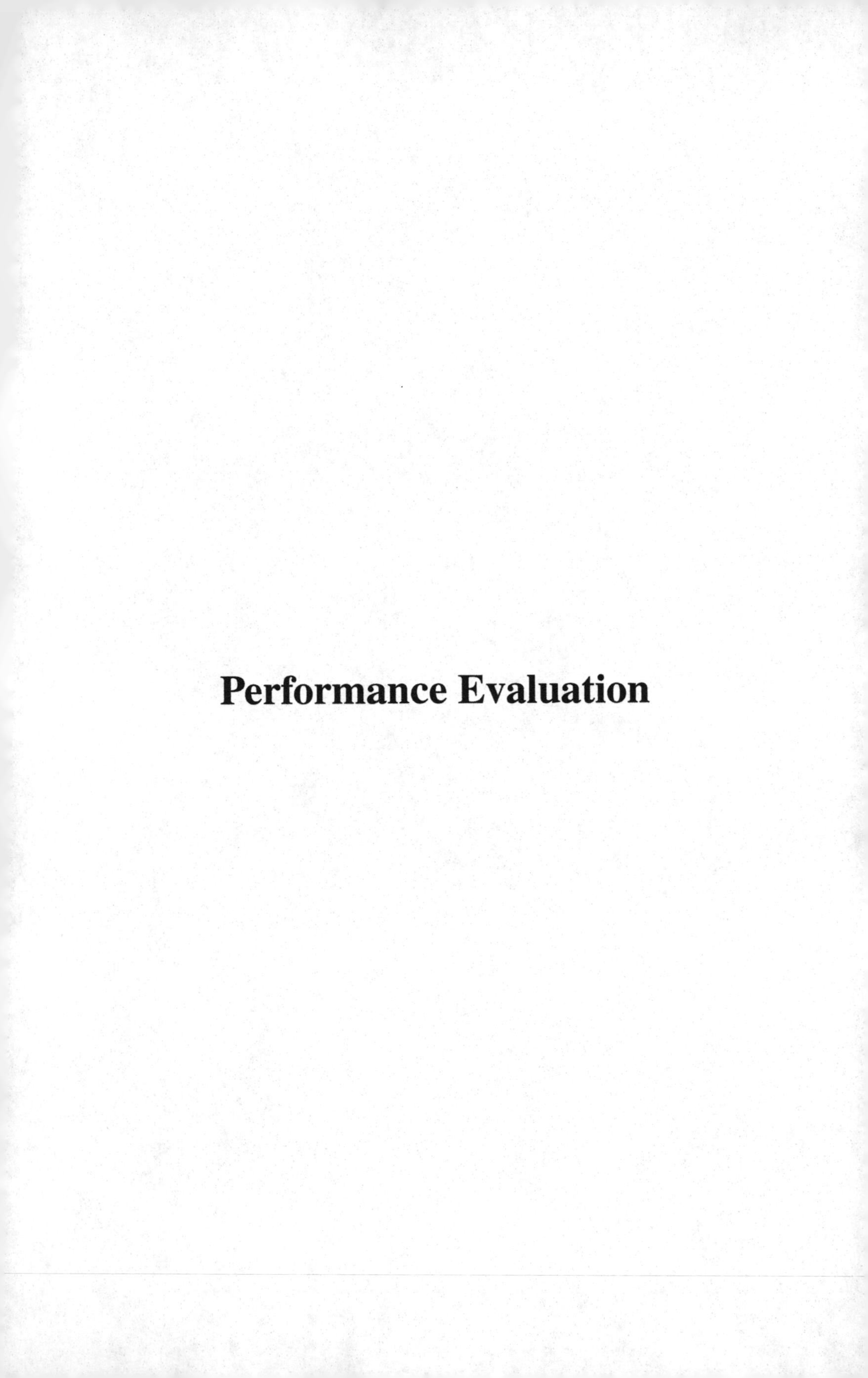

Performance Evaluation

Queue Response Times with Server Speed Controlled by Measured Utilizations

Murray Woodside(✉)

Carleton University, Ottawa, Canada
cmw@sce.carleton.ca

Abstract. Because CPUs use speed control to conserve energy their response times may be greater at low loads, than if they were operating at full speed, giving a flatter response curve against load and trading off longer response time at light loads for energy savings. When conservative control is applied to average utilizations and the averaging time is too long, a different and somewhat "toxic" response-time curve results instead. A numerical investigation was undertaken of open and closed single-server queues with server speed controlled by feedback of CPU utilization measures, which is a common approach for CPUs. With higher target utilizations and longer averaging times for the measured utilization, an undesirable non-monotonic pattern (rising response time, then falling, finally rising again) emerges. This gives unstable behaviour and could disrupt autoscaling strategies that assume monotonically increasing response times. Recommendations have been found for controller parameters, to avoid non-monotonic response times. It is concluded that speed control based on measured utilizations has limited usefulness if performance is a concern, which is in line with industry recommendations. Better speed governors are needed.

Keywords: Performance management · Performance model · Controlled queue

1 Introduction

Modern computer processors have a controllable clock, which can be set by a strategy implemented in the operating system to speed up or slow down the processor, and save energy at lower speeds. The classical queuing models for processor performance need to be adapted because they assume a known constant service rate. This note investigates the response times for a commonly used feedback control strategy and finds that the response time may follow a desirable flattened pattern, or a highly undesirable non-monotonic zig-zag shape which could reduce the effectiveness of some adaptive mechanisms, depending on the speed control parameters. While standard settings may favour the former, more aggressive power-saving will move the parameters towards the non-monotonic behaviour and may cause unexpected system instability, and disrupt adaptive application management. The purpose of this work is to understand the phenomenon and to characterize the parameters that give the desirable and undesirable patterns.

A. Abate and A. Marin (Eds.): QEST 2021, LNCS 12846, pp. 295–309, 2021.
https://doi.org/10.1007/978-3-030-85172-9_16

Speed control has been studied in many different ways, but not for feedback of performance measures. For a general stochastic workload, optimal stochastic control provides policies for the optimal speed as a function of the queue state, as in the work of Lu et al. [10] to optimize a combination of power and response time. These authors also give references to other research in this direction. While this gives an optimal solution to the problem considered here, feedback of queue state is not commonly used in practice, as described below. The purpose here is not to find the ideal control but to model the effect of policies that are used in practice.

Optimal policies have also been found for problems in which the system is more narrowly specified. For example, in [9] Li et al. consider controlling processing speed, memory latency and memory bandwidth on multiple processors to minimize the makespan of a predefined parallel processing job across the cores of a processor. In [13], Rao et al. consider controlling the overheating effect of completing a given total computation in minimum time, leading to an optimal profile of speed values over time. In [12], Mutapcic et al. determine optimal speed values for heterogeneous co-located processors that share a heat sink. In [7] speed control is used to help achieve task deadlines, and also to deliver video frames on time for video streaming (to compensate for the variation in the processing to be done per frame).

The technology of power management is described by Gough, Steiner and Saunders [6]. Processors offer a set of clock speeds (called P-states by Intel [8]) which can be set by the operating system. Linux has a set of standard speed-control policies called governors [5]. Many governors have been implemented (a list of 117 of them with short descriptions is given in [14]), but three are notable:

- "performance" [3, 6], which runs at maximum speed when the CPU is busy and switches to minimum when it is idle,
- "conservative" [3] which steps the speed up and down based on average utilization,
- "schedutil" [3], an evolved version similar to "conservative" which uses moving-average load measures formed by the scheduler, related to utilization.

Here the "conservative" governor is modeled, since it represents the strongest effort to save power. These controllers are for Linux; similar controllers are used by the Android [1] and Windows [11, 6] operating systems. Processors are also capable of short-term additional speed (called "turbo" operation) which cannot be sustained due to overheating; this work does not consider non-sustainable speeds.

A CPU with the "conservative" governor is modeled below as a single server with open and closed Markovian workloads and control via a speed reduction factor S that can be set in the range $S_{min} < S < 1.0$. The controller measures the server utilization U over an interval and then raises S by a step if U is too high, or lowers it if U is too low. No previous study of queues with this family of speed control policy could be found, which motivates this report.

A Naive Motivating Observation

Suppose a queue has a target value of utilization of U^* and a service rate $S\mu$ with $S_{min} < S < 1.0$. Figure 1 shows the two response-time functions for the extreme values of S, and two paths that might be enforced by speed control. A-AA-C-D is a desirable case which maintains a constant response time over a wide range of loads. However a controller might instead give a curve like A-B-C-D, if it enforced the principles:

- maintain S_{min} as long as the utilization is below U^*
- raise the speed to prevent the utilization from exceeding U^* if possible

An adaptive system controller that scales up servers depending on R might be trapped in the segment AB and deploy unnecessary extra servers, rather than exploit the improved response available around C.

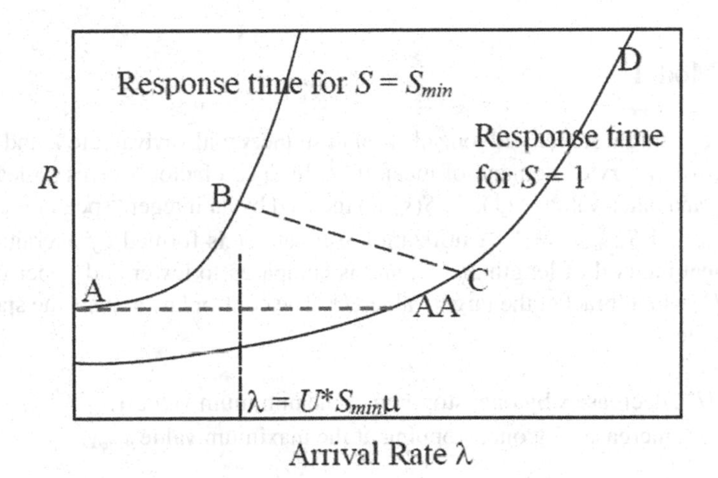

Fig. 1. Naive idea of possible response curves

In real systems the speed is controlled in steps at discrete control instants, and the control is based on past measured utilizations rather than on exact knowledge. This introduces both a control delay and statistical estimation errors. A Markov model with these features is developed below, with numerical solutions. In Sect. 4 the naive view above is formalized as an idealized "perfect knowledge" (PK) approximation to $R(\lambda)$. Both the approximation and the Markov model have the zigzag form seen in Fig. 1.

An Experiment

A simple experiment shows that actual controllers have been adjusted to give desirable flattening, but also confirms the existence of non-monotonic response time functions. A closed concurrent workload with 300 programs was run in parallel on a laptop under the "power saving" power management option which controls the processor speed to a target utilization. The average response times shown in Table 1 initially rise, drop between throughputs of 5000/s and 8700/s, and then rise again. The measured values are accurate to about 2% so the dip is statistically significant, although small.

Table 1. Measured response times on a microsoft windows personal computer

Think time between operations (sec)	Throughput (responses/sec)	Mean response time R (sec)	Confidence interval for R ($\pm$)
0	22561	0.01330	0.00031
0.025	8754	0.00927	0.00024
0.05	5024	0.00971	0.00015
0.075	3547	0.00957	0.00028
0.15	1881	0.00950	0.00027

2 The Model

We consider a single server with one class of customers with arrival rate λ and exponentially distributed service demand of mean μ^{-1}. Its speed factor S gives a mean service rate of $S\mu$, and takes values $S(1), \ldots S(s_{max})$ indexed by an integer "speed index" s, with $S(1) = S_{min}$ and $S(s_{max}) = 1$. A utilization estimate $\widehat{U}$ is formed by averaging over a measurement interval of length Δ sec, and is compared to lower and upper thresholds U^- and U^+, which bracket the target value $U*$. The control law adjusts the speed index s according to:

- if $\widehat{U} < U^-$, decrease s by one, stopping at the minimum value 1,
- if $\widehat{U} > U^+$, increase s by one, stopping at the maximum value s_{max}.

To make the solution numerically practical a discrete-time version was adopted with a time-step of length δ. In each time-step there is an arrival with probability $\lambda\delta$ and a departure (if the server is busy) with probability $\mu\delta$. For small values of δ this approximates Poisson arrivals at rate λ and exponential service at rate μ. To make the state space finite the state n was truncated at N and arrivals in state $n = N$ are lost.

The decision interval was taken as a multiple $K\delta$ of the time-step δ. The utilization is measured by counting the number b of the K steps that have a busy server, and is compared to lower and upper target values KU^-, KU^+. For the decision, if s_0 is the previous index, the new value $s_{new} = s_{new}(s_0, b)$ is given by:

Speed-Control Algorithm:

```
if      b < KU⁻ : s_new(s_0, b) = max(s_0 − 1, 1)
elseif  b > KU⁺ : s_new(s_0, b) = min(s_0 + 1, s_max)
else            : s_new(s_0, b) = s_0.
```

Because s changes only at decision points, the model can be decomposed into two Markov Chain models, a transient chain one to model the states between decision points, and the other embedded at the decision points, to model the transitions in s

(1) The lower level Discrete-Time Markov Chain (DTMC) [2] models the transient behaviour of the queue state and the accumulated busy time over one interval between decision points. It has:

- s_{max} submodels, one for each value of s,
- each with a state (n, b) at substep k, where $n \in [0, ..., N]$ is the number of customers in the queue and $b \in [0, ..., K]$ is the accumulated busy time of the server,
- an initial state $(n_0, 0)$, where n_0 is the value of n at the decision point that begins the interval.

The lower-level DTMC is solved for each combination of s and n_0 to determine the probability $p(n, b; k, n_0, s)$ for state (n, b) at substep k of the interval. The model equations and solution method are conventional [2]; the important result is the final probability $p(n, b; K, n_0, s)$, and the mean number of customers over the transient, denoted as $\overline{n}(n_0, s)$

(2) The upper level DTMC models the joint state (n, s) of the queue and the speed controller, embedded immediately after the decision points for the control algorithm. It has transition probabilities $A(n_0, s_0; n, s)$ for a transition from state (n_0, s_0) immediately after a decision, to the successor state (n, s) immediately after the next decision K substeps later. The steady-state probabilities of this DTMC are denoted $\pi(n, s)$.

The transition probabilities A for the upper level model are found from the solutions of the lower level model. Let $B(s; s_0, b)$ be the set of utilization measures b such that the speed-control algorithm above gives $s_{new}(s_0, b) = s$. Then

$$A(n_0, s_0; n, s) = \sum\nolimits_{b \in B(s, s_0, b)} p(n, b; K, n_0, s) \tag{1}$$

The upper-level DTMC with $(N + 1)s_{max}$ states was solved for the steady-state probabilities $\pi(n, s)$ for the number in queue and the speed level, after a decision point. The equations and the solution are conventional [2]. The steady state mean number in the queue is found by combining the results for the transients, conditioned on the initial state, giving the overall mean number in the queue $\overline{\overline{n}}$ and overall mean arrival rate (without the lost arrivals) of $\overline{\overline{\lambda}}$:

$$\overline{\overline{n}} = \sum_{n_0, s_0} \overline{n}(n_0, s_0)\pi(n_0, s_0)$$

$$\overline{\overline{\lambda}} = \sum_{n_0, s_0} \overline{\lambda}(n_0, s_0)\pi(n_0, s_0)$$

The mean response time for the given arrival rate, service rate, speed factor steps and utilization thresholds is then given by:

$$R = \overline{\overline{n}} / \overline{\overline{\lambda}} \tag{2}$$

3 The Solution

To explore the form of the solution, the major parameters were varied across their possible range of values. The service rate μ was normalized to 1/sec, and.

- $U_0 = \lambda/\mu$ (the high-speed utilization) was varied from 0.1 to 0.9;
- the target server utilization $U*$ was varied from 0.3 to 0.9;
- the steps in speed were set to the values [0.4, 0.6, 0.8, 1.0] and
- the control interval Δ took values [1, 10, 30, 100, 300] sec.

Other parameters were:

- the utilization thresholds $U^- = 0.9$ U* and $U^+ = \min(1.1U*, 0.95)$.
- the time-step $\delta = 0.2$ s. in the lower-level model
- a limit of 60 queue states, giving a limit of N = 59 customers

Some results giving a broad picture of the solution are shown in Fig. 2. The computed response times are plotted as circles, and the boundary cases of the maximum-speed and minimum-speed response times are plotted as solid curves for reference.

The smaller is Δ, the closer R stays to the high-speed asymptote. For higher Δ larger $U*$ gives larger response times, and the response curves are almost flat. For $\Delta = 100$ and 300 the zigzag shape predicted in Fig. 1 emerges and has a very pronounced peak. The "desirable" flat response pattern is obtained for cases with $\Delta \le 30$ and pronounced "undesirable" peaking is obtained for cases with larger Δ and for U* > 0.5.

The cause of the undesirable peaking can be traced to two factors that are both linked to Δ, the control delay. For small Δ the control delay (which is at least Δ) is small but the estimation accuracy is low, while for large Δ the accuracy is better but the control delay is large. Clearly the effect of a large control delay overwhelms the system. Consider this scenario: a combination of stochastic load fluctuations and estimation errors puts the server into a low-speed state which leads to a long unstable transient increase in the queue until the end of the next estimation/control period.

In uncontrolled queueing systems, heavier loads always lead to longer response times, and many adaptive scaling algorithms for computer systems are based on this. The survey in [4] found that the largest number of reported autoscalers use this kind of simple feedback loop. We can imagine a scaling algorithm based on response time, which increases the capacity of a heavily loaded system that is operating to the right of the peak. This would give increased response times as the load on each CPU is reduced, and therefore it would scale up further until it crosses the peak into the less economical regime to the left, and then be stuck there.

Worse, the value of Δ is difficult to tune. Δ is the ratio of the measurement interval to the mean service time of CPU requests, which varies with the program being executed, so a CPU could be driven randomly between the desirable pattern in parts (a)–(c) and the less desirable pattern in parts (d) and (e).

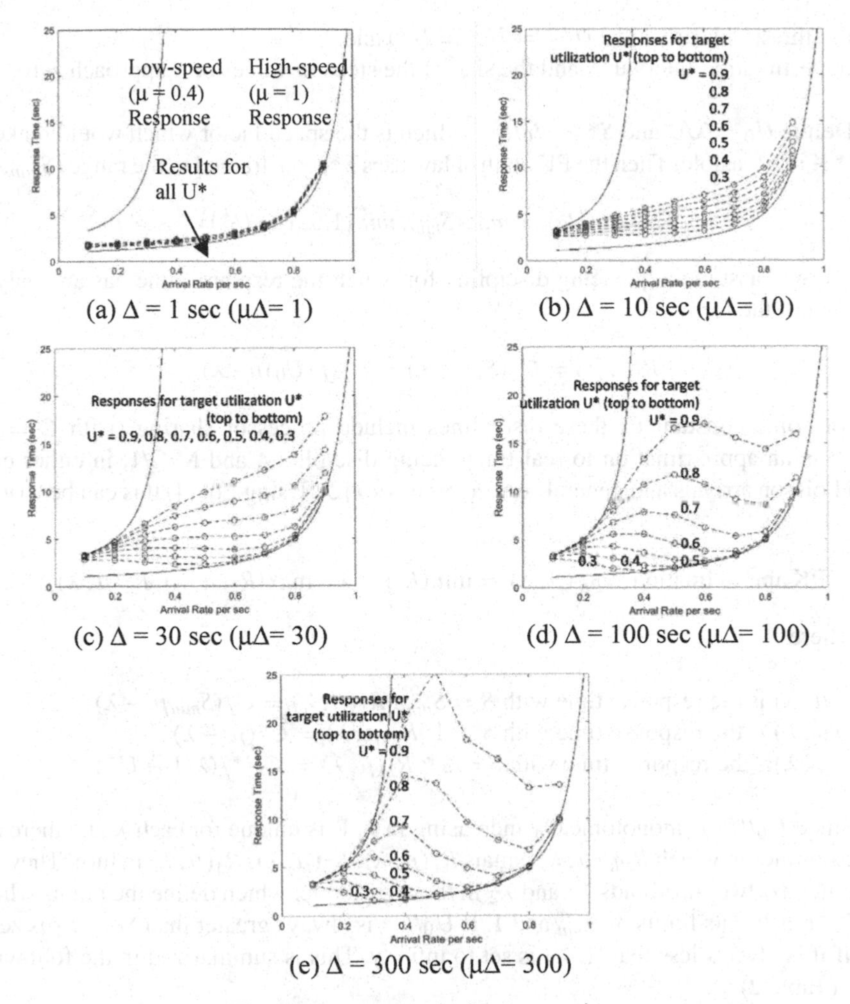

(a) $\Delta = 1$ sec ($\mu\Delta = 1$) (b) $\Delta = 10$ sec ($\mu\Delta = 10$)

(c) $\Delta = 30$ sec ($\mu\Delta = 30$) (d) $\Delta = 100$ sec ($\mu\Delta = 100$)

(e) $\Delta = 300$ sec ($\mu\Delta = 300$)

Fig. 2. Exact (numerical) response times

Two practical recommendations emerge from Fig. 2. First, if U^* is set to a mid-range value such as 0.5 it tends to give a wide plateau at near-constant response time, for a wide range of Δ. Second, if Δ is set at a moderately small value such as 10 times the mean service time ($\Delta = 10$ in Fig. 5)) the response time is attractively flat, while a long averaging time (300 times the mean service time) gives unstable behaviour and possibly very long responses (at the peak). However the longer averaging time also gives lower control overhead and (as shown below) lower average power.

4 An Idealized "Perfect Knowledge" (PK) Analysis

An idealized analytic model was created by assuming that:

1. the controller has perfect knowledge of λ, μ, and U over the next interval ($\widehat{U} = U$)

2. the thresholds are equal ($U^- = U^+ = U^*$) and
3. the estimation interval Δ and the size of the steps in values of S approach zero,

Define $U_0 = \lambda/\mu$ and $S^* = U_0/U^*$, which is the speed factor which would make $U = U^*$ if it is feasible. Then the PK control law uses S^*, constrained to the range (S_{min}, 1):

$$S = f(U_0) = max(S_{min}, min(1.0, U_0/U^*)) \tag{3}$$

We will assume a queueing discipline for which the response time has an analytic solution of the form:

$$R(\lambda, \mu) = C/(S\mu - \lambda) = C/(f(U_0)\mu - \lambda) \tag{4}$$

for some constant C; these disciplines include processor sharing (with $C = 1$), which is an approximation to real time-slicing disciplines, and M/G/1, in either case with Poisson arrivals and general service processes [2]. Using Eq. (4) this can be written as:

$$\text{PK approximation: } R_{PK}(\mu, \lambda) = \min(R_1(\mu, \lambda), \max(R_2(\mu, \lambda), R_3(\mu, \lambda))) \tag{5}$$

where

- $R_1(\mu, \lambda)$ is the response time with $S = S_{min}$: $R_1(\mu, \lambda) = C/(S_{min}\mu - \lambda)$
- $R_3(\mu, \lambda)$ is the response time with $S = 1$: $R_3(\mu, \lambda) = C/(\mu - \lambda)$
- $R_2(\mu, \lambda)$is the response time with $S = S^*$: $R_2(\mu, \lambda) = CU^*/(\lambda(1 - U^*)$

Since U_0/U^* is monotonically increasing in λ, R is unique for each λ and there are three regimes in which $R_{PK}(\mu, \lambda)$ equals $R_1(\mu, \lambda)$, $R_2(\mu, \lambda)$, $R_3(\mu, \lambda)$ in turn. They are separated by two thresholds λ_1 and λ_2 in the arrival rate, which define the points where U_0/U^* reaches its limits of S_{min} and 1. If U_0/U^* is always greater than S_{min}, λ_1 is zero, and if it is always less than 1, λ_2 is set to infinity. This is summarized in the following table (Table 2).

Table 2. Definition of the PK-approximation

Regime	Condition on λ	S given by	Response time $R(\lambda, \mu)$ given by
AB: low speed	$\lambda \le \lambda_1 = \mu U^* S_{min}$	$S = S_{min}$	$R_1 = C/(S_{min}\mu - \lambda)$
BC: controlled speed	$\Lambda_1 < \lambda \le \lambda_2 = \mu U^*$	$S = S^* = U_0/U^* = S_{min} + (\lambda - \lambda_1)/(\mu U^*)$	$R_2 = C/(S\mu - \lambda)$
CD: high speed	$\lambda > \lambda_2$	$S = 1$	$R_3 = C/(\mu - \lambda)$

For an M/M/1 queue (for which $C = 1$) with nominal service rate $\mu = 1.0$ and $S_{min} = 0.3$, the response time approximations for varying λ and some different settings of the

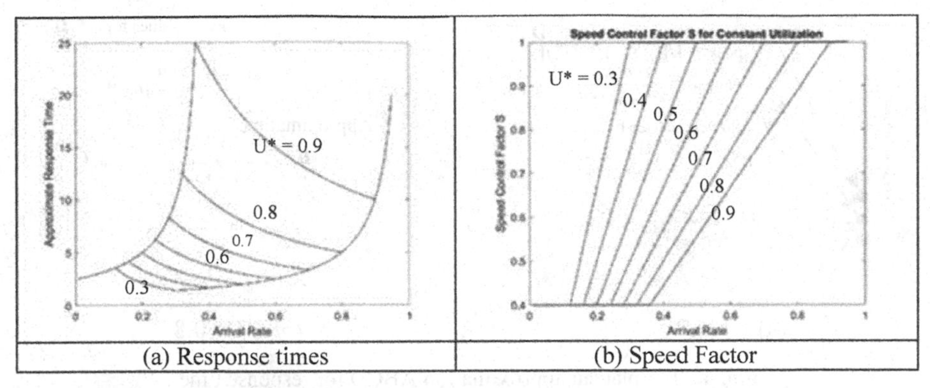

Fig. 3. The PK-Approximation response time and speed control factor for an open single server queue with various utilization targets, with $S_{min} = 0.3$

target utilization U^* are shown in Fig. 3(a). They strongly resemble Fig. 1. Figure 3(b) shows the speed control settings. Since the power used increases with S it is evident that the higher U* is set, the less power is used.

In the limit as $\mu\Delta \rightarrow 0$, $\hat{U}$ will be either zero (if the server is idle) or 1 (if it is busy) and the control will switch between S_{min} for $\hat{U} = 0$. and $S = 1$ for $\widehat{U = 1}$; thus whenever there is a customer the server will run at maximum speed and:

$$\text{Zero} - \text{averaging} - \text{interval asymptote}: R(\mu, \lambda) = R_3(\mu, \lambda) \tag{6}$$

This corresponds to the "performance" governor in Linux [5].

The accuracy of the PK approximation is quite good for some situations and poor in others. Figure 4 compares it to the numerical exact results. For moderate target utilization ($U^* = 0.5$) it is quite accurate for a wide rage of values of the normalized estimation time $\mu\Delta$. For a higher value ($U^* = 0.8$) it is however quite poor. The hump in the response time curve corresponds in positioning and amplitude for intermediate values of $\mu\Delta$, but is more pronounced in the exact results for large U* a large $\mu\Delta$.

In general the exact response time is below the approximation (sometimes much below) for low loads and above it for high loads. This is due to the estimation errors in $\hat{U}$; when the ideal speed is near a boundary (S_{min} or S_{max}) the estimation errors tend to diffuse the speed away from the boundary. The exaggerated hump for large $\mu\Delta$ is probably due to the larger control delay which allows the queue to build up when the system accidentally (due to estimation error) enters a state with low speed and high load. For long averaging times ($\mu\Delta = 100$ and 300) the mean relative absolute error (MRAE) was 9.5%. Over all cases the MRAE was 33%, and it was particularly high for short averaging and large U^*.

A Usable "Plateau" Approximation

For moderate values of U^* up to about 0.5 the flattening in the response time curve can be described by a simpler "plateau approximation" with a constant response in the middle range between the curves for S_{min} and for $S = 1$. Figures 4(a) and 4(b) show it as a bold line BC placed halfway between the response time at zero load ($R = 1/(\mu\ S_{min})$)

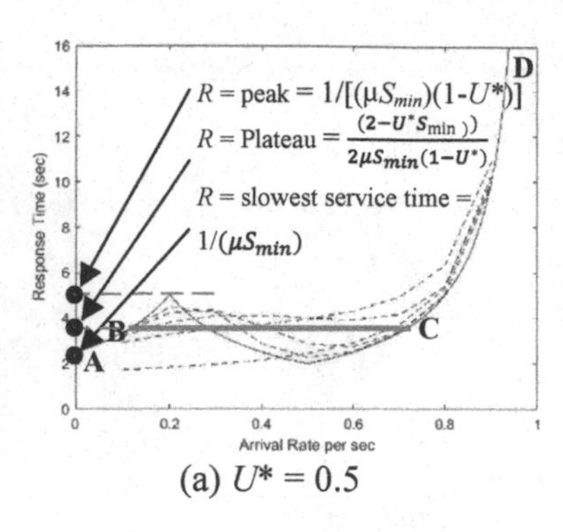

(a) $U^* = 0.5$

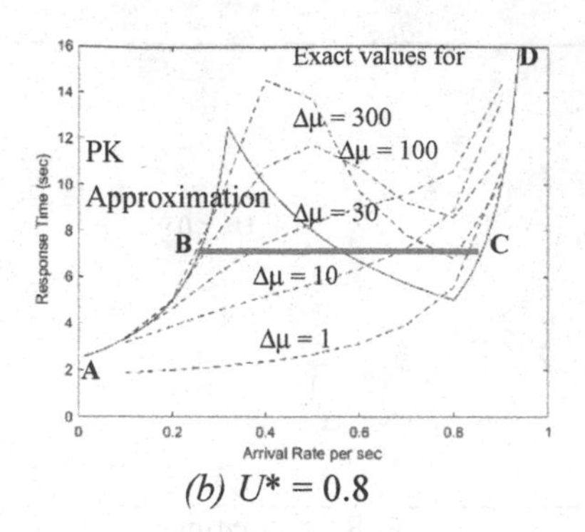

(b) $U^* = 0.8$

Fig. 4. The plateau approximation ABCD for response time

and the response time at the peak of the PK approximation, which can be shown from Eq. (7) and (8) to be $R = 1/[(\mu\ S_{min})(1 - U^*)]$. The plateau is approximated by their average, given by.

$$\text{Plateau}: R_P(\mu, \lambda) = \left(2 - U^* S_{min)}\right)/2\mu S_{min}\left(1 - U^*\right)$$

Combining this with the low and high-speed boundary curves the entire plateau approximation is given by

$$\text{Plateau approximation}: R(\mu, \lambda) = \min(R_1(\mu, \lambda),\ \max\ (R_2(\mu, \lambda), R_P(\mu, \lambda)) \qquad (7)$$

(where R_1 and R_2 are given by Eq. (4)), and is shown as the curve ABCD in Figs. 4(a) and 4(b). For the cases studied that have target utilization $U^* \leq 0.5$ and normalized estimation interval $\mu\Delta \leq 100$, the MRAE was 11.8%.

To summarize the various possible cases, approximate values can be computed, with average errors around 10% in these situations:

- for very small $\mu\Delta$ by using the high-speed response time,
- for moderate $\mu\Delta$ and U^* by the Plateau approximation Eq. (7), and
- for large $\mu\Delta$ by the PK approximation, Eq. (5).

For moderate $\mu\Delta$ and large U^* a useful approximation has not been found.

5 Effectiveness of Utilization Control and Power Saving

The controller is driven by utilization values and a natural question is, how close does it stay to the target U^*? Figure 6 shows utilization results for the same queue with U^* = {0.3, 0.4, 0.5, 0.6 0.7, 0.8, 0.9}. Following the bottom dashed curve for $U^* = 0.3$ we can see that it does not dwell at $U = 0.3$ for any substantial interval of arrival rates, unless $\mu\Delta$ is at least 100. Thus, the utilization is not well controlled.

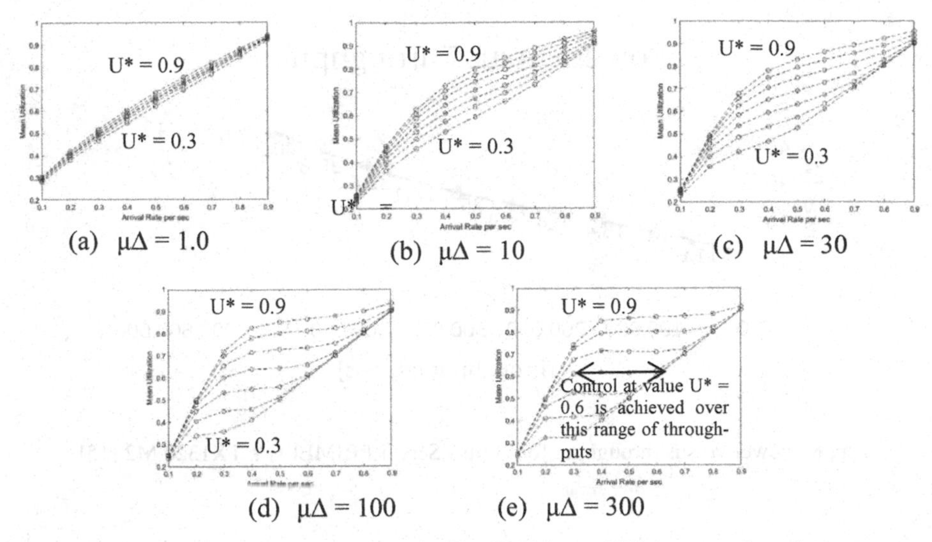

Fig. 5. Utilizations for the same cases as in Fig. 2. Values of U* increase from bottom to top, from 0.3 to 0.9 in steps of 0.1.

Figure 5shows that only for $\mu\Delta = 300$ is there substantial success in achieving the target utilization over a range of throughputs. For $\Delta = 100$ there is some success, and for shorter averaging times very little.

How effective is this form of speed control at reducing power? The SPECPower benchmark measures power consumption as a function of processing speed, as illustrated in Fig. 6. These results will be used, normalized to the maximum speed taken as $S = 1$, to estimate the power consumption based on S. For the processor in Fig. 7, taking a line from the point at zero ops (assumed to correspond to $S = 0$) to the maximum reported speed ($S = 1$) gives as a rough approximation:

$$\text{Power} = 11.6 + 33.1S \tag{8}$$

Using Eq. (8) and the mean value of S found from the model solution gives the power levels shown in Fig. 7. The dashed lines curves are for values of U^* in the range 0.3–0.9 as before, increasing from top to bottom; the solid lines are for $S = 1$. The available power savings are negligible at high loads, about 20% at $U_0 = 0.5$ and about 50% at $U_0 = 0.2$, compared to the solid line. Notice that as long as the power function is linear and increasing, the conclusion that one condition uses less power than another is unaffected by the particular values of the coefficients in Eq. (8).

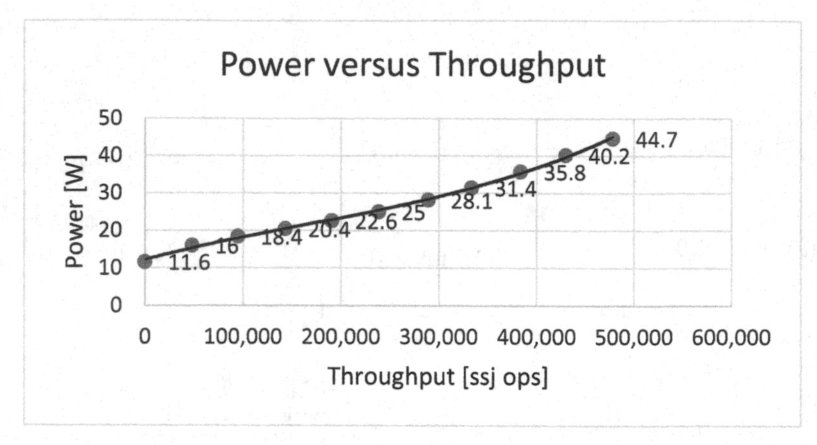

Fig. 6. Power versus throughput for Fujitsu Server PRIMERGY TX1320 M2 [15]

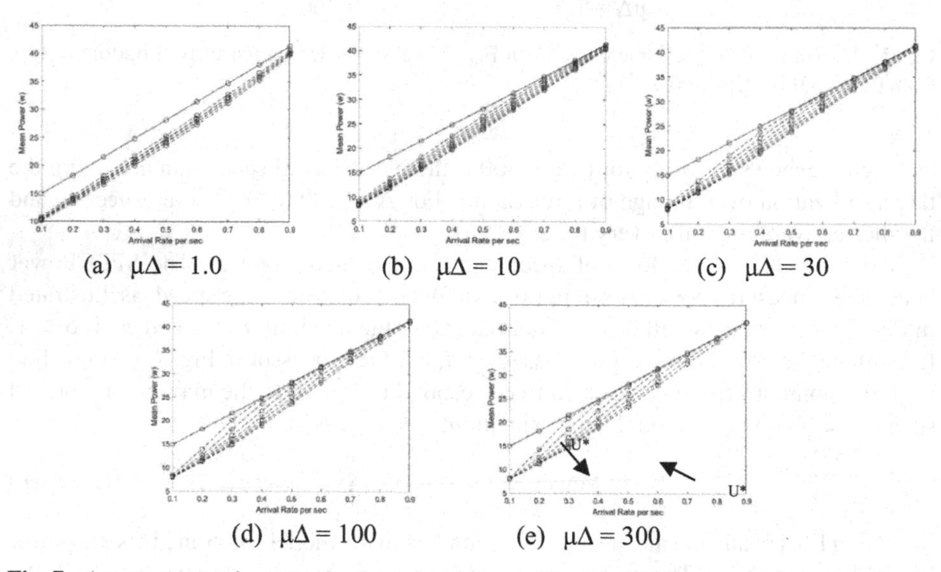

Fig. 7. Average power in watts, for the cases shown in Fig. 2, using the power approximation in Eq. (8). Values of U^* increase from top to bottom.

6 Control of Finite-Population Queues

A finite-population model may be preferred since a processor usually has a finite number of processes or threads that may request to be scheduled. A corresponding PK approximation can be constructed using the well-known solution of the M/M/1//N queue, giving the solutions displayed in Fig. 8. To compare the approximation for the closed and open cases the same example was analyzed with a finite population of 60 customers, and $0.3 < S < 1.0$. The arrival rate r for each customer not at the server ranged from 0 to

0.03 which gave a similar range of throughput values from 0 to 1, as in the open cases examined above.

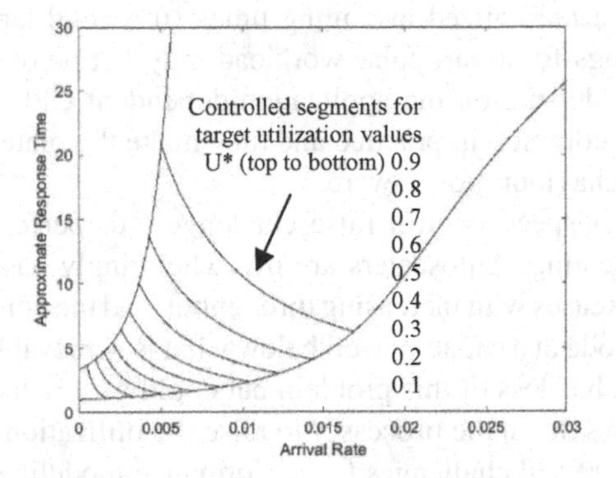

Fig. 8. Approximate "perfect knowledge" response time curves for a closed queue with 60 customers

As the arrival rate increases the response time follows the upper (minimum-speed) bound up to the point where its target utilization $U*$ is reached, then follows the descending curve for that $U*$ until it reaches the lower (high-speed) bound. The curves have the same zig-zag shape as those for the open system shown in Fig. 2(a).

The Markov model for this system follows that of Sect. 3 except for a modified arrival process with rate $(N - n)r$ arrivals/sec in queue state n. Some numerical solutions for four-level control, corresponding to the open arrivals case in Fig. 2, are displayed in Fig. 9. The results are very similar.

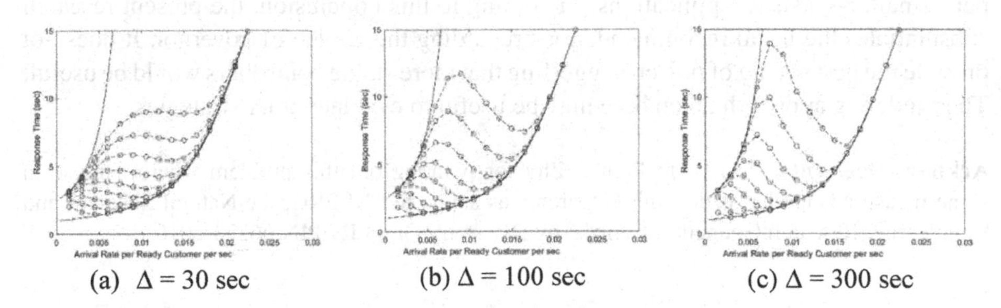

(a) $\Delta = 30$ sec (b) $\Delta = 100$ sec (c) $\Delta = 300$ sec

Fig. 9. Some speed-controlled response times for a closed M/M/1//N queue with N = 60

7 Conclusions

Speed-controlled processors can in some circumstances provide nearly constant response time over a wide range of throughputs, represented by the "plateau" approximation in Sect. 6. However with a power-saving strategy such as the "conservative" governor for

Linux, the response time increases steeply beyond this controlled range and performance collapses with little warning. If the power-management parameters are not ideal, pathological behaviour can set in, in the form of the non-monotonic (humped) response curves found for large normalized averaging times ($\mu\Delta$) and large target utilizations (U^*). The ideal settings for a particular workload may not be obtainable, or be stable over time, since they depend on the application-dependent CPU service times. Large target utilizations are common in practice and they make the plateau range smaller and the non-monotonic behaviour more severe.

These attributes of speed control raise challenges for performance management, particularly for autoscaling. Autoscalers are overwhelmingly based on an assumption that response time increases with increasing throughput, and the non-monotonic response curve could trap the node at a capacity well below what is achievable. Autoscaling based on a target utilization has less of this problem but could give reduced capacity because the speed control slows down the processor to raise the utilization (and save power).

These results also reveal challenges for performance modeling. The exact response time calculation is not tractable for practical solvers. The PK and plateau approximations may be useful but have limited accuracy in important cases. A known problem is the effect of speed control on the measurement of CPU demands for operations; the controller state must be known while measuring. Calibration of models from performance tests or from measurements made "in the wild" will be affected. This affects the usefulness of the models for capacity planning or deployment planning.

Most of these challenges can be offset by always using the "performance" governor, which uses minimum power with zero requests and switches immediately to maximum power when any task is scheduled. The response time is close to that for maximum power. This is the least power-efficient governor, but there seem to be compelling reasons to prefer it.

A natural conclusion is that only the "performance" governor has a useful future in performance-sensitive applications. In coming to this conclusion, the present research substantiates the usual recommendations regarding the choice of governor. It does not provide the best saving of power, suggesting that more stable controllers would be useful. The modeling approach taken here may be useful to evaluate other strategies.

Acknowledgements. Thanks to Wenbo Zhu for pointing out this problem where it arose in some measurements he made. This research was supported NSERC, the Natural Sciences and Engineering Research Council of Canada, by Discovery Grant RGPIN 06274-2016.

References

1. Anonymous: CPU governors explained, 27 June 2012. https://forum.xda-developers.com/showthread.php?t=1736168. Accessed 4 Jan 2020
2. Bolch, G., Greiner, S., de Meer, H., Trivedi, K.S.: Queueing Networks and Markov Chains: Modeling and Performance Evaluation with Computer Science Applications, Wiley (2006)
3. Brodowski, D.: Linux CPUFreq governors. https://www.kernel.org/doc/Documentation/cpu-freq/governors.txt
4. Chen, T., Bahsoon, R., Yao, X.: A survey and taxonomy of self-aware and self-adaptive cloud autoscaling systems. ACM Comput. Surv. **51**(3), 1–40 (2018). Article 61

5. Doleželová, M., Heves, J., East, J., Domingo, D., Landmann, R., Reed, J.: Power management guide for RedHat enterprise Linux 6, section 3.2: using CPUFREQ GOVERNORS. https://access.redhat.com/documentation/en-us/red_hat_enterprise_linux/6/html/power_management_guide/cpufreq_governors. Accessed 4 Jan 2020
6. Gough, C., Steiner, I., Saunders, W.: Energy efficient servers, chapter 2 CPU power management, pp. 21–70, and chapter 8 characterization and optimization, pp. 269–306, Apress (2015)
7. Jadoon, J.K.: Evaluation of power management strategies on actual multiprocessor platforms, Doctoral thesis, Universite de Nice – Sophia Antipolis, March 2013
8. Kidd, T.: Power management states: P-states, C-states, and package C-states, April 2014. http://software.intel.com/en-us/articles/power-management-states-p-states-c-states-and-package-c-states. Accessed Apr 2019
9. Li, B., León, E.A., Cameron, K.W.: COS: a parallel performance model for dynamic variations in processor speed, memory speed, and thread concurrency. In: Proceedings of HPDC 2017, the 26th International Symposium on High-Performance Parallel and Distributed Computing, Washington, pp. 155–166, June 2017
10. Lu, Y., Sharma, M., Squillante, M.S., Zhang, B.: Stochastic optimal dynamic control of Gi/Gi/1 queues with time-varying workloads. Probab. Eng. Inf. Sci. **30**, 470–491 (2016)
11. Microsoft: Processor power management options, 10 April 2017. https://docs.microsoft.com/en-us/windows-hardware/customize/power-settings/configure-processor-power-management-options. Accessed Apr 2019
12. Mutapcic, A., Boyd, S., Murali, S., Atienza, D., De Micheli, G., Gupta, R.: Processor speed control with thermal constraints. IEEE Trans. Circuits Syst. **56**(9), 1994–2008 (2009)
13. Rao, R., Vrudhula, S., Chakrabarti, C., Chang, N.: An optimal analytical solution for processor speed control with thermal constraints. In: Proceedings of International Symposium on Low Power Electronics and Design (ISLPED 2006), Tergensee, Germany, pp. 292–297, October 2006
14. Saber (psuedonym): Collective guide of CPU governors, I/O schedulers and other kernel variables, 8 March 2015. https://forum.xda-developers.com/t/ref-guide-most-up-to-date-guide-on-cpu-governors-i-o-schedulers-and-more.3048957/
15. Standard Performance Evaluation Corporation: SPECpower_ssj2008: Fujitsu FUJITSU Server PRIMERGY TX1320 M2, 25 November 2015. https://www.spec.org/power_ssj2008/results/res2015q4/power_ssj2008-20151110-00704.html. Accessed 7 Dec 2018

Service Demand Distribution Estimation for Microservices Using Markovian Arrival Processes

Runan Wang(✉), Giuliano Casale, and Antonio Filieri

Department of Computing, Imperial College London, London, UK
{runan.wang19,g.casale,a.filieri}@mperial.ac.uk

Abstract. Building performance models for microservices applications in DevOps is costly and error-prone. Accurate service demand distribution estimation is critical to performance model parameterization. However, traditional service demand estimation methods focus on capturing the mean service demand, disregarding higher-order moments of the distribution. To address this limitation, we propose to estimate higher moments of the service demand distribution for a microservice from monitoring traces. We first generate a closed queueing model to abstract a microservice and model the departure process at the queue node as a Markovian arrival process. This allows formulating the estimation of service demand as an optimization problem, which aims to find the optimal parameters of the first multiple moments of the service demand distribution based on the inter-departure times. We then estimate the service demand distribution with a novel maximum likelihood algorithm, and heuristics to mitigate the computational cost of the optimization process for scalability. We apply our method to real traces from a microservice-based application and demonstrate that its estimations lead to greater prediction accuracy than exponential distributions assumed in traditional service demand estimation approaches.

Keywords: Service demand distribution · Markovian arrival process · Maximum likelihood estimation · Queueing models · Performance

1 Introduction

DevOps has been widely adopted in industry, becoming an important part of today's software development methodologies [2]. Compared with traditional software development, DevOps exploits a high degree of automation throughout the whole pipeline to shorten the development life cycle and deliver high-quality applications.

While DevOps provides software engineers with advantages like frequent releases of new features and fast resolution of technical issues, how to keep a speedy pace of delivery to production and ensure the quality of the software at the same time remains an open challenge [3]. Performance models can help to describe the system with a simplified abstraction, further enabling simulation and forecasting for use by both developers and operators. Stochastic models such as queueing networks [14], layered queueing networks [13], Petri nets [20] are widely used to represent web applications. Instead,

A. Abate and A. Marin (Eds.): QEST 2021, LNCS 12846, pp. 310–328, 2021.
https://doi.org/10.1007/978-3-030-85172-9_17

software architecture models are appropriate to describe changes in software structures and resources [21].

To build performance models in the context of DevOps, it is important to consider both architectural and analytical models. Existing methods for generating architecture-level models like UML [28] and Palladio Component Model [7] often rely on manual analysis and domain knowledge, which cannot satisfy the requirement of high-degree automation in DevOps. In addition, the description languages of architectural models in previous works are independent of deployment, which brings complexity to frequent deployment and automatic calibration of performance models responding to new alternatives during this process. Compared to the above models, TOSCA [4] provides a directed topological description of applications that allows deployment and lifecycle management via dedicated orchestrators. As such, TOSCA is increasingly widespread to describe microservice-based applications.

To enable simulation and prediction with TOSCA, model-to-model transformations are required to decompose an architectural model into an analytical model that can be solved with analytical solvers via simulation. Among the parameters of a performance model generated in this way, service demand is a critical aspect that should be specified [8,33]. Service demand generally refers to the cumulative time a request spends receiving service from system resources, such as CPU or disk, accumulated over all visits. The accuracy of service demand specification is decisive for the predictive effectiveness of performance models. Therefore, it is critical to specify accurate service demand distribution in TOSCA models.

Service demand can be estimated with measurements of CPU utilization and response time collected via system monitoring. Several different approaches for service demand estimation have been proposed over the years, such as utilization law [10], response time approximations based on linear regression [26], non-linear optimization [34] and also machine learning methods [12]. However, most of the existing approaches for service demand estimation mainly focus on estimating the *mean* service demand. Restricting attention to the mean can limit the accuracy of service demand estimation. This issue as the higher-order moments can affect the accuracy of critical metrics such as higher percentiles of the response time.

In this paper, we propose to estimate the service demand distribution. To learn the service demand, we first represent a microservice as a closed queueing system, with the finite population representing the maximum parallelism level within the microservice, and in which the service demand for queue nodes is characterized with a general acyclic phase-type (APH) distribution. After generating the continuous-time Markov chain (CTMC) for this model, we filter the departure transitions into a Markovian arrival process (MAP) to characterize the departure process at the queue node. The problem of service demand estimation can then be formulated as an optimization problem to infer the service demand distribution that maximizes the likelihood of the collected trace data with the departure process MAP. The optimal parameters of the service demand distribution can be obtained from matching moments of the APH distribution by a global search with maximum likelihood estimation.

To address the high cost of global optimization, we then propose a heuristic estimation method. In this method, the problem of service demand distribution estimation is divided into sub-problems of fitting different moments, using a collection of estimation methods. The required given data for fitting parameters for estimation with MAP consists of the inter-departure times, response times and the time of departure instant, which can be directly collected with network traffic sniffing from pairs of arrival and departure events. To evaluate our method, we apply it to the analysis of real traces from deploying and monitoring a microservice-based application. The results show that our method can fit the distribution of real traces with a high degree of accuracy.

The rest of the paper is organized as follows. In Sect. 2, we recall necessary background and definitions. In Sect. 3 the inter-departure time model and problem formulation are introduced. In Sect. 4, we discuss our proposed service demand distribution estimation method based on global optimization with maximum likelihood. In Sect. 5, we introduce the heuristic method for service demand distribution estimation. We present our experimental results in Sect. 6. Related work is summarized in Sect. 7. Finally, we draw conclusions in Sect. 8.

2 Preliminaries

Acyclic Phase-Type Distribution. A phase-type (PH) distribution [23] can be defined as the distribution of absorbing time in a continuous-time Markov chain (CTMC) with finite states $\{1, 2, ..., m, m+1\}$, where the first m states are transient and the last state is absorbing. The infinitesimal generator matrix of the CTMC $\boldsymbol{G}$ is

$$\boldsymbol{G} = \begin{bmatrix} \boldsymbol{T} & \boldsymbol{t} \\ \boldsymbol{0}^T & 0 \end{bmatrix}$$

The sub-generator $\boldsymbol{T}$ with dimension $m \times m$ specifies the transition rate from state i to state j. We also define $\boldsymbol{t} = -\boldsymbol{T}\boldsymbol{e}$, where $\boldsymbol{e}$ denotes a column vector of 1 with appropriate dimension. We can further describe the stationary distribution of the transient states with $\boldsymbol{\alpha} = (\alpha_1, \alpha_2, \ldots, \alpha_m)$, subject to $\boldsymbol{\alpha e} = 1, \alpha_i \geq 0$. A PH distribution may thus be compactly specified as $PH(\boldsymbol{\alpha}, \boldsymbol{T})$.

An acyclic PH (APH) distribution [5] is a subset of PH distributions with acyclic underlying Markov chain. This implies that any state in the underlying Markov chain cannot be visited more than once before absorption. If a random variable Y has APH distribution with parameter $\boldsymbol{\alpha}'$ and $\boldsymbol{T}'$, we write $Y \sim APH(\boldsymbol{\alpha}', \boldsymbol{T}')$.

Service Demand Distribution Modeling. In this paper, the service demand distribution is modeled as an APH distribution. The parameters of $APH(\boldsymbol{\alpha}, \boldsymbol{T})$ can be obtained with various PH distribution fitting methods [15]. In this work, we will use the method of moment matching, which can fit the parameters to match an arbitrary number of moments of a reference on empirical distribution. In particular, we consider using the first three moments to study the APH distribution for service demand. The third moment (skewness, S_k) is considered for its characterization on the fitting performance of the end of the tail.

$$S_k = \frac{E[(X - \eta)^3]}{(E[(X - \eta)^2])^{3/2}} \tag{1}$$

In Eq. (1), η denotes the mean value. The first three moments can be described with the mean value (η), the squared coefficient of variance (SCV, c^2) and the skewness (S_k).

$$E[X^2] = (1 + c^2)\eta^2 \tag{2}$$

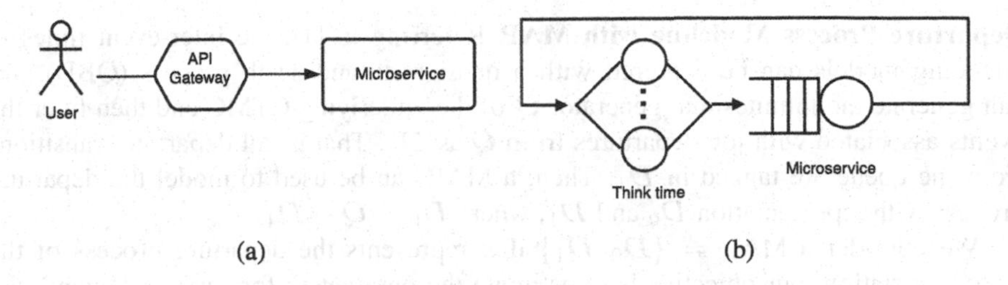

Fig. 1. The structure (a) and the queueing model (b) for of the example applicaton

$$E[X^3] = S_k(c^2)^{3/2}\eta^3 + 3\eta^3c^2 + \eta^3 \tag{3}$$

Thus, we can write the service demand distribution as function of the first three moments $E[X]$, $E[X^2]$ and $E[X^3]$ with parameters η, c^2, and S_k.

Markovian Arrival Processes. MAPs [22] are able to incorporate correlations between successive inter-arrival times. An n-state MAP consists of two stochastic processes, referring to a counting process and a phase process modeled by a finite state (n states) CTMC with infinitesimal generator $\boldsymbol{Q}$. Let $\boldsymbol{D}_0$ be a matrix associated with transitions without arrivals with non-negative off-diagonal elements; $\boldsymbol{D}_0$ and $\boldsymbol{D}_1$ satisfy $\boldsymbol{Q} = \boldsymbol{D}_0 + \boldsymbol{D}_1$ and $(\boldsymbol{D}_0 + \boldsymbol{D}_1)\boldsymbol{e} = \boldsymbol{0}$.

3 Problem Formulation

We propose to observe the departure process of a microservice and determine parameters of service demand distribution, modeled as an APH, using maximum likelihood estimation.

Microservice-based applications can be abstracted as a queueing model. Here we take a simple microservice as an illustrating example[1]. This is a simple microservice exposing a body mass index (BMI) calculation service. The calculation service is a minimalistic microservice that only receives requests and posts responses without external processing. We generate a closed workload to simulate the microservice clients in the system – the structure of the example is given in Fig. 1(a). Figure 1(b) illustrates the analytical model for the application, consisting of a closed queueing network describing the microservice buffer and server, as well as the think time of clients. The model features N concurrent users, each modeled as a job. Scheduling could be either first-come first-served (FCFS) or Processor-sharing (PS) order depending on the implementation details of the web server handling the requests within the microservice. We assume exponentially distributed user think times at the delay station. The problem is to determine the APH service demand distribution in the queueing station. Note that since we focus on a single class of jobs, the model admits a product-form solution for the steady-state distribution, while no specific product-form simplifications are available to analyze the departure process of this queue. As such, the service distribution identification problem does not satisfy a simple analytical closed-form to conduct inference.

[1] https://github.com/go-chassis/go-bmi.

Departure Process Modeling with MAP. Referring to [1], the inter-event times in queueing models can be captured with a quasi birth-and-death process (QBD). We can generate the infinitesimal generator $\boldsymbol{Q}$ of the underlying CTMC and then filter the events associated with job departures from $\boldsymbol{Q}$ as $\boldsymbol{D}_1$. That is, all departure transitions from the queue are tagged in $\boldsymbol{D}_1$. Then, a MAP can be used to model the departure process with representation $\boldsymbol{D}_0$ and $\boldsymbol{D}_1$, where $\boldsymbol{D}_0 = \boldsymbol{Q} - \boldsymbol{D}_1$.

We consider a MAP $= \{\boldsymbol{D}_0, \boldsymbol{D}_1\}$ that represents the departure process of the queueing station, our objective is to estimate the parameters for service demand distribution as $APH(\boldsymbol{\alpha}, \boldsymbol{T})$ from the observable inter-departure times (IDTs). We denote the time between two successive departure events i and $i-1$ as $X_i = a_i - a_{i-1}$. Thus, the IDTs of jobs are $\boldsymbol{X} = [X_1, X_2, \dots, X_{n-1}]$. Since the departure process is modeled as a MAP, the IDTs follows a PH distribution $PH(\boldsymbol{\pi}, \boldsymbol{D}_0)$, where $\boldsymbol{\pi} = \boldsymbol{\pi}(-\boldsymbol{D}_0)^{-1}\boldsymbol{D}_1$ indicating the stationary distribution of the embedded chain. If $\boldsymbol{D}_0$ is acyclic, then the PH distribution specializes into an APH one. This distribution produces an interval stationary initialization for the MAP.

For the MAP described above, the joint probability density function (PDF) of IDTs $\boldsymbol{X} = [X_1, \dots, X_n]$ is

$$f(\boldsymbol{X}) = \boldsymbol{\pi} e^{\boldsymbol{D}_0 X_1} \boldsymbol{D}_1 e^{\boldsymbol{D}_0 X_2} \boldsymbol{D}_1 \dots e^{\boldsymbol{D}_0 X_n} \boldsymbol{D}_1 \boldsymbol{e} \tag{4}$$

For computational convenience, we assume that the given departure events are independent. The logPDF of the IDTs can be approximated as

$$\log f(\boldsymbol{X}) = \sum_{i=1}^{n} \log(\boldsymbol{\pi} e^{\boldsymbol{D}_0 X_i} \boldsymbol{D}_1 \boldsymbol{e}) \tag{5}$$

In general, let $\boldsymbol{\theta}$ be the parameter set of the service demand distribution to be estimated. The log-likelihood for the IDTs is

$$\log f(\boldsymbol{\theta}|\boldsymbol{X}) = \sum_{i=1}^{n} \log(\boldsymbol{\pi} e^{\boldsymbol{D}_0(\boldsymbol{\theta}) X_i} \boldsymbol{D}_1(\boldsymbol{\theta}) \boldsymbol{e}) \tag{6}$$

In (6), $\boldsymbol{D}_0(\boldsymbol{\theta})$ and $\boldsymbol{D}_1(\boldsymbol{\theta})$ describe the functional dependencies between $\boldsymbol{D}_0$, $\boldsymbol{D}_1$ and the service demand distribution parameters $\boldsymbol{\theta}$ such as its moments, e.g., η, c^2, and S_k mentioned in Sect. 2. Then our problem of service demand distribution estimation can be formulated as finding the parameters that maximize the log-likelihood of the IDTs measured from the monitoring traces.

$$f_{obj}(\boldsymbol{\theta}) = \arg\max_{\boldsymbol{\theta} \geq 0} \quad \log f(\boldsymbol{X}|\boldsymbol{\theta}) \tag{7}$$

4 Global Optimization Based Estimation

The parameter estimation of service demand distribution is based on observations of real system trace. In this paper, we consider a finite observation with n samples. Our measured observation consists of the IDTs, the timestamps of each departing instant and the response times.

Data Preprocessing. For a real system, there could be a large number of requests from the users arriving within a very short period. If we directly take all of the samples in the trace, it could be quite time-consuming to calculate the likelihood function in (6), due to the cost of evaluating the matrix exponential.

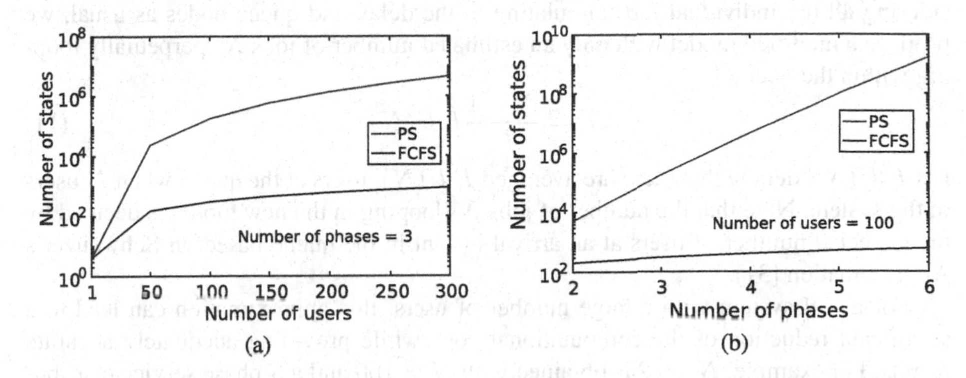

Fig. 2. The number of states in the CTMC state space with PS and FCFS

To address the above issue, we observe that the inter-departure times of jobs can be grouped into different patterns. In order to accelerate the execution times, we apply clustering based on k-means [17] to partition the IDTs to obtain K groups of data with cluster centroids $\boldsymbol{C} = [C_1, C_2, \ldots, C_K]$. Then, the log joint PDF in (5) can be approximated based on the IDT clusters as

$$\log f(\boldsymbol{C}) = \sum_{i=1}^{K} L_i \log(\pi e^{\boldsymbol{D}_0 C_i} \boldsymbol{D}_1 \boldsymbol{e}) \tag{8}$$

where L_i denotes the number of points in cluster i.

CTMC State Space Explosion. Increasing of the number of concurrent users, the state space of the CTMC can easily suffer state-space explosion. Assume a single-server queue where the service demand SD of the queue node is APH distributed and a delay node as shown in the example in Fig. 1, and there are N users in the queueing network. First, we consider the jobs in the queue are processed with a first-come-first-served (FCFS) order. Only one job can be served by the server at one time. Let P denote the number of phases in the service process, i.e., the number of columns in $\boldsymbol{\alpha}$. The number of states in the state space is

$$s = N \cdot P + 1 \tag{9}$$

Instead, if the server follows a processor sharing (PS) scheduling strategy, i.e., multiple jobs can be served simultaneously, the number of states in the state space is

$$s = \sum_{i=0}^{N} (N + 1 - i) \binom{i + P - 2}{i} \tag{10}$$

Compared to FCFS, we can see that the state space for PS grows combinatorially, as shown in Fig. 2, making the analysis of the CTMC intractable.

To mitigate the complexity of dealing with PS scheduling, we propose to capture the behaviours of the original model with a simpler model. We focus on the mean queue length that can be approximated by using mean-value analysis (MVA). Instead of considering all the individual jobs circulating in the delay and queue nodes as usual, we propose a modified model with only an estimated number of jobs $N^{'}$ perpetually looping within the queue.

$$N^{'} = \frac{N-1}{N} E[U(N)] \tag{11}$$

Let $E[U(N)]$ denote that there are averaged $E[U(N)]$ users at the queue when N users in the system. Note that the number of jobs $N^{'}$ looping in the new model is decided by the expected number of users at an arrival instant in the queue based on Schweitzer's Approximation [31].

For a real system with a large number of users, this approximation can lead to a significant reduction of the computational cost, while providing adequately accurate results. For example, $N^{'} = 2$ is obtained with $N = 100$ and a 3-phase service distribution, the number of states in the state space is only 10, which is much smaller compared to the one shown in Fig. 2(a) with $N = 100$ under PS schedule.

MLE for Service Demand Distribution. Our objective is to search optimal parameters for approximating service demand distributions. Given the observed trace data, a common approach for parameter estimation is maximum likelihood estimation (MLE) [24], which casts the estimation as a global optimization problem. We propose an estimation method that combines MLE with simulations of queueing models to approximate the APH distribution for service demand.

Algorithm 1 describes the implementation of this method in details. The algorithm requires a set of clustered IDTs and the searching boundaries. In each iteration, the algorithm generates a set of moments satisfying the bound constraints and then the APH distribution is fitted from the current moments. The conditions of convergence for the algorithm are set as follows. There are two conditional parameters, including the maximum number of evaluation on the objective function ($MaxEvalFunc$) and a last step tolerance for ending iteration (Tol). For moment matching we use the BuTools package [16], pointing to *APHFit* in Line 3. Note that we need to satisfy that the APH distribution is feasible with given parameters, i.e., both $\boldsymbol{\alpha}$ and $\boldsymbol{T}$ are not empty or zero. After obtaining the service demand distribution SD, a queueing model is generated with a queue node of SD. The current queueing model can be solved by analyzing the underlying CTMC, obtaining the infinitesimal generator $\boldsymbol{Q}$. By analyzing the transitions in $\boldsymbol{Q}$, the transition rates of departure events on the queue node can be filtered for $\boldsymbol{D}_1$ as shown in Lines 6–8.

In Algorithm 1, the optimal parameters of service demand distribution are obtained with the maximum likelihood value of the monitoring traces. However, the computation of the infinitesimal generator involves the computation of a matrix exponential, which is computationally expensive and rises numerical instability. To mitigate these issues, we use CTMC uniformization [29] which is well-known to be an effective numerical method for computing transient measures involving matrix exponential. For transient analysis, uniformization techniques can be applied with sub-generator $\boldsymbol{D}_0$ and the initial distribution $\boldsymbol{\pi}$ of the MAP. Since the transient rate in $\boldsymbol{D}_0$ of a real system could be large, to guarantee stable calculations, we adopt the scaling method from [32], involving a scaling factor q to avoid floating-point errors. In Line 10, the scaling CTMC uniformization method is defined as *ctmc_uniform*, which takes π, $\boldsymbol{Q} - \boldsymbol{D}_1$ and the centroid of the cluster as the input. The approximated transient probability is obtained as $\boldsymbol{\beta}$ for $\boldsymbol{\pi} e^{\boldsymbol{D}_0}$. Then the log-likelihood value can be computed using (8) at Line 11.

Algorithm 1. Global optimization based estimation method

Input: C $\leftarrow$ Set of clustered inter-departure times $[c_1, c_2, \ldots, c_n, l_1, l_2, \ldots, l_n]$, where c_i is the centriod value and l_i is the number of points in cluster i.
LB $\leftarrow$searching lower bound
UB $\leftarrow$searching upper bound
Output: $SD \leftarrow$ Estimated service demand with APH distribution
1: Random initialize $M_0 \leftarrow [m_1, m_2, m_3]$, $LB < M_0 < UB$
2: **while** $MaxEvalFunc \leq 10^{10}$ and $Tol \geq 10^{-8}$ **do**
3: Service demand distribution $APH(\boldsymbol{\alpha}, \boldsymbol{T}) \leftarrow APHFit[m_1, m_2, m_3]$
4: **if** $APH(\boldsymbol{\alpha}, \boldsymbol{T})$ is feasible **then**
5: Generate a queueing network model QNM with service demand $APH(\boldsymbol{\alpha}, \boldsymbol{T})$
6: $\boldsymbol{Q} \leftarrow$ solve(QNM)
7: Filter $\boldsymbol{D}_1$ from the infinitesimal generator $\boldsymbol{Q}$
8: MAP $\leftarrow \{\boldsymbol{Q} - \boldsymbol{D}_1, \boldsymbol{D}_1\}$, generate $\boldsymbol{\pi}$
9: **for** $i = 1$ to n **do**
10: $\boldsymbol{\beta} \leftarrow ctmc\_uniform(\boldsymbol{\pi}, \boldsymbol{Q} - \boldsymbol{D}_1, c_i)$
11: $L \leftarrow L + \log(\boldsymbol{\beta} \boldsymbol{D}_1 \boldsymbol{e}) l_i$
12: **end for**
13: **end if**
14: **end while**
15: Get optimal parameter set $[m_1, m_2, m_3]$ with maximum likelihood value
16: **return** $SD \leftarrow APHFit[m_1, m_2, m_3]$

---

5 Heuristics-Based Estimation

As illustrated in Algorithm 1, the global optimization method for service demand distribution estimation needs to consider a large search space. It could be time-consuming to obtain the optimal parameter set maximizing the likelihood value. The method presented in this section estimates the parameters sequentially, rather than jointly, offering a heuristic estimation that trades accuracy for speed.

Mean Service Demand Estimation. The mean value of service demand can be efficiently estimated based on performance measurements from monitoring traces. We refer to the work in [26], which allows estimating the expected value of service demand with queue length and response times. Both queue length and response times are easily measured with system monitoring. Since in this work we target the departure process, the input dataset of the estimation method contains the following data by calculating from system monitoring at departing occurrence.

- The timestamp of a job departing from the queue node (DT)
- The response times from the monitoring traces (R)
- The queue length seen upon arrivals (A).

Considering a single class of jobs in the system, let N be the size of the population in the closed queueing network. Therefore, the mean service demand $E[D]$ for the single-class case can be estimated as [26]:

$$E[D] = \frac{E[R]}{1 + E[A]} \tag{12}$$

where $E[R]$ and $E[A]$ is the expected value of response times and the queue length seen upon arrivals, respectively.

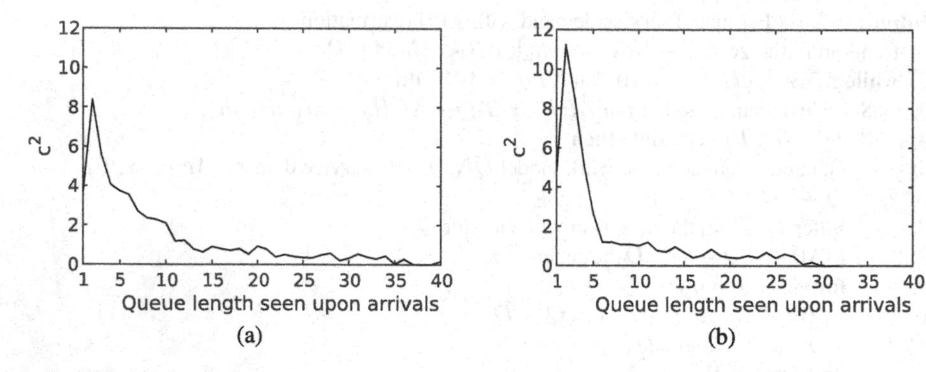

Fig. 3. c^2 conditional on the queue length seen upon arrivals for simulation (a) and the real trace (b)

SCV Estimation. To estimate the second moment of service demand, we investigate the state-dependent behavior of the system. The estimation formulation is derived from the SCV on the mean queue length seen upon arrival. In a queueing system, the response time of a job is related to the number of jobs in the queue node either waiting or receiving service. We propose to estimate c^2 using the following heuristic expression

$$c^2 = \max \frac{E[(R_{ij} - E[R_i])^2]}{E[R_i]^2} \tag{13}$$

where i is a value for the length of queue seen upon arrival, with $i = 0, \ldots, max(A)$. R_i denotes a set of response times of which the queue length seen upon arrival is i, and R_{ij} denotes the jth response time belonging to R_i. In detail, at each departure instant, we can collect the response time of the current completed job and the number of jobs remaining in the queue. Then the queue length is sorted and the response times can be grouped according to different numbers of the queue length as R_{ij}.

To demonstrate the accuracy of the approximation for SCV, we conduct an experiment with $N = 100$. We analyze real trace data from monitoring and compare the c^2 of real traces to the simulated queueing model with estimated parameters based on MLE. We can see from Fig. 3(a) that the c^2 conditional on the queue length first increases to the maximum and then decrease with the growth of queue length. The same pattern can be also observed for the real trace as shown in Fig. 3(b). The simulated c^2 value is 8.3. It can be seen that the $max(c^2)$ of the simulation is close to the one for the real trace with $c^2 = 11.2$.

Heuristic Service Demand Distribution Estimation. Our heuristic method to accelerate the MLE estimation is shown in Algorithm 2. Compared to the global optimization in Algorithm 1, the heuristic-based method is used to estimate the service demand distribution with three separate estimation methods to fit the first three moments, including mean service demand (η) and SCV (c^2) estimation in the beginning of Sect. 5 and a MLE-based skewness (S_k) estimation as shown in Algorithm 2.

Algorithm 2. Heuristics based estimation method

Input: $C \leftarrow$ Set of clustered inter-departure times $[c_1, c_2, \ldots, c_n, l_1, l_2, \ldots, l_n]$, where c_i is the centriod value and l_i is the number of points in cluster i.
$R \leftarrow$ Set of response times $[r_1, r_2, \ldots, r_n]$
$DT \leftarrow$ Set of times on the departure instant $[dt_1, dt_2, \ldots, dt_n]$
$S_k = [S_{k1}, S_{k2}, \ldots, S_{kj}] \leftarrow$ searching set of S_k
Output: $SD \leftarrow$ Estimated service demand with APH distribution
1: Compute A from DT and R
2: $\eta \leftarrow meanEstimate(A, R)$
3: $c^2 \leftarrow scvEstimate(A, R)$
4: **for** $i = 1$ to j **do**
5: Compute the first three moments $[m_1, m_2, m_3]$ from m, c^2 and S_{ki}
6: $APH(\boldsymbol{\alpha}, \boldsymbol{T}) \leftarrow APHFit[m_1, m_2, m_3]$
7: **if** $APH(\boldsymbol{\alpha}, \boldsymbol{T})$ is feasible **then**
8: Repeat execution of lines 5 to 12 in Algorithm 1
9: **end if**
10: **end for**
11: Get optimal skewness set of service demand, $S_k \leftarrow S_{ki}$ with $max(L)$
12: Compute the first three moments $[m_1, m_2, m_3]$ from η, c^2 and S_k
13: **return** $SD \leftarrow APHFit[m_1, m_2, m_3]$

The algorithm requires a set of IDTs, the departure times, the response time of each job, and the queue lengths seen upon arrivals. In Line 1, we first compute the arrival times with the departure times DT and response times R and then calculate the queue length seen upon arrivals for each job. The mean service demand is estimated based on response times and the queue length seen upon arrivals. Then the algorithm estimates c^2 by (13). The only search parameter for our method now is the skewness S_k. Here we perform MLE on the departure process with Eq. (6). As in the search-based global optimization, we first generate a set of candidate skewness values $S_k = [S_{k1}, S_{k2}, \ldots]$. For each S_{ki}, we generate a queueing model with corresponding service demand and then calculate the likelihood value as Lines 5–12 in Algorithm 1. The algorithm will search on all candidates in the set, and the process is repeated with multiple candidate points for robustness. The estimated result of the skewness is finally decided on the max likelihood value. Therefore, the final result is obtained with a collection of three-parameter estimation methods.

6 Evaluation

This section introduces the experimental setup and evaluation metrics, and a comparison of our method against the baseline algorithms.

6.1 Experimental Setup

To evaluate the proposed service demand distribution estimation, we conduct several experiments and compare the results to our baseline algorithm using an open-source microservice-based application called Sock Shop[2]. Sock Shop consists of 13 different services and all services communicate using REST APIs over HTTP. We use Docker Compose for the multi-container orchestration. We then generate closed workloads with different intensity using Locust[3]. In the experiment, we target a service that does not interact with a database, which avoids indirect drifts in the response time due to the state of the database.

Table 1. Workload pattern

CPU level	Low	Medium	High
CPU utilization (U)	33%	43%	95%
Number of users (N)	50	100	300

The experimental environment is as follows. For the deployment of the application, we use a server running Ubuntu 16.04.7, and our target service is pinned to a separate CPU core. Locust is running on 6 different servers, including one host node and 5 distributed nodes to simulate the concurrent users. We experimented with populations of different numbers of users to assess the corresponding CPU utilization level as shown in Table 1.

In all the following experiments, the users' think time is exponentially distributed with a mean of 0.1 s.

We conduct experiments with each population in Table 1 and capture the network traffic with a dockerized `tcpdump` that is triggered over HTTP. During the experiments, we monitor HTTP traffic on the source and destination nodes of our target service. Then the traffic data is parsed to extract the request and response information of each request. For each different population size, we collect 10,000 HTTP request and response pairs to build up the trace dataset. Every dataset consists of the response time of each request and the time of departure instant.

[2] https://microservices-demo.github.io/.
[3] https://locust.io/.

Baseline Algorithm. The service demand is usually modeled as exponential with the estimated mean value [33]. In our experiments, we also fit as baseline an exponential distribution for service demand.

Metrics. Since the real service times of systems are usually difficult to measure, we opt to measure the response time via monitoring and use the complementary cumulative distribution function (CCDF) of the response times as our metrics. We construct the queueing model parameterized with the estimated service demand distribution and use the simulation-based JMT solver in LINE [27] to compute the corresponding response times to compare with. Thus, the only difference between the baseline and our experiments is the service demand distribution at the queueing node.

6.2 Data Preprocessing and Clustering

To demonstrate the trade-off between computational complexity and the approximation accuracy due to the choice of the number of clusters K, we consider a simulated experiment with 50 users in a single-server queueing system. In this experiment, we estimate the service demand distribution with the original data and clustered data for different values of K. Figure 4 shows the simulation results. It can be observed from Fig. 4(b) that with clustered IDTs the execution time of departure process MAP modeling and log-likelihood calculation drops by almost 33% compared to the initial execution time with the whole trace. The accuracy of parameter estimation with the clustered data is evaluated in Fig. 4(a) by means of CCDF diagrams. While small values of K lead to a coarse approximation, increasing K the CCDF for the clustered data rapidly converges to the CCDF estimated from the whole dataset without clustering. As can be noted from Fig. 4(a), the curves become indistinguishable for $K \geq 100$. We can thus conclude that our clustering heuristics does not significantly reduce the accuracy of the estimation for K large enough (K $\geq$ 100 in our experiment), while significantly reducing the computational cost of the estimation.

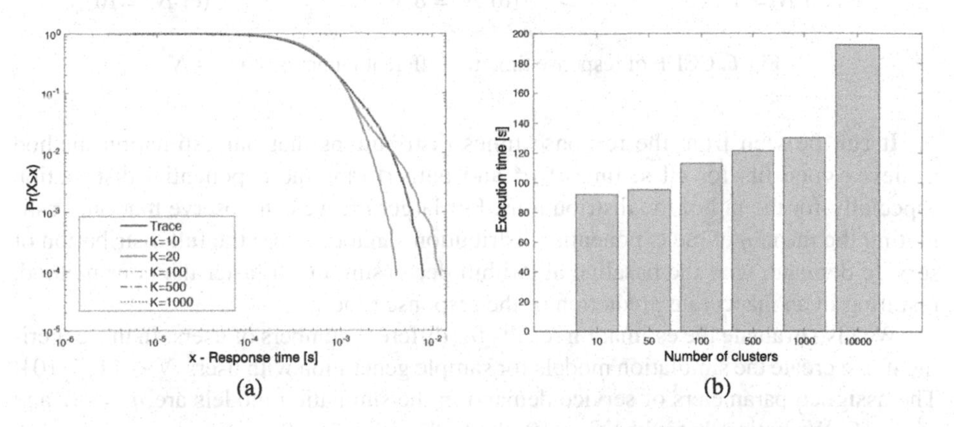

Fig. 4. The CCDF of response times (a) and the execution times (b) of maximum likelihood estimation under different number of clusters

6.3 Numerical Experiment Results

We conduct the following numerical experiments to assess the effectiveness of our global-search based method for the service demand distribution estimation. We first generate single-server queueing models that can simulate the behaviours of a simplistic microservice, setting different known values for service demand at the queue node. Then we generate samples of inter-departure times by simulating. The mean service demand η is fixed at 0.7 and different SCV are selected from $c^2 \in \{0.5, 1, 4, 16\}$. After generating sample traces via simulation, we execute Algorithm 1 and compare the estimated SD with the known values. We also compare with the baseline algorithm on the same simulated traces. The results of fitting with different service demand distributions are plotted as CCDF diagrams in Fig. 5.

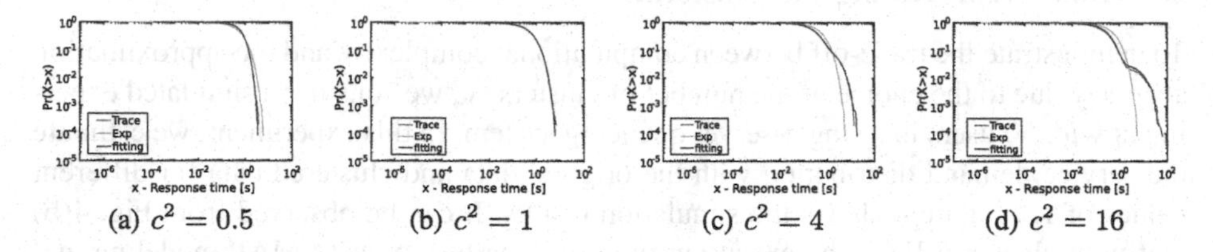

(a) $c^2 = 0.5$ (b) $c^2 = 1$ (c) $c^2 = 4$ (d) $c^2 = 16$

Fig. 5. CCDF of response time for different setting of c^2

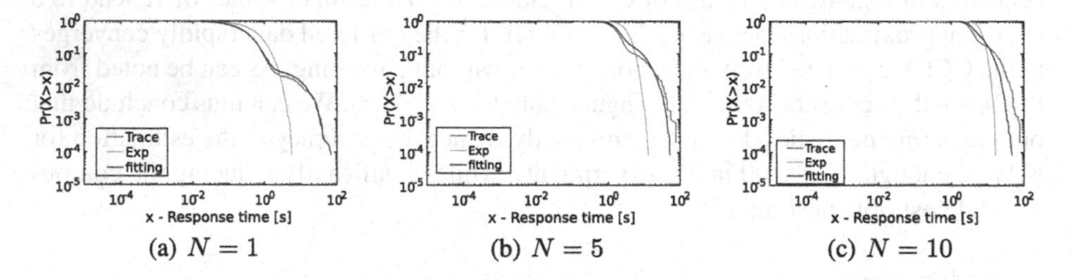

(a) $N = 1$ (b) $N = 5$ (c) $N = 10$

Fig. 6. CCDF of response time for different numbers of users N

It can be seen from the response times distributions that our estimation method achieves good fits for all setting of c^2 and outperforms the exponential distribution especially for the tail of the distributions. For larger c^2, we can observe that only estimating the mean η of the exponential distribution cannot capture the full distribution of service demand, with the baseline algorithm decreasing much faster than our method, resulting in an inaccurate prediction of the response times.

We also evaluate the estimation results for different numbers of users. In this experiment, we create the simulation models for sample generation with users $N \in \{1, 5, 10\}$. The assigned parameters of service demand in the simulation models are $\eta = 0.7$ and $c^2 = 16$. We notice that with $N = 10$, the CPU utilization from simulation can reach over 95% which is close to saturation. However, compared to the baseline, the fitting results of our method under different values of N are much closer to the simulated response time distribution in the tail than the baseline estimation, as shown in Fig. 6.

6.4 Analysis of Results on Measured Traces

Scheduling Policy – FCFS. We now turn our attention to experiments on the real system. First, we present our experimental results for different numbers of users with FCFS scheduling strategy. For a single server system with FCFS, the service times can be obtained from sampling IDTs when the server is not idle. Therefore, for FCFS, we can directly calculate the first three moments for these sampled service times. Here we define this direct measured method as FCFS-single method. We evaluate the simulation results with FCFS-single method, baseline algorithm, and our method. The CCDF of response times for low, medium, and high CPU utilization are plotted in Fig. 7, respectively. The parameter estimation results with FCFS are summarized in Table 2, and the results based on FCFS-single measurement are in Table 3.

We first observe that the FCFS-single method yields a similar accuracy of fitting the tail of the distribution under low and medium utilization level, whereas the curve based on the FCFS-single is farther away than our method for the body of the distribution. For higher CPU utilization, our method has a better fit for the tail of the distribution. We interpret that the real system is not precisely served by FCFS scheduling, while it exhibits some state-dependent degree of CPU sharing, and the heuristic we propose

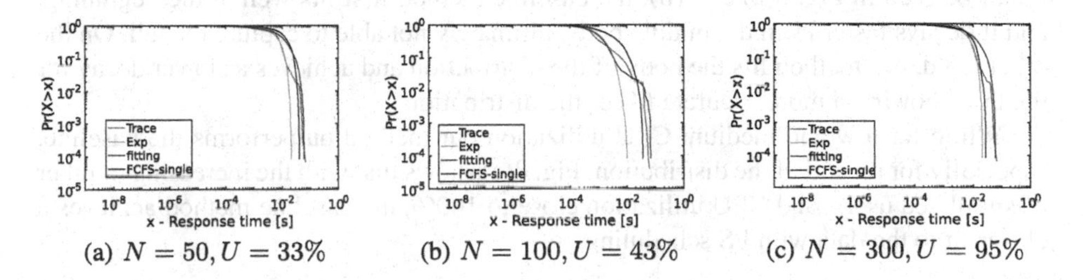

(a) $N = 50, U = 33\%$ (b) $N = 100, U = 43\%$ (c) $N = 300, U = 95\%$

Fig. 7. CCDF of response time with different utilization for the fitted models with FCFS.

Table 2. Service demand distribution parameter estimation results for FCFS

N	$\eta \times (10^{-4})$	c^2	S_k	Likelihood $\times$ (10^4)
50	8.33	8.24	11	6.62
100	6.21	11.22	57	14.31
300	3.50	8.03	12	10.64

Table 3. Service demand distribution parameter estimation results for FCFS with FCFS-single method

N	$\eta \times (10^{-4})$	c^2	S_k
50	13.61	2.31	4
100	7.96	2.85	5
300	3.72	2.87	7

captures instances of large service variance that occur in certain system states, which can be inferred from c^2 in Table 3.

Compared to the baseline algorithm with exponential distribution, it can be seen that for all 3 different N of users our method produces a closer fit. As one can see for the body fitting, both baseline and our proposed model yield good performance. For $N = 50$ and $N = 100$, our model fits the body with better accuracy, whereas the baseline method is outperformed for $N = 300$. However, for the fitting of the tail, the baseline method is worse in terms of accuracy for all values of N. Overall, our estimated distribution achieves a better fit throughout the distribution curve, producing an approximation of the real trace distribution accurately. Compared to the baseline, our proposed method can model the heavy-tail behaviors, which is significant to model different intensive workloads.

Scheduling Policy – PS. To analyze the impact of scheduling policies, we conduct experiments on the queueing model with $N \in \{50, 100, 300\}$ with PS scheduling policy. The parameter estimation results with PS is shown in Table 4 and the fitting comparison is plotted in Fig. 8. We can see from the figure that switching from FCFS to PS impacts the results of the baseline in a significant manner. First, comparing the estimated skewness to the results for FCFS, we can observe smaller values of skewness. As it can be seen in Fig. 8(a) and (b), the baseline method first fits well at the beginning, and it decays faster from the middle body, ultimately not able to capture the tail. On the other hand, our method fits the body of the distribution and achieves a slower decay for the tail, showing a more accurate fit for the distribution.

While for low and medium CPU utilization our method outperforms the baseline, especially for the tail of the distribution, Fig. 8(c) shows that with the increasing number of simulated users, and CPU utilization close to 100%, the baseline method achieves a closer fit to the data with PS scheduling.

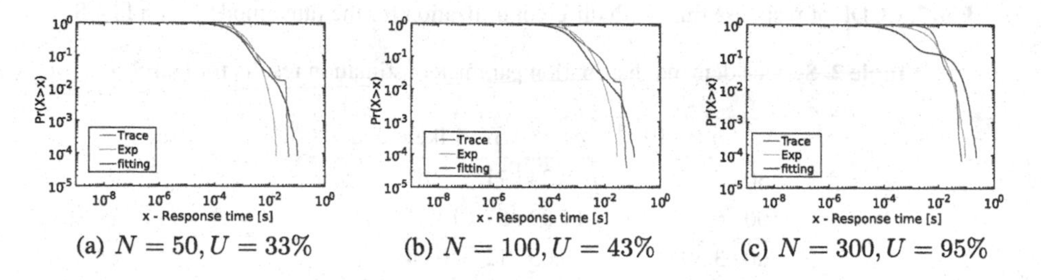

(a) $N = 50, U = 33\%$ (b) $N = 100, U = 43\%$ (c) $N = 300, U = 95\%$

Fig. 8. CCDF of response time with different utilization for the fitted models with PS.

Table 4. Service demand distribution parameter estimation results for PS

N	$\eta \times (10^{-4})$	c^2	S_k	Likelihood $\times (10^4)$
50	8.33	8.24	9	6.11
100	6.21	11.22	10	14.53
300	3.50	8.03	4	10.68

It is not difficult to see from Table 2 and Table 4 that the likelihood values for $N = 50, 100, 300$ with PS are close to the one with FCFS. In reality, the actual system scheduling will factor in several elements such as caching, memory bandwidth, and operating system scheduling, which are neither perfectly PS nor FCFS. Thus, our results indicate that either models provide a reasonable approximation to the observed system behaviour, but FCFS appears more suitable to model heavy loads.

Summarizing, the previous results indicate that in the majority of instances the proposed method is able to fit the service process characteristics of microservices with higher fidelity than simple exponential models, especially capturing tail behaviors with higher accuracy.

7 Related Work

To enable automatic generation for accurate performance models, model parameterization brings out an important problem of resource demand estimation. There are several works using regression methods. Rolia [30] introduce the resource demand estimation problem with linear regression techniques. In general, linear regression has been employed to solve the service demand estimation mostly based on utilization [9,10] and response time [26]. Neural networks like recurrent neural networks (RNN) can be applied to estimate resource demands for the ability to predict time series data like the works in [12]. Machine learning can also help to select the optimal approaches for estimation on account of varieties of existing resource demand approaches [19]. However, most of the mentioned approaches based on regression only enable to obtain the mean value of the resource demand instead of full distributions, lacking higher-order properties of the demand. These time-series based prediction is not suitable for assigning the required parameters for performance models.

There are many existing works on modeling with a queueing system in which the arrivals of users are characterized with a Poisson arrival process and an exponentially distributed service time [6,18]. However, they are not accurate enough to capture and represent the behaviors of real systems, especially for the tail of the service demand distribution. To characterize more accurate service demand, PH distribution provides possibilities to approximate arbitrary distributions, which has been used in studies of modeling and simulation like [11,25].

8 Conclusion

In this paper, we have introduced a service demand distribution estimation method by using Markovian arrival processes for microservices. We have first presented a closed queueing model for a microservice and characterized the service demand of the queue node with an APH distribution. We have then proposed to model the departure process at the queue node of this model with a MAP, in which the parameters of service demand can be exposed. The service demand distribution can be estimated with maximum likelihood based on the departure process MAP. We applied the global search and a heuristic estimation method on solving the optimization problem. The proposed heuristic estimation method can effectively reduce the computational cost compared to the global search approach. We further showed that the proposed estimation method yields a better performance on fitting real traces of microservices compared to the baseline.

As we consider a single-server queue with a single-class workload in this work, our future research aims at generalizing the estimation method with multi-classes models. In the next step, we plan to model the departure process with a marked MAP, which may increase the model complexity and the likelihood of CTMC explosion. To deal with the problems of multi-class queues, multi-class classification methods like the one-against-all algorithm may provide the feasibility to reduce the number of classes. Aggregation methods may be explored to address this issue.

Acknowledgement. The work of Giuliano Casale has been partly funded by the EU's Horizon 2020 program under grant agreement No. 825040.

References

1. Andersen, A.T., Neuts, M.F., Nielsen, B.F.: Lower order moments of inter-transition times in the stationary QBD process. Methodol. Comput. Appl. Probab. **2**(4), 339–357 (2000)
2. Bass, L., Weber, I., Zhu, L.: DevOps: A Software Architect's Perspective. Addison-Wesley Professional, Boston (2015)
3. Bezemer, C.P., et al.: How is performance addressed in devops?. In: Proceedings of the 2019 ACM/SPEC International Conference on Performance Engineering, pp. 45–50 (2019)
4. Binz, T., Breitenbücher, U., Kopp, O., Leymann, F.: TOSCA: portable automated deployment and management of cloud applications. In: Bouguettaya, A., Sheng, Q., Daniel, F. (eds.) Advanced Web Services. Springer, New York (2014). https://doi.org/10.1007/978-1-4614-7535-4_22
5. Bobbio, A., Horváth, A., Telek, M.: Matching three moments with minimal acyclic phase type distributions. Stoch. Models **21**(2–3), 303–326 (2005)
6. Bramson, M., Lu, Y., Prabhakar, B.: Randomized load balancing with general service time distributions. ACM SIGMETRICS Perform. Eval. Rev. **38**(1), 275–286 (2010)
7. Brosig, F., Kounev, S., Krogmann, K.: Automated extraction of palladio component models from running enterprise java applications. In: Proceedings of the Fourth International ICST Conference on Performance Evaluation Methodologies and Tools, pp. 1–10 (2009)
8. Casale, G., et al.: Radon: rational decomposition and orchestration for serverless computing. SICS Softw.-Intensiv. Cyber-Phys. Syst. **35**(1), 77–87 (2020)
9. Casale, G., Cremonesi, P., Turrin, R.: How to select significant workloads in performance models. In: CMG Conference Proceedings, pp. 58–108. Citeseer (2007)
10. Casale, G., Cremonesi, P., Turrin, R.: Robust workload estimation in queueing network performance models. In: 16th Euromicro Conference on Parallel, Distributed and Network-Based Processing (PDP 2008), pp. 183–187. IEEE (2008)
11. Dudin, S., Dudina, O.: Retrial multi-server queuing system with PHF service time distribution as a model of a channel with unreliable transmission of information. Appl. Math. Model. **65**, 676–695 (2019)
12. Duggan, M., Mason, K., Duggan, J., Howley, E., Barrett, E.: Predicting host CPU utilization in cloud computing using recurrent neural networks. In: 2017 12th International Conference for Internet Technology and Secured Transactions (ICITST), pp. 67–72. IEEE (2017)

13. Franks, G., Al-Omari, T., Woodside, M., Das, O., Derisavi, S.: Enhanced modeling and solution of layered queueing networks. IEEE Trans. Softw. Eng. **35**(2), 148–161 (2008)
14. Garetto, M., Cigno, R.L., Meo, M., Marsan, M.A.: A detailed and accurate closed queueing network model of many interacting TCP flows. In: Proceedings IEEE INFOCOM 2001. Conference on Computer Communications. Twentieth Annual Joint Conference of the IEEE Computer and Communications Society (Cat. No. 01CH37213), vol. 3, pp. 1706–1715. IEEE (2001)
15. Horváth, A., Telek, M.: PhFit: a general phase-type fitting tool. In: Field, T., Harrison, P.G., Bradley, J., Harder, U. (eds.) TOOLS 2002. LNCS, vol. 2324, pp. 82–91. Springer, Heidelberg (2002). https://doi.org/10.1007/3-540-46029-2_5
16. Horváth, G., Telek, M.: BuTools 2: a rich toolbox for Markovian performance evaluation. In: VALUETOOLS (2016)
17. Krishna, K., Murty, M.N.: Genetic k-means algorithm. IEEE Trans. Syst. Man Cybern., Part B (Cybern.) **29**(3), 433–439 (1999)
18. Krishnamoorthy, A., Manikandan, R., Lakshmy, B.: A revisit to queueing-inventory system with positive service time. Ann. Oper. Res. **233**(1), 221–236 (2015)
19. Liao, S., Zhang, H., Shu, G., Li, J.: Adaptive resource prediction in the cloud using linear stacking model. In: 2017 Fifth International Conference on Advanced Cloud and Big Data (CBD), pp. 33–38. IEEE (2017)
20. Marsan, M.A., Balbo, G., Conte, G., Donatelli, S., Franceschinis, G.: Modelling with Generalized Stochastic Petri Nets, vol. 292. Wiley, New York (1995)
21. Mazkatli, M., Koziolek, A.: Continuous integration of performance model. In: Companion of the 2018 ACM/SPEC International Conference on Performance Engineering, pp. 153–158 (2018)
22. Neuts, M.F.: Models based on the Markovian arrival process. IEICE Trans. Commun. **75**(12), 1255–1265 (1992)
23. O'Cinneide, C.A.: Characterization of phase-type distributions. Stoch. Models **6**(1), 1–57 (1990)
24. Pan, J.X., Fang, K.T.: Maximum likelihood estimation. In: Growth Curve Models and Statistical Diagnostics. Springer Series in Statistics. Springer, New York (2002). https://doi.org/10.1007/978-0-387-21812-0_3
25. Parini, A., Pattavina, A.: Modelling voice call interarrival and holding time distributions in mobile networks. In: Proceedings of ITC 2005, pp. 729–738 (2005)
26. Pérez, J.F., Pacheco-Sanchez, S., Casale, G.: An offline demand estimation method for multi-threaded applications. In: 2013 IEEE 21st International Symposium on Modelling, Analysis and Simulation of Computer and Telecommunication Systems, pp. 21–30. IEEE (2013)
27. Pérez, J.F., Casale, G.: Line: evaluating software applications in unreliable environments. IEEE Trans. Reliab. **66**(3), 837–853 (2017)
28. Petriu, D.C., Shen, H.: Applying the UML performance profile: graph grammar-based derivation of LQN models from UML specifications. In: Field, T., Harrison, P.G., Bradley, J., Harder, U. (eds.) TOOLS 2002. LNCS, vol. 2324, pp. 159–177. Springer, Heidelberg (2002). https://doi.org/10.1007/3-540-46029-2_10
29. Reibman, A., Trivedi, K.: Numerical transient analysis of Markov models. Comput. Oper. Res. **15**(1), 19–36 (1988)
30. Rolia, J., Vetland, V.: Correlating resource demand information with arm data for application services. In: Proceedings of the 1st International Workshop on Software and Performance, pp. 219–230 (1998)
31. Schweitzer, P.: Approximate analysis of multiclass closed networks of queues. J. ACM **29**(2) (1981)
32. Sidje, R.B., Burrage, K., MacNamara, S.: Inexact uniformization method for computing transient distributions of Markov chains. SIAM J. Sci. Comput. **29**(6), 2562–2580 (2007)

33. Spinner, S., Casale, G., Brosig, F., Kounev, S.: Evaluating approaches to resource demand estimation. Perform. Eval. **92**, 51–71 (2015)
34. Wang, W., Huang, X., Qin, X., Zhang, W., Wei, J., Zhong, H.: Application-level CPU consumption estimation: towards performance isolation of multi-tenancy web applications. In: 2012 IEEE Fifth International Conference on Cloud Computing, pp. 439–446. IEEE (2012)

Performance Analysis of Work Stealing Strategies in Large Scale Multi-threaded Computing

Grzegorz Kielanski and Benny Van Houdt(✉)

University of Antwerp, Middelheimlaan 1, Antwerp 2020, Belgium
{Grzegorz.Kielanski,Benny.Houdt}@uantwerpen.be

Abstract. Distributed systems use randomized work stealing to improve performance and resource utilization. In most prior analytical studies of randomized work stealing, jobs are considered to be sequential and are executed as a whole on a single server. In this paper we consider a homogeneous system of servers where parent jobs spawn child jobs that can feasibly be executed in parallel. When an idle server probes a busy server in an attempt to steal work, it may either steal a parent job or multiple child jobs.

To approximate the performance of this system we introduce a Quasi-Birth-Death Markov chain and express the performance measures of interest via its unique steady state. We perform simulation experiments that suggest that the approximation error tends to zero as the number of servers in the system becomes large. Using numerical experiments we compare the performance of various simple stealing strategies as well as optimized strategies.

Keywords: Performance analysis · Matrix analytic methods · Distributed computing

1 Introduction

Jobs in multithreaded computing systems consist of several threads [2,24]. Upon starting the execution a main thread (which we call a parent job) several other threads are spawned (which we call child jobs). These spawned child jobs are initially stored locally, but can be redistributed at a later stage. One way of redistributing jobs is called "randomized work stealing": servers that become empty start probing other servers at random (uniformly) and if the probed server has pending jobs, some of its jobs are transferred to the probing server [2,5]. Another option is to make use of "randomized work sharing", where servers that have pending jobs probe others to offload some of their work to other servers.

Work stealing solutions have been studied by various authors and are often used in practice. They have been implemented for example in the Cilk programming language [3,6], Intel TBB [19], Java fork/join framework [12], KAAPI [9]

A. Abate and A. Marin (Eds.): QEST 2021, LNCS 12846, pp. 329–348, 2021.
https://doi.org/10.1007/978-3-030-85172-9_18

and .NET Task Parallel Library [13]. Some early studies on work sharing and stealing include [5,16,22]. In [5] the performance of work stealing and sharing is compared for homogenous systems with exponential job sizes. Using similar techniques the work in [5] was generalized to heterogeneous systems in [16]. The key takeaway from these papers is that work stealing clearly outperforms work sharing in system with high load. [22] focused on shared-memory systems and assumes that migrated jobs have a higher service demand and migrating jobs requires some time.

More recent work includes [8,14,15,21,23]. In [8] the authors analyse the system consisting of several homogeneous clusters with exponential job sizes and where half of the jobs are transferred when a probe is successful. A fair comparison between stealing and sharing strategies is given for homogeneous networks and exponential job sizes in [14,15] and for non-exponential job sizes in [23]. Further, the comparison in [15] is extended to heterogeneous networks in [21]. The key difference with the current paper is that in these prior works jobs are considered to be sequential and are always executed as a whole on a single server.

In this paper, we consider a system of homogeneous servers that uses a randomized work stealing policy. We consider a set of policies where if a server with pending child jobs is probed by an idle server, some of its child jobs are transferred. When a server is probed that does not have any pending child jobs, a pending parent job is transferred instead (if available). The work presented in this paper is closely related to [20], where two systems are considered: one system where parent jobs can be stolen and the other system where child jobs can be stolen one at a time. In the current paper we allow that several child jobs can be stolen at once and the main objective is to provide insights on how to determine the number of child jobs that should be transferred in such an event. When several child jobs can be stolen at once, child jobs may be transferred several times before being executed and this considerably complicates the analysis compared to [20]. In [20] we also introduced a mean field model and showed that this mean field model has a unique fixed point given by the steady state vector of a structured Markov chain. For the model considered in this paper a similar type of result can be established (albeit with more effort). However, due to the page limitations, we decided to directly present the structured Markov chain instead.

The main contributions of the paper are the following:

1. We introduce a Quasi-Birth-Death (QBD) Markov chain describing a single server queueing system with negative arrivals that is used to approximate the performance of the work stealing system. We present simulation results that suggest that as the number of servers becomes large, the approximation error tends to zero.
2. We prove that this QBD has a unique stationary distribution for which we provide formulas for the waiting, service, mean waiting and mean service time. These are the main technical results of the paper.
3. We compare the performance of several stealing strategies. Our main insight is that the strategy of stealing half of the child jobs performs well for low loads and/or high probe rates and stealing all child jobs is a good heuristic when the load is high and/or the probe rate is low.

The rest of this paper is organized as follows. In Sect. 2 we describe the system while the Quasi-Birth-Death (QBD) Markov chain is introduced in Sect. 3 and the response time distribution is analyzed in Sect. 4. In Sect. 5 we describe the work stealing strategies considered and present the performance of these strategies using numerical examples. Section 6 contains some concluding remarks and possible future work. The QBD approximation is validated using simulation in Appendix A.

2 System Description and Strategies

We consider a system with N homogeneous servers each with an infinite buffer to store jobs. Parent jobs arrive in each server according to a local Poisson arrival process with rate λ. Upon entering service a parent job spawns $i \in \{0, 1, \ldots, m\}$, with $m \geq 1$, child jobs, the number of which follows a general distribution with finite support p_i (i.e., $p_i > 0$ for every i and $\sum_{i=0}^{m} p_i = 1$). These child jobs are stored locally and have priority over any parent jobs (either already present or yet to arrive), while the spawning parent job continues service. Thus, when a (parent or child) job completes service the server first checks to see whether it has any waiting child jobs, if so it starts service on a child job. If there are no child jobs present, service on a waiting parent job starts (if any are present). We assume that parent and child jobs have exponentially distributed service requirements with rates μ_1 and μ_2 respectively.

When a server is idle, it probes other servers at random at rate $r > 0$, where r is a system parameter. Note that r determines the amount of communication between the servers and increasing r should improve performance at the expense of a higher communication overhead. When a server is probed (by an idle server) and it has waiting (parent or child) jobs, we state that the probe is successful. When a successful probe reaches a server without waiting child jobs, a parent job is transferred to the idle server. Note that such a transferred parent job starts service and spawns its child jobs at the new server.

When a successful probe reaches a server with pending/waiting child jobs, several child jobs can be transferred at once. If the probed server is serving a parent job and there are i child jobs in the buffer of the probed server, $j \leq i$ child jobs are stolen with probability $\phi_{i,j}$ (i.e., for every i we have $\sum_{j=1}^{i} \phi_{i,j} = 1$). On the other hand if a child job is being processed by the probed server and there are i child jobs waiting in the buffer of the probed server, $j \leq i$ child jobs are stolen with probability $\psi_{i,j}$ (i.e., for every i we have $\sum_{j=1}^{i} \psi_{i,j} = 1$). For ease of notation we set $\phi_{i,j} = \psi_{i,j} = 0$ if $j > i$. Probes and job transfers are assumed to be instantaneous.

The main objective of this paper is to study how the probabilities $\phi_{i,j}$ and $\psi_{i,j}$ influence the response time of a job, where the response time is defined as the time between the arrival of a parent job and the completion of the parent and all its spawned child jobs. Given the above description, it is clear that we get a Markov process if we keep track of the number of parent and child job in each of the N servers. This Markov process however does not appear to have a product form, making its analysis prohibitive.

Instead we use an approximation method, the accuracy of which is investigated in Appendix A. The idea of the approximation exists in focusing on a single server and assuming that the queue lengths at any other server are independent and identically distributed as in this particular server. Within the context of load balancing, this approach is known as the cavity method [4]. In fact all the analytical models used in [5,8,14–16,20–23] can be regarded as cavity method approximations. A common feature of such an approximation is that it tends to become more accurate as the number of servers tends to infinity, as we demonstrate in Appendix A for our model. The cavity method typically involves iterating the so-called cavity map [4]. However, in our case the need for such an iteration is avoided by deriving expressions for the rates at which child and parent jobs are stolen.

3 Quasi-Birth-Death Markov Chain

In this section we introduce a Quasi-Birth-Death (QBD) Markov chain to approximate the system from the viewpoint of a single server. Let $\lambda_p(r)$ denote the rate at which parent jobs are stolen when the server is idle. Let $\lambda_{c,1}(r), \ldots, \lambda_{c,m}(r)$ denote respectively the rates at which $1, \ldots, m$ child jobs are stolen. We provide formulas for these rates further on. The evolution of a single server has the following characteristics, where the negative arrivals correspond to steal events:

1. When the server is busy, arrivals of parent jobs occur according to a Poisson process with rate λ. When the server is idle, parent jobs arrive at the rate $\lambda + \lambda_p(r)$, while a batch of i child jobs arrives at rate $\lambda_{c,i}(r)$ for $i = 1, \ldots, m$.
2. Upon entering service, a parent job spawns $i \in \{0, 1, \ldots, m\}, m \geq 1$, child jobs with probability p_i. Child jobs are stored locally.
3. Child jobs have priority over any parent jobs *waiting* in the queue and are thus executed immediately after their parent job completes when executed on the same server.
4. Parent and child jobs have exponentially distributed service requirements with rates μ_1 and μ_2, respectively.
5. If there are parent jobs and no child jobs waiting in the buffer of the server then a negative parent arrival occurs at the rate rq, where $q = 1 - \rho$ is the probability that a queue is idle (where ρ is defined in (1)).
6. If a parent job is in service and there are $i \in \{1, \ldots, m\}$ child jobs in the buffer of the server, a batch of j negative child job arrivals occurs at the rate $rq\phi_{i,j}$, for all $j \in \{1, \ldots, i\}$.
7. If a child job is in service and there are $i \in \{1, \ldots, m-1\}$ child jobs pending in the buffer of the server, a batch of j negative child job arrivals occurs at the rate $rq\psi_{i,j}$, for all $j \in \{1, \ldots, i\}$.

Note that the load of the system can be expressed as

$$\rho = \lambda \left(\frac{1}{\mu_1} + \frac{\sum_{n=1}^{m} n p_n}{\mu_2} \right). \tag{1}$$

Table 1. Transitions for the QBD in Sect. 3

	From To	Rate	For
1	$(0,0,0) \to (0,j,0)$	$\lambda_{c,j}(r)$	$j = 1,\ldots,m,$
2	$(0,0,0) \to (0,j,1)$	$(\lambda + \lambda_p(r))p_j$	$j = 0,1,\ldots,m,$
3	$(X,Y,Z) \to (X+1,Y,Z)$	λ	$X+Y+Z \geq 1,$
4	$(X,Y,1) \to (X,Y,0)$	μ_1	$X \geq 0, Y \geq 1$ or $X = 0, Y = 0,$
5	$(X,Y,0) \to (X,Y-1,0)$	μ_2	$X \geq 0, Y \geq 2$ or $X = 0, Y = 1,$
6	$(X,1,0) \to (X-1,j,1)$	$\mu_2 p_j$	$X \geq 1, j = 0,1,\ldots,m,$
7	$(X,0,1) \to (X-1,j,1)$	$\mu_1 p_j$	$X \geq 1, j = 0,1,\ldots,m,$
8	$(X,Y,Z) \to (X-1,Y,Z)$	rq	$X \geq 1, Y+Z = 1,$
9	$(X,Y,1) \to (X,Y-j,1)$	$rq\phi_{Y,j}$	$X \geq 0, Y \geq j, j = 1,\ldots,m,$
10	$(X,Y,0) \to (X,Y-j,0)$	$rq\psi_{Y-1,j}$	$X \geq 0, Y \geq j+1, j = 1,\ldots,m-1$

Denote by $X \geq 0$ the number of parent jobs waiting, by $Y \in \{0,1,\ldots,m\}$ the number of child jobs in the server (either in service or waiting), and by $Z \in \{0,1\}$ whether a parent job is currently in service ($Z = 1$) or not ($Z = 0$). The possible transitions of the QBD Markov chain are listed in Table 1, corresponding to: 1. a batch of j child jobs arriving at an idle queue and the first child job proceeding directly into service, 2. a parent job arriving at an idle queue and proceeding directly into service, spawning j child jobs, 3. a parent arriving to a non-idle queue, 4. completion of a parent in service, not succeeded by another parent job, 5. child service completion, succeeded by either another child job or no job, 6. child service completion, succeeded by a parent job that enters service and spawns j child jobs, 7. parent service completion, succeeded by a parent job that enters service and spawns j child jobs, 8. negative parent job arrival, 9. a parent is in service and a batch of negative child job arrivals occurs, 10. a child job is in service and a batch of negative child job arrivals occurs.

The three dimensional process $\{X_t(r), Y_t(r), Z_t(r) : t \geq 0\}$ is an irreducible, aperiodic Quasi-Birth-Death process. We state that the *level* $\ell = *$ when the chain is in state $(0,0,0)$, while for any state with $X = \ell$ different from $(0,0,0)$, we state that the chain is in level ℓ (for $\ell \geq 0$). When the level $\ell \geq 0$, the *phase* of the QBD is two dimensional and given by (Y,Z). The $2m+1$ phases of level $\ell \geq 0$ are ordered such that the j-th phase corresponds to $(Y,Z) = (j,0)$, for $j = 1,\ldots,m$ and phase $m+1+j$ to $(Y,Z) = (j,1)$ for $j = 0,\ldots,m$.

As explained below, the generator of the process is

$$Q(r) = \begin{bmatrix} -\lambda_0(r) & \sum_{j=1}^{m} \lambda_{c,j}(r) e_j + (\lambda + \lambda_p(r))\alpha & & \\ \mu & B_0(r) & A_1 & \\ & A_{-1}(r) & A_0(r) & A_1 \\ & & \ddots & \ddots & \ddots \end{bmatrix}$$

with $\lambda_0(r) = \sum_{j=1}^{m} \lambda_{c,j}(r) + \lambda + \lambda_p(r)$, with e_j a row vector with 1 in its j-th entry and zeros elsewhere. The initial probability vector α records the distribution of child jobs upon a parent job entering service and is given by $\alpha = [0'_m \; p_0 \; p_1 \; \dots \; p_m]$, where 0_i is a column vector of zeros of length i. Indeed, at rate $\lambda_{c,j}(r)$ a batch of j child jobs arrives in an idle server, causing a jump to level 0 and phase j, while at rate $\lambda + \lambda_p(r)$ a parent job arrives that spawns j child jobs with probability p_j causing a jump to phase $m+1+j$ of level 0.

Define

$$S(r) = \begin{bmatrix} S_{00}(r) & 0 \\ S_{10} & S_{11}(r) \end{bmatrix},$$

where $S_{00}(r)$ is an $m \times m$ matrix and $S_{11}(r)$ is an $(m+1) \times (m+1)$ matrix,

$$S_{00}(r) = rq \begin{bmatrix} & & \\ \psi_{1,1} & & \\ \vdots & \ddots & \\ \psi_{m-1,m-1} & \dots & \psi_{m-1,1} & \end{bmatrix} + \begin{bmatrix} -\mu_2 & & & \\ \mu_2 & \ddots & & \\ & \ddots & \ddots & \\ & & \mu_2 & -\mu_2 \end{bmatrix},$$

$$S_{10} = \begin{bmatrix} 0 & \dots & \\ \mu_1 & & \\ & \mu_1 & \\ & & \ddots \end{bmatrix}, \quad S_{11}(r) = rq \begin{bmatrix} & & \\ \phi_{1,1} & & \\ \vdots & \ddots & \\ \phi_{m,m} & \dots & \phi_{m,1} & \end{bmatrix} + \begin{bmatrix} -\mu_1 & & \\ & \ddots & \\ & & \ddots \\ & & & -\mu_1 \end{bmatrix}.$$

The matrix $A_0(r)$ contains the possible transitions for which the level $\ell > 0$ remains unchanged, this is when child jobs are stolen, or when a waiting child moves into service. Hence

$$A_0(r) = S(r) - \lambda I - rqI.$$

Note that even when there are no child jobs waiting, the rate rq appears on the main diagonal due to the negative parent arrivals. When $\ell = 0$ there are no parent jobs waiting and therefore the negative parent arrivals that occur in phase 1 and $m+1$ have no impact. This implies that

$$\begin{aligned} B_0(r) &= A_0(r) + rqV_0 \\ &= S(r) - \lambda I - rq(I - V_0), \end{aligned}$$

where $V_0 = \text{diag}([1 \; 0'_{m-1} \; 1 \; 0'_m])$. The level ℓ can only decrease by one due to a service completion from a phase with no pending child jobs, that is, from phase 1 and $m+1$. To capture these events define $\mu = [\mu_2 \; 0'_{m-1} \; \mu_1 \; 0'_m]'$. The level can also decrease due to a negative parent arrival when $\ell > 0$. The matrix $A_{-1}(r)$ records the transitions for which the level decreases and therefore equals

$$A_{-1}(r) = \mu\alpha + rqV_0.$$

Finally, parent job arrivals always increase the level by one:

$$A_1 = \lambda I.$$

Denote by $A(r) = A_{-1}(r) + A_0(r) + A_1$, the generator of the phase process, then

$$A(r) = S(r) + \mu\alpha - rq(I - V_0).$$

Define

$$\pi_*(r) = \lim_{t\to\infty} P[X_t(r) = 0, Y_t(r) = 0, Z_t(r) = 0],$$

and for $\ell \geq 0$,

$$\pi_\ell(r) = (\pi_{\ell,1,0}(r), \dots \pi_{\ell,m,0}(r), \pi_{\ell,0,1}(r), \dots, \pi_{\ell,m,1}(r))$$

where

$$\pi_{\ell,j,k}(r) = \lim_{t\to\infty} P[X_t(r) = \ell, Y_t(r) = j, Z_t(r) = k].$$

Due to the QBD structure [17], we have

$$\pi_0(r) = \pi_*(r)R_0(r), \tag{2}$$

where $R_0(r)$ is a row vector of size $2m + 1$ and for $\ell \geq 1$,

$$\pi_\ell(r) = \pi_0(r)R(r)^\ell, \tag{3}$$

where $R(r)$ is a $(2m+1) \times (2m+1)$ matrix and by [11] the smallest nonnegative solution to

$$A_1 + R(r)A_0(r) + R(r)^2A_{-1}(r) = 0.$$

Also, due to the balance equations with $\ell = 0$, we have

$$\sum_{j=1}^{m} \lambda_{c,j}(r)e_j + (\lambda + \lambda_p(r))\alpha + R_0(r)B_0(r) + R_0(r)R(r)A_{-1}(r) = 0$$

and due to [11, Chapter 6]

$$A_1G(r) = R(r)A_{-1}(r),$$

where $G(r)$ is the smallest nonnegative solution to

$$A_{-1}(r) + A_0(r)G(r) + A_1G(r)^2 = 0.$$

Combining the above yields the following expression:

$$R_0(r) = -\left(\sum_{j=1}^{m} \lambda_{c,j}(r)e_j + (\lambda + \lambda_p(r))\alpha\right)(B_0(r) + \lambda IG(r))^{-1}, \tag{4}$$

where $B_0(r) + \lambda IG(r)$ is a subgenerator matrix and is therefore invertible. We note that $R(r)$ and $G(r)$ are independent of $\lambda_{c,1}(r), \dots, \lambda_{c,m}(r)$ and $\lambda_p(r)$ and can be computed easily using the toolbox presented in [1]. To fully characterize the QBD in terms of λ, μ_1, μ_2 and the probabilities $p_i, \phi_{i,j}$ and $\psi_{i,j}$, we need to specify $\lambda_{c,1}(r), \dots, \lambda_{c,m}(r)$ and $\lambda_p(r)$.

To determine these rates we use the following observation: as all parent and child jobs are executed on some server, $q = 1 - \rho$ should be the probability that the QBD is in state $(0, 0, 0)$. In this state batches of j child jobs arrive at rate $\lambda_{c,j}(r)$. Therefore $q\lambda_{c,j}(r)$ should equal the parent arrival rate λ times the expected number of times that a batch of j child jobs is stolen per parent job. The main difficulty in using this equality lies in the fact that we must also take into account that a child job can be stolen several times before it is executed.

To this end and as a preparation for Proposition 1, we define recursively $p_{0,i}(r)$, $i = 1, \ldots, m$, as the probability that the QBD visits phase $(i, 0)$ during the service of a job and similarly $p_{1,i'}(r)$, $i' = 0, \ldots, m$ that phase $(i', 1)$ is visited. By conditioning on whether we first have a service completion or steal event, we have

$$\begin{aligned} p_{1,m}(r) &= p_m, \\ p_{1,i}(r) &= p_i + \frac{rq}{rq + \mu_1} \sum_{j>i} p_{1,j}(r)\phi_{j,j-i}, \end{aligned}$$

for $i \in \{0, \ldots, m-1\}$, and

$$p_{0,i}(r) = \frac{\mu_1}{rq + \mu_1} p_{1,i}(r) + \frac{\mu_2}{rq + \mu_2} p_{0,i+1}(r) + \frac{rq}{rq + \mu_2} \sum_{j>i} p_{0,j}(r)\psi_{j-1,j-i},$$

for $i \in \{1, \ldots, m\}$, with $p_{0,m+1} = 0$. Note that

$$p_{1,0}(r) + p_{0,1}(r) = 1, \tag{5}$$

as phase $(1, 0)$ or $(0, 1)$ is visited before any job completes service.

We also define $p_i^j(r)$, for $1 \leq i \leq j \leq m$, as the probability that the QBD visits phase $(i, 0)$ given that it is in the phase $(j, 0)$ before a job completes service. We have

$$\begin{aligned} p_j^j(r) &= 1, \\ p_i^j(r) &= \frac{\mu_2}{rq + \mu_2} p_{i+1}^j(r) + \frac{rq}{rq + \mu_2} \sum_{k=i+1}^{j} \psi_{k-1,k-i} p_k^j(r), \end{aligned}$$

for $i \in \{1, \ldots, j-1\}$. Note that we have $p_1^j(r) = 1$, for $1 \leq j \leq m$, as the QBD visits phase $(0, 1)$ before completing service if it is in phase $(0, j)$. We are now in a position to define $\lambda_{c,i}(r)$ recursively as:

$$\begin{aligned} \lambda_{c,m}(r) &= \frac{\lambda}{q} p_{1,m}(r) \frac{rq}{rq + \mu_1} \phi_{m,m} \\ \lambda_{c,i}(r) &= \frac{\lambda}{q} \frac{rq}{rq + \mu_1} \sum_{j \geq i} p_{1,j}(r)\phi_{j,i} + \frac{\lambda}{q} \frac{rq}{rq + \mu_2} \sum_{j>i} p_{0,j}(r)\psi_{j-1,i} \\ &\quad + \sum_{j=i+1}^{m} \lambda_{c,j}(r) \sum_{k=i+1}^{j} p_k^j(r)\psi_{k-1,i} \frac{rq}{rq + \mu_2} \end{aligned} \tag{6}$$

for $i \in \{1,\ldots,m-1\}$. Note that $p_{1,m}(r)rq\phi_{m,m}/(rq+\mu_1)$ indeed equals the expected number of batches of size m that are stolen per parent job (as the job must spawn m child jobs and these must be stolen as a batch before the parent completes service). For $i < m$, the first two sums represent the expected number of size i batches that are stolen from the original server, while the double sum counts the expected number of such steals that occur on a server different from the original server.

It remains to define $\lambda_p(r)$, for this we demand that $\pi_*(r) = q$ and that

$$\pi_*(r) + \sum_{\ell\geq 0} \pi_\ell(r)e = 1,$$

where e is a column vector of ones. Then from Eqs. (2) and (3),

$$q\Big(1 + R_0(r)(I - R(r))^{-1}e\Big) = 1, \tag{7}$$

where the inverse of $I - R(r)$ exists due to Proposition 1. Using (4) and (7) we get:

$$\lambda_p(r) = \frac{(1-q) - q(\sum_{j=1}^{m} \lambda_{c,j}(r)e_j + \lambda\alpha)w}{q\alpha w}, \tag{8}$$

with $w = -(B_0(r) + \lambda I G(r))^{-1}(I - R(r))^{-1}e$. Note that $\lambda_p(r)$ is well-defined for $q > 0$, i.e. $\rho < 1$. This completes the description of the QBD Markov chain.

Proposition 1. *The QBD process* $\{X_t(r), Y_t(r), Z_t(r) : t \geq 0\}$ *has a unique stationary distribution for any* $r \geq 0$ *if* $\rho < 1$.

Proof. The positive recurrence of the QBD process only depends on the matrices $A_{-1}(r)$, $A_0(r)$ and A_1 [17]. These three matrices are the same three matrices as those of the QBD characterizing the M/MAP/1 queue where the MAP service process is characterized by $(S_0(r), S_1(r))$ with $S_0(r) = S(r) - rqI$ and $S_1(r) = \mu\alpha + rqV_0$. As such the QBD process is positive recurrent if and only if the arrival rate λ is less than the service completion intensity of the MAP $(S_0(r), S_1(r))$. This intensity equals $\theta^{(r)}S_1(r)e/\theta^{(r)}e$, where the vector $\theta^{(r)}$ is such that $\theta^{(r)}(S_0(r) + S_1(r)) = 0$.

We note that $S_0(r) + S_1(r) = A_{-1}(r) + A_0(r) + A_1 = A(r)$ and define

$$\theta^{(r)}_{(0,1)} = \frac{1}{\mu_2} p_{0,1}(r),$$

$$\theta^{(r)}_{(0,i')} = \frac{1}{rq+\mu_2} p_{0,i'}(r),$$

$$\theta^{(r)}_{(1,0)} = \frac{1}{\mu_1} p_{1,0}(r),$$

$$\theta^{(r)}_{(1,i)} = \frac{1}{rq+\mu_1} p_{1,i}(r),$$

for $i' = 2, \ldots, m$ and for $i = 1, \ldots, m$. Define $v^{(r)} = \theta^{(r)} A(r)$. Then, using (5),

$$v_i^{(r)} = p_{i-m-1} - p_{1,i-m-1}(r) + \frac{rq}{rq+\mu_1} \sum_{j>i-m-1} p_{1,j}(r)\phi_{j,j-i-m-1} = 0,$$

for $i = m+1, \ldots, 2m+1$, and

$$\begin{aligned} v_{i'}^{(r)} &= -p_{0,i'}(r) + 1[i<m]\frac{\mu_2}{rq+\mu_2} p_{0,i'+1}(r) \\ &+ \frac{rq}{rq+\mu_2}\sum_{j>i} p_{0,j}(r)\psi_{j-1,j-i'} + \frac{\mu_1}{rq+\mu_1} p_{1,i'}(r) = 0, \end{aligned}$$

for $i' = 1, \ldots, m$. Hence $\theta^{(r)} A(r) = \theta^{(r)}(S_0(r) + S_1(r)) = 0$. As

$$\begin{aligned} \frac{\theta^{(r)} S_1(r) e}{\theta^{(r)} e} &= \frac{1}{\theta^{(r)} e}\left(\frac{\mu_2 + rq}{\mu_2} p_{0,1}(r) + \frac{\mu_1 + rq}{\mu_1} p_{1,0}(r)\right) \\ &\geq \frac{1}{\theta^{(r)} e}(p_{0,1}(r) + p_{1,0}(r)) = \frac{1}{\theta^{(r)} e}, \end{aligned}$$

it suffices that $\lambda < 1/\theta^{(r)} e$ for the chain to be positive recurrent. For $r = 0$ we have $p_{1,i}(r) = p_i$ and $p_{0,i'} = \sum_{j \geq i'} p_j$, which implies that $\theta^{(0)} e = \rho/\lambda$. Therefore $\lambda < 1/\theta^{(0)} e$ is equivalent to demanding that $\rho < 1$. As $\theta^{(r)} e$ is the mean time between two service completions of the MAP process where the state is reset according to the vector α, we have that $\theta^{(r)} e$ decreases in r. This completes the proof as $\rho < 1$ implies that $\lambda < 1/\theta^{(0)} e \leq 1/\theta^{(r)} e$.

4 Response Time Distribution

We define $T(r)$ as the response time of a job in a system with probe rate r. The response time is defined as the length of the time interval between the arrival of a parent job and the completion of this parent job and all of its spawned child jobs. $T(r)$ can be expressed as the sum of the waiting time $W(r)$ and the service time $J(r)$. The waiting time is defined as the amount of time that the parent job waits in the queue before its service starts. Clearly, the waiting and the service time of a job are independent in our QBD model.

Theorem 1. *The distribution of the waiting time is given by*

$$P[W(r) > t] = (e' \otimes \pi_0 (I - R(r))^{-1}) e^{\mathbb{W}t} vec\langle I \rangle$$

with $\mathbb{W} = ((A_0(r) + A_1)' \otimes I) + ((A_{-1}(r))' \otimes R(r))$ *and where* $vec\langle\cdot\rangle$ *is the column stacking operator. The mean waiting time is*

$$E[W(r)] = \int_0^\infty P[W(r) > t]\, dt = (e' \otimes \pi_0 (I - R(r))^{-1})(-\mathbb{W})^{-1} vec\langle I \rangle.$$

Proof. We repeat the arguments of the proof of [20, Theorem 6.1]. Let $(N(k,t))_{j,j'}$ be the probability that there are exactly k transitions that decrease the level by one in $(0,t)$ and the phase at time t equals j' given that the level never decreased below 1 and the phase was j at time 0. Due to the PASTA property we have

$$P[W(r) > t] = \sum_{n=1}^{\infty} \pi_{n-1} \sum_{k=0}^{n-1} N(k,t)e,$$

as $(\pi_{n-1})_j$ is the probability that a tagged parent job is the nth parent job waiting in the queue immediately after it arrived and the service phase equals j. In such case there can be at most $n-1$ events that decrease the level otherwise $W(r) < t$. Thus,

$$\begin{aligned} P[W(r) > t] &= \sum_{k=0}^{\infty} \pi_0 \sum_{n=k+1}^{\infty} R(r)^{n-1} N(k,t)e \\ &= \pi_0 (I - R(r))^{-1} \sum_{k=0}^{\infty} R(r)^k N(k,t)e. \end{aligned}$$

Using the same arguments as in [18] or [10] one finds that

$$vec\left\langle \sum_{k=0}^{\infty} R(r)^k N(k,t) \right\rangle = e^{\mathbb{W}t} vec\left\langle I \right\rangle.$$

The proof is completed by noting that $vec\langle ABC\rangle = (C' \otimes A) vec\langle B\rangle$.

The service time distribution $J(r)$ is more difficult to compute compared to the model in [20]. This is due to the fact that child jobs can be stolen multiple times before finally going into service.

We define $J_{0,k}(r)$ as the distribution of the time that it takes for k child jobs in a server to be completed ($k = 1, \ldots, m$). Similarly, we define $J_{1,k}(r)$ as the distribution of the time that it takes for a parent job and k child jobs in a server to be completed ($k = 0, \ldots, m$). The service time distribution can then be expressed as

$$P[J(r) \le t] = \sum_{k=0}^{m} p_k P[J_{1,k}(r) \le t].$$

Clearly, $P[J_{0,1}(r) \le t] = 1 - e^{-\mu_2 t}$ and $P[J_{1,0}(r) \le t] = 1 - e^{-\mu_1 t}$. For $k > 1$, we can condition on the first service completion or steal event to find that

$$\begin{aligned} P[J_{0,k}(r) \le t] = \int_0^t \Bigg(& rq \sum_{j=1}^{k-1} \psi_{k-1,j} P[J_{0,k-j}(r) \le t-s] P[J_{0,j}(r) \le t-s] \\ & + \mu_2 P[J_{0,k-1}(r) \le t-s] \Bigg) e^{-(rq+\mu_2)s} ds, \end{aligned} \tag{9}$$

and for $k > 0$ this yields

$$P[J_{1,k}(r) \le t] = \int_0^t \left(rq \sum_{j=1}^{k} \phi_{k,j} P[J_{1,k-j}(r) \le t-s] P[J_{0,j}(r) \le t-s] \right.$$
$$\left. + \mu_1 P[J_{0,k}(r) \le t-s] \right) e^{-(rq+\mu_1)s} ds. \tag{10}$$

While the above formulas recursively determine the service time, they are less suited for numerical computations, we therefore also develop a recursive scheme for the mean service time.

Consider a set of s servers, where the k-th server contains i_k child jobs, where $s \ge 1$, $0 \le i_1 + \cdots + i_s \le m$ and $i_k \ge 0$ for $k = 1, \ldots, s$. Let $E_{i_1,\ldots,i_s}(r)$ be the expected time until all these child jobs have completed service. Define similarly $E^p_{i_1,\ldots,i_s}(r)$, except that the first server contains i_1 child jobs and a parent job (that is in service).

By definition, we can drop i_k's that are zero (expect i_1 in $E^p_{i_1,\ldots,i_s}(r)$) and can permute the indices of $E_{i_1,\ldots,i_s}(r)$ and all indices except the first one of $E^p_{i_1,\ldots,i_s}(r)$. We have for $s \ge 1$

$$E_{1'_s}(r) = \frac{1}{\mu_2} \sum_{k=1}^{s} \frac{1}{k}. \tag{11}$$

We now define recursively, assuming $i_k \ge 1$ for $k = 1, \ldots, s$:

$$E_{i_1,\ldots,i_s}(r) = \frac{1}{s\mu_2 + rq \sum_{k=1}^{s} 1[i_k \ge 2]} \left(1 + \mu_2 \sum_{k=1}^{s} E_{i_1,\ldots,i_{k-1},i_k-1,i_{k+1},\ldots,i_s}(r) \right.$$
$$\left. + rq \sum_{k=1}^{s} \sum_{n=1}^{i_k-1} \psi_{i_k-1,n} E_{i_1,\ldots,i_{k-1},i_k-n,i_{k+1},\ldots,i_s,n}(r) \right).$$

We have $E^p_0(r) = 1/\mu_1$ and we define recursively for $s \ge 1$

$$E^p_{1'_s}(r) = \frac{1}{\mu_1 + (s-1)\mu_2} \left(1 + \mu_1 E_{1'_s}(r) + (s-1)\mu_2 E^p_{1'_{s-1}}(r) \right)$$
$$= \frac{1}{\mu_1 + (s-1)\mu_2} \left(1 + \frac{\mu_1}{\mu_2} \sum_{k=1}^{s-1} \frac{1}{k} + (s-1)\mu_2 E^p_{1'_{s-1}}(r) \right),$$

where we have used (11) in the last equality. Finally, we define recursively, assuming $i_k \ge 1$ for $k = 2, \ldots, s$:

$$E^p_{i_1,\ldots,i_s}(r) = \frac{1}{\mu_1 + (s-1)\mu_2 + rq1[i_1 \ge 1] + rq \sum_{k=2}^{s} 1[i_k \ge 2]} \left(1 \right.$$
$$+ \mu_1 E_{i_1,\ldots,i_s}(r) + rq \sum_{n=1}^{i_1} \phi_{i_1,n} E^p_{i_1-n,i_2,\ldots,i_s,n}(r)$$

$$+ \mu_2 \sum_{k=2}^{s} E^p_{i_1,\ldots,i_{k-1},i_k-1,i_{k+1},\ldots,i_s}(r)$$
$$+ rq \sum_{k=2}^{s} \sum_{n=1}^{i_k-1} \psi_{i_k-1,n} E^p_{i_1,\ldots,i_{k-1},i_k-n,i_{k+1},\ldots,i_s,n}(r) \Bigg).$$

The expectation $E[J(r)]$ can now be computed as:

$$E[J(r)] = \sum_{k=0}^{m} p_k E^p_k(r).$$

Note that when $r \to \infty$, children spawned by a parent job get immediately distributed amongst empty servers. Therefore, as $r \to \infty$, we get

$$E[J(r)] \to \sum_{k=0}^{m} p_k E^p_{0,1'_k}(r). \tag{12}$$

5 Numerical Experiments

In this section we perform numerical experiments to compare the performance of several stealing strategies. Duc to the lack of space, we present only a subset of the experiments performed. The main conclusions in these additional experiments (e.g., different μ_2 values) are in agreement with the results presented. Define Ψ as the matrix where $[\Psi]_{i,j} = \psi_{i,j}$ and define Φ similarly. Note that a strategy is fully characterized by Ψ and Φ. The strategies considered are as follows:

1. **Steal one:** The strategy of always stealing one child job, that is $\phi_{i,1} = \psi_{i,1} = 1$ for every i.
2. **Steal half:** The strategy of always stealing half of the pending child jobs. If n, the number of pending child jobs, is uneven, there is a fifty percent chance that $\lfloor n/2 \rfloor$ child jobs get stolen and $\lceil n/2 \rceil$ jobs otherwise;
3. **Steal all:** The strategy of stealing all of the pending child jobs, that is $\phi_{i,i} = \psi_{i,i} = 1$ for every i.

Note that these strategies do not rely on any knowledge on the (mean) job sizes or system load.

We compare the mean response time for these strategies with the optimal monotone deterministic strategy. A strategy is called deterministic if for every $i \leq m$ there exists a $j \leq i$ such that $\phi_{i,j} = 1$ and a $k \leq i$ such that $\psi_{i,k} = 1$. It is called monotone deterministic (MD) if in addition having $\psi_{i,j} = 1$ and $\psi_{i',j'} = 1$ with $i < i'$ implies that $j \leq j'$ for all i, j, i', j' and the same holds for Φ. Experiments not included in the paper suggest that the optimal strategy, that is, the optimal Ψ and Φ matrices, corresponds to an MD strategy. The optimal MD strategy is determined using brute-force and its mean response time is denoted as $T_{MD}(r)$. Let $\mathbf{p} = [p_0, p_1, \ldots, p_m]$.

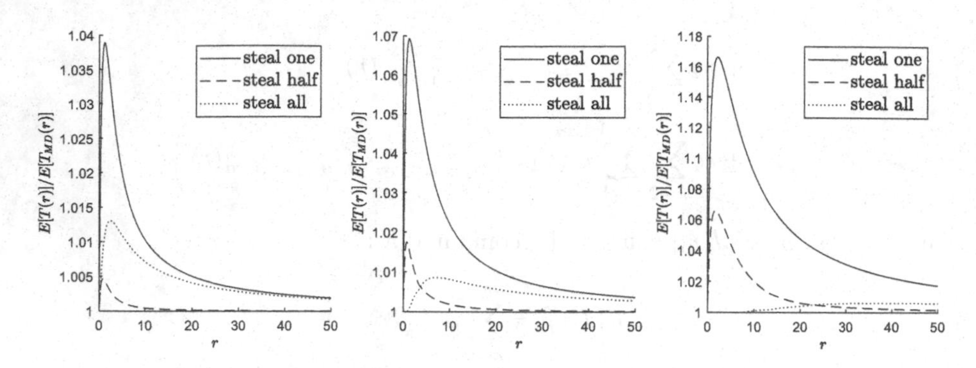

Fig. 1. Example 1 with $\rho = 0.15$ (left), $\rho = 0.5$ (mid) and $\rho = 0.85$ (right).

Example 1. In Fig. 1 we examine the effect of increasing the steal rate on how well the three strategies perform compared to the optimal MD strategy. More specifically, we plot the mean response time $E[T(r)]$ of our three policies normalized by the mean response time $E[T_{MD}(r)]$ of the optimal MD policy. We do this for $\rho \in \{0.15, 0.5, 0.85\}, \mu_1 = 1, \mu_2 = 2, \mathbf{p} = \mathbf{1}_5'/5$ and $r \in [0.05, 50]$. We note that there exists no universal best strategy. The strategy of stealing one job performs the worst. This is due to the fact that relatively very little work of the pending jobs is transferred. When $\mu_2 < \mu_1$, examples can be constructed where the strategy of stealing a single child outperforms the others. For moderately high values of r or for low loads the strategy where half of the child jobs get stolen is close to the optimal MD strategy. This is intuitively clear as in such case there is a small chance that there are pending parent jobs in a queue, so stealing half of the child jobs more or less balances the work. In fact, it seems that as r becomes large enough the optimal strategy for systems where $\mu_1 \leq \mu_2$ is stealing $\lfloor i/2 \rfloor + 1$ out of i children. For low values of r the strategy of stealing all child jobs performs well, as there is a fair chance that there are pending parents in the queue and it can take a long time until the server is probed again.

For $\rho = 0.85$ the matrices Ψ, Φ of the optimal MD strategy change as follows: for low values of r the best strategy is the one of stealing all jobs, that is Ψ and Φ are identity matrices of size $m-1$ and m respectively. Then, at approximately $r = 7.6$, $\psi_{3,2}$ becomes one. Around $r = 13.5$, the $\phi_{4,3}$ becomes one and finally $\phi_{3,2} = 1$ around $r = 20.35$. For $\rho = 0.5$ we see a similar evolution: for low values of r the best strategy is stealing all child jobs. Then, at approximately $r = 0.85$, $\psi_{3,2}$ becomes one. Around $r = 1.55$, the $\phi_{4,3}$ becomes one and finally $\phi_{3,2} = 1$ around $r = 3.35$.

The number of MD strategies grows quickly in function of m (in fact one can prove that for a given m there exist $C(m)C(m+1)$ such strategies, where $C(k)$ denotes the k-th Catalan number). This implies that it can take a long time to determine the optimal MD strategy for systems with larger m values.

We therefore introduce a smaller family of strategies and compare our three strategies with the optimal strategy in this smaller family to limit the brute-force search. We call a strategy bounded monotone deterministic (BMD) if it is monotone deterministic and $\psi_{i,j} = 1$ implies $\psi_{i+1,j} = 1$ or $\psi_{i+1,j+1} = 1$ for every i and the same holds for Φ. Note that there are $2^{m-2}2^{m-1} = 2^{2m-3}$ BMD strategies for a given $m \geq 2$. The optimal BMD strategy is determined using brute-force and we denote its mean response time by $T_{BMD}(r)$. The mean response time of the optimal BMD strategy may exceed that of the optimal MD strategy as indicated in the next example.

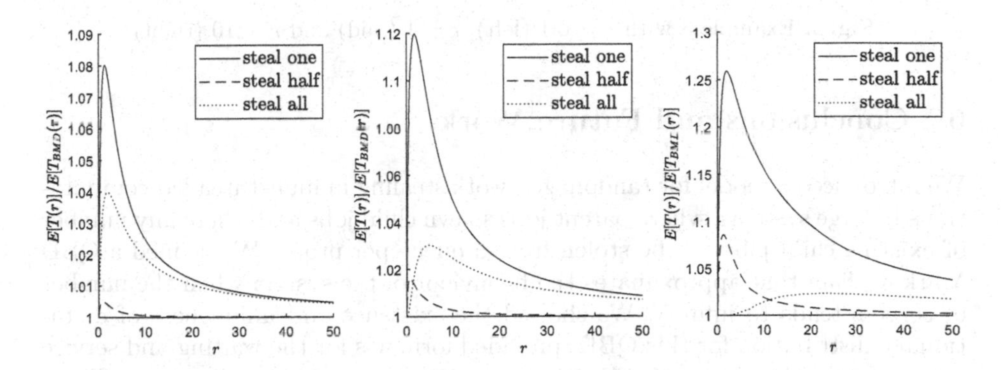

Fig. 2. Example 2 with $\rho = 0.15$ (left), $\rho = 0.5$ (mid) and $\rho = 0.85$ (right).

Example 2. In Fig. 2 we examine the effect of increasing the steal rate on how well the three strategies perform compared to the optimal BMD strategy when $m = 6$ instead of $m = 4$ as in the previous example. We do this for $\rho \in \{0.15, 0.5, 0.85\}, \mu_1 = 1, \mu_2 = 2, \mathbf{p} = \mathbf{1}'_7/7$ and $r \in [0.05, 50]$. It is clear that the main insights are similar as in the $m = 4$ case, except that more substantial gains can be achieved by optimizing Ψ and Φ. We also performed some experiments to compare the performance of the optimal MD and BMD strategies and noted that for $r \in [6.9, 7.4]$, the optimal MD strategy has $\psi_{3,2} = 1$ and $\psi_{4,4} = 1$, which is not BMD. The reduction in the mean response time was however very limited.

Example 3. In Fig. 3 we illustrate the effect of increasing the load ρ on the mean response time. We do this for $\rho \in [0.05, 0.95], \mu_1 = 1, \mu_2 = 2, m = 4, \mathbf{p} = \mathbf{1}'_5/5$ and $r \in \{0.1, 1, 10\}$. These result confirm that stealing all is best when the load is sufficiently high, while stealing half of the child jobs is good for systems with a limited load.

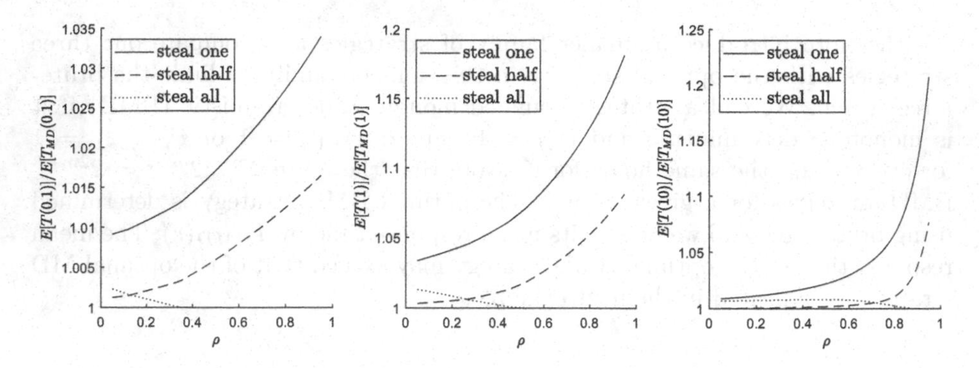

Fig. 3. Example 3 with $r = 0.1$ (left), $r = 1$ (mid) and $r = 10$ (right).

6 Conclusions and Future Work

We introduced a model for randomized work stealing in multithreaded computations in large systems, where parent jobs spawn child jobs and where any number of existing child jobs can be stolen from a queue per probe. We defined a QBD Markov chain that approximates the behaviour of the system when the number of servers tends to infinity. We showed the existence and uniqueness of a stationary distribution for this QBD, provided formulas for the waiting and service times and provided a practical way of calculating expected service times. These are the main technical contributions of the paper. Using numerical experiments we examined the effect of changing the load ρ and the steal rate r. We concluded that the stealing policy where the half of child jobs gets stolen every time is in general a good stealing policy for higher values of r, while the strategy of stealing all children performs best for low values of r. We concluded further that stealing only one child performs the worst in most of the cases. Finally, using simulation, we validated the accuracy of the QBD model.

Possible generalizations include stealing multiple parent jobs (up to some finite amount) per probe and systems where offspring of a job can spawn further offspring (multigenerational multithreading). One can also attempt to relax the exponential service time requirements for child and/or parent jobs. This may be challenging as this complicates several aspects of the model such as determining the rates $\lambda_{c,j}(r)$.

A Model Validation

Based on numerical experiments in the Sect. 5, we see that stealing all or half of the children are good stealing policies: stealing all works best for low values of r, while stealing half of the children works well for higher values. Therefore, we validate the mean field model for the policy of stealing all or half of the children. We always start the simulations from an empty system and simulate the behaviour for $T = 10^5$ with a warm up period of 33% of T.

In Fig. 4 we focus on the case where all children are stolen. The 95% confidence intervals were computed based on 5 runs with $N = 500$ servers, $m = 4$, $\mu_1 = 1, \mu_2 = 2, \rho = 0.75$, $\mathbf{p} = (1, 1, 1, 1, 1)/5$ and $r \in \{1, 5\}$. We see that there is an excellent match between the simulated waiting and service times and those of the QBD model (calculated using Sect. 4).

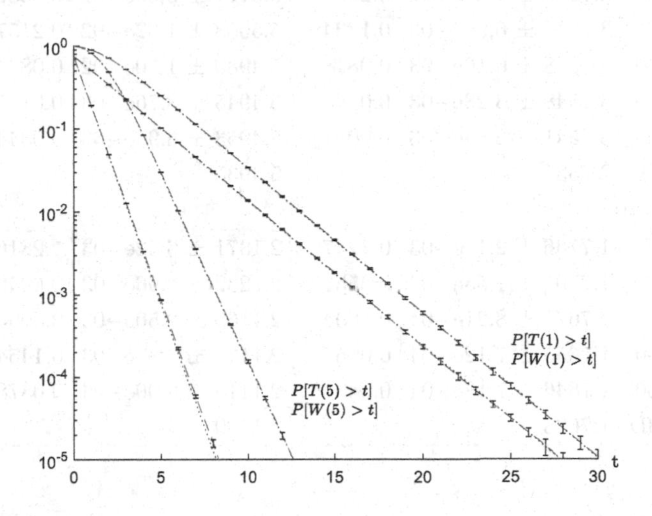

Fig. 4. Waiting and response times from the QBD (blue dots) and simulations (red dashed line) with confidence intervals for 5 runs. (Color figure online)

In Table 2 we compare the relative error of the simulated mean response time, based on 20 runs, to the one obtained from Section 4. We do this for $\mu_1 = 1, \mu_2 = 2$, $\mathbf{p} = (1, 1, 1, 1, 1)/5$, $\rho \in \{0.75, 0.85\}$, $r \in \{1, 10\}$ and $N \in \{250, 500, 1000, 2000, 4000\}$.
The relative error in all cases is below 1.5% and tends to increase with the steal rate r. Further, the relative error seems roughly to halve when doubling N, which is in agreement with the results in [7].

Next we validate the model for the strategy of stealing half of the children using the same simulation settings. In Fig. 5, we see that there is an excellent match between the simulated waiting and service times and those of the QBD model. Similarly to Table 2, we see in Table 3 that the relative error is below 1.5% in all cases, tends to increase with the steal rate r and seems about halved when doubling N.

Table 2. Relative error of simulation results for $E[T(r)]$, based on 20 runs

N	$\rho = 0.75$		$\rho = 0.85$	
	sim. ± conf.	rel.err.%	sim. ± conf.	rel.err.%
$r = 1$				
250	3.7650 ± 1.08e−02	0.2986	5.5121 ± 3.08e−02	0.3386
500	3.7588 ± 6.98e−03	0.1334	5.5053 ± 1.62e−02	0.2157
1000	3.7568 ± 6.16e−03	0.0818	5.4980 ± 1.60e−02	0.0821
2000	3.7548 ± 3.28e−03	0.0283	5.4945 ± 8.76e−03	0.0197
4000	3.7541 ± 2.53e−03	0.0091	5.4953 ± 6.96e−03	0.0344
QBD	3.7537		5.4935	
$r = 10$				
250	1.7766 ± 2.11e−03	0.7247	2.1371 ± 6.32e−03	1.2816
500	1.7701 ± 1.53e−03	0.3553	2.1232 ± 3.96e−03	0.6249
1000	1.7671 ± 8.21e−04	0.1894	2.1165 ± 2.50e−03	0.3090
2000	1.7655 ± 7.12e−04	0.0957	2.1131 ± 1.88e−03	0.1454
4000	1.7646 ± 7.18e−04	0.0437	2.1119 ± 8.00e−04	0.0878
QBD	1.7638		2.1100	

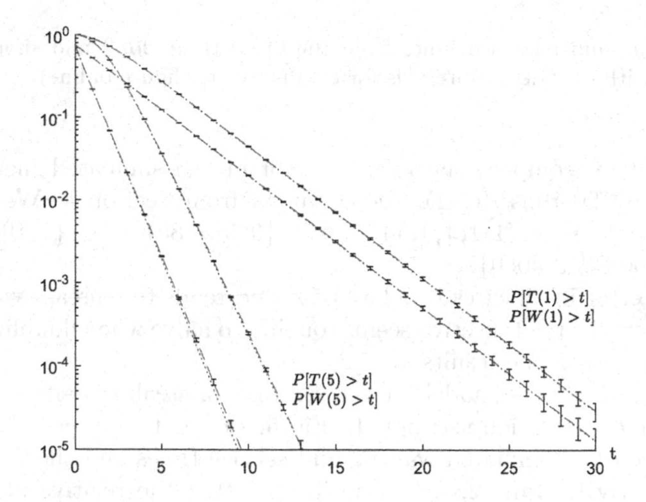

Fig. 5. Waiting and response times from the QBD (blue dots) and simulations (red dashed line) with confidence intervals for 5 runs. (Color figure online)

Table 3. Relative error of simulation results for $E[T(r)]$, based on 20 runs

N	$\rho = 0.75$		$\rho = 0.85$	
	sim. ± conf.	rel.err.%	sim. ± conf.	rel.err.%
$r = 1$				
250	3.9305 ± 1.45e−02	0.2392	5.8435 ± 2.91e−02	0.2830
500	3.9261 ± 1.26e−02	0.1271	5.8331 ± 1.68e−02	0.1045
1000	3.9231 ± 5.55e−03	0.0506	5.8288 ± 1.04e−02	0.0307
2000	3.9225 ± 4.63e−03	0.0353	5.8281 ± 1.33e−02	0.0187
4000	3.9219 ± 2.71e−03	0.0200	5.8279 ± 9.34e−03	0.0153
QBD	3.9211		5.8270	
$r = 10$				
250	1.7822 ± 2.34e−03	0.7748	2.1782 ± 5.92e−03	1.3017
500	1.7752 ± 2.00e−03	0.3790	2.1642 ± 3.21e−03	0.6506
1000	1.7720 ± 1.25e−03	0.1965	2.1576 ± 2.82e−03	0.3437
2000	1.7703 ± 8.84e−04	0.1007	2.1537 ± 1.82e−03	0.1623
4000	1.7695 ± 3.83e−04	0.0567	2.1520 ± 1.65e−03	0.0832
QBD	1.7685		2.1502	

References

1. Bini, D., Meini, B., Steffé, S., Van Houdt, B.: Structured Markov chains solver: software tools. In: Proceeding From the 2006 Workshop on Tools for Solving Structured Markov Chains, pp. 1–14 (2006)
2. Blumofe, R., Leiserson, C.: Scheduling multithreaded computations by work stealing. J. ACM (JACM) **46**(5), 720–748 (1999)
3. Blumofe, R., Joerg, C., Kuszmaul, B., Leiserson, C., Randall, K., Zhou, Y.: Cilk: an efficient multithreaded runtime system. J. Parallel Distrib. Comput. **37**(1), 55–69 (1996)
4. Bramson, M., Lu, Y., Prabhakar, B.: Randomized load balancing with general service time distributions. In: ACM SIGMETRICS 2010, pp. 275–286 (2010). https://doi.org/10.1145/1811039.1811071, http://doi.acm.org/10.1145/1811039.1811071
5. Eager, D., Lazowska, E., Zahorjan, J.: A comparison of receiver-initiated and sender-initiated adaptive load sharing. Perform. Eval. **6**(1), 53–68 (1986)
6. Frigo, M., Leiserson, C.E., Randall, K.H.: The implementation of the cilk-5 multithreaded language. In: Proceedings of the SIGPLAN 1998 Conference on Program Language Design and Implementation, pp. 212–223 (1998)
7. Gast, N.: Expected values estimated via mean-field approximation are 1/n-accurate. In: Proceedings of the ACM on Measurement and Analysis of Computing Systems, vol. 1, no. 1, p. 17 (2017)
8. Gast, N., Gaujal, B.: A mean field model of work stealing in large-scale systems. ACM SIGMETRICS Perform. Eval. Rev. **38**(1), 13–24 (2010)
9. Gautier, T., Besseron, X., Pigeon, L.: Kaapi: A thread scheduling runtime system for data flow computations on cluster of multi-processors. In: Proceedings of the 2007 International Workshop on Parallel Symbolic Computation, pp. 15–23 (2007)

10. Horváth, G., Van Houdt, B., Telek, M.: Commuting matrices in the queue length and sojourn time analysis of map/map/1 queues. Stoch. Model. **30**(4), 554–575 (2014)
11. Latouche, G., Ramaswami, V.: Introduction to matrix analytic methods in stochastic modeling, vol. 5. SIAM (1999)
12. Lea, D.: A java fork/join framework. In: Proceedings of the ACM 2000 Conference on Java Grande, JAVA 2000, New York, NY, USA, pp. 36–43. Association for Computing Machinery (2000). https://doi.org/10.1145/337449.337465
13. Leijen, D., Schulte, W., Burckhardt, S.: The design of a task parallel library. In: Proceedings of the 24th ACM SIGPLAN Conference on Object Oriented Programming Systems Languages and Applications. OOPSLA 2009, New York, NY, USA, pp. 227–242. Association for Computing Machinery (2009). https://doi.org/10.1145/1640089.1640106
14. Minnebo, W., Hellemans, T., Van Houdt, B.: On a class of push and pull strategies with single migrations and limited probe rate. Perform. Eval. **113**, 42–67 (2017)
15. Minnebo, W., Van Houdt, B.: A fair comparison of pull and push strategies in large distributed networks. IEEE/ACM Trans. Networking (TON) **22**(3), 996–1006 (2014)
16. Mirchandaney, R., Towsley, D., Stankovic, J.: Adaptive load sharing in heterogeneous distributed systems. J. Parallel Distrib. Comput. **9**(4), 331–346 (1990)
17. Neuts, M.: Matrix-Geometric Solutions in Stochastic Models: An Algorithmic Approach. John Hopkins University Press, Baltimore (1981)
18. Ozawa, T.: Sojourn time distributions in the queue defined by a general QBD process. Queue. Syst. Appl. **53**(4), 203–211 (2006)
19. Robison, A., Voss, M., Kukanov, A.: Optimization via reflection on work stealing in TBB. In: 2008 IEEE International Symposium on Parallel and Distributed Processing, pp. 1–8. IEEE (2008)
20. Sonenberg, B., Kielanski, G., Van Houdt, B.: Performance analysis of work stealing in large scale multithreaded computing. ACM ToMPECS (2021, to appear)
21. Spilbeeck, I.V., Houdt, B.V.: Performance of rate-based pull and push strategies in heterogeneous networks. Perform. Eval. **91**, 2–15 (2015)
22. Squillante, M., Nelson, R.: Analysis of task migration in shared-memory multiprocessor scheduling. SIGMETRICS Perform. Eval. Rev. **19**(1), 143–155 (1991). http://doi.acm.org/10.1145/107972.107987
23. Van Houdt, B.: Randomized work stealing versus sharing in large-scale systems with non-exponential job sizes. IEEE/ACM Trans. Networking **27**, 2137–2149 (2019)
24. Wirth, N.: Tasks versus threads: an alternative multiprocessing paradigm. Software Concepts Tools **17**, 6–12 (1996)

Abstractions and Aggregations

Abstraction-Guided Truncations for Stationary Distributions of Markov Population Models

Michael Backenköhler[1(✉)], Luca Bortolussi[2,3], Gerrit Großmann[1], and Verena Wolf[1,3]

[1] Saarbrücken Graduate School of Computer Science, Saarland University, Saarland Informatics Campus E1 3, Saarbrücken, Germany
michael.backenkoehler@uni-saarland.de
[2] University of Trieste, Trieste, Italy
[3] Saarland University, Saarland Informatics Campus E1 3, Saarbrücken, Germany

Abstract. To understand the long-run behavior of Markov population models, the computation of the stationary distribution is often a crucial part. We propose a truncation-based approximation that employs a state-space lumping scheme, aggregating states in a grid structure. The resulting approximate stationary distribution is used to iteratively refine relevant and truncate irrelevant parts of the state-space. This way, the algorithm learns a well-justified finite-state projection tailored to the stationary behavior. We demonstrate the method's applicability to a wide range of non-linear problems with complex stationary behaviors.

Keywords: Long-run behavior · State-space aggregation · Lumping · Truncation

1 Introduction

In many areas of science, stochastic models of interacting populations can describe systems in which the discrete population sizes evolve stochastically in continuous time. Such problems naturally occur in a wide range of areas such as chemistry [16], systems biology [42,45], epidemiology [35] as well as queuing systems [9] and finance [37].

Interactions between agents, commonly referred to as *reactions*, happen at exponentially distributed random times. Their rate depends on the current system state, i.e. the population sizes. This results in a continuous-time Markov chain semantics [4]. An important part of the analysis of such models concerns their long-run behavior. Given an ergodic underlying Markov chain, the chain's stationary distribution characterizes this behavior. For some special model classes, such as zero-deficiency networks [3], analytical solutions for the stationary distribution are known. However, most models require numerical approaches, often based on some form of approximation to guarantee tractability.

A. Abate and A. Marin (Eds.): QEST 2021, LNCS 12846, pp. 351–371, 2021.
https://doi.org/10.1007/978-3-030-85172-9_19

Those approaches can be based on stochastic simulation [16] (which for steady-state analysis tends to be slow and inaccurate) or moment-bounds via mathematical programming [23]. Here, we draw on numerical approaches based on state-space truncation, which represent a viable option to approximate stationary distributions [24]. Truncation-based approaches have the benefit of describing the complete dynamics within a finite subset of the typically very large or infinite state-space. As such, they enable the approximation of complex distributions that are not well-described by low-order moments.

The main step in the computation of such an approximation is the identification of a suitable truncation, i.e. a subset of the state-space encompassing most of the stationary probability mass. Existing methods typically rely on Foster-Lyapunov drift conditions to define such subsets [12]. While these truncations come with bounds on the contained stationary probability mass, they typically are far larger than necessary. The truncation is usually strongly constrained by the form of the chosen Lyapunov function [12,17]. Optimizing over possible functions to identify efficient truncations is technically challenging and, to our knowledge, has not been demonstrated for general reaction networks [34].

In this work, we address the identification of suitable truncations by using an aggregation-refinement scheme. Initially, a Lyapunov analysis yields a set containing at least $1 - \epsilon$ of the stationary probability mass. On this subset of the state-space, we apply an aggregation scheme that groups together states in hypercube macro-states. Throughout each of these macro-states, we assume a uniform distribution among its constituent micro-states. This allows us to roughly analyze large portions of the state-space with exponentially fewer variables. We then iteratively truncate and refine the approximation based on the stationary distribution of this aggregated Markov chain. We keep only the most relevant macro-states and continue this scheme until the macro-states contain a single original state. In this way, we arrive at an effective truncation to compute an approximation of the stationary distribution.

We investigate the approximation results on case studies with known stationary distributions and complex models with intricate stationary distributions. We evaluate the truncation quality by assessing the stationary probability mass captured. To this end, we use analytical solutions and bounds given by a Lyapunov analysis. Further, we explore the control of the truncation size through the truncation parameter. Finally, we demonstrate the method on the p53 oscillator model exhibiting a complex stationary distribution.

The rest of the paper is organized as follows: Sect. 2 discusses related work, Sect. 3 introduces background material, Sect. 4 is devoted to the description of our method, Sect. 5 presents an experimental validation, and finally Sect. 6 contains a final discussion.

2 Related Work

For some specific models, analytical solutions for the stationary distribution have been found [26,29]. For the class of zero-deficiency networks, the stationary distribution is known to have a Poisson product form [2]. Monomolecular reaction networks can be solved explicitly, as well [21].

The analysis of countably infinite-sized state-spaces is often handled by predefined truncations [27]. Sophisticated state-space truncations for the (unconditioned) forward analysis have been developed that give lower bounds. They typically provide a trade-off between computational load and tightness of the bound [5,20,28,33,36]. Such methods cannot be directly applied to the estimation of stationary distributions because the approximation usually introduces a sink-state.

Truncations for stationary distributions often involve re-direction schemes for transitions leaving and entering the subset. A comprehensive survey of such state-space truncation methods can be found in Kuntz et al. [25]. A popular method of identifying truncations is the construction of a suitable Lyapunov function. Beyond their use for establishing ergodicity [12,17,30], these functions can be used to obtain truncations, guaranteed to contain a certain amount of stationary probability mass [12]. Using Lyapunov functions for the construction of truncations often leads to very conservative sets [34]. Different approaches have been employed to find truncations: In Gupta et al. [18] SSA estimates are used to set up an increasing family of truncations.

Apart from approaches based on state-space truncations, moment-based approaches have been particularly popular recently [13,15,23,38]. Such approaches are based on the fact that particular matrices of distributional moments such as mean and variance are positive semi-definite. Along with linear constraints stemming from the Kolmogorov equations [7], a semi-definite program can be formulated and solved using existing tools. While this method is suited to compute bounds on both moments and subsets of the state-space, its application is limited, due to numerical issues inherent in the formulation [13].

An approach where quantities are only described in terms of their magnitude has been proposed by Ceska and Kretínský [11]. This allows for an efficient qualitative analysis of both dynamic and transient behavior.

An aggregation scheme similar to the one used here has been previously proposed in [6] to analyze the bridging problem on Markov population models. This is the problem of analyzing process dynamics under both initial and terminal constraints.

Aggregation-based numerical methods for computing the stationary distribution of discrete or continuous-time Markov chains have been studied in previous work. Popular approaches rely on an alternation of aggregation and disaggregation of the state-space [39,41]. In the case of stiff chains, such aggregations are typically based on a separation of time-scales [10]. However, these methods have been developed for finite chains with arbitrary structure and are motivated by numerical issues of standard methods such as the power method or Jacobi iteration [41]. They do not consider a truncation of irrelevant states, while here our aggregation approach is used to determine the most relevant states under stationary conditions in large or infinite chains with population structure.

3 Preliminaries

3.1 Markov Population Models

A Markov population model (MPM) describes the stochastic interactions among agents of distinct types in a well-stirred system. This assumes that all agents are equally distributed in space, which allows us to keep track only of the overall copy number of agents for each type. Therefore the state-space is $\mathcal{S} \subseteq \mathbb{N}^{n_S}$ where n_S denotes the number of agent types or populations. Interactions between agents are expressed as *reactions*. These reactions have associated gains and losses of agents, given by non-negative integer vectors v_j^- and v_j^+ for reaction j, respectively. The overall change by a reaction is given by the vector $v_j = v_j^+ - v_j^-$. A reaction between agents of types $S_1, \ldots, S_{n_S}$ is specified in the following form:

$$\sum_{\ell=1}^{n_S} v_{j\ell}^- S_\ell \xrightarrow{\alpha_j(x)} \sum_{\ell=1}^{n_S} v_{j\ell}^+ S_\ell \,. \tag{1}$$

The propensity function α_j gives the rate of the exponentially distributed firing time of the reaction as a function of the current system state $x \in \mathcal{S}$. In population models, *mass-action* propensities are most common. In this case the firing rate is given by the product of the number of reactant combinations in x and a *rate constant* c_j, i.e.

$$\alpha_j(x) := c_j \prod_{\ell=1}^{n_S} \binom{x_\ell}{v_{j\ell}^-} \,. \tag{2}$$

In this case, we give the rate constant in (1) instead of the function α_j. For a given set of n_R reactions, we define a stochastic process $\{X_t\}_{t \geq 0}$ describing the evolution of the population sizes over time t. Due to the assumption of exponentially distributed firing times[1], X is a continuous-time Markov chain (CTMC) on $\mathcal{S}$ with infinitesimal generator matrix Q, where the entries of Q are

$$Q_{x,y} = \begin{cases} \sum_{j: x+v_j = y} \alpha_j(x) \,, & \text{if } x \neq y, \\ -\sum_{j=1}^{n_R} \alpha_j(x) \,, & \text{otherwise.} \end{cases} \tag{3}$$

The probability distribution over time is given by an initial value problem. Given an initial state x_0, the distribution[2]

$$\pi(x_i, t) = \Pr(X_t = x_i \mid X_0 = x_0), \quad t \geq 0 \tag{4}$$

evolves according to the Kolmogorov forward equation

$$\frac{d}{dt}\pi(t) = \pi(t) Q \,, \tag{5}$$

[1] Note that in addition mild regularity assumptions are necessary for the existence of a unique CTMC X, such as non-explosiveness [4]. These assumptions are typically valid for realistic reaction networks.

[2] In the sequel, we assume an enumeration of all states in $\mathcal{S}$. We simply write x_i for the state with index i and drop this notation for entries of a state x.

where $\pi(t)$ is an arbitrary vectorization $(\pi(x_1,t),\pi(x_2,t),\ldots,\pi(x_{|\mathcal{S}|},t))$ of the states.

Example. Consider a birth-death process as a simple example. This model is used to describe a wide variety of phenomena and often constitutes a sub-module of larger models. For example, it represents an M/M/1 queue with service rates being linearly dependent on the queue length. Note that even for this simple model, the state-space is countably infinite.

Model 1 (Birth-Death Process). *The model consists of exponentially distributed arrivals and service times proportional to queue length. It can be expressed using two mass-action reactions:*

$$\varnothing \xrightarrow{\mu} S \quad \textit{and} \quad S \xrightarrow{\gamma} \varnothing .$$

The initial condition $X_0 = 0$ holds with probability one.

3.2 Stationary Distribution

Assuming ergodicity of the underlying chain, a stationary distribution π_∞ is an invariant distribution, namely a fixed point of the Kolmogorov forward equation (5). Let π_∞ be the vector description of a stationary distribution. It then satisfies

$$0 = \pi_\infty Q \quad \text{and} \quad 1 = \sum_{x\in\mathcal{S}} \pi_\infty(x) \tag{6}$$

as a fixed point of the Kolmogorov equation (5). Stationary distributions are connected to the *long-run* behavior of an MPM [12], as the system's distribution will converge to the (unique) stationary distribution. The connection of the stationary distribution to the long-run behavior becomes clear when considering the ergodic theorem. For some $A \subseteq \mathcal{S}$,

$$\lim_{T\to\infty} \frac{1}{T}\int_0^T 1_A(X_t)\,dt = \sum_{x\in A} \pi_\infty(x) . \tag{7}$$

Thus, the mean occupation time for set A over infinite trajectories is the stationary measure for A. Equation (7) shows that we can assess long-run behavior using the stationary distribution and vice-versa.

Example. Returning to the example of Model 1 it is obvious that the state-space is irreducible. Further, we can easily show, that the stationary distribution is Poissonian with rate μ/γ:

$$\pi_\infty(x) = \frac{(\mu/\gamma)^x \exp(-\mu/\gamma)}{x!} .$$

For simplicity, we assume throughout that the state-space is composed of a single communicating class. Checking ergodicity given a countably infinite number of states is achieved by providing a suitable Foster-Lyapunov function [31]. Some automated techniques have been proposed for this task [12,17,34].

3.3 Truncation-Based Approximation of π_∞

In many relevant cases, the state-space is huge or infinite and therefore the stationary solution cannot be computed directly. To make such a computation possible we have to restrict ourselves to a finite manageable subset of the state-space and assume the majority of the probability mass is concentrated within that finite subset. The main problem is to deal with the transitions leading to and from the truncated set (cf. Fig. 1). In forward analysis, the outgoing transitions are simply redirected into a sink-state. This way, a forward analysis provides lower bounds since mass leaving the truncation does not re-enter. This approach, however, is unsuitable for the computation of stationary distributions because mass would accumulate in the sink-state leading to a distribution assigning all mass to it. Therefore, transitions leaving the truncation need to be redirected back into the truncation.

The process' dynamics outside the truncation are defined by the *stochastic complement* [40]. If its behavior was known, one could redirect outgoing to incoming transitions optimally and preserve the correct stationary distribution. However, this reentry distribution is typically unknown in most relevant cases. Many different reentry distributions have been used, such as redirecting to some internal state or states with incoming transition from outside the truncation. Reference [24] provides a comprehensive review of such methods.

The most natural choice is to pick a reentry distribution that redirects mass to states with incoming transitions from truncated states (cf. Fig. 1 (center)).

Using varying redirections, we can compute bounds on the stationary probability conditioned on a truncation [40, (Thm. 14)]. To do this, one has to compute the stationary distribution for every possible way of connecting all outgoing to a single incoming transition. Naturally, such an algorithm is rather expensive since one has to solve a linear system for each combination. Therefore this method of computing bounds is costly on very large truncations, often given by Lyapunov functions.

When computing an approximation instead of bounds, we employ a uniform redirection scheme: Outgoing transitions are split uniformly among incoming transitions. Due to the threshold-based truncation scheme, we are likely to end up with a somewhat uniform distribution over in-boundary states (see Sect. 4.3).

The identification of good truncations remains a major task in such approximations. Using approaches such as Lyapunov functions (Sect. 3.4) [12] or moment-bounds [24] can provide a good initial estimate, but typically the resulting truncations are far larger than necessary. This leads to dramatically increased computational costs, especially when bounding methods mentioned above are performed. Until a system for a larger truncation is solved, the precise location of most of the probability mass is often unknown. Instead of solving the full system for such a large space, we employ an aggregation scheme to cover large areas of the state-space with exponentially fewer variables.

Error bounds have been derived for increasing truncation sets in the case of linear Lyapunov functions [18]. However, until now it has not been shown that these bounds are applicable in practice [32]. Alternatively, one can monitor

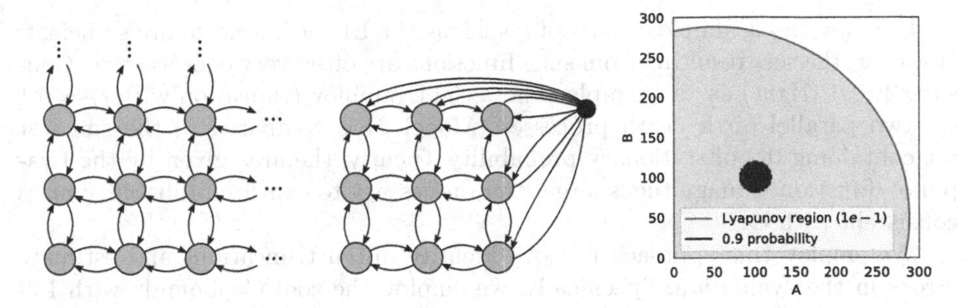

Fig. 1. (left) A countably infinite state-space. (center) Outgoing transitions are redirected (according to the reentry distribution) to states that have incoming transitions from outside the truncation. (right) A comparison of the area prescribed by a Lyapunov analysis using Geobound and threshold 0.1 and the minimal area containing 0.9 stationary probability mass. The model is a parallel birth death process (Model 2).

the product of the probability-outflow rate and the maximum L1-norm, which bounds the approximation error up to a constant $M > 0$, assuming a linear Lyapunov function exists [18].

3.4 Lyapunov Bounds

It is well-known that for a CTMC X, ergodicity can be proven by a Lyapunov function $g : \mathcal{S} \to \mathbb{R}_+$ [12,30]. Given the g, we define its *drift* d as its average infinitesimal change, which is obtained applying the generator Q to g.

$$d(x) = \sum_{j=1}^{n_R} \alpha_j(x)(g(x + v_j) - g(x)) \tag{8}$$

Usually, such a function g grows in all directions on the positive orthant, while its drift $d(x)$ decreases in all directions. More formally, g is characterized by having finite level sets $\{x \in \mathcal{S} \mid g(x) < l\}$ for all $l > 0$. At the same time,

$$\mathcal{C}_{\epsilon_\ell} = \{x \in \mathcal{S} \mid \frac{\epsilon_\ell}{c} d(x) > \epsilon_\ell - 1\} \tag{9}$$

should be finite, where $\infty > c \geq \sup_{x \in \mathcal{S}} d(x)$. In this case, $\mathcal{C}_{\epsilon_\ell}$ contains at least $1 - \epsilon_\ell$ of stationary probability mass for any $\epsilon_\ell \in (0, 1)$ [40, Thm. 8]. Given that $\mathcal{C}_{\epsilon_\ell}$ is finite, the chain is ergodic and

$$\sum_{x \in \mathcal{C}_{\epsilon_\ell}} \pi(x) > 1 - \epsilon_\ell \tag{10}$$

bounding the stationary probability mass contained within $\mathcal{C}_{\epsilon_\ell}$.

In many cases, simple choices of g such as the L1- or L2- norm are sufficient. However, the sets resulting from such functions are often very conservative. Consider Fig. 1 (right) as an example, where the Lyapunov truncation with $\epsilon_\ell = 0.1$ for two parallel birth death processes (Model 2) is compared to the smallest set containing 0.9 of stationary probability. Clearly, the area given by the Lyapunov function is magnitudes larger than necessary to capture probability mass consistent with ϵ_ℓ.

We employ this approach to both identify initial truncations and estimate errors in the evaluation. Specifically, we employ the tool Geobound[3] with L2-norm as function g implementing techniques presented in [12].

4 Method

In this work, we propose a method to identify a truncation that optimizes the trade-off between the size of the considered state-space and the approximation error due to the finite state-space projection. To this end, we start with a very coarse-grained model abstraction that we refine iteratively. The coarse-grained model is based on an grid-shaped aggregation (i.e., lumping) scheme that identifies a set of macro-states. These macro-states can be used to compute an interim model solution that guides the refinement in the next step. We perform refinements until the approximation arrives at the resolution of the original model (i.e., each macro-state has only one constituent) such that the aggregation introduces no approximation error.

We explain the construction of macro-states in Sect. 4.1 and their initialization in Sect. 4.2. We present the iterative refinement algorithm in Sect. 4.3.

4.1 State-Space Aggregation

A macro-state is a collection of micro-states (or simply states) treated as one state in the aggregated model, which can be seen as an abstraction of the original model. The aggregation scheme defines a partitioning of the state-space. We choose a scheme based on a grid structure. That is, each macro-state is a hypercube in $\mathbb{N}^{n_S}$.

Hence, each macro-state $\bar{x}_i(\ell^{(i)}, u^{(i)})$ (denoted by $\bar{x}_i$ for notational ease) can be identified using two vectors $\ell^{(i)}$ and $u^{(i)}$. The vector $\ell^{(i)}$ gives the corner closest to the origin, while $u^{(i)}$ gives the corner farthest from the origin. Formally,

$$\bar{x}_i = \bar{x}_i(\ell^{(i)}, u^{(i)}) = \{x \in \mathbb{N}^{n_S} \mid \ell^{(i)} \leq x \leq u^{(i)}\}, \tag{11}$$

where '$\leq$' denotes element-wise comparison.

In order to solve the aggregated model, we need to define transition rates between macro-states. Therefore, we assume that, given that the system is in a particular macro-state, all constituent states are equally likely (uniformity

[3] https://mosi.uni-saarland.de/tools/geobound.

assumption). This assumption is the reason why the aggregated model provides only a coarse-grained approximation.

The uniformity assumption is a modeling choice yielding significant advantages. Firstly, it eases the computation of the rates between macro-states and, therefore, makes a fast solution of the aggregated model possible. Secondly, even though it induces an approximation error, it provides suitable guidance as uniformity assumption spreads out the probability mass conservatively. Hence, it becomes less likely that regions of interest are disregarded. Lastly, the uniformity assumption is theoretically well-founded, as it stems from the maximum entropy principle: In the absence of concrete knowledge about the probability distribution inside a macro-state, we assume the distribution with the highest uncertainty, i.e., the uniform distribution.

The grid structure makes the computation of transition rates between macro-states particularly convenient and computationally simple. Mass-action reaction rates can be given in a closed-form, due to the Faulhaber formulae [22] and more complicated rate functions such as Hill-functions can often be handled as well by taking appropriate integrals [6].

Suppose, we are interested in the transition rate from macro-state $\bar{x}_i$ to macro-state $\bar{x}_k$ according to reaction j. Using the uniformity assumption, this is simply the mean rate of the states in $\bar{x}_i$ that go to $\bar{x}_k$ using j. However, only a small subset of constituents in $\bar{x}_i$ are actually relevant for this transition. Hence, we identify the subset of states of $\bar{x}_i$ that lie at the border to $\bar{x}_k$ and in such a way that applying reaction j shifts them to a state in $\bar{x}_k$. Then, we sum up the corresponding rates of these states. Lastly, we normalize according to the number of states inside of $\bar{x}_i$.

It is easy to see that the relevant set of border states is itself an interval-defined macro-state $\bar{x}_{i \xrightarrow{j} k}$. To compute this macro-state we can simply shift $\bar{x}_i$ by v_j, take the intersection with $\bar{x}_k$ and project this set back. Formally,

$$\bar{x}_{i \xrightarrow{j} k} = ((\bar{x}_i + v_j) \cap \bar{x}_k) - v_j\,, \tag{12}$$

where the additions are applied element-wise to all states making up the macro-states. For ease of notation, we also define a general exit state

$$\bar{x}_{i \xrightarrow{j}} = ((\bar{x}_i + v_j) \backslash \bar{x}_i) - v_j. \tag{13}$$

This state captures all micro-states inside $\bar{x}_i$ that can leave the state via reaction j.

This uniformity assumption gives rise to the following Q-matrix of the aggregated model:

$$\bar{Q}_{\bar{x}_i, \bar{x}_k} = \begin{cases} \sum_{j=1}^{n_R} \bar{\alpha}_j \left(\bar{x}_{i \xrightarrow{j} k}\right) / \, |\bar{x}_i|\,, & \text{if } \bar{x}_i \neq \bar{x}_k \\ -\sum_{j=1}^{n_R} \bar{\alpha}_j \left(\bar{x}_{i \xrightarrow{j}}\right) / |\bar{x}_i|\,, & \text{otherwise} \end{cases} \tag{14}$$

where

$$\bar{\alpha}_j(\bar{x}) = \sum_{x \in \bar{x}} \alpha_j(x). \tag{15}$$

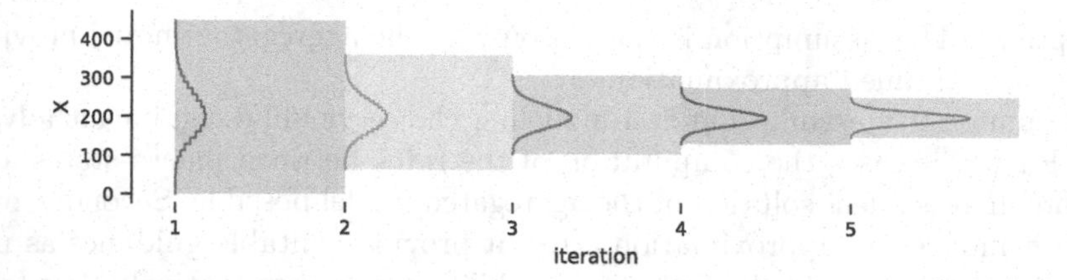

Fig. 2. The state-space refinement algorithm on a birth-death process. From left to right the state size is halved and states with low probability are removed from the truncation. The final truncation is a typical truncation with states of size 1 and the initial states are of size 2^4.

is the sum of all rates belonging to reaction j in $\bar{x}$.

Under the assumption of polynomial rates, as is the case for mass-action systems, we can compute the sum of rates over this transition set efficiently using Faulhaber's formula. As an example consider the following mass-action reaction $2X \xrightarrow{c} \varnothing$. For macro-state $\bar{x} = \{0, \dots, n\}$ we can compute the corresponding lumped transition rate

$$\bar{\alpha}(\bar{x}) = \frac{c}{2} \sum_{i=1}^{n} i(i-1) = \frac{c}{2} \sum_{i=1}^{n} (i^2 - i) = \frac{c}{2} \left(\frac{2n^3 + 3n^2 + n}{6} - \frac{n^2 + n}{2} \right)$$

eliminating the explicit summation in the lumped propensity function.

4.2 Initial Aggregation

The initial aggregated space $\hat{\mathcal{S}}^{(0)}$ should encompass all regions of the state-space that could contain significant mass because states outside this initial area will not be refined. In principle, multiple approaches could be used to identify such a region. One possibility is the computation of moment bounds for the stationary distribution [13,15]. Based on these bounds on expectations and covariances, an initial truncation could be fixed. The approach we use here is to identify such a region by a Lyapunov analysis [12]. This way, we obtain a polynomial describing a semi-algebraic subset of the entire state-space containing $1 - \epsilon_\ell$ of the mass, where $\epsilon_\ell > 0$ can be fixed arbitrarily. These sets usually are far larger than a minimal set containing $1 - \epsilon_\ell$ of stationary probability mass would be. As an initial aggregation, we build an aggregation on a subset $[0..n]^{n_S} \subset \mathcal{S}$ containing the set prescribed by the Lyapunov analysis.

4.3 Iterative Refinement Algorithm

The refinement algorithm (Algorithm 1) starts with a set of large macro-states that are iteratively refined, based on approximate stationary distributions. We start by constructing square macro-states of size 2^m in each dimension for some $m \in \mathbb{N}$ such that they form a large-scale grid $\mathcal{S}^{(0)}$. Hence, each initial macro-state has a volume of $(2^m)^{n_S}$. This choice of grid size is convenient because we

Algorithm 1: Lumping to approximate the stationary distribution

input : Initial partitioning $\mathcal{S}^{(0)}$, truncation threshold ϵ
output: approximate stationary distribution $\hat{\pi}_\infty$

1 **for** $i = 1, \dots, m$ **do**
2 $\quad \hat{\pi}_\infty^{(i)} \leftarrow$ solve approximate stationary distribution on $\mathcal{S}^{(i)}$;
3 $\quad \mathcal{R} \leftarrow$ choose smallest $\mathcal{R}' \subseteq \mathcal{S}^{(i)}$ such that $\sum_{\bar{x} \in \mathcal{R}'} \hat{\pi}_\infty^{(i)}(\bar{x}) \geq 1 - \epsilon$;
4 $\quad \mathcal{S}^{(i+1)} \leftarrow \bigcup_{\bar{x} \in \mathcal{R}} \text{split}(\bar{x})$;
5 $\quad$ update $\hat{Q}$-matrix;
6 **return** $\hat{\pi}_\infty^{(m)}$;

can halve states in each dimension. Moreover, this choice ensures that all states have an equal volume and we end up with unit-sized macro-states, equivalent to a truncation of the original non-lumped state-space.

An iteration of the state-space refinement starts by computing the stationary distribution, using the lumped $\hat{Q}$-matrix. Based on a threshold parameter $\epsilon > 0$ states are either removed or split (line 4), depending on the mass assigned to them by the approximate stationary probabilities $\hat{\pi}_\infty^{(i)}$. Thus, each macro-state is either split into 2^{n_S} new states or removed entirely. The result forms the next lumped state-space $\mathcal{S}^{(i+1)}$. The $\hat{Q}$-matrix is updated (line 5) using 14 to calculate the transition rates of the next aggregated truncation $\mathcal{S}^{(i+1)}$. Entries of truncated states are removed from the updated transition matrix. Transitions leading to them are re-directed according to the re-entry matrix (see Sect. 3.3). After m iterations (we started with states of side lengths 2^m) we have a standard finite state projection scheme on the original model tailored to computing an approximation of the stationary distribution.

This way, the refinement algorithm focuses only on those parts of the state-space contributing most to the stationary distribution. For instance, in Fig. 2 the stationary probability mass mostly concentrates around $\#S = 200$. Therefore, states that are further away from this area can be dropped in further refinement. This filtering (line 3 in Algorithm 1) ensures that states contributing significantly to $\hat{\pi}_\infty^{(i)}$ will be kept and refined in the next iteration. The selection of states is done by sorting states in descending order according to their approximate probability mass. This ensures the construction of the smallest possible subset chosen for refinement according to the approximation. Then states are collected until their overall approximate mass is above $1 - \epsilon$.

An interesting feature of the aggregation scheme is that the distribution tends to spread out more. This is due to the assumption of a uniform distribution inside macro-states. To gain an intuition, consider a macro-state that encompasses a peak of the stationary distribution. If we re-distribute the actual probability mass inside this macro-state uniformly, a higher probability is assigned to states at the macro-state's border. When plugging such macro-states together, this increased mass away from the peak will increase the mass assigned to adjacent macro-states. This effect is illustrated by the example of a birth-death process in Fig. 2. Due to this effect, an iterative refinement typically keeps an over-approximation in terms of state-space area. This is a desirable feature since relevant regions are less likely to be pruned due to lumping approximations.

5 Results

A prototype was implemented in Rust 1.50 and Python 3.8. The linear systems were solved either using Numpy [19] for up to 5000 states, or the sparse linear solver as available through Scipy [43], or the iterative biconjugate gradient stabilized algorithm [44] (up to 10,000 iterations and absolute tolerance 10^{-16}).

The examples that we consider in the sequel are typical benchmarks for the analysis of MPMs. For most of them, appropriate Lyapunov functions have been determined using Geobound [40]. However, the corresponding Lyapunov sets containing at least $1 - \epsilon_\ell$ of the stationary probability mass are very large for typical choices of ϵ_ℓ (e.g. $\epsilon_\ell \in \{0.1, 0.05, 0.001\}$). Even for extremely large ϵ_ℓ, say $\epsilon_\ell = 0.8$, the remaining state-space may still be huge (e.g., 15,198 states).

5.1 Parallel Birth-Death Process

We first examine the algorithm on the simple example of two parallel, uncoupled birth-death processes.

Model 2 (Parallel Birth-Death Process). *Two uncoupled parallel birth-death processes result in a simple stationary distribution that is given by a product of two Poisson distributions.*

$$\varnothing \xrightarrow{\rho} A \qquad A \xrightarrow{\delta} \varnothing \qquad \varnothing \xrightarrow{\rho} B \qquad B \xrightarrow{\delta} \varnothing$$

As a parameterization we choose $\rho = 100$ *and* $\delta = 1$.

For this model, the stationary distribution is known to be the product of two Poisson distributions with rate ρ/δ.

According to the Lyapunov analysis with a 1e-4 bound, we fix the initial truncation to a 70×70 grid of macro-states with size 2^7 in each dimension. This implies 8 iterations of the algorithm to arrive at a truncation with the original granularity. In Fig. 3, we illustrate the truncations of different iterations. Over the iterations, the covered area decreases, while the aggregation granularity increases. The final truncation distribution approximation is also depicted and covers $1 - 1.27\text{e-}2$ of the true stationary distribution (cf. Table 1).

For this case study, we also compute state-wise bounds on the probabilities conditioned on the truncation as discussed in Sect. 3.3. In Fig. 6 (right), we present the difference between upper and lower bound for $\epsilon = 0.1$. We observe intervals that are narrowest in the truncation's interior near the distribution's mode. The largest intervals or the largest absolute uncertainty is present in the boundary states. This indicates, that the specific reentry distribution has little effect on the main approximate stationary mass. More detailed results on the intervals' magnitudes are given in Table 1.

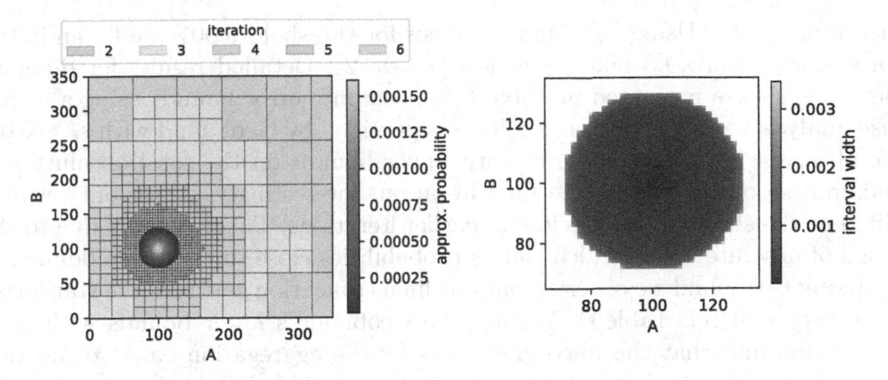

Fig. 3. Results for Model 2 with truncation threshold $\epsilon = 0.1$. (left) Truncations of different iterations are layered on top of each other. At higher iterations, truncations cover less area but increase in detail, due to the refinement of macro-states. The final approximation is indicated by its approximate probabilities. (right) The difference between the upper and lower bounds on the probability conditioned on the truncation.

5.2 Exclusive Switch

The exclusive switch [8] has three different modes of operation, depending on the DNA state, i.e. on whether a protein of type one or two is bound to the DNA.

Model 3 (Exclusive Switch). *The exclusive switch model consists of a promoter region that can express both proteins P_1 and P_2. Both can bind to the region, suppressing the expression of the other protein. For certain parameterizations, this leads to a bi-modal or even tri-modal behavior.*

$$D \xrightarrow{\rho_1} D + P_1 \qquad D \xrightarrow{\rho_2} D + P_2 \qquad P_1 \xrightarrow{\lambda} \varnothing \qquad P_2 \xrightarrow{\lambda} \varnothing$$

$$D + P_1 \xrightarrow{\beta} D.P_1 \qquad D.P_1 \xrightarrow{\gamma_1} D + P_1 \qquad D.P_1 \xrightarrow{\rho_1} D.P_1 + P_1$$

$$D + P_2 \xrightarrow{\beta} D.P_2 \qquad D.P_2 \xrightarrow{\gamma_2} D + P_2 \qquad D.P_2 \xrightarrow{\rho_2} D.P_2 + P_2$$

We choose parameter values $\rho_1 = 0.7$, $\rho_2 = 0.6$, $\lambda = 0.02$, $\beta = 0.005$, $\gamma_1 = 0.06$, and $\gamma_2 = 0.05$.

Since the exclusive switch models mutually exclusive binding of proteins to a single genetic locus, we know a priori that there are exactly three distinct operating modes. In particular are D, $D.P_1$, and $D.P_2$ mutually exclusive such that $X_D(t) + X_{D.P_1}(t) + X_{D.P_2}(t) = 1$, $\forall t \geq 0$. This model characteristic often leads to bi-modal stationary distributions, where one or the other protein is more abundant depending on the genetic state.

Accordingly, we adjust the initial truncation: The state-space for the DNA states is not lumped. Instead we "stack" lumped approximations of the P_1–P_2 plane upon each other. Such special treatment of DNA states is common for such models [28]. Using Lyapunov analysis for threshold 0.001, we fix an initial state-space of 63×63 macro-states with size 2^7. Detailed results for different parameters ϵ are presented in Table 3. We compute error bounds using a worst-case analysis based on reference solutions provided by Geobound with $\epsilon_\ell = 0.01$. We observe a strong decrease in both upper bounds on the total absolute and maximal absolute error in the final iteration. Interestingly, the errors between different thresholds are very close in earlier iterations. This is mainly due to the usage of absolute errors which causes probabilities close to the mode dominate.

Using Geobound we observe that our final truncation captures the stationary mass very well (cf. Table 1). We use the Geobound's lower bounds with $\epsilon_\ell = 1e-2$ and find that the uncovered mass by the aggregation-based truncation is magnitudes lower than ϵ or close to it (for $\epsilon = 0.1$). While they capture the mass well, they are much smaller than the Geobound truncation ($\epsilon_\ell = 0.1$) with 16,780 states, regardless of the threshold parameter ϵ.

In Fig. 4 (left), we show the effect of the threshold parameter ϵ on the size of the final truncation. We observe a roughly linear increase in size with an exponential decrease of ϵ.

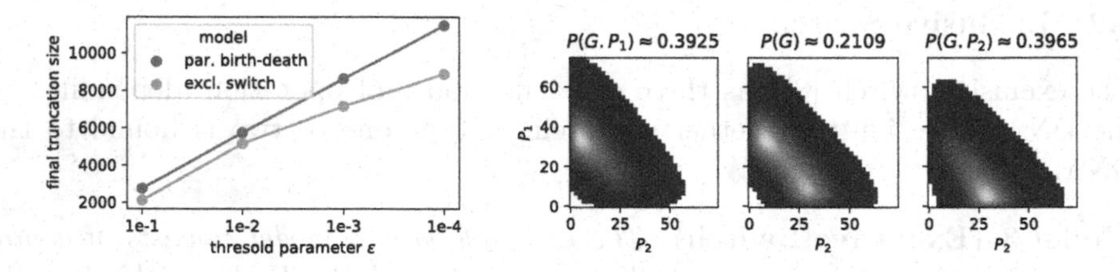

Fig. 4. (left) The sizes of the final truncation vs. the threshold parameter ϵ. (right) The approximate stationary distribution of the exclusive switch (Model 3) obtained with ϵ = 1e-4.

Table 1. Results for Model 2 and Model 3: The characteristics of the lower-upper bound intervals on the conditional probability and the (upper bound on) mass not contained in the truncation are given.

Model		Threshold parameter ϵ			
		1e-1	1e-2	1e-3	1e-4
2	Total width	1.2336	3.0938e-02	5.3916e-04	8.1249e-06
	Max. width	3.4752e-03	9.2954e-05	4.0400e-07	4.6521e-09
	Outside mass	1.2708e-02	1.0568e-04	1.0500e-06	1.0617e-08
3	Total width	5.5171	1.5559	2.8946e-02	3.7161e-04
	Max. width	1.5898e-01	3.3089e-03	3.4733e-05	3.8412e-07
	Outside mass $\leq$	1.5274e-01	1.2973e-03	2.0249e-05	2.7280e-07

5.3 P53 Oscillator

We now consider a model of the interactions of the tumor suppressor p53 [14]. The system describes the negative feedback loop between p53 and the oncogene Mdm2. Species pMdm2 models a precursor to Mdm2. This model is particularly interesting due to its complex three-dimensional oscillatory behavior. The model is ergodic with a unique stationary distribution [17].

Model 4 (p53 Oscillator).

$$\varnothing \xrightarrow{k_1} \text{p53} \qquad \text{p53} \xrightarrow{k_2} \varnothing \qquad \text{p53} \xrightarrow{k_4} \text{p53} + \text{pMdm2}$$

$$\text{p53} \xrightarrow{\alpha_4(\cdot)} \varnothing \qquad \text{pMdm2} \xrightarrow{k_5} \text{Mdm2} \qquad \text{Mdm2} \xrightarrow{k_6} \varnothing$$

The non-polynomial degradation reaction rate

$$\alpha_4(x) = k_3 x_{\text{Mdm2}} \frac{x_{\text{p53}}}{x_{\text{p53}} + k_7} .$$

The parameterization based on [1] is $k_1 = 90$, $k_2 = 0.002$, $k_3 = 1.7$, $k_4 = 1.1$, $k_5 = 0.93$, $k_6 = 0.96$, *and* $k_7 = 0.01$.

With the exception of propensity function α_4, we can compute the transition rates $\bar{\alpha}_i$ using the Faulhaber formulae, as discussed in Sect. 4.1. We consider α_4 separately, because it is non-polynomial and therefore, we have to make an approximation. The fraction occurring in the non-linear propensity function α_4 can roughly be characterized as an activation function: Due to the low value of parameter $k_7 = 0.01$ we can approximate

$$\frac{x_{\text{p53}}}{x_{\text{p53}} + k_7} \approx \begin{cases} 0 & \text{if } x_{\text{p53}} = 0 \\ 1 & \text{otherwise} \end{cases}$$

We use this approximation at the coarser levels of aggregation to efficiently compute the approximate transition rate $\bar{\alpha}_4$. At the finest granularity we switch back to exact propensity function α_4.[4]

Due to the exponential increase stemming from the three-dimensional nature of this model, we only evaluated with parameter $\epsilon = 0.1$. According to a Lyapunov analysis (Sect. B), the area covered by an $6 \times 6 \times 6$ macro-states with size 2^{20}, covers 0.9 of stationary mass. A truncation of this same area would consist of 226,492,416 states instead of the 216 macro-states. The model has a striking oscillatory behavior (cf. Fig. 5 (top right)) that is reflected in its stationary distribution. This feature is well-captured in the approximate distribution, where the oscillatory behavior leads to a complex stationary distribution (cf. Fig. 5 (bottom right)). This distribution leads to a non-trivial truncation (357,488 states) which is tailored to the main stationary mass (Fig. 5 (left)).

[4] We note, that $\sum_{i=0}^{n} i/(i+k_7)$ can be solved analytically. However, the approximation presented above is much simpler to compute.

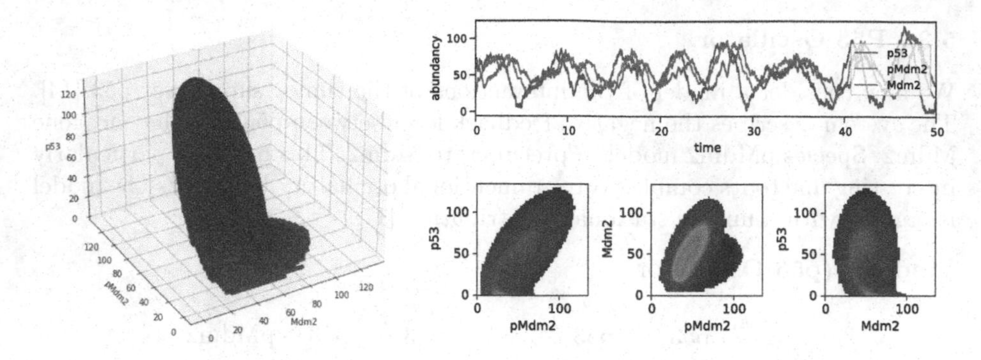

Fig. 5. (left) The final truncation at original granularity derived for the p53 oscillator. (top right) A sample trajectory illustrating the oscillatory long-run behavior. (bottom right) The approximate marginal distributions of the stationary distribution based on the truncation derived with $\epsilon = 0.1$.

6 Conclusion

State-of-the-art methods for numerically calculating the stationary distribution of Markov Population Models rely on coarse truncations of irrelevant parts of large or infinite discrete state-spaces. These truncations are either obtained from the stationary statistical moments of the process or from Lyapunov theory. They are limited in shape because these methods do not take into account the detailed steady-state flow within the truncated state-space but only consider the average drift or stationary moments.

Here, we propose a method to find a tight truncation that is not limited in its shape and iteratively optimizes the set based on numerically cheap solutions of abstract intermediate models. It captures the main portion of probability mass even in the case of complex behaviors efficiently. In particular, the method represents another option, where Lyapunov analysis leads to forbiddingly large truncations.

Acknowledgements. This work is supported by the DFG project "MULTIMODE".

A Detailed Results

See Tables 2, 3 and Fig. 6.

Table 2. Detailed results for Model 2. The errors are computed wrt. the reference Poissonian product. The total absolute error and the maximum absolute errors are given.

		Iteration i							
ϵ		1	2	3	4	5	6	7	8
1e-1	$\lvert\mathcal{S}^{(i)}\rvert$	4,900	28	52	112	232	472	960	1,932
	Tot. error	1.91	1.84	1.73	1.55	1.29	9.35e-1	4.88e-1	3.54e-2
	Max. error	3.15e-3	3.13e-3	3.08e-3	2.98e-3	2.77e-3	2.38e-3	1.57e-3	6.04e-5
1e-2	$\lvert\mathcal{S}^{(i)}\rvert$	4,900	52	104	208	464	988	2,008	4,052
	Tot. error	1.91	1.84	1.73	1.56	1.30	9.46e-1	5.01e-1	6.22e-4
	Max. error	3.15e-3	3.13e-3	3.08e-3	2.98e-3	2.78e-3	2.39e-3	1.59e-3	8.33e-7
1e-3	$\lvert\mathcal{S}^{(i)}\rvert$	4,900	84	152	300	652	1,440	2,996	6,068
	Tot. error	1.91	1.83	1.73	1.56	1.30	9.46e-1	5.01e-1	9.83e-6
	Max. error	3.15e-3	3.13e-3	3.08e-3	2.98e-3	2.78e-3	2.39e-3	1.59e-3	1.14e-8
1e-4	$\lvert\mathcal{S}^{(i)}\rvert$	4,900	116	212	400	848	1,872	3,960	8,060
	Tot. error	1.91	1.83	1.73	1.56	1.30	9.46e-1	5.01e-1	9.83e-6
	Max. error	3.15e-3	3.13e-3	3.08e-3	2.98e-3	2.78e-3	2.39e-3	1.59e-3	1.83e-10

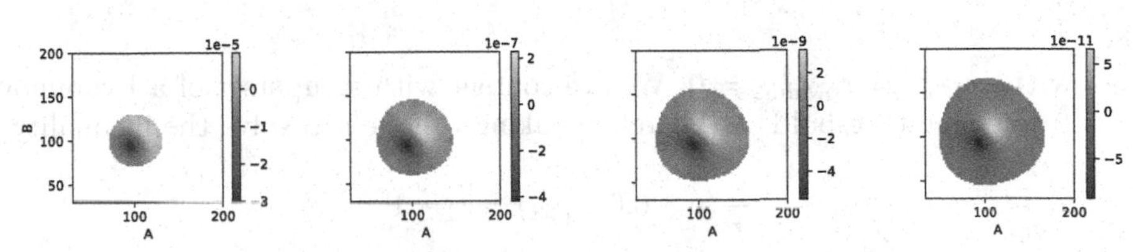

Fig. 6. The error over the truncation wrt. the analytical solution

B Lyapunov Analysis of the p53 Oscillator

We now derive Lyapunov-sets for the p53 oscillator case study (Model 4). Let the Lyapunov function

$$g(x) = 120x_{\mathrm{p53}} + 0.2x_{\mathrm{pMdm2}} + 0.1x_{\mathrm{Mdm2}}\,. \tag{16}$$

Then the drift

$$\begin{aligned} d(x) &= -\frac{k_3 x_{\mathrm{Mdm2}} x_{\mathrm{p53}}}{x_{\mathrm{p53}} + k_7} - 0.1k_6 x_{\mathrm{Mdm2}} + 120k_1 \\ &\quad - 120k_2 x_{\mathrm{p53}} + 0.2k_4 x_{\mathrm{p53}} - 0.1k_5 x_{\mathrm{pMdm2}} \\ &= -\frac{204 x_{\mathrm{Mdm2}} x_{\mathrm{p53}}}{x_{\mathrm{p53}} + 0.01} - 0.096x_{\mathrm{Mdm2}} - 0.02x_{\mathrm{p53}} \\ &\quad - 0.0093x_{\mathrm{pMdm2}} + 10800\,. \end{aligned} \tag{17}$$

Clearly, $c = \sup_{x \in S} d(x) = 10800$. In particular, the supremum c is at the origin since all non-constant terms are negative. The slowest rate of decrease for (17) is

Table 3. Detailed results for Model 3. Upper bounds on the total absolute error and the maximum absolute error are given. The worst-case errors are computed wrt. the reference Geobound solution with $\epsilon_\ell = 1e-2$.

		Iteration i							
ϵ		1	2	3	4	5	6	7	8
1e-1	$\lvert\mathcal{S}^{(i)}\rvert$	11907	20	32	60	140	340	840	2116
	Tot. error $\leq$	1.86e0	1.85e0	1.45e0	1.18e0	9.31e-1	6.41e-1	4.67e-1	4.89e-1
	Max. error $\leq$	1.63e-3	1.63e-3	1.55e-3	1.40e-3	1.22e-3	9.36e-4	8.40e-4	1.40e-3
1e-2	$\lvert\mathcal{S}^{(i)}\rvert$	11907	48	112	148	300	720	1892	5156
	Tot. error $\leq$	1.86e0	1.84e0	1.44e0	1.21e0	9.56e-1	6.65e-1	3.41e-1	3.31e-2
	Max. error $\leq$	1.63e-3	1.62e-3	1.53e-3	1.39e-3	1.20e-3	9.59e-4	5.86e-4	5.37e-5
1e-3	$\lvert\mathcal{S}^{(i)}\rvert$	11907	84	192	244	488	1084	2692	7152
	Tot. error $\leq$	1.86e0	1.83e0	1.46e0	1.22e0	9.63e-1	6.67e-1	3.37e-1	8.01e-4
	Max. error $\leq$	1.63e-3	2.95e-2	1.54e-3	1.39e-3	1.20e-3	9.51e-4	5.79e-4	1.09e-6
1e-4	$\lvert\mathcal{S}^{(i)}\rvert$	11907	124	324	352	672	1436	3408	8864
	Tot. error $\leq$	1.86e0	1.83e0	1.46e0	1.22e0	9.63e-1	6.67e-1	3.37e-1	1.12e-5
	Max. error $\leq$	1.63e-3	3.19e-2	1.54e-3	1.39e-3	1.20e-3	9.51e-4	5.79e-4	1.28e-8

x_{p53} with $x_{\text{Mdm2}} = x_{\text{pMdm2}} = 0$. We are content with a superset of a Lyapunov set (9) for some threshold ϵ_ℓ. Therefore taking (9), we can solve the inequality

$$\frac{\epsilon_\ell}{c}(c - 0.02x_{\text{p53}}) > \epsilon_\ell - 1$$

for x_{p53} and

$$\frac{c}{0.02\epsilon_\ell} < x_{\text{p53}}\,. \tag{18}$$

Therefore

$$\pi_\infty\left(\left\{x \in \mathcal{S} \mid \frac{c}{0.2\epsilon_\ell} < \|x\|\right\}\right) > 1 - \epsilon_\ell\,. \tag{19}$$

References

1. Ale, A., Kirk, P., Stumpf, M.P.: A general moment expansion method for stochastic kinetic models. J. Chem. Phys. **138**(17), 174101 (2013)
2. Anderson, D.F., Craciun, G., Kurtz, T.G.: Product-form stationary distributions for deficiency zero chemical reaction networks. Bull. Math. Biol. **72**(8), 1947–1970 (2010)
3. Anderson, D.F., Kurtz, T.G.: Continuous time Markov chain models for chemical reaction networks. In: Koeppl, H., Setti, G., di Bernardo, M., Densmore, D. (eds.) Design and Analysis of Biomolecular Circuits, pp. 3–42. Springer, New York (2011)
4. Anderson, W.J.: Continuous-Time Markov Chains: An Applications-Oriented Approach. Springer, New York (2012). https://doi.org/10.1007/978-1-4612-3038-0

5. Andreychenko, A., Mikeev, L., Spieler, D., Wolf, V.: Parameter identification for Markov models of biochemical reactions. In: Gopalakrishnan, G., Qadeer, S. (eds.) CAV 2011. LNCS, vol. 6806, pp. 83–98. Springer, Heidelberg (2011). https://doi.org/10.1007/978-3-642-22110-1_8
6. Backenköhler, M., Bortolussi, L., Großmann, G., Wolf, V.: Analysis of Markov jump processes under terminal constraints. arXiv preprint arXiv:2010.10096 (2020)
7. Backenköhler, M., Bortolussi, L., Wolf, V.: Generalized method of moments for stochastic reaction networks in equilibrium. In: Bartocci, E., Lio, P., Paoletti, N. (eds.) CMSB 2016. LNCS, vol. 9859, pp. 15–29. Springer, Cham (2016). https://doi.org/10.1007/978-3-319-45177-0_2
8. Barzel, B., Biham, O.: Calculation of switching times in the genetic toggle switch and other bistable systems. Phys. Rev. E **78**(4), 041919 (2008)
9. Breuer, L.: From Markov Jump Processes to Spatial Queues. Springer, New York (2003). https://doi.org/10.1007/978-94-010-0239-4
10. Cao, W.L., Stewart, W.J.: Iterative aggregation/disaggregation techniques for nearly uncoupled Markov chains. J. ACM (JACM) **32**(3), 702–719 (1985)
11. Češka, M., Křetínský, J.: Semi-quantitative abstraction and analysis of chemical reaction networks. In: Dillig, I., Tasiran, S. (eds.) CAV 2019. LNCS, vol. 11561, pp. 475–496. Springer, Cham (2019). https://doi.org/10.1007/978-3-030-25540-4_28
12. Dayar, T., Hermanns, H., Spieler, D., Wolf, V.: Bounding the equilibrium distribution of Markov population models. Numer. Linear Algebra Appl. **18**(6), 931–946 (2011)
13. Dowdy, G.R., Barton, P.I.: Bounds on stochastic chemical kinetic systems at steady state. J. Chem. Phys. **148**(8), 084106 (2018)
14. Geva-Zatorsky, N., et al.: Oscillations and variability in the p53 system. Mol. Syst. Biol. **2**(1) (2006). 2006.0033
15. Ghusinga, K.R., Vargas-Garcia, C.A., Lamperski, A., Singh, A.: Exact lower and upper bounds on stationary moments in stochastic biochemical systems. Phys. Biol. **14**(4), 04LT01 (2017)
16. Gillespie, D.T.: Exact stochastic simulation of coupled chemical reactions. J. Phys. Chem. **81**(25), 2340–2361 (1977)
17. Gupta, A., Briat, C., Khammash, M.: A scalable computational framework for establishing long-term behavior of stochastic reaction networks. PLoS Comput. Biol. **10**(6), e1003669 (2014)
18. Gupta, A., Mikelson, J., Khammash, M.: A finite state projection algorithm for the stationary solution of the chemical master equation. J. Chem. Phys. **147**(15), 154101 (2017)
19. Harris, C.R., et al.: Array programming with NumPy. Nature **585**, 357–362 (2020). https://doi.org/10.1038/s41586-020-2649-2
20. Henzinger, T.A., Mateescu, M., Wolf, V.: Sliding window abstraction for infinite Markov chains. In: Bouajjani, A., Maler, O. (eds.) CAV 2009. LNCS, vol. 5643, pp. 337–352. Springer, Heidelberg (2009). https://doi.org/10.1007/978-3-642-02658-4_27
21. Jahnke, T., Huisinga, W.: Solving the chemical master equation for monomolecular reaction systems analytically. J. Math. Biol. **54**(1), 1–26 (2007)
22. Knuth, D.E.: Johann faulhaber and sums of powers. Math. Comput. **61**(203), 277–294 (1993)
23. Kuntz, J., Thomas, P., Stan, G.B., Barahona, M.: Rigorous bounds on the stationary distributions of the chemical master equation via mathematical programming. arXiv preprint arXiv:1702.05468 (2017)

24. Kuntz, J., Thomas, P., Stan, G.B., Barahona, M.: Approximations of countably infinite linear programs over bounded measure spaces. SIAM J. Optim. **31**(1), 604–625 (2021)
25. Kuntz, J., Thomas, P., Stan, G.B., Barahona, M.: Stationary distributions of continuous-time Markov chains: a review of theory and truncation-based approximations. SIAM Rev. **63**(1), 3–64 (2021)
26. Kurasov, P., Lück, A., Mugnolo, D., Wolf, V.: Stochastic hybrid models of gene regulatory networks-a PDE approach. Math. Biosci. **305**, 170–177 (2018)
27. Kwiatkowska, M., Norman, G., Parker, D.: PRISM 4.0: verification of probabilistic real-time systems. In: Gopalakrishnan, G., Qadeer, S. (eds.) CAV 2011. LNCS, vol. 6806, pp. 585–591. Springer, Heidelberg (2011). https://doi.org/10.1007/978-3-642-22110-1_47
28. Lapin, M., Mikeev, L., Wolf, V.: SHAVE: stochastic hybrid analysis of Markov population models. In: Proceedings of the 14th International Conference on Hybrid Systems: Computation and Control, pp. 311–312 (2011)
29. Mélykúti, B., Hespanha, J.P., Khammash, M.: Equilibrium distributions of simple biochemical reaction systems for time-scale separation in stochastic reaction networks. J. R. Soc. Interface **11**(97), 20140054 (2014)
30. Meyn, S.P., Tweedie, R.L.: Stability of Markovian processes III: Foster-Lyapunov criteria for continuous-time processes. Adv. Appl. Probab. **25**, 518–548 (1993)
31. Meyn, S.P., Tweedie, R.L.: Markov Chains and Stochastic Stability. Springer, London (2012). https://doi.org/10.1007/978-1-4471-3267-7
32. Meyn, S.P., Tweedie, R.L., et al.: Computable bounds for geometric convergence rates of Markov chains. Ann. Appl. Probab. **4**(4), 981–1011 (1994)
33. Mikeev, L., Neuhäußer, M.R., Spieler, D., Wolf, V.: On-the-fly verification and optimization of DTA-properties for large Markov chains. Formal Methods Syst. Des. **43**(2), 313–337 (2013)
34. Milias-Argeitis, A., Khammash, M.: Optimization-based Lyapunov function construction for continuous-time Markov chains with affine transition rates. In: 53rd IEEE Conference on Decision and Control, pp. 4617–4622. IEEE (2014)
35. Mode, C.J., Sleeman, C.K.: Stochastic Processes in Epidemiology: HIV/AIDS, Other Infectious Diseases, and Computers. World Scientific (2000)
36. Munsky, B., Khammash, M.: The finite state projection algorithm for the solution of the chemical master equation. J. Chem. Phys. **124**(4), 044104 (2006)
37. Pardoux, E.: Markov Processes and Applications: Algorithms, Networks, Genome and Finance, vol. 796. Wiley (2008)
38. Sakurai, Y., Hori, Y.: A convex approach to steady state moment analysis for stochastic chemical reactions. In: 2017 IEEE 56th Annual Conference on Decision and Control (CDC), pp. 1206–1211. IEEE (2017)
39. Schweitzer, P.J.: A survey of aggregation-disaggregation in large Markov chains. Numer. Solution Markov Chains **8**, 63–88 (1991)
40. Spieler, D.: Numerical analysis of long-run properties for Markov population models. Ph.D. thesis, Saarland University (2014)
41. Stewart, W.J.: Introduction to the Numerical Solution of Markov Chains. Princeton University Press (1994)
42. Ullah, M., Wolkenhauer, O.: Stochastic Approaches for Systems Biology. Springer, New York (2011). https://doi.org/10.1007/978-1-4614-0478-1
43. Virtanen, P., et al.: SciPy 1.0: fundamental algorithms for scientific computing in Python. Nat. Methods **17**, 261–272 (2020). https://doi.org/10.1038/s41592-019-0686-2

44. Van der Vorst, H.A.: Bi-CGSTAB: a fast and smoothly converging variant of BI-CG for the solution of nonsymmetric linear systems. SIAM J. Sci. Stat. Comput. **13**(2), 631–644 (1992)
45. Wilkinson, D.J.: Stochastic Modelling for Systems Biology. CRC Press, Boca Raton (2018)

Reasoning About Proportional Lumpability

Carla Piazza[1] and Sabina Rossi[2(✉)]

[1] Università di Udine, Udine, Italy
carla.piazza@uniud.it
[2] Università Ca' Foscari Venezia, Venice, Italy
sabina.rossi@unive.it

Abstract. In this paper we reason about the notion of *proportional lumpability*, that generalizes the original definition of lumpability to cope with the state space explosion problem inherent to the computation of the performance indices of large stochastic models. Lumpability is based on a state aggregation technique and applies to Markov chains exhibiting some structural regularity.

Proportional lumpability formalizes the idea that the transition rates of a Markov chain can be altered by some factors in such a way that the new resulting Markov chain is lumpable. It allows one to derive exact performance indices for the original process.

We prove that the problem of computing the coarsest proportional lumpability which refines a given initial partition is well-defined, i.e., it has always a unique solution. Moreover, we introduce a polynomial time algorithm for solving the problem. This provides us further insights on both the notion of proportional lumpability and on generalizations of partition refinement techniques.

Keywords: Markov chains · Lumpability · Algorithms

1 Introduction

Markov chains constitute the basic underlying semantics model of a plethora of modelling formalism for reliability analysis and performance evaluation of complex systems, such as Stochastic Petri nets [22], Stochastic Automata Networks [24], queuing networks [3] and Markovian process algebras [10,11].

Although the use of high-level specification formalisms highly simplifies the design of compositional/hierarchical quantitative models, the stochastic process underlying even a very simple model may have a large number of states that makes its analysis a difficult, sometimes impossible, task. In order to study models with a very large state space without resorting to approximation or simulation techniques we can attempt to reduce the state space of the underlying Markov chain by aggregating states with equivalent behaviours (according to a notion of equivalence that captures our concept of behaviour). An interesting class of these aggregation methods that can be decided by the structural

A. Abate and A. Marin (Eds.): QEST 2021, LNCS 12846, pp. 372–390, 2021.
https://doi.org/10.1007/978-3-030-85172-9_20

analysis of the original Markov chain is known as *lumping*. In the literature, several notions of lumping have been introduced: strong and weak lumping [15], exact lumping [25], and strict lumping [4]. The lumpability method allows one to efficiently compute the exact values of the performance indices when the model is actually lumpable. However, it is well known that not all Markov chains are lumpable. Indeed, Markov chains arising in real-life applications are, in general, not lumpable. To cope with this problem, in [7] the notion of *quasi-lumpability* has been introduced. The idea is that a quasi-lumpable Markov chain can be altered in such a way that the new resulting Markov chain is lumpable and steady state probability bounding methods [5,7,8] can be applied to the new lumpable Markov chain in order to obtain bounds on the performance indices of the original model.

In [19], the notion of *proportional lumpability* has been introduced. It extends the original definition of lumpability but, differently than the general definition of quasi-lumpability, it allows one to derive exact performance indices for the original process. In [20] we extended the work presented in [19] by comparing the notion of proportional lumpability with other definitions of lumping such as weak lumpability [15,17] and the notion of exact lumpability for ordinary differential equations (ODEs) [16,18].

The definition of proportional lumpability requires to find a function that assigns a positive coefficient to each state of the system. Being the set of all possible such functions infinite, the existence of an efficient algorithmic technique to either check or compute proportional lumpability is not an immediate consequence of the definition.

In this paper we study the properties of proportional lumpability and present two alternative characterizations of it. The first characterization has been proved in [20] and allows one to efficiently verify whether a partition of the state space of a Markov chain is induced by an equivalence relation which is a proportional lumpability. The second charaterization is a novel contribution and it is exploited to design a polynomial time algorithm to compute the coarsest proportional lumpability of a given Markov chain. Indeed, in the case of the classical notion of strong lumpability, partition refinement algorithms are at the basis of the efficient computation of the coarsest lumpability included in a given initial partition. In the same spirit, we prove that the problem of computing the coarsest proportional lumpability which refines a given initial partition is well-defined, i.e., it has always a unique solution. Moreover, we introduce a polynomial time algorithm for solving the problem. This provides us further insights on both the notion of proportional lumpability and on generalizations of partition refinement techniques.

Structure of the Paper. The paper is structured as follows: In Sect. 2 we review the theoretical background on continuous-time Markov chains and recall the concept of strong lumpability. The notion of proportional lumpability is introduced in Sect. 3 and one novel characterization of it is proved. In Sect. 4 an algorithm for proportional lumpability is presented and both its correctness and its complexity are proved. Section 5 concludes the paper.

2 Background

In this section we rapidly review the fundamentals of continuous-time Markov chains and the concept of lumpability.

Continuous-Time Markov Chains. A Continuous-Time Markov Chain (CTMC) is a stochastic process $X(t)$ for $t \in \mathbb{R}^+$ taking values into a discrete state space $\mathcal{S}$ such that the *Markov property* holds, i.e., the conditional (on both past and present states) probability distribution of its future behaviour is independent of its past evolution until the present state:

$$Prob(X(t_{n+1}) = s_{n+1} \mid X(t_1) = s_1, X(t_2) = s_2, \ldots, X(t_n) = s_n) = \\ Prob(X(t_{n+1}) = s_{n+1} \mid X(t_n) = s_n).$$

A stochastic process $X(t)$ is said to be *stationary* if the collection of random variables $(X(t_1), X(t_2), \ldots, X(t_n))$ has the same distribution as the collection $(X(t_1+\tau), X(t_2+\tau), \ldots, X(t_n+\tau))$ for all $t_1, t_2, \ldots, t_n, \tau \in \mathbb{R}^+$. A CTMC $X(t)$ is said to be *time-homogeneous* if the conditional probability $Prob(X(t+\tau) = s \mid X(t) = s')$ does not depend upon t, and is *irreducible* if every state in $\mathcal{S}$ can be reached from every other state. A state in a Markov process is called *recurrent* if the probability that the process will eventually return to the same state is one. A recurrent state is called *positive-recurrent* if the expected return time is finite. A CTMC is *ergodic* if it is irreducible and all its states are positive-recurrent. In the case of finite Markov chains, irreducibility is sufficient for ergodicity. Henceforth, we consider ergodic CTMCs.

An ergodic CTMC possesses an *equilibrium* (or *steady-state*) *distribution*, that is the *unique* collection of positive real numbers $\pi(s)$ with $s \in \mathcal{S}$ such that

$$\lim_{t \to \infty} Prob(X(t) = s \mid X(0) = s') = \pi(s).$$

Notice that the above equation for $\pi(s)$ is independent of s'. We denote by $q(s, s')$ the transition rate out of state s to state s', with $s \neq s'$, and by $q(s)$ the sum of all transition rates out of state s to any other state in the chain. A state s for which $q(s) = \infty$ is called an instantaneous state since when entered it is instantaneously left. Whereas such states are theoretically possible, we shall assume throughout that $0 < q(s) < \infty$ for each state s. The infinitesimal generator matrix $\mathbf{Q}$ of a CTMC $X(t)$ with state space $\mathcal{S}$ is the $|\mathcal{S}| \times |\mathcal{S}|$ matrix whose off-diagonal elements are the $q(s, s')$'s and whose diagonal elements are the negative sum of the extra diagonal elements of each row, i.e., $q(s, s) = -\sum_{s' \in \mathcal{S},\ s' \neq s} q(s, s')$. For the sake of simplicity, we use $q(s, s')$ to denote the components of matrix $\mathbf{Q}$. For $s \in \mathcal{S}$ and $S \subseteq \mathcal{S}$ we write $q(s, S)$ to denote $\sum_{s' \in S} q(s, s')$.

Any non-trivial vector of positive real numbers $\boldsymbol{\mu}$ satisfying the system of global balance equations (GBEs) $\boldsymbol{\mu}\mathbf{Q} = \mathbf{0}$ is called *invariant measure* of the CTMC. For an irreducible CTMC $X(t)$, if $\boldsymbol{\mu}_1$ and $\boldsymbol{\mu}_2$ are two invariant measures of $X(t)$, then there exists a constant $k > 0$ such that $\boldsymbol{\mu}_1 = k\boldsymbol{\mu}_2$. If the CTMC is ergodic, then there exists a unique invariant measure $\boldsymbol{\pi}$ whose components sum to unity, i.e., $\sum_{s \in \mathcal{S}} \pi(s) = 1$. In this case $\boldsymbol{\pi}$ is the *equilibrium* or *steady-state distribution* of the CTMC.

Strong Lumpability. In the context of performance and reliability analysis, the notion of *lumpability* provides a model aggregation technique that can be used for generating a Markov chain that is smaller than the original one but allows one to determine exact results for the original process.

The concept of lumpability can be formalized in terms of equivalence relations over the state space of the Markov chain. Any such equivalence induces a *partition* on the state space of the Markov chain and aggregation is achieved by clustering equivalent states into macro-states, thus reducing the overall state space. If the partition can be shown to satisfy the so-called *strong* lumpability condition [2,15], then the equilibrium solution of the aggregated process may be used to derive an exact solution of the original one.

The notion of strong lumpability has been introduced in [15] and further studied in [1,4,21,26].

Definition 1 (Strong lumpability). *Let $X(t)$ be a CTMC with state space $\mathcal{S}$ and $\sim$ be an equivalence relation over $\mathcal{S}$. We say that $X(t)$ is* strongly lumpable *with respect to $\sim$ (resp., $\sim$ is a* strong lumpability *for $X(t)$) if $\sim$ induces a partition on the state space of $X(t)$ such that for any equivalence class $S_i, S_j \in \mathcal{S}/\sim$ with $S_i \neq S_j$ and $s, s' \in S_i$,*

$$q(s, S_j) = q(s', S_j)\,.$$

Thus, an equivalence relation over the state space of a Markov process is a strong lumpability if it induces a partition into equivalence classes such that for any two states within an equivalence class their aggregated transition rates to any other class are the same. Notice that every Markov process is strongly lumpable with respect to the identity relation, and also with respect to the trivial relation having only one equivalence class.

In [15] the authors prove that for an equivalence relation $\sim$ over the state space of a Markov process $X(t)$, the aggregated process is a Markov process for every initial distribution if, and only if, $\sim$ is a strong lumpability for $X(t)$. Moreover, the transition rate between two aggregated states $S_i, S_j \in \mathcal{S}/\sim$ is equal to $q(s, S_j)$ for any $s \in S_i$.

Proposition 1 (Aggregated process for strong lumpability). *Let $X(t)$ be a CTMC with state space $\mathcal{S}$, infinitesimal generator $\mathbf{Q}$ and equilibrium distribution $\boldsymbol{\pi}$. Let $\sim$ be a strong lumpability for $X(t)$ and $\widetilde{X}(t)$ be the aggregated process with state space $\mathcal{S}/\sim$ and infinitesimal generator $\widetilde{\mathbf{Q}}$ defined by: for any equivalence class $S_i, S_j \in \mathcal{S}/\sim$,*

$$\widetilde{q}(S_i, S_j) = q(s, S_j)$$

for any $s \in S_i$. Then the equilibrium distribution $\widetilde{\boldsymbol{\pi}}$ of $\widetilde{X}(t)$ is such that for any equivalence class $S \in \mathcal{S}/\sim$,

$$\widetilde{\pi}(S) = \sum_{s \in S} \pi(s)\,.$$

3 Proportional Lumpability

The notion of *proportional lumpability* has been introduced in [19]. As the notion of *quasi-lumpability* [7], also called *near-lumpability* in [4], proportional lumpability extends the original definition of strong lumpability but, differently from the general definition of quasi-lumpability, it allows one to derive an exact solution of the original process.

Definition 2 (Proportional lumpability). *Let $X(t)$ be a CTMC with state space $\mathcal{S}$ and $\sim$ be an equivalence relation over $\mathcal{S}$. We say that $X(t)$ is* proportionally lumpable *with respect to $\sim$ (resp., $\sim$ is a* proportional lumpability *for $X(t)$) if there exists a function κ from $\mathcal{S}$ to $\mathbb{R}^+$ such that $\sim$ induces a partition on the state space of $X(t)$ satisfying the property that for any equivalence classes $S_i, S_j \in \mathcal{S}/\sim$ with $S_i \neq S_j$ and $s, s' \in S_i$,*

$$\frac{q(s, S_j)}{\kappa(s)} = \frac{q(s', S_j)}{\kappa(s')}.$$

*We say that $X(t)$ is κ-*proportionally lumpable *with respect to $\sim$ (resp., $\sim$ is a κ-*proportional lumpability *for $X(t)$) if $X(t)$ is* proportionally lumpable *with respect to $\sim$ and function κ.*

The following theorem [19] proves that proportional lumpability allows one to compute an exact solution for the original model.

Theorem 1 (Aggregated process for proportional lumpability). *Let $X(t)$ be a CTMC with state space $\mathcal{S}$, infinitesimal generator $\mathbf{Q}$ and equilibrium distribution $\boldsymbol{\pi}$. Let κ be a function from $\mathcal{S}$ to $\mathbb{R}^+$, $\sim$ be a κ-proportional lumpability for $X(t)$ and $\widetilde{X}(t)$ be the aggregated process with state space $\mathcal{S}/\sim$ and infinitesimal generator $\widetilde{\mathbf{Q}}$ defined by: for any equivalence classes $S_i, S_j \in \mathcal{S}/\sim$*

$$\widetilde{q}(S_i, S_j) = \frac{q(s, S_j)}{\kappa(s)}$$

for any $s \in S_i$. Then the invariant measure $\widetilde{\boldsymbol{\mu}}$ of $\widetilde{X}(t)$ is such that for any equivalence class $S \in \mathcal{S}/\sim$,

$$\widetilde{\mu}(S) = \sum_{s \in S} \pi(s)\kappa(s) . \tag{1}$$

The next Definition 3 introduces a way to perturb a proportionally lumpable CTMC in order to obtain a strongly lumpable one. In contrast with previous perturbation-based approaches, Theorem 2 gives a way to compute the stationary probabilities of a proportionally lumpable chain given those of the perturbed lumpable one. The proof of Theorem 2 is given in [19].

Definition 3 (Perturbed Markov chains). *Let $X(t)$ be a CTMC with state space $\mathcal{S}$, and infinitesimal generator $\mathbf{Q}$. Let κ be a function from $\mathcal{S}$ to $\mathbb{R}^+$. We*

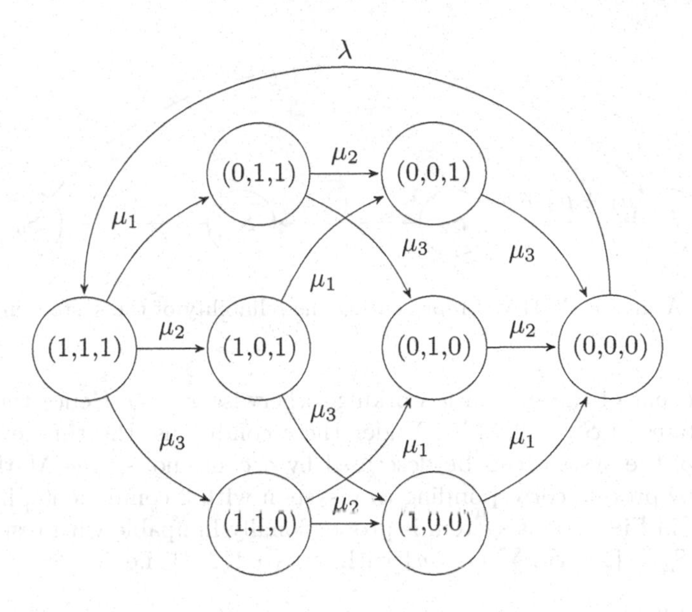

Fig. 1. CTMC representing the reliability of a system with 3 components.

say that a CTMC $X'(t)$ with infinitesimal generator $\mathbf{Q}'$ is a perturbation of $X(t)$ with respect to κ if $X'(t)$ is obtained from $X(t)$ by perturbing its rates such that for all $s, s' \in \mathcal{S}$ with $s \neq s'$,

$$q'(s, s') = \frac{q(s, s')}{\kappa(s)} .$$

Theorem 2 (Equilibrium distribution for proportional lumpability). *Let $X(t)$ be a CTMC with state space $\mathcal{S}$, infinitesimal generator $\mathbf{Q}$ and equilibrium distribution $\boldsymbol{\pi}$. Let κ be a function from $\mathcal{S}$ to $\mathbb{R}^+$. Then, for any perturbation $X'(t)$ of the original chain $X(t)$ with respect to κ according to Definition 3 with infinitesimal generator $\mathbf{Q}'$ and equilibrium distribution $\boldsymbol{\pi}'$, the equilibrium distribution $\boldsymbol{\pi}$ of $X(t)$ satisfies the following property: let $K = \sum_{s \in \mathcal{S}} \pi'(s)/\kappa(s)$ then, for all $s \in \mathcal{S}$*

$$\pi(s) = \frac{\pi'(s)}{K\ \kappa(s)} .$$

Example 1. Consider the standard reliability problem for a system consisting of N components. The time to failure of each component $i \in \{1, \ldots, N\}$ is exponentially distributed with rate μ_i and it is independent of the state of the other components. This type of system has been studied in several works, like, e.g., [9,12–14,27]. As in [14], we assume that when the system fails it is restored to a new "good" state and the time it takes for this restoration is exponentially distributed with rate λ. At any point in time, the state of the system can be represented as a boolean vector of size N, $\bar{x} = (x_1, \ldots, x_N)$, where $x_i = 1$ if the

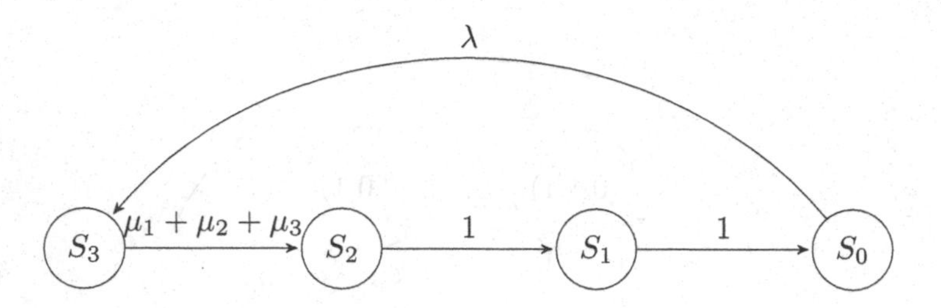

Fig. 2. Aggregated CTMC representing the reliability of the system in Fig. 1,

i-th component of the system is working, otherwise $x_i = 0$. Hence the set of all possible states is $\mathcal{S} = \{0,1\}^N$. Under these conditions, the time evolution of the state of the system can be described by a continuous time Markov chain. The Markov process corresponding to a system with 3 components, i.e., $N = 3$, is depicted in Fig. 1. This system is proportionally lumpable with respect to the partition: $S_n = \{\bar{x} \in \mathcal{S} : \sum x_i = n\}$ with $n \in \{0, 1, 2, 3\}$, i.e.,

$S_0 = \{(0,0,0)\}$
$S_1 = \{(1,0,0),(0,1,0),(0,0,1)\}$
$S_2 = \{(1,1,0),(1,0,1),(0,1,1)\}$
$S_3 = \{(1,1,1)\}$

and the function κ such that for each state $s \in S_1 \cup S_2$, $\kappa(s) = q(s)$, while for $s \in S_0 \cup S_3$, $\kappa(s) = 1$.

Thus, we can analyze the aggregated Markov chain represented in Fig. 2 and, by Theorems 1 and 2 we can compute the exact solution to the original model.

3.1 Alternative Characterizations of Proportional Lumpability

We present two alternative characterizations of proportional lumpability. The first characterization has been proved in [20] and allows one to efficiently verify whether a partition of the state space of a Markov chain is induced by a proportional lumpability. The second charaterization is a novel contribution and is exploited in the next section to design a polynomial time algorithm to compute the coarsest proportional lumpability of a given Markov chain.

First, for a given equivalence relation $\sim$ over the state space of a CTMC, we denote by $q_\sim(s)$ the sum of all transition rates from the state s to any state t such that $s \not\sim t$, i.e., for all $s \in \mathcal{S}$,

$$q_\sim(s) = \sum_{t \not\sim s} q(s,t).$$

The following theorem shows that proportional lumpability can be characterized in terms of $q_\sim(s)$ by replacing $\kappa(s)$ with $q_\sim(s)$ in the original definition.

Theorem 3 (Characterization 1 of proportional lumpability [20]). *Let $X(t)$ be an ergodic CTMC with state space $\mathcal{S}$ and $\sim$ be an equivalence relation over $\mathcal{S}$. The relation $\sim$ is a proportional lumpability for $X(t)$ if and only if for any equivalence classes $S_i, S_j \in \mathcal{S}/\sim$ with $S_i \neq S_j$ and $s, s' \in S_i$,*

1. $q_\sim(s) \neq 0$ *if and only if* $q_\sim(s') \neq 0$
2. *if* $q_\sim(s) \neq 0$ *then*
$$\frac{q(s, S_j)}{q_\sim(s)} = \frac{q(s', S_j)}{q_\sim(s')}.$$

While the above characterization can be exploited to efficiently check whether a given relation is a proportional lumpability, it is not immediate to guess how to use it within an algorithm for the computation of the proportional lumpability that refines a given initial relation. As we will see in Sect. 4, if the relation changes during the computation, $q_\sim$ also changes. So it could be the case that one of the equalities of item 2 which is not true at the current step will become true later. On the other hand, the following characterization of proportional lumpability is easier to use to define a partition refinement algorithm for proportional lumpability.

Theorem 4 (Characterization 2 of proportional lumpability). *Let $X(t)$ be an ergodic CTMC with state space $\mathcal{S}$ and $\sim$ be an equivalence relation over $\mathcal{S}$. The relation $\sim$ is a proportional lumpability for $X(t)$ if and only if for any equivalence classes $S_i, S_j, S_k \in \mathcal{S}/\sim$ with $S_i \neq S_j$, $S_i \neq S_k$, and $s, s' \in S_i$,*

1. $q(s, S_k) \neq 0$ *if and only if* $q(s', S_k) \neq 0$ *and*
2. *if* $q(s, S_k) \neq 0$, *then*
$$\frac{q(s, S_j)}{q(s, S_k)} = \frac{q(s', S_j)}{q(s', S_k)}$$

Proof. $\Rightarrow$) Suppose that $\sim$ is a κ-proportional lumpability for a function $\kappa : \mathcal{S} \rightarrow \mathbb{R}^+$, i.e., for any equivalence classes $S_i, S_j \in \mathcal{S}/\sim$ with $S_i \neq S_j$ and $s, s' \in S_i$,

$$\frac{q(s, S_j)}{\kappa(s)} = \frac{q(s', S_j)}{\kappa(s')}. \tag{2}$$

Item 1. follows by the definition of proportional lumpability. Moreover, if $q(s, S_k) \neq 0$ we have that also $q(s', S_k) \neq 0$ and

$$\frac{q(s, S_j)}{q(s, S_k)} = \frac{q(s, S_j)}{\kappa(s)} \frac{\kappa(s)}{q(s, S_k)} = \frac{q(s', S_j)}{\kappa(s')} \frac{\kappa(s')}{q(s', S_k)} = \frac{q(s', S_j)}{q(s', S_k)}.$$

$\Leftarrow$) Suppose that $\sim$ is an equivalence relation such that for any equivalence classes $S_i, S_j, S_k \in \mathcal{S}/\sim$ with $S_i \neq S_j$, $S_i \neq S_k$, and $s, s' \in S_i$,

1. $q(s, S_k) \neq 0$ if and only if $q(s', S_k) \neq 0$ and
2. if $q(s, S_k) \neq 0$, then
$$\frac{q(s, S_j)}{q(s, S_k)} = \frac{q(s', S_j)}{q(s', S_k)}$$

For each $S \in \mathcal{S}/\sim$ such that there exists $s \in S$ with $q_{\sim}(s) \neq 0$ we choose a class $B_S \neq S$ of $\mathcal{S}/\sim$ such that $q(s, B_S) \neq 0$. We define $\kappa : \mathcal{S} \longrightarrow \mathbb{R}^+$ as follows:

- if $q_{\sim}(s) = 0$, then $\kappa(s) = 1$ otherwise
- if $s \in S$, then $\kappa(s) = q(s, B_S)$.

We prove that $\sim$ is a κ-proportional lumpability. Let $S_i, S_j \in \mathcal{S}/\sim$ with $S_i \neq S_j$ and $s, s' \in S_i$

$$\frac{q(s, S_j)}{\kappa(s)} = \frac{q(s, S_j)}{q(s, B_{S_i})} = \frac{q(s', S_j)}{q(s', B_{S_i})} = \frac{q(s', S_j)}{\kappa(s')}.$$

□

3.2 Comparison with Lumpability of the Embedded Markov Chain

We compare proportional lumpability with lumpability of the embedded Markov chain [20]. The following Examples 2 and 3 are novel.

One standard approach for computing the stationary probability distribution of an ergodic continuous-time Markov chain $X(t)$ is by analyzing its embedded Markov chain $X^E(t)$. Strictly speaking, the embedded Markov chain is a regular discrete-time Markov chain (DTMC), sometimes referred to as its jump process. Given $X(t)$ with state space $\mathcal{S}$, each element of the one-step transition probability matrix of the corresponding embedded Markov chain is denoted by $p(s, s')$, and represents the conditional probability of the transition from state s into state s', defined by:

$$p(s, s') = \frac{q(s, s')}{q(s)} \qquad \text{for } s \neq s'$$

while $p(s, s) = 0$. Assuming that $X^E(t)$ is aperiodic, let $\boldsymbol{\pi}^*$ be its steady-state distribution. Then, one may derive the distribution $\boldsymbol{\pi}$ of $X(t)$ as follows: let $W = \sum_{s \in \mathcal{S}} \pi^*(s)/q(s)$, then

$$\pi(s) = \frac{\pi^*(s)}{Wq(s)}.$$

Notice that, in general, our definition of $q_{\sim}(s)$ is different from that of $q(s)$, hence the fact that $X(t)$ is proportionally lumpable does not imply that the corresponding embedded Markov chain $X^E(t)$ is lumpable.

On the other hand, if $X^E(t)$ is lumpable then $X(t)$ is proportionally lumpable with respect to function κ from $\mathcal{S}$ to $\mathbb{R}^+$ such that $\kappa(s) = q(s)$ for all $s \in \mathcal{S}$. In conclusion, we can say that if $X(t)$ has a strongly lumpable embedded process, then it is also proportional lumpable but the opposite does not hold.

Example 2. Consider again the problem of reliability for a system consisting of N components. Suppose that we are now interested in the number of components working at any point time. Thus the state space $\mathcal{S} = \{S_i : 0 \leq i \leq N\}$

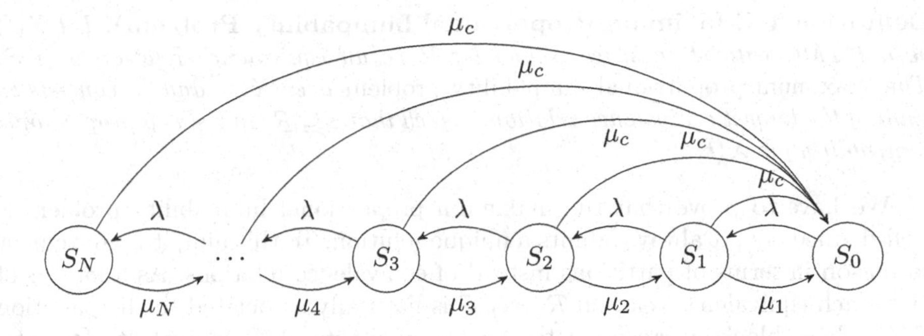

Fig. 3. CTMC for system repair model with common cause failures.

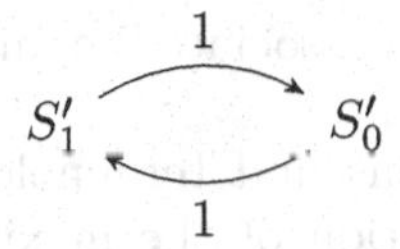

Fig. 4. Aggregated CTMC for system repair model with common cause failures.

where S_i denotes the state of the system where i components are working. We assume that in each state S_i, the time to failure of a component is exponentially distributed with rate μ_i. Each componenet can be restored with rate λ. In some cases, the system fails due to the simultaneous failure of components due to common factors. Common cause failures may arise due to the failure of common power supply, environmental conditions (e.g., earthquake, flood, humidity, etc.), common maintenance problems, etc. Simultaneous failure due to common cause may occur with failure rate μ_c. The state transition diagram for system repair model is depicted in Fig. 3.

This model is proportionally lumpable with respect to the relation $\sim$ over $\mathcal{S}$ given by the reflexive, symmetric and transitive closure of $\{(S_i, S_j) : 1 \leq i, j \leq N\}$, and the function κ such that $\kappa(S_i) = q_\sim(S_i)$ for $i \in \{0, \dots, N\}$. This relation induces two equivalence classes, $C_0 = \{S_0\}$ and $C_1 = \{S_1, \dots, S_N\}$, and the model in Fig. 3 is proportionally lumpable to the one depicted in Fig. 4.

In this case the model in Fig. 3 has not a strongly lumpable embedded process due to the fact that $q(S_i) \neq q_\sim(S_i)$ for each $i \in \{0, \dots, N\}$.

Example 3. Consider the model described in Example 1. We showed that the CTMC depicted in Fig. 1 is proportionally lumpable. It is easy to see that this model has also a strongly lumpable embedded process. Indeed, this trivially follows by Theorem 3 and the fact that $q(s) = q_\sim(s)$ for all $s \in \mathcal{S}$ where $\sim$ is the relation inducing the partition $S_n = \{\bar{x} \in \mathcal{S} : \sum x_i = n\}$ with $n \in \{0, 1, 2, 3\}$.

4 Computing Proportional Lumpability

In this section we consider the maximum proportional lumpability problem.

Definition 4 (Maximum Proportional Lumpability Problem). *Let $X(t)$ be a CTMC with state space $\mathcal{S}$ and let $\mathcal{R}$ be an equivalence relation over $\mathcal{S}$. The* maximum proportional lumpability problem *over $X(t)$ and $\mathcal{R}$ consists in finding the largest equivalence relation $\sim$ such that $\sim \subseteq \mathcal{R}$ and $\sim$ is a proportional lumpability for $X(t)$.*

We have to prove that the maximum proportional lumpability problem is well-defined, i.e., it always admits a unique solution. To this aim, it is convenient to reason in terms of partitions instead of equivalence relations. As a matter of fact, each equivalence relation $\mathcal{R}$ over $\mathcal{S}$ is naturally associated to the partition $\mathcal{S}/\mathcal{R}$ whose blocks correspond to the maximal sets of $\mathcal{R}$-equivalent elements, and vice-versa. This allows us to talk about proportional lumpabilities as both equivalence relations and partitions. In particular, a partition $\mathcal{P}$ is said to be a *proportional partition* when it is associated to an equivalence relation which is a proportional lumpability.

We introduce some notations and terminologies over partitions useful for providing an alternative definition of the maximum proportional lumpability problem.

Given two partitions $\mathcal{P}_1$ and $\mathcal{P}_2$ over $\mathcal{S}$ we say that $\mathcal{P}_1$ is *finer* than $\mathcal{P}_2$, denoted by $P_1 \sqsubseteq P_2$, if and only if for each block B_1 of $\mathcal{P}_1$ there exists a block B_2 of $\mathcal{P}_2$ such that $B_1 \subseteq B_2$. This is equivalent to say that the blocks of $\mathcal{P}_2$ are unions of blocks of $\mathcal{P}_1$. Equivalently we say that $\mathcal{P}_2$ is coarser than $\mathcal{P}_1$ (also $\mathcal{P}_1$ refines $\mathcal{P}_2$) if $\mathcal{P}_1$ is finer than $\mathcal{P}_2$.

Let $\mathcal{R}_1$ and $\mathcal{R}_2$ be two equivalence relations over $\mathcal{S}$. It holds that $\mathcal{R}_1 \subseteq \mathcal{R}_2$ if and only if the partition $\mathcal{P}_1 \equiv \mathcal{S}/\mathcal{R}_1$ associated to $\mathcal{R}_1$ is finer than the partition $\mathcal{P}_2 \equiv \mathcal{S}/\mathcal{R}_2$ associated to $\mathcal{R}_2$, i.e., $\mathcal{P}_1 \sqsubseteq \mathcal{P}_2$.

Definition 5 (Maximum Proportional Partition Problem). *Let $X(t)$ be a CTMC with state space $\mathcal{S}$, let $\mathcal{P}$ be a partition over $\mathcal{S}$. The* maximum proportional partition problem *over $X(t)$ and $\mathcal{P}$ consists in finding the coarsest proportional partition $\mathcal{P}_\sim$ refining $\mathcal{P}$.*

Proposition 2 (Equivalence of the two problems). *Let $X(t)$ be a CTMC with state space $\mathcal{S}$. Let $\mathcal{R}$ be an equivalence relation over $\mathcal{S}$ and $\mathcal{S}/\mathcal{R}$ be the partition associated to $\mathcal{R}$. $\sim$ is the solution of the maximum proportional lumpability problem over $X(t)$ and $\mathcal{R}$ if and only if the partition $\mathcal{S}/\sim$ is the solution of the maximum proportional partition problem over $X(t)$ and $\mathcal{S}/\mathcal{R}$.*

Proof. This is an immediate consequence of the definitions. □

As a consequence, from now on we will focus on the maximum proportional partition problem.

Notice that the partition $\mathcal{S}/Id$, where Id is the identity relation, is associated to the proportional lumpability Id and it is finer than any other partition $\mathcal{P}$. In other terms the set of proportional partitions that refine a given partition $\mathcal{P}$ is always not empty. However, it could be the case that for a given partition $\mathcal{P}$ such set contains different elements which are maximal with respect to the partial order $\sqsubseteq$. The following property will allow us to prove that this is never the case, i.e., that the maximum proportional partition problem has always a unique solution. The proofs of the following lemma and theorem are reported in the Appendix.

Lemma 1. *Let $X(t)$ be a CTMC with state space $\mathcal{S}$ and let $\mathcal{P}_1$ and $\mathcal{P}_2$ be two proportional partitions over $\mathcal{S}$. Let $\mathcal{P}$ be the smallest partition that is coarser than both $\mathcal{P}_1$ and $\mathcal{P}_2$. $\mathcal{P}$ is a proportional partition.*

Theorem 5 (Uniqueness). *The maximum proportional partition problem has always a unique solution.*

Partition refinement algorithms already defined in the context of bisimulation [23] and lumpabilities [1,28] are based on the following idea: at every step each existing block B is split into B_1, B_2 using a reference block S, called *splitter*, which witnesses that the elements of B_1 and B_2 are not equivalent, no matter how S will be split during the next steps. In such framework the correctness of the algorithm is proved by proving that:

ST. Step Correctness: at each step the current partition is refined into a new one that is coarser than the solution;
FC. Final Convergence: the final partition is a proportional partition.

In order to be able to proceed along the same lines, we first need to prove that the maximum proportional partition problem has a chance be solved by iteratively applying refinement steps.

Proposition 3 (Iterative Refinements). *Let $X(t)$ be a CTMC with state space $\mathcal{S}$, let $\mathcal{P}$ be a partition over $\mathcal{S}$. Let $\mathcal{P}_\sim$ be the solution of the maximum proportional partition problem over $X(t)$ and $\mathcal{P}$. If $\mathcal{P}'$ is finer than $\mathcal{P}$ and coarser than $\mathcal{P}_\sim$, i.e., $\mathcal{P}_\sim \sqsubseteq \mathcal{P}' \sqsubseteq \mathcal{P}$, then the solution of the maximum proportional partition problem over $X(t)$ and $\mathcal{P}'$ is $\mathcal{P}_\sim$.*

Proof. This is an immediate consequence of the definition of maximum proportional partition problem. □

We now focus on Step Correctness, i.e., we define splitting strategies that approaches the current partition to the result. To this aim we deeply analyse the characterization provided in Theorem 4.

Notice that if $\mathcal{P}$ has a unique class, then $\mathcal{P}$ is a proportional partition, i.e., no refinement is needed. In the case of partitions with only two classes the second condition of Theorem 4 is trivially satisfied, so we get the following characterization for such simple partitions.

Lemma 2. *Let $X(t)$ be a CTMC with state space $\mathcal{S}$, let $\mathcal{P}$ be a partition over $\mathcal{S}$ with $|\mathcal{P}| = 2$. $\mathcal{P}$ is a proportional partition if and only if for all $S_i, S_k \in \mathcal{P}$ with $S_i \neq S_k$, and $s, s' \in S_i$ it holds that*

$$q(s, S_k) \neq 0 \quad \textit{iff} \quad q(s', S_k) \neq 0$$

Algorithm 1. Fix point computation of BISIMSPLIT

```
1: function BisimSplit(X(t), P)
2:     repeat
3:         Bool = True
4:         for S, B ∈ P  with  S ≠ B do                 ▷ S splits B
5:             B₁ = {s ∈ B | q(s, S) ≠ 0}
6:             if B₁ ≠ B ∧ B₁ ≠ ∅ then
7:                 P = (P \ {B}) ∪ {B₁, B \ B₁}
8:                 Bool = False
9:     until Bool                                       ▷ Exit when Bool is True
10:    return P
```

Proof. This is an immediate consequence of Theorem 4. □

In the general case only the left to right direction of the above result continues to hold and provides us a first splitting strategy.

Lemma 3. *Let $X(t)$ be a CTMC with state space $\mathcal{S}$, let $\mathcal{P}$ be a partition over $\mathcal{S}$. If $\mathcal{P}$ is a proportional partition, then for all $S_i, S_k \in \mathcal{P}$ with $S_i \neq S_k$, and $s, s' \in S_i$ it holds that*

$$q(s, S_k) \neq 0 \qquad \textit{iff} \qquad q(s', S_k) \neq 0$$

Proof. It immediately follows from the definition of proportional lumpability. □

Hence, we can split blocks exploiting the above condition. If s and s' in S_i are such that $q(s, S_k) \neq 0$ while $q(s', S_k) = 0$, no matter how S_k will be split during the computation s and s' will always violate the condition with respect to at least one new class $S'_k \subseteq S_k$. So, we split S_i separating the elements reaching S_k from those that do not reach S_k. We call such splits BISIMSPLIT, since they are exactly the splits performed in classical strong bisimulation algorithms [23]. In Algorithm 1 we describe the function that computes these splits until a fix-point is reached.

Proposition 4 (BisimSplit Correctness). *Let $X(t)$ be a CTMC with state space $\mathcal{S}$, let $\mathcal{P}$ be a partition over $\mathcal{S}$. Let $\mathcal{P}_\sim$ be the solution of the maximum proportional partition problem over $X(t)$ and $\mathcal{P}$. Let $\mathcal{P}'$ be the partition returned by* BISIMSPLIT*$(X(t), \mathcal{P})$. $\mathcal{P}'$ is finer than $\mathcal{P}$ and coarser than $\mathcal{P}_\sim$.*

Proof. This is a consequence of Lemma 3. □

At this point we focus on the second condition of Theorem 4 and we translate it in a splitting strategy. If s and s' satisfy the first condition of Theorem 4, but not the second one, then one could believe that next splits on S_j and S_k could avoid the problem. In other terms it could be possible for s and s' to remain in the same block thanks to changes in S_j and S_k. The following result proves that this is never the case. The proof is reported in the Appendix.

Lemma 4. *Let $X(t)$ be a CTMC with state space $\mathcal{S}$, let $\mathcal{P}$ be a partition over $\mathcal{S}$. Let $\mathcal{P}_\sim$ be the solution of the maximum proportional partition problem over $X(t)$ and $\mathcal{P}$. If there exist $S_i, S_j, S_k \in \mathcal{P}$ with $S_i \neq S_j$, $S_i \neq S_k$, and $s, s' \in S_i$ such that $q(s, S_k) \neq 0$, $q(s', S_k) \neq 0$, and*

$$\frac{q(s, S_j)}{q(s, S_k)} \neq \frac{q(s', S_j)}{q(s', S_k)}$$

then s and s' belong to different blocks in $\mathcal{P}_\sim$.

As a consequence we get the splitting strategy described in Algorithm 2.

Algorithm 2. Fix point computation of PROPSPLIT

```
1: function PROPSPLIT(X(t), P)
2:     repeat
3:         Bool = True
4:         for S, T ∈ P with S ≠ T do
5:             for B ∈ P with B ≠ S, B ≠ T and ∀s ∈ B it is q(s,T) ≠ 0 do
6:
7:                 ℬ = {B_1, ..., B_n} such that B_f ⊆ B and
8:                         for all s, s' ∈ B_f it is q(s,S)/q(s,T) = q(s',S)/q(s',T)
9:                 if |ℬ| > 1 then
10:                    P = (P \ {B}) ∪ ℬ
11:                    Bool = False
12:    until Bool                                        ▷ Exit when Bool is True
13:    return P
```

Proposition 5 (PropSplit Correctness). *Let $X(t)$ be a CTMC with state space $\mathcal{S}$, let $\mathcal{P}$ be a partition over $\mathcal{S}$. Let $\mathcal{P}_\sim$ be the solution of the maximum proportional partition problem over $X(t)$ and $\mathcal{P}$. Let $\mathcal{P}'$ be the partition returned by* PROPSPLIT*$(X(t), \mathcal{P})$. $\mathcal{P}'$ is finer than $\mathcal{P}$ and coarser than $\mathcal{P}_\sim$.*

Proof. This is a consequence of Lemma 4. □

The algorithm we propose for solving the maximum proportional partition problem alternatively applies the two above described splitting strategies until a fix point is reached. It is described in Algorithm 3.

Since in Proposition 3 we proved that the problem can be solved through an iterative algorithm and in Propositions 4 and 5 we provided the Step Correctness, it only remains to prove that the final result is a proportional partition and to analyse the complexity of the procedure.

Theorem 6 (Correctness and Complexity). *Let $X(t)$ be a CTMC with state space $\mathcal{S}$, let $\mathcal{P}$ be a partition over $\mathcal{S}$.* MAXPROP*$(X(t), \mathcal{P})$ returns the solution of the maximum proportional partition problem over $X(t)$ and $\mathcal{P}$ in time $O(|\mathcal{S}|^4)$.*

Algorithm 3. Fix point computation of the Maximum Proportional Partition

```
1: function MaxProp(X(t), P)
2:     repeat
3:         P' = P
4:         P = PropSplit(X(t),BisimSplit(X(t), P))
5:     until P = P'
6:     return P
```

Proof. As far as correctness is concerned, in virtue of Propositions 3, 4, and 5 we only have to prove that the output of the algorithm is a proportional partition. The output of the algorithm is a fix-point for the function PropSplit($X(t)$,BisimSplit($X(t)$, _)). We have that BisimSplit implements the first condition of Theorem 4 and PropSplit implement the second condition of Theorem 4. So, since Theorem 4 is a characterization for proportional lumpability, the output of the algorithm is a proportional partition.

During the computation $O(|\mathcal{S}|)$ splits will be performed by either BisimSplit or PropSplit, since in the worst case the final partition has $\Theta(|\mathcal{S}|)$ blocks. Each split performed by BisimSplit can be computed in time $O(|\mathcal{S}|^2)$, e.g., by exploiting [23]. As for the splits performed by PropSplit, from the infinitesimal generator of $X(t)$ and the blocks of the current partition in time $\Theta(|\mathcal{S}|^2)$ we can compute a matrix in which for each state s and each class S we store $q(s, S)$. This matrix has size $O(|\mathcal{S}^2|)$. Each block T of the current partition corresponds to a column t in the matrix. For each column t we compute a new matrix in which for each row s having $q(s, t) \neq 0$ we normalize all the row dividing by $q(s, t)$. This take time $O(|\mathcal{S}|^2)$ and allow us to split each class B with respect to all other classes S, through a single complete scan of the matrix. Hence, for each normalizer T we need time $O(|\mathcal{S}|^2)$. Since T has $O(|\mathcal{S}|)$ possible values, one split of PropSplit requires $O(|\mathcal{S}|^3)$. □

Notice that the above complexity result can be refined by exploiting adjacency lists, hence replacing a factor $|\mathcal{S}|^2$ by the number of non-null elements of the infinitesimal generator of $X(t)$.

5 Conclusion

In this paper we recall the notion of proportional lumpability and present two characterizations of it. These characterizations allow us to develop a computational method for proportional lumpability. More precisely, the first characterization has been proved in [20] and can be exploited to efficiently check whether a given relation is a proportional lumpability, while the second characterization is a novel contribution and allows us to develop an algorithm for the computation of the proportional lumpability that refines a given initial relation.

The algorithm we presented for proportional lumpability at the moment does not exploit any ad-hoc technique for reducing the computational complexity, such as the process the smallest half policy presented in [23] for bisimulation computation and extended to lumpability in [6,28]. As future work we plan to investigate along this direction.

A Appendix

Proof of Lemma 1

First notice that each block $A \in \mathcal{P}$ can be written both as a union of blocks of $\mathcal{P}_1$ and as a union of blocks of $\mathcal{P}_2$, i.e.,

$$A = A_{11} \cup A_{12} \cup \cdots \cup A_{1k_1} = A_{21} \cup A_{22} \cup \cdots \cup A_{2k_2}$$

with $A_{ij} \in \mathcal{P}_i$.

Since $\mathcal{P}_1$ and $\mathcal{P}_2$ are proportional partitions, there exist two functions κ_1, κ_2 from $\mathcal{S}$ to $\mathbb{R}^+$ that witness this fact. This implies that if we take two states s and s' which are not in A and are in a block B_i of $\mathcal{P}_i$, it holds that:

$$\frac{q(s,A)}{\kappa_i(s)} = \frac{\sum_{j=1}^{k_i} q(s, A_{ij})}{\kappa_i(s)} = \frac{\sum_{j=1}^{k_i} q(s', A_{ij})}{\kappa_i(s')} = \frac{q(s',A)}{\kappa_i(s')}$$

This last can be rewritten as:

$$q(s,A) = \frac{\kappa_i(s)}{\kappa_i(s')} q(s',A)$$

For each block $B \in \mathcal{P}$ we fix a representative element $b \in B$. For each $b' \in B$ there exists at least one finite sequence $b_0, b_1, \ldots, b_m$ such that $b_0 = b$, $b_m = b'$ and for each $h = 0, \ldots, m-1$ there exists B_h such that $b_h, b_{h+1} \in B_h$ and either $B_h \in \mathcal{P}_1$ or $B_h \in \mathcal{P}_2$. For each $b' \in B$ we fix one of such sequences. For the sake of clarity, let us consider a simple case where $b, b_1 \in B_0 \in \mathcal{P}_1$, $b_1, b_2 \in B_1 \in \mathcal{P}_2$, and $b_2, b' \in B_2 \in \mathcal{P}_1$. Let $A \in \mathcal{P}$ with $A \neq B$. In virtue of the last equation, we have:

$$q(b,A) = \frac{\kappa_1(b_1)}{\kappa_1(b)} q(b_1,A) = \frac{\kappa_1(b)}{\kappa_1(b_1)} \frac{\kappa_2(b_1)}{\kappa_2(b_2)} q(b_2,A) = \frac{\kappa_1(b)}{\kappa_1(b_1)} \frac{\kappa_2(b_1)}{\kappa_2(b_2)} \frac{\kappa_1(b_2)}{\kappa_1(b')} q(b',A)$$

In the general case we obtain:

$$q(b,A) = K(b,b') q(b',A)$$

where $K(b,b')$ is a product of fractions involving values of κ_1 and κ_2 that depends on the sequence that we have fixed from b to b'. Since both b and the sequence have been fixed we can define $\overline{K}(b') = K(b,b')$. As a consequence, if $b', b'' \in B$ we obtain that for each $A \in \mathcal{P}$ with $A \neq B$ it holds

$$\overline{K}(b') q(b',A) = q(b,A) = \overline{K}(b'') q(b'',A)$$

This means that $\mathcal{P}$ is a proportional partition. □

Proof of Lemma 4

Let $S_j = A_1 \cup \cdots \cup A_n$ and $S_k = B_1 \cup \ldots B_m$ with $A_f, B_h \in \mathcal{P}_\sim$. Let κ be a function witnessing that $\mathcal{P}_\sim$ is a proportional lumpability. If by contradiction there exists a block $C \in \mathcal{P}_\sim$ such that $s, s' \in C$, then we would have

$$\frac{q(s, A_f)}{\kappa(s)} = \frac{q(s', A_f)}{\kappa(s')}$$

for each $f = 1, \ldots, n$ and

$$\frac{q(s, B_h)}{\kappa(s)} = \frac{q(s', B_h)}{\kappa(s')}$$

for each $h = 1, \ldots, m$. As a consequence by summing for $f = 1, \ldots, n$ and $h = 1, \ldots, m$ we have

$$\frac{q(s, S_j)}{\kappa(s)} = \frac{q(s', S_j)}{\kappa(s')} \text{ and } \frac{q(s, S_k)}{\kappa(s)} = \frac{q(s', S_k)}{\kappa(s')}$$

Since by hypothesis it holds $q(s, S_k) \neq 0$ and $q(s', S_k) \neq 0$ we get

$$\frac{q(s, S_j)}{q(s, S_k)} = \frac{q(s', S_j)}{q(s', S_k)}$$

which contradicts the hypothesis. □

Proof of Theorem 5

The existence of at least one solution is trivial, since the identity relation is a proportional lumpability.

As far as the uniqueness is concerned, let us consider the maximum proportional partition problem over $X(t)$ and $\mathcal{P}$. Let us assume by contradiction that the set of proportional partitions that refines $\mathcal{P}$ has at least two different maximal elements. This means that there are two different partitions $\mathcal{Q}_1$ and $\mathcal{Q}_2$ such that:

a. $\mathcal{Q}_i$ is a proportional partition;
b. $\mathcal{Q}_i$ refines $\mathcal{P}$;
c. each $\mathcal{Q}'$ coarser than $\mathcal{Q}_i$ and refining $\mathcal{P}$ is not a proportional partition.

By Lemma 1 the smallest partition $\mathcal{Q}$ that is coarser than both $\mathcal{Q}_1$ and $\mathcal{Q}_2$ is a proportional partition. Moreover, since both $\mathcal{Q}_1$ and $\mathcal{Q}_2$ refine $\mathcal{P}$, it holds that $\mathcal{Q}$ refines $\mathcal{P}$. This contradicts item c. □

References

1. Alzetta, G., Marin, A., Piazza, C., Rossi, S.: Lumping-based equivalences in Markovian automata: algorithms and applications to product-form analyses. Inf. Comput. **260**, 99–125 (2018). https://doi.org/10.1016/j.ic.2018.04.002
2. Baarir, S., Beccuti, M., Dutheillet, C., Franceschinis, G., Haddad, S.: Lumping partially symmetrical stochastic models. Perform. Eval. **68**(1), 21–44 (2011). https://doi.org/10.1016/j.peva.2010.09.002
3. Balsamo, S., Marin, A.: Queueing networks. In: Bernardo, M., Hillston, J. (eds.) SFM 2007. LNCS, vol. 4486, pp. 34–82. Springer, Heidelberg (2007). https://doi.org/10.1007/978-3-540-72522-0_2
4. Buchholz, P.: Exact and ordinary lumpability in finite Markov chains. J. Appl. Probab. **31**, 59–75 (1994). https://doi.org/10.1017/S0021900200107338
5. Courtois, P.J., Semal, P.: Computable bounds for conditional steady-state probabilities in large Markov chains and queueing models. IEEE J. Sel. Areas Commun. **4**(6), 926–937 (1986)
6. Derisavi, S., Hermanns, H., Sanders, W.H.: Optimal state-space lumping in Markov chains. Elsevier Inf. Process. Lett. **87**(6), 309–315 (2003)
7. Franceschinis, G., Muntz, R.: Bounds for quasi-lumpable Markov chains. Perform. Eval. **20**(1–3), 223–243 (1994). https://doi.org/10.1016/0166-5316(94)90015-9
8. Franceschinis, G., Muntz, R.: Computing bounds for the performance indices of quasi-lumpable stochastic well-formed nets. IEEE Trans. Software Eng. **20**(7), 516–525 (1994). https://doi.org/10.1109/32.297940
9. Frostig, E.: Jointly optimal allocation of a repairman and optimal control of service rate for machine repairman problem. Eur. J. Oper. Res. **116**(2), 274–280 (1999)
10. Hermanns, H.: Interactive Markov Chains. LNCS, vol. 2428. Springer, Heidelberg (2002). https://doi.org/10.1007/3-540-45804-2
11. Hillston, J.: A Compositional Approach to Performance Modelling. Cambridge Press (1996). https://doi.org/10.1017/CBO9780511569951
12. Hooghiemstra, G., Koole, G.: On the convergence of the power series algorithm. Perform. Eval. **42**(1), 21–39 (2000)
13. Katehakis, M., Derman, C.: Optimal repair allocation in a series system. Math. Oper. Res. **9**(4), 615–623 (1984)
14. Katehakis, M., Smit, L.: A successive lumping procedure for a class of Markov chains. Probab. Eng. Inf. Sci. **26**(4), 483–508 (2012)
15. Kemeny, J.G., Snell, J.L.: Finite Markov Chains. Springer, New York (1976)
16. Kuo, J., Wei, J.: Lumping analysis in monomolecular reaction systems. Analysis of approximately Lumpable system. Ind. Eng. Chem. Fundam. **8**(1), 124–133 (1969)
17. Ledoux, J.: A necessary condition for weak lumpability in finite Markov processes. Oper. Res. Lett. **13**(3), 165–168 (1993)
18. Li, G., Rabitz, H.: A general analysis of exact lumping in chemical kinetics. Chem. Eng. Sci. **44**(6), 1413–1430 (1989)
19. Marin, A., Piazza, C., Rossi, S.: Proportional Lumpability. In: André, É., Stoelinga, M. (eds.) FORMATS 2019. LNCS, vol. 11750, pp. 265–281. Springer, Cham (2019). https://doi.org/10.1007/978-3-030-29662-9_16
20. Marin, A., Piazza, C., Rossi, S.: Proportional lumpability and proportional bisimilarity. Acta Informatica (2021). https://doi.org/10.1007/s00236-021-00404-y
21. Marin, A., Rossi, S.: On the relations between Markov chain lumpability and reversibility. Acta Informatica **54**(5), 447–485 (2017). https://doi.org/10.1007/s00236-016-0266-1

22. Molloy, M.K.: Performance analysis using stochastic petri nets. IEEE Trans. Comput. **31**(9), 913–917 (1982). https://doi.org/10.1109/TC.1982.1676110
23. Paige, R., Tarjan, R.E.: Three partition refinement algorithms. SIAM J. Comput. **16**(6), 973–989 (1987)
24. Plateau, B.: On the stochastic structure of parallelism and synchronization models for distributed algorithms. SIGMETRICS Perf. Eval. Rev. **13**(2), 147–154 (1985). https://doi.org/10.1145/317795.317819
25. Schweitzer, P.: Aggregation methods for large Markov chains. In: Proceedings of the International Workshop on Computer Performance and Reliability, pp. 275–286. North Holland (1984)
26. Sumita, U., Rieders, M.: Lumpability and time-reversibility in the aggregation-disaggregation method for large Markov chains. Commun. Stat. Stoch. Models **5**, 63–81 (1989). https://doi.org/10.1080/15326348908807099
27. Ungureanu, V., Melamed, B., Katehakis, M., Bradford, P.: Deferred assignment scheduling in cluster-based servers. Clust. Comput. **9**(1), 57–65 (2006)
28. Valmari, A., Franceschinis, G.: Simple $O(m \log n)$ time Markov chain lumping. In: Esparza, J., Majumdar, R. (eds.) TACAS 2010. LNCS, vol. 6015, pp. 38–52. Springer, Heidelberg (2010). https://doi.org/10.1007/978-3-642-12002-2_4

Lumpability for Uncertain Continuous-Time Markov Chains

Luca Cardelli[1], Radu Grosu[2], Kim G. Larsen[3], Mirco Tribastone[4], Max Tschaikowski[3(✉)], and Andrea Vandin[5,6]

[1] University of Oxford, Oxford, UK
[2] TU Wien, Vienna, Austria
[3] Aalborg University, Aalborg, Denmark
tschaikowski@cs.aau.dk
[4] IMT School for Advanced Studies Lucca, Lucca, Italy
[5] Sant'Anna School of Advanced Studies, Pisa, Italy
[6] DTU Technical University of Denmark, Lyngby, Denmark

Abstract. The assumption of perfect knowledge of rate parameters in continuous-time Markov chains (CTMCs) is undermined when confronted with reality, where they may be uncertain due to lack of information or because of measurement noise. In this paper we consider uncertain CTMCs, where rates are assumed to vary non-deterministically with time from bounded continuous intervals. This leads to a semantics which associates each state with the reachable set of its probability under all possible choices of the uncertain rates. We develop a notion of lumpability which identifies a partition of states where each block preserves the reachable set of the sum of its probabilities, essentially lifting the well-known CTMC ordinary lumpability to the uncertain setting. We proceed with this analogy with two further contributions: a logical characterization of uncertain CTMC lumping in terms of continuous stochastic logic; and a polynomial time and space algorithm for the minimization of uncertain CTMCs by partition refinement, using the CTMC lumping algorithm as an inner step. As a case study, we show that the minimizations in a substantial number of CTMC models reported in the literature are *robust* with respect to uncertainties around their original, fixed, rate values.

1 Introduction

Motivation. Continuous-time Markov chains (CTMCs) are a fundamental tool for describing a wide range of natural and engineered systems and serve as the underlying semantics for several formalisms such as stochastic Petri Nets [13], stochastic process algebra (e.g., [29,30]), and chemical reaction networks [22]. A CTMC is typically characterized by a number of parameters such as arrival and service rates in a queuing network [45], transmission and infection rates of epidemic processes [43], and the kinetic rates of a chemical reaction. In essentially all practical situations, however, knowing the values of all parameters *precisely* is unlikely. This may be due to measurement noise when parameters are to be

A. Abate and A. Marin (Eds.): QEST 2021, LNCS 12846, pp. 391–409, 2021.
https://doi.org/10.1007/978-3-030-85172-9_21

estimated from observations, as well as to our inability to accurately observe events at certain spatio-temporal scales—a well-known problem notably arising in computational systems biology [11]. In addition, sometimes the modeler wishes to be deliberately imprecise about the value of certain parameters in order to explicitly account for the disagreement between the real system and its model.

These motivations have stimulated a vigorous line of research into quantitative modeling frameworks where *uncertainty* is a first-class citizen, with the basic idea to replace known constants with *sets* of values which can be nondeterministically assigned to parameters. A prominent instance is Jonnson and Larsen's interval specification systems [31] (equivalent to interval-valued finite Markov chains [35]), where the probability of making a transition between two states of a discrete-time Markov chain is assumed to be taken from a continuous interval of possible values, later generalized to polynomial constraints [7].

Contributions. In this paper we consider uncertain CTMCs (UCTMCs). They allow time-varying nondeterministic uncertainty in the values of the rate parameters within given bounded intervals. This is essentially the continuous-time analogue of the model of nondeterminism in [20,42], and can be seen as an over-approximation for a time-invariant interpretation of uncertainty which underlies a family of CTMCs, one for each possible choice of rate parameter values [31].

Here we study minimization of UCTMCs, motivated by the appeal to work with models of smaller size that still preserve quantities of interest for analysis and verification purposes. We proceed by means of analogies with the well-known CTMCs counterpart of ordinary lumpability [4,34] (reviewed in Sect. 2):

- CTMC ordinary lumpability identifies a partition of the state space which induces a lumped chain where each macro-state represents a partition block; the probability of being in each macro-state at all time points is equal to the sum of the probabilities of the states of the original CTMC belonging to that block [4]. The semantics of UCTMCs associates each state with the reachable set of the probabilities of that state under all possible values of the uncertain transition rates at any time point. *Mutatis mutandis*, our notion of lumpability is such that the lumped UCTMC preserves reachable sets of sums of the states in each block. In fact, UCTMC lumpability turns out to be a conservative extension of CTMC lumpability.
- We study the logical characterization of UCTMC lumpability. Similarly to the characterization of continuous stochastic logic (CSL) [1] by F-bisimulation [1], a notion closely related to ordinary lumpability, we prove that UCTMC lumpability preserves a conservative extension of CSL to UCTMCs, where a CSL formula is satisfied by a UCTMC if it is true for all possible rate values.
- CTMCs enjoy an efficient minimization algorithm based on partition refinement which computes the coarsest ordinarily lumpable partition that refines a given initial partition of states [17,46]. Here we develop an analogous algorithm for UCTMCs where the CTMC lumping algorithm is used as an inner step: the coarsest UCTMC lumpable partition is the coarsest one that refines both of the two time-homogeneous CTMCs derived by choosing the lower and

upper bounds for all uncertainty intervals, respectively. Thus, the minimization algorithm takes $\mathcal{O}(r s \log s)$ steps in the worst case, where r is the number of transitions and s is the number of states of the UCTMC.

As an application, we consider the problem of analyzing the "robustness" of CTMC lumping, i.e., to what extent the minimization depends on the specific choice of rate parameters of a model. Using a prototype implementation, we study how adding uncertainty intervals around the constant values of the rates of a CTMC model preserves the original CTMC lumping.

Further Related Work. A UCTMC can be seen as a continuous-time Markov decision process (MDP). Indeed, we formally show in Sect. 3.3 that the UCTMC can be alternatively given as a time-inhomogeneous continuous-time MDP with an uncountable action space, which represents the values within the uncertainty intervals, see [42] and [23, Section 2.2]. This model of uncertainty is different from the state of the art concerned with MDPs where the action space is finite and/or policies are time-independent (alternatively, untimed or time-invariant), see for instance [5,6,25,39]. Another related model is that of parametric Markov chains and parametric MDPs [14,26,36], where certain transition probabilities have symbolic parameters. A parametric model underlies an (infinite) family of Markov models, one for each possible evaluation of the parameters. However, each member of this family is time-invariant because the instantiation of the parameters is assumed fixed throughout the time course evolution of the process.

Most notions of lumpability and bisimulation for these models of uncertainty impose constraints that must hold for all actions (in the case of MDPs [26, 39,44]) or, analogously, for all parameter evaluations (for parametric Markov chains [26]). Instead, our notion of lumping can aggregate states even when realizations of the uncertain transition rates make the resulting time-inhomogeneous Markov chain not lumpable. In order to clarify this difference, let us consider the simple graph structure in the right inset. If $q_{2,1}$ and $q_{3,1}$ are constant values, then the graph represents a continuous-time Markov chain. In this case, states 2 and 3 can be aggregated by ordinary lumpability if $q_{2,1} = q_{3,1}$. In the case of a parametric Markov chain, $q_{2,1}$ and $q_{3,1}$ can be expressions over parameters; yet, parametric Markov chain lumping requires these two expressions to be equal for all possible assignments of the parameters [26]—hence, each member of the family of Markov chains will be ordinarily lumpable. A similar remark applies to lumpability of parametric MDPs. Indeed, if $q_{i,j}(a)$ denotes the transition rate from state i into state j in the case of any action a, the lumpability condition requires that $q_{2,1}(a) = q_{3,1}(a)$. Instead, a UCTMC has bounded intervals as transitions. Applied to this simple example, our proposed notion of lumpability will require that the intervals of both transitions be equal; however, according to the semantic interpretation of a UCTMC, this model underlies behavior in the form of (time-varying) CTMCs which have different transition rates when the uncertainty is resolved.

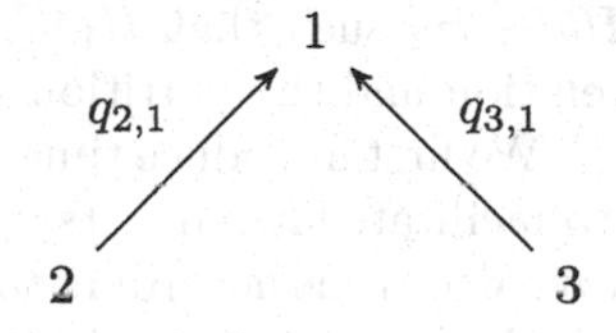

The closest notion to UCTMC lumping is the alternating probabilistic bisimulation considered in [27] for discrete-time interval MDPs. Similarly to us, alternating probabilistic bisimulation: (i) does not require that realizations of the uncertain transition probabilities make the discrete-time Markov chain lumpable; (ii) can be computed in polynomial time; (iii) preserves quantitative and logical properties; however, in [27] it is not proved that the bisimulation is indeed necessary for the preservation of such properties. We relate UCTMC lumping to alternating probabilistic bisimulation by defining an approximation for the continuous-time MDP interpretation of the UCTMC that discretizes both time and the action space using an MDP with a probabilistic scheduler. On this discretized model, we show that a UCTMC lumping does correspond to a probabilistic alternating bisimulation, see Sect. 3.4.

Paper Organization. Sect. 2 provides the background, while Sect. 3 introduces UCTMCs and discusses techniques for their analysis. Section 4, instead, introduces UCTMC lumpability, its quantitative and logical characterization, and an algorithm for the computation of the coarsest UCTMC lumpability. Section 5 continues with an evaluation of UCTMC lumpability on a set of benchmarks from the literature, while Sect. 6 concludes the paper.

2 Preliminaries

In this section we fix the notation and briefly recall the definitions of CTMCs and lumpability that will be used throughout the paper.

Notation. We use ∂_t to denote derivative with respect to time t, while x^T is the transpose of a vector x. Pointwise equivalence of functions is denoted by $\equiv$, while $:=$ signifies a definition. Given two partitions $\mathcal{H}_1$ and $\mathcal{H}_2$ of a set $\mathcal{V}$, we say that $\mathcal{H}_1$ is a refinement of $\mathcal{H}_2$ if for any $H_1 \in \mathcal{H}_1$ there exists a (unique) $H_2 \in \mathcal{H}_2$ such that $H_1 \subseteq H_2$. We shall not distinguish among an equivalence relation and the partition induced by it.

We first introduce time-inhomogeneous (alternatively, time-varying) CTMCs. To facilitate later results, throughout this paper we assume that transition rates vary with time according to *uniformly piecewise analytic* functions, i.e., functions which are analytic and bounded on all intervals $[kh; (k+1)h)$, where $k \geq 0$ is an integer and $h > 0$ is a given fixed time step.

Definition 1 (CTMC). *A time-varying CTMC is a tuple* $(\mathcal{V}, Q)$ *where* $\mathcal{V}$ *is a set of states* $\mathcal{V} = \{1, \ldots, n\}$*, while* $Q = (q_{i,j})_{i,j}$ *is a time-varying transition rate matrix such that* $q_{i,j} : \mathbb{R}_{\geq 0} \to \mathbb{R}_{\geq 0}$ *is a uniformly piecewise analytic transition rate function from* i *into* j*.* □

The following result relates the (transient) probability distributions of $(\mathcal{V}, Q)$ to the Kolmogorov equations for time-varying transition rates [23, Section 2.2].

Theorem 1. *Given a CTMC $(\mathcal{V}, Q)$ and an initial probability distribution $\pi[0]$, the probability distributions $\pi(t)$ exist and satisfy, for all $t \in \mathbb{R}_{\geq 0}$, the Kolmogorov equation*[1]

$$\partial_t \pi(t)^T = \pi(t)^T Q(t), \qquad \text{where } \pi(0) = \pi[0]. \tag{1}$$

Thanks to Theorem 1, ordinary lumpability for time-varying CTMCs is a straightforward generalization of ordinary lumpability for time-homogeneous CTMCs (e.g., [4]). The next well-known result provides a quantitative characterization of ordinary lumpability.

Theorem 2 (Ordinary Lumpability). *Given a CTMC $(\mathcal{V}, Q)$, a partition $\mathcal{H}$ of the set of states $\mathcal{V}$ is an ordinary lumping if*

$$\sum_{j \subset H'} q_{i_1,j} \equiv \sum_{j \subset H'} q_{i_2,j}, \text{ for all } H, H' \in \mathcal{H} \text{ and } i_1, i_2 \in H,$$

The lumped CTMC $(\hat{\mathcal{V}}, \hat{Q})$ is given by

- *States $\hat{\mathcal{V}} := \{i_H \mid H \in \mathcal{H}\}$, where $i_H \in H$ is an arbitrary representative of H.*
- *Transition rate matrix $\hat{Q} = (\hat{q}_{i_H,i_{H'}})_{H,H'}$, where*

$$\hat{q}_{i_H,i_{H'}} := \sum_{j \in H'} q_{i_H,j} \qquad \text{forall } H, H' \in \mathcal{H}.$$

If the initial probability distribution of $(\hat{\mathcal{V}}, \hat{Q})$ is defined by $\hat{\pi}[0]_{i_H} = \sum_{i \in H} \pi[0]_i$ for all $H \in \mathcal{H}$ and the transient probability distributions of $(\hat{\mathcal{V}}, \hat{Q})$ are denoted by $\hat{\pi}$, the following holds.

- *If $\mathcal{H}$ is an ordinary lumping, then $\hat{\pi}_{i_H} \equiv \sum_{i \in H} \pi_i$ for all $H \in \mathcal{H}$ and $\pi[0]$.*
- *If $\mathcal{H}$ is such that $\hat{\pi}_{i_H} \equiv \sum_{i \in H} \pi_i$ for all $H \in \mathcal{H}$ and $\pi[0]$, then $\mathcal{H}$ is an ordinary lumping.*

3 Uncertain Continuous-Time Markov Chains

UCTMCs allow transition rates to vary non-deterministically with time within bounded continuous intervals. After the formal introduction of the model (Sect. 3.1), we provide the semantics of UCTMCs both in terms of reachable sets of their probability distributions using the Kolmogorov equations (Sect. 3.2) and by means of an encoding into a time-inhomogeneous continuous-time MDP (Sect. 3.3). The time and action-space discrete approximation of the latter semantics is presented in Sect. 3.4.

---

[1] Proofs are given in the extended version available at doi.org/10.5281/zenodo.4699211.

3.1 Model Definition

Definition 2 (Uncertain CTMC). *An uncertain CTMC* $(\mathcal{V}, m, M)$ *is a set of states* $\mathcal{V} = \{1, \ldots, n\}$ *and non-negative matrices* $m = (m_{i,j})_{i,j}$ *and* $M = (M_{i,j})_{i,j}$, *with* $m \leq M$, *describing the lower and upper bounds of the transition rates, respectively.* □

According to the above definition, a UCTMC $(\mathcal{V}, m, M)$ induces two *extremal* (time-homogeneous) CTMCs $(\mathcal{V}, m)$ and $(\mathcal{V}, M)$ by fixing all lower and upper bounds, respectively, for each transition rate.

Example. Throughout of this paper, we will use the UCTMC depicted in Figure 1 as a running example. To favor intuition, it can be interpreted as a symmetric model of two components (e.g., two virtual machines) with a binary state (e.g., down/0 and up/1). Assuming independent events, each UCTMC state tracks a possible configuration of the two machines. Each transition is labeled with the interval within which the rates can vary; we use distinct symbols α, β, γ to indicate different activities of an hypothetical system under study, such as start-up, shut-down or machine migration, respectively. When all parameters are precisely known, i.e., $\underline{\alpha} = \overline{\alpha}$, $\underline{\beta} = \overline{\beta}$, and $\underline{\gamma} = \overline{\gamma}$, there is the ordinary lumping consisting of blocks {11}, {01, 10}, and {00}. In this paper we will develop the theory to capture such symmetry for UCTMCs. For this, here we also observe that the aforementioned ordinary lumping carries over to the two extremal time-homogeneous CTMCs. Indeed, it turns out that these are the only two CTMCs needed to consider for UCTMC lumpability, although the UCTMC admits time-varying behaviors that do not satisfy the conditions of CTMC ordinary lumpability stated in Theorem 2 and also mentioned in Sect. 1.

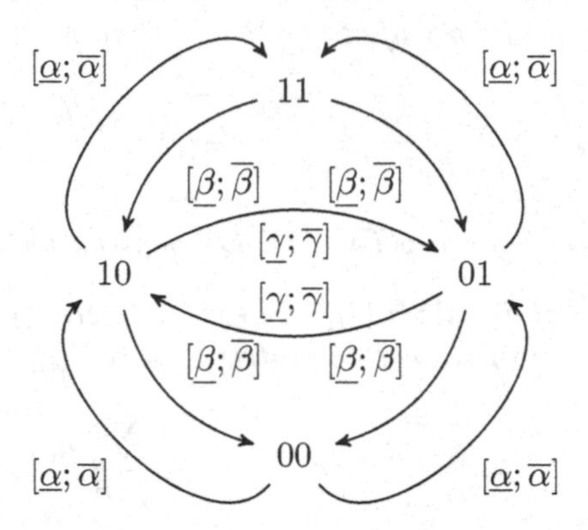

Fig. 1. Running example.

3.2 Reachable-Set Semantics

Analogously to the probability distribution of a CTMC obeying the Kolmogorov equations, the semantics of a UCTMC is given by the set of reachable probability distributions under all possible time-varying values of the transition rate matrix.

Definition 3 (UCTMC reachable-set semantics). *The semantics of a given UCTMC* $\mathcal{U} = (\mathcal{V}, m, M)$ *is provided by the reachable sets*

$$\mathcal{R}_{\mathcal{U}}(H, \tau, \pi[0]) = \Big\{ \sum_{i \in H} \pi_i(\tau) \mid \partial_t \pi(t)^T = \pi(t)^T Q(t)$$
$$\text{such that } \pi(0) = \pi[0] \text{ and } Q \text{ is admissible} \Big\},$$

where $\tau \geq 0$ *and* $H \subseteq \mathcal{V}$*, while* $Q = (q_{i,j})_{i,j}$ *is admissible if, for all* $t \geq 0$ *and* $i \neq j$*,* $q_{i,j}(t) \in [m_{i,j}; M_{i,j}]$ *and* $q_{i,j}$ *is uniformly piecewise analytic function of time.* □

Remark 1. The common notion of reachable sets is recovered by restricting H to singleton blocks only, i.e., $\{\{i\} \mid i \in \mathcal{V}\}$. We allow for general blocks because our ultimate goal is to relate sums of reachable probability distributions of a UCTMC to the reachable probability distributions of a lumped UCTMC.

The reachable-set semantics gives a concrete operational view of the model. Indeed, reachable sets can be analyzed in two ways. The first is by determining, by means of some analytical approach, the transient probabilities for all possible time-varying transition rates satisfying $m_{i,j} \leq q_{i,j}(t) \leq M_{i,j}$ that obey the ODE in Eq. 1. The second way is to compute reachable intervals by formal under- and over-approximation, using well-established techniques for uncertain dynamical systems, of which Eq. 1 are an instance, such as those implemented in C2E2, Flow* or SpaceEx, see [2,12,19] and references therein. We remark however that the aforementioned methods apply to nonlinear dynamical systems of which UCTMCs are a specific instance. While time-varying transition rates can be interpreted as control inputs that are steering the transient probabilities towards certain values, it is also possible to approximate UCTMCs by carefully chosen discrete-time MDPs (DTMDP). As discussed further below, in both cases the presence of time-varying uncertainty may result in computationally challenging problems, thus further motivating the development of efficient reduction techniques.

3.3 CTMDP Semantics

As introduced in Sect. 1, a UCTMC can also be seen as an instance of a time-inhomogeneous continuous-time MDP (CTMDP). To see this, we consider a CTMDP with the scheduler model as in [23, Section 2.2], which can be intuitively described as follows. For a sufficiently small time step $h > 0$, a CTMDP that is in state $i \in \mathcal{V}$ at some time $kh \geq 0$, where $k \geq 0$ is an integer, may choose an action a_i from $\mathcal{A}(i)$, the set of available actions in state i. With this, the CTMDP remains in state i on $[kh; kh + h)$, while at time $kh + h$ the state is:

- $j \neq i$, with probability $q_{i,j}(kh, a_i)h + o(h)$, where $o(h)$ refers to the standard small-o notation, while $q_{i,j}(kh, a_i)$ denotes the transition rate from state i into state j at time kh under action a_i;
- i, with probability $1 + q_{i,i}(kh, a_i)h + o(h)$.

Note that $q_{i,j}(kh, a_i)h + o(h)$ and $1 + q_{i,i}(kh, a_i)h + o(h)$ can be interpreted as transition probabilities of the embedded DTMC under action a_i at step k. Indeed, in the special case when the transition rates are time-invariant, the *time-homogeneous* CTMDP admit a characterization in terms of sojourn times and an embedded discrete time Markov chain, according to which the choice of action $a_i \in \mathcal{A}(i)$ upon entering state $i \in \mathcal{V}$ gives a sojourn time in state i that is exponentially distributed with rate $-q_{i,i}(a_i)$, and the probability to move into a state $j \neq i$ equal to $-q_{i,j}(a_i)/q_{i,i}(a_i)$ (see Theorem 2.8.2 in [40]).

Under this model, the discussion in [23, Section 2.2] yields the following relationship between a UCTMC and a CTMDP, where, essentially the uncountable many actions of the latter encode the uncertainty intervals of the former.

Theorem 3. *For a given UCTMC $(\mathcal{V}, m, M)$, consider the CTMDP $(\mathcal{V}, \mathcal{A}, \mathcal{M})$ where an action taken at time t in state i, denoted by $a_i(t)$, is a row vector such that each component $a_{i,j}(t)$ determines the transition rate from i into j at time t. More formally:*

- *The set of actions in state $i \in \mathcal{V}$ be given by $\mathcal{A}(i) = \prod_{j \neq i}[m_{i,j}; M_{i,j}]$;*
- *The transition rate from state i to state j at time t under action $a_i \in \mathcal{A}(i)$ is denoted by $q_{i,j}(t, a_i)$ and is given by $a_{i,j} \in [m_{i,j}; M_{i,j}]$, where $a_{i,j}$ is the j-th entry of a_i;*
- *The policies form the set $\mathcal{M}$ and are given by uniformly piecewise analytic functions $a : [0; \infty) \to \prod_{i \in \mathcal{V}} \mathcal{A}(i)$.*

For such a CTMDP, the maximization (respectively, minimization) of the probability of reaching a state in block H at the time τ corresponds to the computation of the maximal (respectively, minimal) value of the reachable set from Definition 3.

3.4 Discrete-Time Approximation of the CTMDP Semantics

In this appendix we present a discrete-time approximation of the CTMDP semantics which is of interest for a two-fold purpose. First, we show that the resulting DTMDP can be analyzed to obtain approximations of the maximal and minimal reachable probabilities for each state using dynamic programming. Second, in the proof of Theorem 9, this approximate DTMDP is used to relate the notion UCTMC lumping with the alternating probabilistic bisimulation of [27].

Instrumental to the DTMDP approximation is an alternative CTMDP encoding which uses finite action spaces, at the expense of probabilistic (instead of deterministic) policies. Before giving this encoding, we convey the main underlying idea on an illustrative example. Let us assume that we are given a CTMDP that can move from state i only into state j and that the corresponding time-dependent deterministic transition rate function is $q_{i,j}(t, a(t)) = a_{i,j}(t)$, where $m_{i,j} = 1$, $M_{i,j} = 2$ and $a_{i,j}(t) = 2 - e^{-t}$. With this, we first replace the continuous interval $[1; 2]$ with the discrete action set $\{\mathbf{m}_{i,j}, \mathbf{M}_{i,j}\}$, where the symbols $\mathbf{m}_{i,j}$ and $\mathbf{M}_{i,j}$ represent the boundary values $m_{i,j} = 1$ and $M_{i,j} = 2$, respectively. Then, the idea is to choose suitable probability functions $\mu_{\mathbf{m}_{i,j}}(t)$ and

$\mu_{\mathbf{M}_{i,j}}(t)$ such that the average transition rate from state i into state j at time t, given by $1\mu_{\mathbf{m}_{i,j}}(t) + 2\mu_{\mathbf{M}_{i,j}}(t)$, is identical to $a_{i,j}(t)$. It can be easily verified that $\mu_{\mathbf{m}_{i,j}}(t) = e^{-t}$ and $\mu_{\mathbf{M}_{i,j}}(t) = 1 - e^{-t}$ induce $a_{i,j}$.

Following [23, Section 2.2], the foregoing example can ge generalized as follows.

Proposition 1. For a given UCTMC $(\mathcal{V}, m, M)$, consider the CTMDP $(\mathcal{V}, \mathcal{A}', \mathcal{M}')$ where an action in state i at time t is taken randomly, is denoted by $a_i(t)$, and is a row vector such that each row entry $a_{i,j}(t) \in \{\mathbf{m}_{i,j}; \mathbf{M}_{i,j}\}$ determines the transition rate from i into j at time t accordingly. Formally, we have the following.

- The set of actions in state $i \in \mathcal{V}$ is given by $\mathcal{A}'(i) = \prod_{j \neq i} \{\mathbf{m}_{i,j}, \mathbf{M}_{i,j}\}$.
- The transition rate of from i into j at time t under action $a_i \in \mathcal{A}'(i)$ is $q_{i,j}(t, a_i) = v(a_{i,j})$, where $v(a_{i,j}) = m_{i,j}$ if $a_{i,j} = \mathbf{m}_{i,j}$ and $v(a_{i,j}) = M_{i,j}$ when $a_{i,j} = \mathbf{M}_{i,j}$.
- The set of policies, $\mathcal{M}'$, constitutes non-negative uniformly piecewise analytic functions μ satisfying $\sum_{a_i \in \mathcal{A}'(i)} \mu_{a_i}(t) = 1$ for all $i \in \mathcal{V}$ and $t \geq 0$. In particular, with $\mathcal{D}(X)$ denoting the set of probability measures on a set X, it holds that $\mathcal{M}'$ is a proper subset of $[0; \infty) \to \prod_{i \in \mathcal{V}} \mathcal{D}(\mathcal{A}'(i))$.

Then, the policy sets $\mathcal{M}$ and $\mathcal{M}'$, where $\mathcal{M}$ refers to the policy set given in Theorem 3, induce the same set of time-inhomogeneous CTMCs.

For a policy $\mu \in \mathcal{M}'$, the Kolmogorov equations $\partial_t \pi(t)^T = \pi(t)^T Q(t, \mu(t))$ describing the transient probabilities of the time-inhomogeneous CTMC can be solved numerically by invoking the Euler method [21], a classic approach for the numeric solution of systems of differential equations. Specifically, by discretizing time into the set $\{0, h, 2h, \ldots\}$, the probability distribution at time kh, denoted by $\pi(kh)$, is approximated by $\pi[k]$, where

$$\pi[k+1]^T := \pi[k]^T \big(I + hQ(kh, \mu(kh))\big),$$

$\pi[0] := \pi(0)$ and I is the identity matrix. Additionally to the known fact that the approximation error is $\mathcal{O}(h)$, we make the key observation that the Euler method defines a time-inhomogeneous DTMC. Indeed, similarly to the discussion in Sect. 3.3, $I + hQ(kh, \mu(kh))$ describes the transition probability matrix of the embedded time-inhomogeneous DTMC.

Together with Theorem 3 and Proposition 1, the next result allows us to formally relate UCTMCs to time-inhomogeneous DTMDPs.

Theorem 4. *Given UCTMC $(\mathcal{V}, m, M)$, set*

$$\Lambda = \max_{i \in \mathcal{V}} \Big(\sum_{j \neq i} M_{i,j} + \sum_{j \neq i} M_{j,i} \Big)$$

and fix $h \leq 1/\Lambda$. Then, $I + hQ(kh, \mu(kh))$ is a stochastic matrix for all $\mu \in \mathcal{M}'$ and $k \geq 0$. With this, consider the DTMDP $(\mathcal{V}, \mathcal{A}', \mathcal{M}'_h)$ given as:

- *The states are* $\mathcal{V}$, *while the actions in state* $i \in \mathcal{V}$ *are given by* $\mathcal{A}'(i) = \prod_{j \neq i}\{m_{i,j}; M_{i,j}\}$.
- *The transition probability from state* i *into state* j *at step* $k \geq 0$ *for* $a_i \in \mathcal{A}'(i)$ *is*

$$p_{i,j}(k, a_i) = \begin{cases} hv(a_{i,j}) & , j \neq i \\ 1 - h\sum_{j \neq i} v(a_{i,j}) & , j = i \end{cases}$$

- *The set of policies is* $\mathcal{M}'_h = \{\nu \mid \nu : \mathbb{N}_0 \to \prod_{i \in \mathcal{V}} \mathcal{D}(\mathcal{A}'(i))\}$. *In particular, for a given policy* ν, *the transition probability from state* i *into state* j *at step* $k \geq 0$ *is given by* $p_{i,j}(k, \nu(k)) = \sum_{a_i \in \mathcal{A}'(i)} \nu_{a_i}(k) p_{i,j}(k, a_i)$.

Then, for any time $\tau > 0$ *and policy* $a \in \mathcal{M}$ *such that the modulus of the derivative of each* $a_{i,j}$ *is bounded by* $\lambda \geq 0$ *almost everywhere, there exits a policy* $\nu \in \mathcal{M}'_h$ *such that*

$$\max_{i \in \mathcal{V}} |\pi_i[k] - \pi_i(\tau)| \leq h \left[\frac{3\Lambda}{2} + \frac{\lambda}{\Lambda} \max_{i \in \mathcal{V}} \deg(i)\right] (e^{\Lambda\tau} - 1) = \mathcal{O}(h),$$

where $\deg(i) = |\{j \neq i \mid m_{i,j} < M_{i,j}\}| + |\{j \neq i \mid m_{j,i} < M_{j,i}\}|$ *are the incoming and outgoing non-deterministic transitions of* i, *while* $k \geq 0$ *minimizes* $|kh - \tau|$.

Theorem 4 ensures that any $\tau \geq 0$ and any admissible transition rate matrix of the UCTMC can be matched by an approximate DTMDP such that the transition probabilities of both, the so-induced DTMC and the so-induced CTMC, are matching up to an ε at $\tau \geq 0$.

We state our first major result which relates reachability- and MDP-semantics.

Theorem 5. *For* $\tau > 0$, *a UCTMC* $\mathcal{U} = (\mathcal{V}, m, M)$ *and* $H \subseteq \mathcal{V}$, *let* k *be such that* $\tau = kh$. *Then, the maximal (minimal) probability for reaching a block* H *at* τ *coincides, by Theorem 3, with the maximum (minimum) of* $\mathcal{R}_{\mathcal{U}}(H, \tau, \pi[0])$ *from Definition 3 and can be computed in*

$$\mathcal{O}\Big(k\big(\sum_{i \in \mathcal{V}} \deg_o^{all}(i)\big)\big(\sum_{i \in \mathcal{V}} 2^{\deg_o(i)}\big)\Big),$$

where $\deg_o^{all}(i) = |\{j \neq i \mid 0 < M_{i,j}\}|$ *is the number of outgoing transitions from state* i, *while* $\deg_o(i) = |\{j \neq i \mid m_{i,j} < M_{i,j}\}|$ *is the number of outgoing non-deterministic transitions from state* i.

The complexity bound from Theorem 5 is polynomial in the number of states and exponential in $\max_i \deg_o(i)$, i.e., the maximal number of outgoing non-deterministic transitions of the approximate DTMDP.

4 UCTMC Lumpability

In Sect. 4.1 we prove that UCTMC lumpability characterizes the preservation of sums of reachable probability distributions. The logical characterization of UCTMC lumpability with respect to continuous stochastic logic is presented in Sect. 4.2. The UCTMC lumping algorithm is discussed in Sect. 4.3.

4.1 UCTMC Lumpability

Definition 4 (UCTMC Lumpability). *A partition $\mathcal{H}$ of $\mathcal{V}$ is a UCTMC lumping of UCTMC $(\mathcal{V}, m, M)$ if it is an ordinary lumping of both CTMCs $(\mathcal{V}, m)$ and $(\mathcal{V}, M)$.* □

For instance, $\mathcal{H} = \{\{00\}, \{01, 10\}, \{11\}\}$ is a UCTMC lumping of the UCTMC from Fig. 1. The lumped UCTMC is obtained in a similar way as for ordinary lumpability.

Definition 5 (Lumped UCTMC). *Assume that $\mathcal{H}$ is a UCTMC lumping of $(\mathcal{V}, m, M)$ and fix, for each $H \in \mathcal{H}$, some representative $i_H \in H$. The lumped UCTMC has states $\hat{\mathcal{V}} := \{i_H \mid H \in \mathcal{H}\}$ and bounds $\hat{m}_{i_H, i_{H'}} := \sum_{j \in H'} m_{i_H, j}$ and $\hat{M}_{i_H, i_{H'}} := \sum_{j \in H'} M_{i_H, j}$.* □

Example. In the case of the UCTMC from Fig. 1, the UCTMC lumping $\mathcal{H} = \{\{11\}, \{01, 10\}, \{11\}\}$ induces the lumped UCTMC in Fig. 2. Each state is labeled with a representative of the corresponding partition block. It is interesting to note that the transitions between states 01 and 10 in the original UCTMC correspond to self-loops in the lumped UCTMC. However, since self-loops induce self canceling terms at the level of forward Kolmogorov equations, they do not have an impact on system's dynamics and can be ignored.

Sums of reachable probability distributions of a UCTMC coincide with the reachable probability distributions of the corresponding lumped UCTMC.

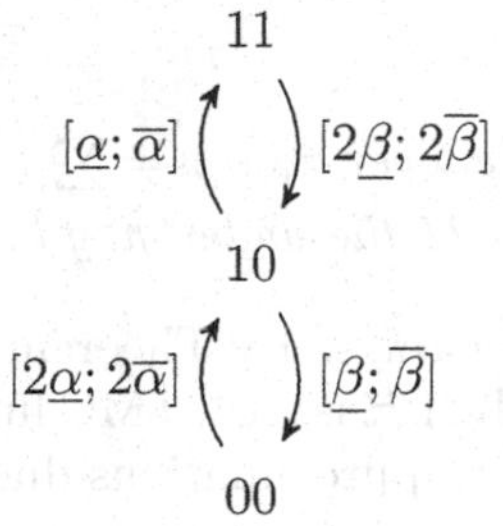

Fig. 2. Lumped UCTMC.

Theorem 6 (Preservation of Reachability). *Assume that $\mathcal{H}$ is a UCTMC lumping of $\mathcal{U} = (\mathcal{V}, m, M)$. Then, for any time $\tau \geq 0$, block $H \in \mathcal{H}$ and initial probability distribution $\pi[0]$, it holds that*

$$\mathcal{R}_{\mathcal{U}}(H, \tau, \pi[0]) = \mathcal{R}_{\hat{\mathcal{U}}}(\{i_H\}, \tau, \hat{\pi}[0]),$$

where $\hat{\mathcal{U}}$ refers to the lumped UCTMC induced by $\mathcal{H}$ and $\hat{\pi}[0]_{i_H} = \sum_{i \in H} \pi[0]_i$ for all $H \in \mathcal{H}$.

Example. In the case of the running example, Theorem 6 ensures, for instance, that $\mathcal{R}_{\mathcal{U}}(\{10, 01\}, t, \pi[0]) = \mathcal{R}_{\hat{\mathcal{U}}}(\{10\}, t, \hat{\pi}[0])$ for all $t \geq 0$ and $\pi[0]$.

We next present a modification of Theorem 6 that allows one to over-approximate sums of reachable probability distributions when $\mathcal{H}$ is not a UCTMC. It resembles [33] which provides over-approximations of uniformized CTMCs.

Theorem 7 (Over-Approximation). *For a given UCTMC $\mathcal{U} = (\mathcal{V}, m, M)$ and partition $\mathcal{H}$ of $\mathcal{V}$, assume that*

– *$m' \leq m$ such that $\mathcal{H}$ is an ordinary lumping of the CTMC $(\mathcal{V}, m')$;*
– *$M \leq M'$ such that $\mathcal{H}$ is an ordinary lumping of the CTMC $(\mathcal{V}, M')$.*

Then, $\mathcal{H}$ is a UCTMC lumping of $\mathcal{U}' = (\mathcal{V}, m', M')$ and for any initial probability distribution $\pi[0]$, the lumped UCTMC $\hat{\mathcal{U}}'$ induced by $\mathcal{U}'$ and $\mathcal{H}$ satisfies

$$\mathcal{R}_{\mathcal{U}}(H, \tau, \pi[0]) \subseteq \mathcal{R}_{\hat{\mathcal{U}}'}(\{i_H\}, \tau, \hat{\pi}'[0])$$

for all $\tau \geq 0$ and $H \in \mathcal{H}$, provided that $\hat{\pi}'[0]_{i_H} = \sum_{i \in H} \pi[0]_i$ for all $H \in \mathcal{H}$.

Our next result is the converse of Theorem 6. Together with Theorem 6, it provides a quantitative characterization of UCTMC lumpability.

Theorem 8 (Quantitative Characterization). *Let $\mathcal{U} = (\mathcal{V}, m, M)$ be some UCTMC and $\hat{\mathcal{U}} = (\hat{\mathcal{V}}, \hat{m}, \hat{M})$ a UCTMC with $\hat{\mathcal{V}} = \{i_H \mid H \in \mathcal{H}\}$ where $\mathcal{H}$ is a partition of $\mathcal{V}$ and for any time $\tau \geq 0$, block $H \in \mathcal{H}$ and initial probability distribution $\pi[0]$, it holds that*

$$\mathcal{R}_{\mathcal{U}}(H, \tau, \pi[0]) = \mathcal{R}_{\hat{\mathcal{U}}}(\{i_H\}, \tau, \hat{\pi}[0])$$

whenever $\hat{\pi}[0]_{i_H} = \sum_{i \in H} \pi[0]_i$ for all $H \in \mathcal{H}$. Then, $\mathcal{H}$ is a UCTMC lumping and $\hat{\mathcal{U}}$ the underlying lumped UCTMC.

Remark 2. By Theorem 8, the subset relation of Theorem 7 becomes an identity only if $\mathcal{H}$ is a UCTMC lumping of $(\mathcal{V}, m, M)$ and $m = m'$, $M = M'$. In particular, over-approximations due to Theorem 7 are proper in general.

We end this section by relating UCTMC lumpability to other notions. First, we observe that UCTMC lumpability is a conservative generalization of ordinary lumpability.

Lemma 1 (Generalization). *Assume that $\mathcal{H}$ is a UCTMC lumping of a UCTMC $(\mathcal{V}, m, M)$ which is deterministic, i.e., $m = M$. Then, $\mathcal{H}$ is an ordinary lumping.*

Second, we prove that any UCTMC admits a DTMDP approximation that discretizes time and action spaces and for which the notions of UCTMC lumpability and alternating probabilistic bisimulation (cf. [27]) coincide.

Theorem 9. *Fix a UCTMC $(\mathcal{V}, m, M)$, an equivalence relation $\mathfrak{R} \subseteq \mathcal{V} \times \mathcal{V}$ and let $\mathcal{H} = \mathcal{V}/\mathfrak{R}$. Then $\mathcal{H}$ is a UCTMC lumpability of $(\mathcal{V}, m, M)$ if and only if $\mathfrak{R}$ is an alternating probabilistic bisimulation of the DTMDP from Theorem 4.*

As mentioned earlier, alternating probabilistic bisimulation only preserves logical and quantitative properties [27] on the domain of DTMDPs, while UCTMC lumping characterizes these on the domain of UCTMCs.

4.2 Logical Characterization

We extend CSL to UCTMCs by defining a formula to be true when it is satisfied by all admissible $Q = (q_{i,j})_{(i,j)}$. This allows one to study safety properties in presence of uncertainty, aligning with [39], which considers CSL for CTMDPs with finite action spaces.

Definition 6 (CSL for UCTMCs). *Given a UCTMC $(\mathcal{V}, m, M)$, the CSL syntax is*

$$\phi ::= a \mid \phi \wedge \phi \mid \neg\phi \mid \mathcal{P}^{\forall}_{\bowtie p}(\boldsymbol{X}^{[t_0;t_1]}\phi) \mid \mathcal{P}^{\forall}_{\bowtie p}(\phi\, \boldsymbol{U}^{[t_0;t_1]}\phi)$$

For an arbitrary small but fixed time step $h > 0$, let $\underline{t}$ denote the smallest grid point in $\{0, h, 2h, \ldots\}$ that minimizes the distance to $t \geq 0$, i.e., $\underline{t} = h \cdot \lfloor t/h \rfloor$, where $\lfloor \cdot \rfloor$ is the floor function. For a given labeling function $\mathcal{L} : \mathcal{V} \to 2^{\mathcal{V}}$ and initial probability distribution $\pi[0]$, the satisfiability operator is defined by induction:

- *$i, t \models a$ iff $a \in \mathcal{L}(i)$;*
- *$i, t \models \phi_1 \wedge \phi_2$ iff $i, t \models \phi_1$ and $i, t_0 \models \phi_2$;*
- *$i, t \models \neg\phi$ iff not $i, t \models \phi$;*
- *$i, t \models \mathcal{P}^{\forall}_{\bowtie p}(\boldsymbol{X}^{[t_0;t_1]}\phi_1)$ iff $i, \underline{t} \models \mathcal{P}_{\bowtie p}(\boldsymbol{X}^{[t_0;t_1]}\phi)$ for all admissible q;*
- *$i, t \models \mathcal{P}^{\forall}_{\bowtie p}(\phi_1\, \boldsymbol{U}^{[t_0;t_1]}\phi_2)$ iff $i, \underline{t} \models \mathcal{P}_{\bowtie p}(\phi_1\, \boldsymbol{U}^{[t_0;t_1]}\phi_2)$ for all admissible q.* □

Similarly to [39], existential quantification is given by $\mathcal{P}^{\exists}_{\bowtie p}(\Phi) := \neg\mathcal{P}^{\forall}_{\neg\bowtie p}(\Phi)$, where $\neg\bowtie$ is defined in the obvious manner (e.g., $\neg \leq$ is $>$). Likewise, $\vee$, $\to$ are defined using $\wedge$, $\neg$.

Theorem 10 (Preservation of CSL). *Let $\mathcal{H}$ be a UCTMC lumping of UCTMC $\mathcal{U}$ and let $\hat{\mathcal{U}}$ be the underlying lumped UCTMC. Further, assume that $\mathcal{L}(i) = \mathcal{L}(j)$ for all $H \in \mathcal{H}$ and $i, j \in H$. With this, define $\hat{\mathcal{A}} := \mathcal{A}$ and $\hat{\mathcal{L}}(i_H) := \mathcal{L}(i_H)$ for all $H \in \mathcal{H}$. Then*

$$i, t \models_{\mathcal{U}} \phi \iff i_H, t \models_{\hat{\mathcal{U}}} \phi$$

for any $t \geq 0$, $h > 0$, block $H \in \mathcal{H}$, state $i \in H$, $\boldsymbol{X}$-operator free CSL formula ϕ and initial probability distribution $\pi[0]$.

The next result is a converse of Theorem 10 and establishes a logical characterization of UCTMC lumpability.

Theorem 11 (Logical Characterization). *Fix a UCTMC $(\mathcal{V}, m, M)$, a partition $\mathcal{H}$ of $\mathcal{V}$ and let $\mathcal{L}$, $\hat{\mathcal{A}}$ and $\hat{\mathcal{L}}$ be as in Theorem 10. Assume further that there exists a UCTMC $(\hat{\mathcal{V}}, \hat{m}, \hat{M})$ such that $\hat{\mathcal{V}} = \{i_H \mid H \in \mathcal{H}\}$ and*

$$i, t \models_{\mathcal{V},m,M} \phi \iff i_H, t \models_{\hat{\mathcal{V}},\hat{m},\hat{M}} \phi$$

for any $t \geq 0$, $h > 0$, $H \in \mathcal{H}$, $i \in H$ and $\boldsymbol{X}$-operator free CSL formula ϕ. Then, $\mathcal{H}$ is a UCTMC lumping and $(\hat{\mathcal{V}}, \hat{m}, \hat{M})$ the underlying lumped UCTMC.

Algorithm 1. Partition refinement algorithm for the computation of the coarsest UCTMC lumping $\mathcal{H}$ from the proof of Theorem 12.

Require: Uncertain CTMC $(\mathcal{V}, m, M)$ and initial partition $\mathcal{H}$
1: **while true do**
2: $\quad\mathcal{H}' \longleftarrow$ coarsest ordinary lumping of CTMC $(\mathcal{V}, m)$ that refines $\mathcal{H}$
3: $\quad\mathcal{H}'' \longleftarrow$ coarsest ordinary lumping of CTMC $(\mathcal{V}, M)$ that refines $\mathcal{H}'$
4: $\quad$**if** $\mathcal{H}'' = \mathcal{H}$ **then**
5: $\qquad$**return** $\mathcal{H}''$
6: $\quad$**else**
7: $\qquad\mathcal{H} \longleftarrow \mathcal{H}''$
8: $\quad$**end if**
9: **end while**

4.3 UCTMC Lumping Algorithm

We next present an algorithm for the efficient computation of the coarsest UCTMC lumping that refines a given partition $\mathcal{H}$. Its steps are as follows.

A1 With $\mathcal{H}$ being the current partition, compute the coarsest ordinary lumping $\mathcal{H}'$ of the CTMC $(\mathcal{V}, m)$ that refines $\mathcal{H}$;
A2 Compute the coarsest ordinary lumping $\mathcal{H}''$ of the CTMC $(\mathcal{V}, M)$ that refines $\mathcal{H}'$;
A3 If $\mathcal{H}'' = \mathcal{H}$, return $\mathcal{H}''$; Otherwise, set $\mathcal{H} := \mathcal{H}''$ and go to A1.

Obviously, if A1 does not refine $\mathcal{H}$ and A2 does not refine $\mathcal{H}'$, then $\mathcal{H}$ is a UCTMC lumping of $(\mathcal{V}, m, M)$. The algorithm terminates because $\mathcal{V}$ is finite. Moreover, it can be shown that the algorithm indeed computes the coarsest UCTMC partition because each refinement produces a partition which, itself, is still refined by the coarsest UCTMC lumping.

The next result summarizes the above discussion. The complexity statement follows thanks to the fact that A1 and A2 can be processed via efficient CTMC lumping algorithms such as [17,46].

Theorem 12. *Given a UCTMC $(\mathcal{V}, m, M)$, let $\mathcal{H}$ be a partition of $\mathcal{V}$. Then, the following can be shown.*

1) *Algorithm 1 computes the coarsest UCTMC lumping refining $\mathcal{H}$.*
2) *The time and space complexity required for one while loop iteration does not exceed $\mathcal{O}(r\log(s))$, where $r := |\{(i,j) \mid m_{i,j} > 0 \text{ or } M_{i,j} > 0\}|$ and $s := |\mathcal{V}|$. The number of while loop iterations, instead, is at most s.*

We conclude this section with two remarks regarding the lumping algorithm. First, we note that it simplifies to the CTMC lumping algorithm if applied to a deterministic UCTMC $(\mathcal{V}, m, M)$, i.e., a UCTMC that satisfies $m = M$. Second, using the correspondence between UCTMC lumpability and probabilistic alternating bisimulation from Theorem 9, it would be possible to apply the algorithm for the largest alternating probabilistic bisimulation [27] of the approximate

DTMDP from Theorem 9. However, in contrast to the UCTMC algorithm, such an approach would incur an exponential dependence on the maximal number of outgoing non-deterministic transitions of the approximate DTMDP.

5 Evaluation

Here we assess UCTMC lumpability in terms of both its computational tractability and reduction power with respect to ordinary lumpability. To this end, we consider uncertain variants of CTMCs of increasing size generated from benchmark models in PRISM [36].

Tool-Support and Replicability. In our experiments we used a prototype implementation of our algorithm based on the tool ERODE [9]. ERODE supports CTMC minimization as a special case of lumping algorithms for nonlinear ordinary differential equations [10]. Given that CTMC ordinary lumpability is the most important inner step of our algorithm, other tools implementing CTMC lumping could have been used, such as MRMC [32], STORM [15], and CoPaR [16,18]. All runtimes reported refer to the execution of ERODE on a common desktop machine with 8 GB RAM. All the material to replicate the experiments is available at https://www.erode.eu/examples.html.

Set-Up. For this evaluation we used CTMCs in the MRMC format [32], generated from PRISM models. We considered CTMCs which describe: a dependable cluster of workstations [28]; a protocol for wireless group communication [3,38]; a model of the cell cycle control in eukaryotes [37,41]. Similarly to [24,27], we considered uncertain relaxations of such CTMCs by adding uncertainty to the transition rates. In particular, in each model we replaced every transition rate value with an interval of fixed length (arbitrarily fixed equal to 20% of the smallest transition rate in the model) centered at the original rate value itself.

Results. The results are provided in Table 1. We report the number of transitions and states of the obtained CTMCs in the second and third column, respectively, as a function of the scaling parameter N. The initial input partition of states, denoted by $\mathcal{H}_0$, was induced by the original model specification by creating blocks of states characterized by the same atomic propositions. The comparison of the runtimes of the minimization algorithms provides an indication of the increased overhead for the reduction (which is proportional to the number of states in the worst case). In all our tests, UCTMC lumpability had, up to a factor of two, the same runtime as the CTMC version. This is because in all models at most two iterations of our algorithm were necessary. The effectiveness of UCTMC lumping can be evaluated by comparing the size of the coarsest UCTMC lumpings with their corresponding CTMC counterparts. Notably, CTMC and UCTMC lumpability coincide on the first two families of models, while in the third family UCTMC lumpability leads to finer (at most 18% more blocks) partitions than the CTMC counterpart.

Table 1. Quantitative comparison of CTMC and UCTMC lumpability. Entries *identical* denote cases with identical CTMC and UCTMC lumpable partitions.

Original model (CTMC)				CTMC Lumpability		UCTMC Lumpability	
N	r	s	$\lvert\mathcal{H}_0\rvert$	*Red.(s)*	$\lvert\mathcal{H}\rvert$	*Red.(s)*	$\lvert\mathcal{H}\rvert$
Workstation cluster							
128	2 908 192	597 012	4	2.21E+1	298 893	2.64E+1	*identical*
192	6 52 4960	1 33 7876	4	6.78E+1	669 517	8.04E+1	*identical*
256	11 583 520	2 373 652	4	1.55E+2	1 187 597	1.85E+2	*identical*
320	18 083 872	3 704 340	4	2.81E+2	1 853 133	3.58E+2	*identical*
384	26 026 016	5 329 940	4	*out of memory*		*out of memory*	
Wireless group communication protocol							
16	686 153	103 173	2	2.26E+0	4 846	2.97E+0	*identical*
24	3 183 849	453 125	2	1.34E+1	20 476	1.61E+1	*identical*
32	10 954 382	1 329 669	2	4.49E+1	58 906	5.50E+1	*identical*
40	22 871 849	3 101 445	2	1.35E+2	135 752	1.61E+2	*identical*
48	46 574 793	6 235 397	2	*out of memory*		*out of memory*	
Cell cycle control in eukaryotes							
2	18 342	4 666	3	1.76E–1	3 514	1.97E–1	4 000
3	305 502	57 667	3	8.21E–1	40 667	9.81E–1	48 147
4	2 742 012	431 101	3	6.45E+0	282 956	7.80E+0	33 9368
5	16 778 785	2 326 666	3	4.58E+1	1 424 935	9.15E+1	1 712 322
6	78 768 799	9 960 861	3	*out of memory*		*out of memory*	

Table 2. Summary of results. UCTMC lumpability generalizes the well-known dynamical, logical and algorithmic properties of ordinary lumpability for CTMCs (in statements concerning complexity, s refers to the numbers of states, while r denotes the number of transitions).

CTMC lumpability		*UCTMC lumpability*
$\Updownarrow$ [4,8,34]	Dynamics	$\Updownarrow$ Theorem 6, 8
$\sum_{i\in H}\pi_i(t)=\hat{\pi}_{i_H}(t)$		$\mathcal{R}(\sum_{i\in H}\pi_i,t,\pi[0])=\mathcal{R}(\hat{\pi}_{i_H},t,\hat{\pi}[0])$
$\Updownarrow$ [1]	Logics	$\Updownarrow$ Theorem 10, 11
$i,t\models_{\mathcal{V}}\phi\Longleftrightarrow i_H,t\models_{\hat{\mathcal{V}}}\phi$		$i,t\models_{\mathcal{V},m,M}\phi\Longleftrightarrow i_H,t\models_{\hat{\mathcal{V}},\hat{m},\hat{M}}\phi$
$\mathcal{O}(r\log(s))$ [17,46]	Complexity	$\mathcal{O}(sr\log(s))$ Theorem 12

6 Conclusion

Uncertain continuous-time Markov chains (UCTMCs) generalize continuous-time Markov chains (CTMCs) by allowing transition rates to non-deterministically take values within given bounded intervals. UCTMC lumpability enjoys a polynomial time and space algorithm for the computation of the largest UCTMC lumping.

Similarly to CTMC lumping that characterizes the preservation of sums of probability distributions, UCTMC lumping characterizes the preservations of reachable sets of sums of probability distributions. We have provided a logical characterization of UCTMC lumpability to uncertain time-varying parameters. Overall, the results in this paper can be put in analogy with the corresponding CTMC counterparts, as summarized in Table 2. The applicability of UCTMC lumpability has been established by presenting substantial reductions in benchmark models. The discretization of a UCTMC as a DTMDP has offered the means to relating UCTMC lumpability to bisimulations for DTMDPs. Future work will consider model-checking algorithms for UCTMCs.

Acknowledgement. Luca Cardelli is supported by a Royal Society Research Professorship. The work has been partially supported by the ERC Advanced Grant LASSO, the Villum Investigator Grant S4OS, the EU-Ecsel project iDev40, the FFG project Adex, the PRIN project SEDUCE, no. 2017TWRCNB, the FWF project COCO no. M-2393-N32, the Poul Due Jensen Foundation grant no. 883901, and by the DFF RP1 project REDUCTO no. 9040-00224B.

References

1. Baier, C., Haverkort, B.R., Hermanns, H., Katoen, J.-P.: Model-checking algorithms for continuous-time Markov Chains. IEEE Trans. Software Eng. **29**(6), 524–541 (2003)
2. Bogomolov, S., Frehse, G., Grosu, R., Ladan, H., Podelski, A., Wehrle, M.: A box-based distance between regions for guiding the reachability analysis of SpaceEx. In: CAV, pp. 479–494 (2012)
3. Bondavalli, A., Coccoli, A., Giandomenico, F.D.: QoS analysis of group communication protocols in wireless environment. In: Ezhilchelvan, P., Romanovsky, A. (eds.) Concurrency in Dependable Computing, pp. 169–188. Springer, Boston (2002). https://doi.org/10.1007/978-1-4757-3573-4_9
4. Buchholz, P.: Exact and ordinary lumpability in finite Markov chains. J. Appl. Prob. **31**(1), 59–75 (1994)
5. Buchholz, P., Hahn, E.M., Hermanns, H., Zhang, L.: Model checking algorithms for CTMDPs. In: CAV, pp. 225–242 (2011)
6. Butkova, Y., Hatefi, H., Hermanns, H., Krcál, J.: Optimal continuous time Markov decisions. In: ATVA, pp. 166–182 (2015)
7. Caillaud, B., Delahaye, B., Larsen, K.G., Legay, A., Pedersen, M.L., Wasowski, A.: Constraint Markov chains. Theoret. Comput. Sci. **412**(34), 4373–4404 (2011)
8. Cardelli, L., Tribastone, M., Tschaikowski, M., Vandin, A.: Symbolic computation of differential equivalences. In: POPL, pp. 137–150 (2016)
9. Cardelli, L., Tribastone, M., Tschaikowski, M., Vandin, A.: ERODE: a tool for the evaluation and reduction of ordinary differential equations. In: TACAS (2017)
10. Cardelli, L., Tribastone, M., Tschaikowski, M., Vandin, A.: Maximal aggregation of polynomial dynamical systems. Proc. Natl. Acad. Sci. **114**(38), 10029–10034 (2017)
11. Češka, M., Dannenberg, F., Paoletti, N., Kwiatkowska, M., Brim, L.: Precise parameter synthesis for stochastic biochemical systems. Acta Informatica **54**(6), 589–623 (2017)

12. Chen, X., Ábrahám, E., Sankaranarayanan, S.: Flow*: an analyzer for non-linear hybrid systems. In: CAV, pp. 258–263 (2013)
13. David, R., Alla, H.: Discrete, Continuous, and Hybrid Petri Nets. Springer, Heidelberg (2005). https://doi.org/10.1007/978-3-642-10669-9
14. Dehnert, C., et al.: PROPhESY: A PRObabilistic ParamEter SYnthesis tool. In: CAV, pp. 214–231 (2015)
15. Dehnert, C., Junges, S., Katoen, J.-P., Volk, M.: A storm is coming: a modern probabilistic model checker. In: Computer Aided Verification - 29th International Conference, CAV 2017, Heidelberg, Germany, July 24–28, 2017, Proceedings, Part II, pp. 592–600 (2017)
16. Deifel, H.-P., Milius, S., Schröder, L., Wißmann, T.: Generic partition refinement and weighted tree automata. In: FM (2019, to Appear)
17. Derisavi, S., Hermanns, H., Sanders, W.H.: Optimal state-space lumping in Markov chains. Inf. Process. Lett. **87**(6), 309–315 (2003)
18. Dorsch, U., Milius, S., Schröder, L., Wißmann, T.: Efficient coalgebraic partition refinement. In: CONCUR, pp. 32:1–32:16 (2017)
19. Fan, C., Qi, B., Mitra, S., Viswanathan, M., Duggirala, P.S.: Automatic reachability analysis for nonlinear hybrid models with C2E2. In: CAV, pp. 531–538 (2016)
20. Fecher, H., Leucker, M., Wolf, V.: Don't Know in probabilistic systems. In: SPIN, pp. 71–88 (2006)
21. William Gear, C.: Numerical Initial Value Problems in Ordinary Differential Equations. Prentice Hall PTR (1971)
22. Gillespie, D.T.: Exact stochastic simulation of coupled chemical reactions. J. Phys. Chem. **81**(25), 2340–2361 (1977)
23. Guo, X., Hernandez-Lerma, O.: Continuous-Time Markov Decision Processes. Springer, Heidelberg (2009). https://doi.org/10.1007/978-3-642-02547-1
24. Hahn, E.M., Hashemi, V., Hermanns, H., Turrini, A.: Exploiting robust optimization for interval probabilistic bisimulation. In: Agha, G., Van Houdt, B. (eds.) QEST 2016. LNCS, vol. 9826, pp. 55–71. Springer, Cham (2016). https://doi.org/10.1007/978-3-319-43425-4_4
25. Hahn, E.M., Hermanns, H., Wimmer, R., Becker, B.: Transient reward approximation for continuous-time Markov chains. IEEE Trans. Reliability **64**(4), 1254–1275 (2015)
26. Hahn, E.M., Hermanns, H., Zhang, L.: Probabilistic reachability for parametric Markov models. STTT **13**(1), 3–19 (2011)
27. Hashemi, V., Turrini, A., Hahn, E.M., Hermanns, H., Elbassioni, K.M.: Polynomial-time alternating probabilistic bisimulation for interval MDPs. In: SETTA, pp. 25–41 (2017)
28. Haverkort, B.R., Hermanns, H., Katoen, J.-P.: On the use of model checking techniques for dependability evaluation. In: SRDS, pp. 228–237 (2000)
29. Hermanns, H., Rettelbach, M.: Syntax, semantics, equivalences, and axioms for MTIPP. In: Proceedings of Process Algebra and Probabilistic Methods, pp. 71–87, Erlangen (1994)
30. Hillston, J.: A Compositional Approach to Performance Modelling. Cambridge University Press, Cambridge (1996)
31. Jonsson, B., Larsen, K.G.: Specification and refinement of probabilistic processes. In: LICS, pp. 266–277 (1991)
32. Katoen, J.-P., Khattri, M., Zapreev, I.S.: A Markov reward model checker. In: QEST, pp. 243–244 (2005)
33. Katoen, J.-P., Klink, D., Leucker, M., Wolf, V.: Three-valued abstraction for continuous-time Markov chains. In: CAV, pp. 311–324 (2007)

34. Kemeny, J.G., Snell, J.L.: Finite Markov Chains. Springer, New York (1976)
35. Kozine, I.O., Utkin, L.V.: Interval-valued finite Markov chains. Reliable Comput. **8**(2), 97–113 (2002)
36. Kwiatkowska, M., Norman, G., Parker, D.: PRISM 4.0: verification of probabilistic real-time systems. In: CAV, pp. 585–591 (2011)
37. Lecca, P., Priami, C.: Cell cycle control in eukaryotes: a biospi model. Electr. Notes Theor. Comput. Sci. **180**(3), 51–63 (2007)
38. Massink, M., Katoen, J.-P., Latella, D.: Model checking dependability attributes of wireless group communication. In: DSN, pp. 711–720 (2004)
39. Neuhäußer, M.R., Katoen, J.-P.: Bisimulation and logical preservation for continuous-time Markov decision processes. In CONCUR, pages 412–427, 2007
40. Norris, J.R.: Markov Chains. Cambridge University Press, Cambridge (1998)
41. Novak, B., Csikasz-Nagy, A., Gyorffy, B., Nasmyth, K., Tyson, J.J.: Model scenarios for evolution of the eukaryotic cell cycle. Philosophical Trans. Roy. Soc. Lond. Ser. B: Biol. Sci. **353**(1378), 2063–2076 (1998)
42. Sen, K., Viswanathan, M., Agha, G.: Model-checking Markov chains in the presence of uncertainties. In: TACAS, pp. 394–410 (2006)
43. Simon, P.L., Taylor, M., Kiss, I.Z.: Exact epidemic models on graphs using graph-automorphism driven lumping. J Math. Biol. **62**(4), 479–508 (2010)
44. Song, L., Zhang, L., Godskesen, J.Chr.: Bisimulations and logical characterizations on continuous-time Markov decision processes. In: VMCAI, pp. 98–117 (2014)
45. Stewart, W.J.: Probability, Markov Chains, Queues, and Simulation. Princeton University Press, PrincetonPrinceton (2009)
46. Valmari, A., Franceschinis, G.: Simple $O(m \log n)$ time Markov chain lumping. In: TACAS, pp. 38–52 (2010)

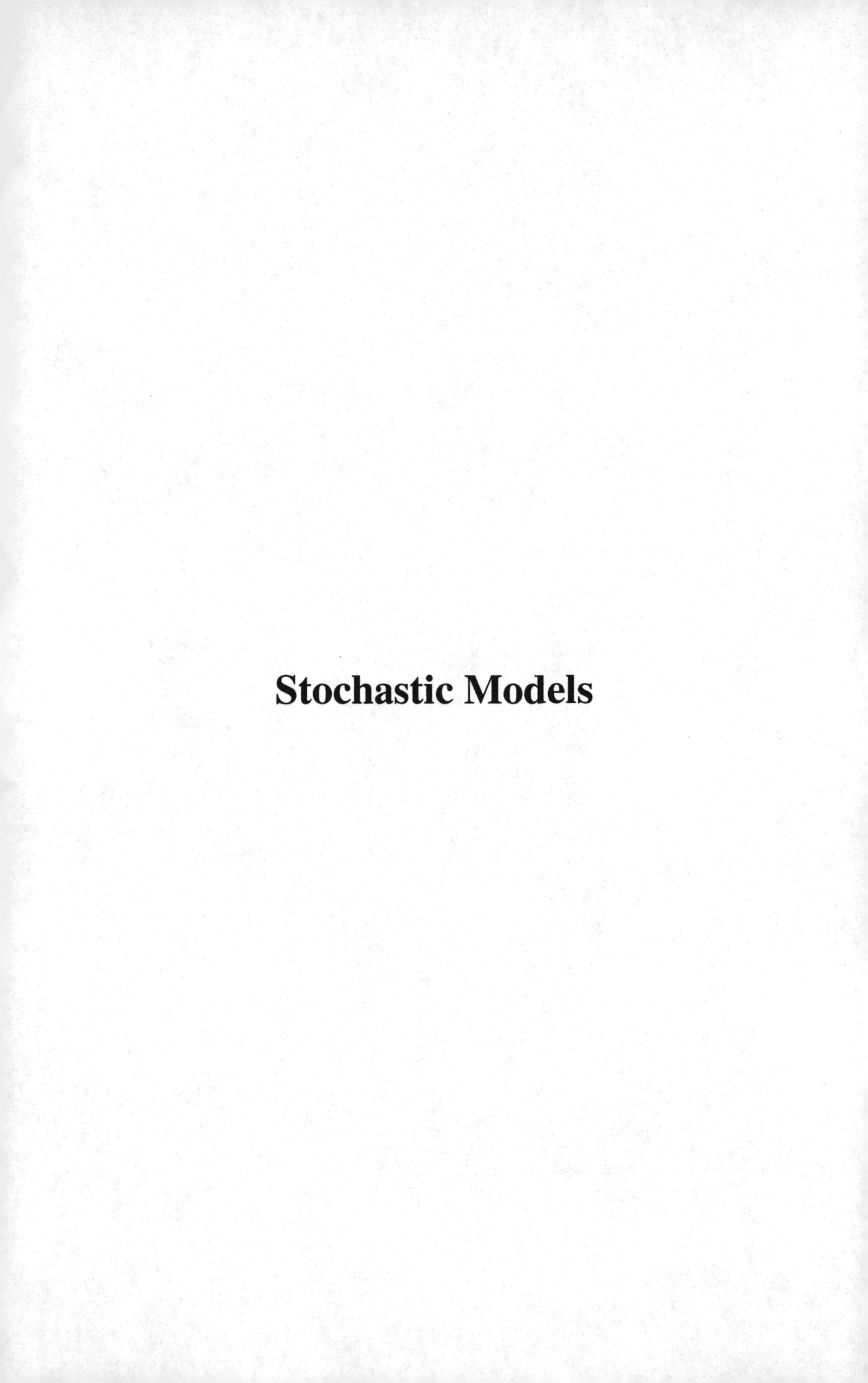

Stochastic Models

Accurate Approximate Diagnosis of (Controllable) Stochastic Systems

Engel Lefaucheux(✉)

Max Planck Institute for Software Systems, Saarland Informatics Campus, Saarbrücken, Germany
elefauch@mpi-sws.org

Abstract. Diagnosis of partially observable stochastic systems prone to faults was introduced in the late nineties. Diagnosability may be specified in different ways: exact diagnosability requires that almost surely a fault is detected and that no fault is erroneously claimed; approximate diagnosability tolerates a small error probability when claiming a fault; last, accurate approximate diagnosability guarantees that the error probability can be chosen arbitrarily small. While all three notions were studied for passive systems such as observable Markov chains, only the exact notion was considered for systems equipped with a controller. As the approximate notion of diagnosability was shown to be undecidable in passive systems, in this article, we complete the picture by deciding the accurate approximate diagnosability for controllable observable Markov chains. More precisely, we show how to adapt the accurate approximate notion to the active setting and establish EXPTIME-completeness of the associated decision problem. We also show how to measure the set of faulty paths that are detected under the accurate approximate notion in the passive setting.

Keywords: Stochastic systems · Partial observation · Control · Diagnosis

1 Introduction

Diagnosis and Diagnosability. There has been an increasing use of software systems for critical operations. When designing such systems, one aims at eliminating faults that could trigger unwanted behaviours. However, for embedded systems interacting with an unpredictable environment, the absence of faults is not a reasonable hypothesis. Thus diagnosis, whose goal consists to detect faults from the observations of the runs of the system, is a crucial task. One of the approaches frequently used to analyse *diagnosability* consists in modelling the system by a transition system whose states (depending on the internal part of the system) are unobservable and events may, depending on their nature, be observable or not. One of the proposed approaches consists in modelling these systems by partially observable labelled transition systems (poLTS) [24]. In such a framework, diagnosability requires that the occurrence of unobservable faults can be deduced accurately from the previous and subsequent observable events.

A. Abate and A. Marin (Eds.): QEST 2021, LNCS 12846, pp. 413–434, 2021.
https://doi.org/10.1007/978-3-030-85172-9_22

In other words, defining the *disclosure set* of a system as the set of faulty paths of the system that can be detected, a system is diagnosable if every faulty path belongs to the disclosure set. Diagnosability for poLTS was shown to be decidable in PTIME [18]. Diagnosis has since been extended to numerous models (Petri nets [12], pushdown systems [20], etc.) and settings (centralized, decentralized, distributed), and have had an impact on important application areas, *e.g.* for telecommunication network failure diagnosis.

Diagnosability for Stochastic Passive Systems. In transition systems, the unpredictable behaviours of the environment are modelled by a nondeterministic choice between the possible events from the current state. However, in order to quantify the risks induced by the faults of the systems, the designer often substitutes the nondeterministic choice by a random choice or equivalently by a weighted one. Then the model becomes a discrete time observable Markov chain (oMC) in the *passive* case (*i.e.* without control). In these models, one can define a probability measure over infinite runs. In that context, the accuracy required to claim a path is faulty can be relaxed. There are three natural variants: (1) exact disclosure, which, as in the non-stochastic case, requires that every path sharing the given observation sequence is faulty in order to claim a fault occurred, (2) ε-disclosure for $\varepsilon > 0$ which tolerates small errors, allowing to claim the failure of a path if the conditional probability that the path is faulty exceeds $1 - \varepsilon$, and (3) Accurate Approximate disclosure (AA-disclosure) which is satisfied when the accuracy of the guess can be chosen arbitrarily high. Diagnosability with exact disclosure has been studied extensively for oMC [6,8,25]. In particular, various exact notions of diagnosability have been shown to be PSPACE-complete for oMC. Due to the quantitative requirement, diagnosability with ε-disclosure was shown to be undecidable while diagnosability with AA-disclosure was surprisingly shown to be in PTIME [7].

Active Diagnosability. Embedded systems are often equipped with one (or more) controller(s) in order to maintain some functionalities of the system in case of a pathological behaviour of the environment. It is thus tempting to add to the controller a diagnosis task. Formally some of the observable events are controllable and considering its current observation, the controller chooses which subset of actions should be allowed to make the system diagnosable. As such, a controller only has access to the observations produced by the system to make his choice. This represents the idea that the control is realised by the same entity as the diagnosis. A system is said *actively diagnosable* if there exists a controller ensuring diagnosability [13,14,17,23,26]. In [17], the authors designed an exponential time algorithm and proved the optimality of this complexity. In stochastic systems, diagnosability has only been considered with exact disclosure and has been proven EXPTIME-complete [5].

Contribution. In this paper, we study diagnosability in stochastic systems under AA-disclosure.

- we introduce an alternative definition of AA-disclosure and establish its equivalence with the notion introduced in [25] (Proposition 1)
- we show that measuring the set of AA-disclosing paths for oMC is PSPACE-complete (Theorem 3);
- we establish that diagnosability with AA-disclosure for controllable oMC is EXPTIME-complete (Theorem 4).

For space concerns, some technical proofs are deferred to the appendix.

2 Diagnosis of Markov Chains

2.1 Observable Markov Chains

For a finite alphabet Σ, we denote by Σ^* (resp. Σ^ω) the set of finite (resp. infinite) words over Σ, $\Sigma^\infty = \Sigma^* \cup \Sigma^\omega$ and ε the empty word. The length of a word w is denoted by $|w| \in \mathbb{N} \cup \{\infty\}$ and for $n \in \mathbb{N}$, Σ^n is the set of words of length n. A word $u \in \Sigma^*$ is a prefix of $v \in \Sigma^\infty$, written $u \leq v$, if $v = uw$ for some $w \in \Sigma^\infty$. The prefix is strict if $w \neq \varepsilon$. For $n \leq |w|$, we write $w_{\downarrow n}$ for the prefix of length n of w. Given a countable set S, a distribution on S is a mapping $\mu : S \to [0,1]$ such that $\sum_{s \in S} \mu(s) = 1$. The support of μ is $\mathsf{Supp}(\mu) = \{s \in S \mid \mu(s) > 0\}$. If $\mathsf{Supp}(\mu) = \{s\}$ is a single element, μ is a Dirac distribution on s written $\mathbf{1}_s$. We denote by $\mathsf{Dist}(S)$ the set of distributions on S.

For the purpose of partially observable problems, the model must be equipped with an *observation function* describing what an external observer can see. The observation function can be obtained via a labelling of states or transitions, both options being known to be equivalent. We thus define observable Markov chains (see Fig. 1).

Definition 1 (Observable Markov chains). *An observable Markov chain (oMC) over alphabet Σ is a tuple $\mathcal{M} = (S, p, \mathsf{O})$ where S is a countable set of states, $p : S \to \mathsf{Dist}(S)$ is the transition function, and $\mathsf{O} : S \to \Sigma$ is the observation function.*

We write $p(s'|s)$ instead of $p(s)(s')$ to emphasise the probability of going to state s' conditioned by being in state s. Given a distribution $\mu_0 \in \mathsf{Dist}(S)$, we denote by $\mathcal{M}(\mu_0)$ the oMC with initial distribution μ_0. For decidability and complexity results, we assume that all probabilities occurring in the model (transition probabilities and initial distribution) are rationals. A (finite or infinite) path of $\mathcal{M}(\mu_0)$ is a sequence of states $\rho = s_0 s_1 \ldots \in S^\infty$ such that $\mu_0(s_0) > 0$ and for each $i \geq 0$, $p(s_{i+1}|s_i) > 0$. For a finite path, $\rho = s_0 s_1 \ldots s_n$, we call n its length and denote its ending state by $\mathsf{last}(\rho) = s_n$. A finite path ρ_1 prefixes a finite or infinite path ρ if there exists a path ρ_2 such that $\rho = \rho_1 \rho_2$. The set $\mathsf{Cyl}(\rho)$ represents the cylinder of infinite paths prefixed by ρ. We denote by $\mathsf{Path}(\mathcal{M}(\mu_0))$ (resp. $\mathsf{fPath}(\mathcal{M}(\mu_0))$) the set of infinite (finite) paths of $\mathcal{M}(\mu_0)$. The *observation sequence* of the path $\rho = s_0 s_1 \ldots$ is the word $\mathsf{O}(\rho) = \mathsf{O}(s_0)\mathsf{O}(s_1)\ldots \in \Sigma^\infty$. For a set R of paths, $\mathsf{O}(R) = \{\mathsf{O}(\rho) \mid \rho \in R\}$ and for a set W of observation sequences, $\mathsf{O}^{-1}(W) = \{\rho \in \mathsf{Path}(\mathcal{M}(\mu_0)) \cup \mathsf{fPath}(\mathcal{M}(\mu_0)) \mid \mathsf{O}(\rho) \in W\}$.

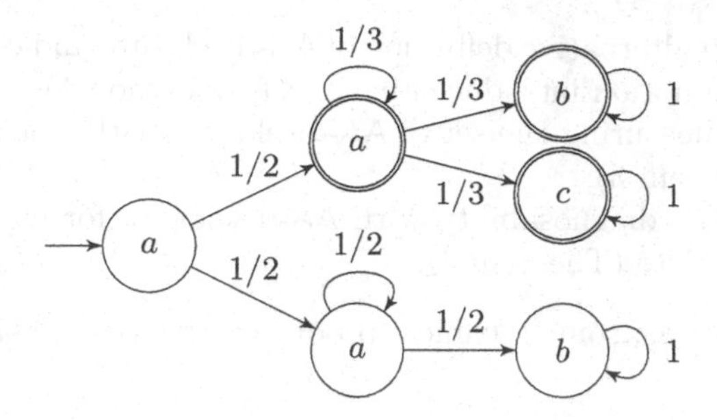

Fig. 1. An observable Markov chain. The arrow entering the leftmost state means that the initial distribution is a Dirac on this state. Faulty states are circled twice.

Forgetting the labels, an oMC with an initial distribution μ_0 becomes a discrete time Markov chain (DTMC). In a DTMC, the set of infinite paths is the support of a probability measure extended from the probabilities of the cylinders by the Caratheodory's extension theorem:

$$\mathbf{P}_{\mathcal{M}(\mu_0)}(\mathsf{Cyl}(s_0 s_1 \dots s_n)) = \mu_0(s_0)p(s_1|s_0)\dots p(s_n|s_{n-1}) \ .$$

When $\mathcal{M}(\mu_0)$ is clear from context, we will sometimes omit the subscript, and write $\mathbf{P}$ for $\mathbf{P}_{\mathcal{M}(\mu_0)}$. Let $\rho \in \mathsf{fPath}(\mathcal{M})$, $w \in \Sigma^*$ and $E \subseteq \Sigma^\omega$, with a small abuse of notation we write $\mathbf{P}(\rho)$ for $\mathbf{P}(\mathsf{Cyl}(\rho))$, $\mathbf{P}(w)$ instead of $\mathbf{P}(\cup_{\rho \in \mathsf{O}^{-1}(w)} \mathsf{Cyl}(\rho))$ and $\mathbf{P}(E)$ instead of $\mathbf{P}(\{\rho \in \mathsf{Path}(\mathcal{M}(\mu_0)) \mid \rho \in \mathsf{O}^{-1}(E)\})$.

2.2 Faulty Paths and Notions of Disclosure

In this paper we are interested in the study of diagnosis, a problem in which one wants to detect whether the current path correspond to a faulty behaviour of the system. We focus on the particular case where the faulty behavior of the system is given by a subset of states $\mathsf{S}^\mathsf{F} \subseteq S$, called *faulty states*, of the model: a (finite or infinite) path $s_0 s_1 \dots$ is *faulty* if $s_i \in \mathsf{S}^\mathsf{F}$ for some i. The set of finite (resp. infinite) faulty paths is denoted F (resp. F_∞). A path that is not faulty is called *correct*. Remark that without loss of generality, we can assume that the set of faulty states is absorbing, *i.e.* if a path visits S^F, it forever remains in S^F.

In non-stochastic systems, a faulty path discloses its failure if it does not share its observation sequence with any correct path, *i.e.* given a path $\rho \in \mathsf{S}^\mathsf{F}$, it discloses its failure iff $\mathsf{O}^{-1}(\mathsf{O}(\rho)) \subseteq \mathsf{S}^\mathsf{F}$. When adding probabilities, one could keep the same definition of disclosure, this is what we call exact disclosure. Denoting $Disc^{exact}$ the set of infinite disclosing faulty paths, the exact diagnosability problem for oMC asks whether $\mathbb{P}(Disc^{exact}) = \mathbb{P}(\mathsf{F}_\infty)$. This problem is known to be PSPACE-complete for oMC [8].

However, one could also weaken the requirement by allowing potential false claims. In this case, a faulty path is disclosing if, based on its observation, the likelihood of the path to be faulty is high. To formalise this likelihood, we define the failure proportion as the conditional probability that a path is faulty, given its observation sequence. Formally, given an oMC $\mathcal{M} = (S, p, \mathsf{O})$, an initial distribution μ_0, $\mathsf{S^F} \subseteq S$ and an observation sequence $w \in \Sigma^*$, the failure proportion associated with the observation sequence w is:

$$\mathsf{Fprop}_{\mathcal{M}(\mu_0)}(w) = \frac{\mathbf{P}(\{\rho \in \mathsf{O}^{-1}(w) \mid \rho \in \mathsf{F}\})}{\mathbf{P}(w)}.$$

This proportion is undefined if $\mathbf{P}(w) = 0$.

Example 1. Consider the oMC of Fig. 1 and the observation sequences a^k, $a^k b^n$ and $a^k c^m$. The observation sequence a^k, for $k > 1$, can be produced by a correct path with probability $1/2^{k-1}$ and by a faulty path with probability $1/2 \times 1/3^{k-2}$. Therefore, $\mathsf{Fprop}_{\mathcal{M}(\mu_0)}(a^k) = \frac{1/3^{k-2}}{1/2^{k-2}+1/3^{k-2}}$ which converges to 0 when k grows to infinity. The failure proportion of the observation $a^k b^n$ with $k > 1$ and $n \geq 1$ is similarly $\mathsf{Fprop}_{\mathcal{M}(\mu_0)}(a^k b^n) = \frac{1/3^{k-1}}{1/2^{k-1}+1/3^{k-1}}$ which remains constant for extensions of $a^k b^n$ as it does not depend on n. Finally, if $m \geq 1$, $\mathsf{Fprop}_{\mathcal{M}(\mu_0)}(a^k c^m) = 1$ as no correct path can produce a 'c'.

Let $\mathcal{M} = (S, p, \mathsf{O})$ be an oMC, μ_0 be an initial distribution and $\mathsf{S^F} \subseteq S$. Given $\varepsilon > 0$ representing the confidence threshold expected for the detection, we can define the approximate notion of disclosure: an observation sequence $w \in \Sigma^*$ is called ε-disclosing if $\mathsf{Fprop}_{\mathcal{M}(\mu_0)}(w) > 1 - \varepsilon$. Moreover, it is ε-min-disclosing if it is ε-disclosing and no strict prefix of w is ε-disclosing. Writing $D^\varepsilon_{\min}$ for the set of ε-min-disclosing observation sequences, the ε-disclosure is defined by

$$Disc^\varepsilon(\mathcal{M}(\mu_0)) = \mathbf{P}(\{\rho \in \mathsf{F} \mid \exists \rho' \leq \rho, \mathsf{O}(\rho') \in D^\varepsilon_{\min}\}).$$

$Disc^\varepsilon$ is thus the probability that a path of the oMC will be faulty and disclose its failure with sufficiently low doubt. The ε-diagnosability problem consists then in deciding whether $Disc^\varepsilon(\mathcal{M}(\mu_0)) = \mathbf{P}(\mathsf{F}_\infty)$. Unfortunately, it is known that this problem is undecidable for $\varepsilon \neq 0$:

Theorem 1 ([7]). *Given $0 < \varepsilon < 1$, the ε-diagnosability problem is undecidable for oMCs.*

In order to regain decidability one can consider a slightly more qualitative notion of approximate information control, that is called accurate approximate. Instead of deeming the failure of a path to be revealed when the proportion of faulty paths goes above a given threshold, an infinite observation sequence is AA-*disclosing* if this proportion converges toward 1. In other words, when observing an AA-disclosing observation sequence, one can get an arbitrarily high confidence that the path is faulty. Formally, an observation sequence $w \in \Sigma^\omega$ is AA-disclosing if $\lim_{n\to\infty} \mathsf{Fprop}_{\mathcal{M}(\mu_0)}(w_{\downarrow n}) = 1$. Writing D^{AA} for the set of AA-disclosing observation sequences, the AA-disclosure is defined by

$$Disc^{\mathsf{AA}}(\mathcal{M}(\mu_0)) = \mathbf{P}(\{\rho \in \mathsf{F} \mid \mathsf{O}(\rho) \in D^{\mathsf{AA}}\})$$

As before, the AA-diagnosability problem consists in deciding if $Disc^{\mathsf{AA}}(\mathcal{M}(\mu_0)) = \mathbf{P}(\mathsf{F}_\infty)$. When an oMC is not AA-diagnosable, it is interesting to measure the probability of undetected faulty paths. This motivates the AA-*disclosure problem* which consists in, given $\lambda \in [0;1]$ and $\bowtie \in \{>, \geq\}$, deciding whether $Disc^{\mathsf{AA}}(\mathcal{M}(\mu_0)) \bowtie \lambda$.

AA-diagnosability was in fact initially defined in [25] slightly differently: a system was then called AA-diagnosable if it was ε-diagnosable for all $\varepsilon > 0$. However, the two definitions are in fact equivalent for oMC.

Proposition 1. *An oMC is* AA*-diagnosable iff it is ε-diagnosable for all $\varepsilon > 0$.*

Proof. Let $\mathcal{M}$ be an oMC and μ_0 an initial distribution.

Suppose that $\mathcal{M}(\mu_0)$ is AA-diagnosable. By definition, given an AA-disclosing observation sequence w, for all $\varepsilon > 0$ there exists $n \in \mathbb{N}$ such that $w_{\downarrow n}$ is ε-disclosing. Therefore for all $\varepsilon > 0$, $Disc^{\mathsf{AA}}(\mathcal{M}(\mu_0)) \leq Disc^{\varepsilon}(\mathcal{M}(\mu_0))$. Moreover, as $\mathcal{M}$ is AA-diagnosable, $Disc^{\mathsf{AA}}(\mathcal{M}(\mu_0)) = \mathbf{P}(\mathsf{F})$. Thus, $Disc^{\varepsilon}(\mathcal{M}(\mu_0)) \geq \mathbf{P}(\mathsf{F})$. Finally, as only faulty paths are disclosing, for all $\varepsilon > 0$ $Disc^{\varepsilon}(\mathcal{M}(\mu_0)) \leq \mathbf{P}(\mathsf{F})$. Thus $Disc^{\varepsilon}(\mathcal{M}(\mu_0)) = \mathbf{P}(\mathsf{F})$ and $\mathcal{M}(\mu_0)$ is ε-diagnosable.

Conversely, suppose that $\mathcal{M}(\mu_0)$ is not AA-diagnosable. Let us consider the set of infinite words $D = \cap_{\varepsilon>0} D^{\varepsilon}_{min} \Sigma^{\omega} \setminus D^{\mathsf{AA}}$. Let us show that $\mathbf{P}(D) = 0$. Let $w \in D$, we have (1) for all $\varepsilon > 0$ there exists $n \in \mathbb{N}$ such that $\mathsf{Fprop}(w_{\downarrow n}) > 1 - \varepsilon$ and (2) $(\mathsf{Fprop}(w_{\downarrow n}))_{n \in \mathbb{N}}$ does not converge toward 1. Given $\varepsilon > 0$, due to (1) we have

$$\begin{aligned} \mathbf{P}(\{\rho \in \mathsf{O}^{-1}(D) \setminus \mathsf{F}\}) &< \sum_{w \in D^{\varepsilon}_{min}} \mathbf{P}(\{\rho \in \mathsf{O}^{-1}(w) \setminus \mathsf{F}\}) \\ &< \sum_{w \in D^{\varepsilon}_{min}} \mathbf{P}(\{\rho \in \mathsf{O}^{-1}(w) \cap \mathsf{F}\}) \frac{\varepsilon}{1-\varepsilon} \\ &< \frac{\varepsilon}{1-\varepsilon}. \end{aligned}$$

As this holds for all $\varepsilon > 0$, $\mathbf{P}(\{\rho \in \mathsf{O}^{-1}(D) \setminus \mathsf{F}\}) = 0$. Moreover, due to (2), there exists $\varepsilon > 0$ such that for infinitely many $n \in \mathbb{N}$ we have $\mathsf{Fprop}(w_{\downarrow n}) < 1 - \varepsilon$. For all $k \in \mathbb{N}$, we denote by E_k the set of prefixes w of words of D such that $\mathsf{Fprop}_{\mathcal{M}(\mu_0)}(w) < 1 - \varepsilon$ for the k'th time. We then have for all k:

$$\begin{aligned} \mathbf{P}(\{\rho \in \mathsf{O}^{-1}(E_k) \setminus \mathsf{F}\}) &= \sum_{w \in E_k} \mathbf{P}(\{\rho \in \mathsf{O}^{-1}(w) \setminus \mathsf{F}\}) \\ &> \sum_{w \in E_k} \mathbf{P}(\{\rho \in \mathsf{O}^{-1}(w) \cap \mathsf{F}\}) \frac{\varepsilon}{1-\varepsilon} \\ &> \frac{\varepsilon}{1-\varepsilon} \mathbf{P}(\{\rho \in \mathsf{O}^{-1}(D) \cap \mathsf{F}\}) \end{aligned}$$

As $(\mathbf{P}(\{\rho \in \mathsf{O}^{-1}(E_k) \setminus \mathsf{F}\}))_{k \in \mathbb{N}}$ converges toward $\mathbf{P}(\{\rho \in \mathsf{O}^{-1}(D) \setminus \mathsf{F}\})$ which is equal to 0, this implies that $\mathbf{P}(\{\rho \in \mathsf{O}^{-1}(D) \cap \mathsf{F}\}) = 0$ and thus that $\mathbf{P}(D) = 0$. As a consequence, $\lim_{\varepsilon \to 0} \mathbf{P}(D^{\varepsilon}_{min}) = \mathbf{P}(D^{\mathsf{AA}})$. As $\mathcal{M}(\mu_0)$ is not AA-diagnosable by assumption, there thus exists $\varepsilon > 0$ such that $\mathcal{M}(\mu_0)$ is not ε-diagnosable. □

The alternative definition of AA-diagnosability was introduced for two reasons. First, through Proposition 1 it helps build a better understanding of this notion, often misunderstood (see for instance the uniform/non-uniform discussion on AA-diagnosability in [8]). Second, it helps clarify and analyse the notion in a controllable framework: as we will see later, we aim to build a single strategy achieving arbitrary high confidence, not a family of strategies each achieving ε-diagnosability for increasingly small ε.

With the accurate approximate approach to diagnosability, one regains decidability. Indeed, the AA-diagnosability problem for finite oMC was shown to be in PTIME in [7]. This result relies on the notion of distance between two oMC introduced in [16] and defined in the following way: the distance between two oMC $\mathcal{M}_1$ and $\mathcal{M}_2$ with initial distribution μ_1 and μ_2 is[1]

$$d(\mathcal{M}_1(\mu_1), \mathcal{M}_2(\mu_2)) = \max_{E \subseteq \Sigma^\omega} |\mathbf{P}_{\mathcal{M}_1(\mu_1)}(E) - \mathbf{P}_{\mathcal{M}_2(\mu_2)}(E)|.$$

The authors of [16] show how to decide in PTIME if the distance between two oMC is 1 thanks to the following characterisation.

Proposition 2 ([16]). *Given two oMC $\mathcal{M}_1$ and $\mathcal{M}_2$ and two initial distributions μ_1 and μ_2, $d(\mathcal{M}_1(\mu_1), \mathcal{M}_2(\mu_2)) < 1$ iff there exists $w \in \Sigma^*$ and two distributions π_1 and π_2 such that, denoting for $i \in \{1,2\}, \mu_i^w(s) = \mathbf{P}_{\mathcal{M}_i(\mu_i)}(\{\rho s \in S^* \mid \mathsf{O}(\rho s) = w\})$, we have, $\mathsf{Supp}(\pi_i) \subseteq \mathsf{Supp}(\mu_i^w)$ and $d(\mathcal{M}_1(\pi_1), \mathcal{M}_2(\pi_2)) = 0$ (i.e. $\forall w' \in \Sigma^*, \mathbf{P}_{\mathcal{M}_1(\pi_1)}(w') = \mathbf{P}_{\mathcal{M}_2(\pi_2)}(w')$).*

Finally, the link between the distance 1 of two oMC and AA-diagnosability was established in [7], giving the PTIME algorithm:

Theorem 2 ([7]). *Let $\mathcal{M}$ be a finite oMC and μ_0 be an initial distribution. $\mathcal{M}(\mu_0)$ is not AA-diagnosable iff there exist two states $s \in \mathsf{S}^\mathsf{F}$ and $s' \in S \setminus \mathsf{S}^\mathsf{F}$ with s' belonging to a bottom strongly connected component (BSCC)[2] of $\mathcal{M}$ and there exist two finite paths ρ and ρ' of $\mathsf{fPath}(\mathcal{M}(\mu_0))$ such that $\mathsf{last}(\rho) = s$, $\mathsf{last}(\rho') = s'$, $\mathsf{O}(\rho) = \mathsf{O}(\rho')$ and $d(\mathcal{M}(\mathbf{1}_s), \mathcal{M}(\mathbf{1}_{s'})) < 1$.*

From the above theorem, one deduces that AA-diagnosability can be tested by checking the distance 1 of an at most quadratic number of oMC, leading to the PTIME algorithm. The results of this paper also study AA-diagnosability by establishing links to the distance 1 problem. These results however go farther than the characterisation of Theorem 2. In particular, when studying controllable systems, we will need to consider infinite oMC. To that end, we can already note that, speaking of the sufficiency condition only, a more general result was in fact proven in [7]:

Proposition 3 ([7]). *Let $\mathcal{M}$ be an oMC, μ_0 be an initial distribution, two states $s \in \mathsf{S}^\mathsf{F}$ and $s' \in S \setminus \mathsf{S}^\mathsf{F}$ with s' such that no faulty state can be reached from s'*

[1] Note that the absolute values are technically not necessary as $\mathbf{P}_{\mathcal{M}_1(\mu_1)}(E) = 1 - \mathbf{P}_{\mathcal{M}_1(\mu_1)}(\Sigma^\omega \setminus E)$.

[2] A BSCC is a strongly connected component that cannot be escaped from.

and two finite paths ρ and ρ' of $\mathsf{fPath}(\mathcal{M}(\mu_0))$ *such that* $\mathsf{last}(\rho) = s$, $\mathsf{last}(\rho') = s'$, $\mathsf{O}(\rho) = \mathsf{O}(\rho')$. *Then* $\mathcal{M}(\mu_0)$ *is* AA-*diagnosable implies that* $d(\mathcal{M}(\mathbf{1}_q), \mathcal{M}(\mathbf{1}_{q'})) = 1$.

While AA-diagnosability can be decided in polynomial time, the AA-disclosure problem is a bit more complicated. This is not surprising as AA-diagnosability consists in testing whether $Disc^{\mathsf{AA}}(\mathcal{M}(\mu_0))$ is equal to $\mathbf{P}(\mathsf{F}_\infty)$ (the latter being easy to compute as it is solely a reachability property) while the AA-disclosure requires to measure precisely $Disc^{\mathsf{AA}}(\mathcal{M}(\mu_0))$.

Theorem 3. *The* AA-*disclosure problem for finite oMC is* PSPACE-*complete.*

Proof (Sketch of proof). In order to solve the AA-disclosure problem in PSPACE. We first build an exponential size oMC which contains additional information compared to the original one. Then we show that there are two kinds of BSCC in this new oMC: the ones that are reached by paths that almost surely have an AA-disclosing observation sequence, and the ones that are reached by paths that almost surely do not correspond to AA-disclosing observation sequences. We then use the existing results for the AA-diagnosability problem to determine the status of each BSCC. Finally, computing the AA-disclosure of the oMC is equivalent to computing the probability to reach the "AA-disclosing" BSCC, which can be done in NC in the size of the oMC, thus giving an overall PSPACE algorithm.

The hardness is obtained by reduction from the universality problem for non-deterministic finite automaton (NFA), which is known to be PSPACE-complete [19]. □

3 Diagnosis of Controllable Systems

3.1 Controllable Observable Markov Chains

An extension of the oMC formalism allowing us to express control requires us to fix at least two features of this formalism: the nature of the control and the distribution of probabilities of the controlled system. *Controllable weighted Observable Markov chains* (CoMC) are an extension of oMC equivalent to the model of controllable weighted labelled transition systems (CLTS) which were introduced for diagnosis in [5] (the difference between the two models lies in whether the states or the transitions are labelled by an observation). CoMC can also be compared to partially observable Markov decision processes (POMDP): the two classes of models are as expressive, but CoMC can be exponentially more succinct.

In order to specify the control in a CoMC, a subset of observable events is considered as controllable. The control strategy forbids a subset of controllable events depending on the sequence of observations it has received so far. The transitions of the system are no longer labelled by (rational) probabilities but rather by (integer) weights which represent their relative probabilities. Given a state and a set of allowed events, in order to obtain a probability distribution on the allowed transitions, the weights of the outgoing transitions labelled by uncontrollable or allowed controllable actions are normalised. Provided that the control strategy does not create any deadlock, the controlled CoMC is an oMC.

Definition 2 (CoMC). *A Controllable weighted Observable Markov chains (CoMC) over alphabet Σ is a tuple $\mathbb{M} = (S, T, \mathsf{O})$ where S is a finite set of states, $T : S \times S \to \mathbb{N}$ is the transition function labelling transitions with integer weights and $\mathsf{O} : S \to \Sigma$ is the observation function.*

The alphabet is partitioned into controllable and uncontrollable events $\Sigma = \Sigma_c \uplus \Sigma_e$. A set $\Sigma_s \subseteq \Sigma$ of *allowed events* in a state $s \in S$ is a set of observations such that $\Sigma_e \subseteq \Sigma_s$ and $\{s' \in S \mid T(s,s') > 0 \wedge \mathsf{O}(s') \in \Sigma_s\} \neq \emptyset$. Given a state s and a set of allowed events Σ_s, we define the transition probability $p(s, \Sigma_s)$ such that for all s' with $\mathsf{O}(s') \in \Sigma_s$, $p(s, \Sigma_s)(s') = \frac{T(s,s')}{\sum_{s'', \mathsf{O}(s'') \in \Sigma_s} T(s,s'')}$. As before, we write $p(s'|s, \Sigma_s)$ instead of $p(s, \Sigma_s)(s')$. Given an initial distribution μ_0, an infinite path of a CoMC $\mathbb{M}(\mu_0)$ is a sequence $\rho = s_0 \Sigma_0 s_1 \Sigma_1 \ldots$ where $\mu_0(s_0) > 0$ and $p(s_{i+1}|s_i, \Sigma_i) > 0$, for $s_i \in S$ and Σ_i is a set of allowed events in s_i, for all $i \geq 0$. As for oMC, we define finite paths, and we use similar notations for the various sets of paths. A sequence of observations and sets of allowed events $b \in (\Sigma \times 2^{\Sigma})^* \Sigma$ is called a *knowledge sequence.* The knowledge sequence of a path of a CoMC $\rho = s_0 \Sigma_0 s_1 \Sigma_1 \ldots s_i$ is $K(\rho) = \mathsf{O}(s_0) \Sigma_0 \mathsf{O}(s_1) \Sigma_1 \ldots \mathsf{O}(s_i)$.

The nondeterministic choice of the set of allowed events is resolved by strategies.

Definition 3 (Strategy for CoMC). *A* strategy *of CoMC $\mathbb{M}$ with initial distribution μ_0 is a mapping $\sigma : (\Sigma \times 2^{\Sigma})^* \Sigma \to \mathsf{Dist}(2^{\Sigma})$ associating to any knowledge sequence a distribution on sets of events.*

We will only consider here strategies that do not generate a deadlock, i.e. strategies σ such that for all state s reached after a knowledge b, $\sigma(b)$ is a distribution on sets of allowed events for s. Given a strategy σ, a path $\rho = s_0 \Sigma_0 s_1 \Sigma_1 \ldots$ of $\mathbb{M}(\mu_0)$ is *σ-compatible* if for all i, $\Sigma_i \in \mathsf{Supp}(\sigma(K(s_0 \Sigma_0 s_1 \Sigma_1 \ldots s_i)))$. A strategy σ is *deterministic* if $\sigma(b)$ is a Dirac distribution for each knowledge sequence b. In this case, we denote by $\sigma(b)$ the set of allowed actions $\Sigma_a \in 2^{\Sigma}$ such that $\sigma(b) = \mathbf{1}_{\Sigma_a}$. Let b be a knowledge sequence. We define $B_{\mathbb{M}(\mu_0)}(b)$ the *belief* about states corresponding to b as follows:

$$B_{\mathbb{M}(\mu_0)}(b) = \{s \mid \exists \rho \in \mathsf{fPath}(\mathbb{M}(\mu_0)),\ K(\rho) = b \wedge s = \mathsf{last}(\rho)\}$$

A strategy σ is *belief-based* if for all b, $\sigma(b)$ only depends on its belief $B_{\mathbb{M}(\mu_0)}(b)$ (*i.e.* given two knowledge sequence b and b' if $B_{\mathbb{M}(\mu_0)}(b) = B_{\mathbb{M}(\mu_0)}(b')$ then $\sigma(b) = \sigma(b')$). For belief-based strategies, we will sometimes write $\sigma(B)$ for the choice of the strategy made for knowledge sequences producing the belief B.

As for oMC, the failure of a path is defined by the reachability of a set $\mathsf{S}^\mathsf{F} \in S$ of faulty states of the CoMC and we assume again that this set is absorbing.

A strategy σ on $\mathbb{M}(\mu_0)$ defines an infinite oMC $\mathbb{M}_\sigma(\mu_0)$ where the set of states is the finite σ-compatible paths, the observation function associates $\Sigma_{n-1}\mathsf{O}(s_n)$ with the state corresponding to the finite path $\rho = s_0 \Sigma_0 \ldots \Sigma_{n-1} s_n$ (Σ_{n-1} being omitted if $n = 0$) and the transition function p_σ is defined for ρ a σ-compatible path and $\rho' = \rho \Sigma_a s'$ by $p_\sigma(\rho'|\rho) = \sigma(K(\rho))(\Sigma_a) p(s'|s, \Sigma_a)$. We denote by

$\mathbf{P}_{\mathbb{M}_\sigma(\mu_0)}$ the probability measure induced by this oMC. When the strategy possesses some good regularity properties, this oMC is equivalent to a finite one (*i.e.* there is a one-to-one correspondence between the paths of each oMC, it preserves the knowledge sequence and the probability. The two oMC have therefore the same disclosure properties). For instance given a deterministic belief based strategy σ, one can define the oMC $\mathbb{M}'_\sigma$ with set of states $S \times 2^\Sigma \times 2^S$, observation $\mathsf{O}'_\sigma(s, \Sigma^\bullet, B) = (\mathsf{O}(s), \Sigma^\bullet)$, initial distribution $\mu_0^\sigma(s, \emptyset, \mathsf{Supp}(\mu_0) \cap \mathsf{O}^{-1}(\mathsf{O}(s))) = \mu_0(s)$ and transition function $p'_\sigma((s_1, \Sigma_1, B_1) \mid (s_2, \Sigma_2, B_2)) = p(s_1 \mid s_2, \Sigma_2)$ if $\sigma(B_1) = \Sigma_2$ and $B_2 = B_{\mathbb{M}(\mu_1)}(\mathsf{O}(s_2))$ for μ_1 a distribution of support B_1, $p'_\sigma((s_1, \Sigma_1, B_1) \mid (s_2, \Sigma_2, B_2)) = 0$ otherwise. The oMC $\mathbb{M}'_\sigma$ is exponential in the size of $\mathbb{M}$ and is equivalent to $\mathbb{M}_\sigma$. When considering belief-based strategies, we will call $\mathbb{M}_\sigma$ the finite equivalent oMC.

Writing $\mathcal{V}_{\mathbb{M}_\sigma(\mu_0)}$ for the set of infinite paths corresponding to AA-disclosing observation sequences in $\mathbb{M}_\sigma(\mu_0)$, we have $Disc^{\mathsf{AA}}(\mathbb{M}_\sigma(\mu_0)) = \mathbf{P}_{\mathbb{M}_\sigma(\mu_0)}(\mathcal{V}_{\mathbb{M}_\sigma(\mu_0)})$. The control of the system is assumed to support the diagnosis. Therefore, the AA-*diagnosability problem* for CoMC consists in, given a CoMC $\mathbb{M}$ and an initial distribution μ_0, deciding whether there exists a strategy σ such that $\mathbb{M}_\sigma(\mu_0)$ is AA-diagnosable (aka, such that $\mathbf{P}_{\mathbb{M}_\sigma(\mu_0)}(\mathcal{V}_{\mathbb{M}_\sigma(\mu_0)}) = \mathbf{P}_{\mathbb{M}_\sigma(\mu_0)}(\mathsf{F}_\infty)$).

Example 2. Consider the CoMC on the left of Fig. 2. Without any control (*i.e.* with a strategy permanently allowing every event), one obtains the oMC of Fig. 1, which is not AA-diagnosable. However, assuming 'b' is a controllable event, the strategy that always forbids it induces the oMC on the right of Fig. 2 which is AA-diagnosable: every faulty path almost surely contains a 'c' that can not be generated by a correct path. This oMC is in fact exactly diagnosable as once a 'c' occurs the failure proportion becomes equal to 1.

Remark that an observation sequence of the oMC induced by a CoMC and a strategy contains both the observation of the state of the CoMC and the choices of allowed events done by the strategy. The observation sequence of a path in the induced oMC is therefore equal to the knowledge sequence of the corresponding path in the CoMC and as such, we will only speak of observation sequences in the following. This choice of observation was done to express that the choices made by the strategy are known to the observer. An important consequence of this decision is that the strategy does not modify which observation sequences are AA-disclosing.

Lemma 1. *Given* $\mathbb{M}$ *a CoMC,* μ_0 *an initial distribution,* $\mathsf{S}^\mathsf{F} \subseteq S$*,* σ, σ' *two strategies and* w *an observation sequence produced by at least one path of* $\mathbb{M}_\sigma(\mu_0)$ *and at least one path of* $\mathbb{M}_{\sigma'}(\mu_0)$*, then* $\mathsf{Fprop}_{\mathbb{M}_{\sigma'}(\mu_0)}(w) = \mathsf{Fprop}_{\mathbb{M}_\sigma(\mu_0)}(w)$.

Proof. Let $\mathbb{M}$ be a CoMC, μ_0 be an initial distribution, σ be a strategy and $w = o_0\Sigma_0 \dots \Sigma_{n-1}o_n$ be an observation sequence produced by at least one path of $\mathbb{M}_\sigma(\mu_0)$. By definition of w, $\mathsf{Fprop}_{\mathbb{M}_\sigma(\mu_0)}(w)$ is defined and in particular $\prod_{i=0}^{n-1} \sigma(\mathsf{O}(w_{\downarrow 2i+1}))(\Sigma_i) \neq 0$. We have

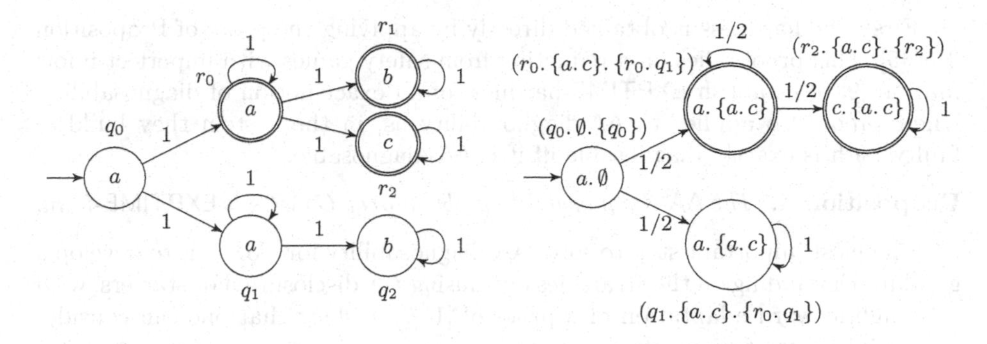

Fig. 2. A CoMC (left) and the finite oMC (right) induced by this CoMC and the strategy that always allow $\{a, c\}$. The observation of a state of the oMC is the pair composed of the observation of the associated state in the CoMC and of the set of allowed events that lead to it. Its name is the triple composed of the associated state in the CoMC, the set of allowed event leading to it and the belief about states that hold in the CoMC when entering this state. The probability in the induced oMC to loop on $(r_0 \cdot \{a \cdot c\} \cdot \{r_0 \cdot q_1\})$ is obtained by dividing the weight $T(r_0, r_0)$ by the weights $T(r_0, r_0)$ and $T(r_0, r_2)$, thus 1/2. The weight $T(r_0, r_1)$ is ignored as b is forbidden.

$$\begin{aligned}
\mathsf{Fprop}_{\mathbb{M}_\sigma(\mu_0)}(w) &= \frac{\mathbf{P}_{\mathbb{M}_\sigma(\mu_0)}(\{\rho \in \mathsf{O}^{-1}(w) \mid \rho \in \mathsf{F}\})}{\mathbf{P}_{\mathbb{M}_\sigma(\mu_0)}(w)} \\
&= \frac{\sum_{\rho \in \mathsf{O}^{-1}(w) \mid \rho \in \mathsf{F}} \mathbf{P}_{\mathbb{M}_\sigma(\mu_0)}(\rho)}{\sum_{\rho \in \mathsf{O}^{-1}(w)} \mathbf{P}_{\mathbb{M}_\sigma(\mu_0)}(\rho)} \\
&= \frac{\sum_{\rho = s_0 \Sigma_0 \dots s_n \in \mathsf{O}^{-1}(w) \mid \rho \in \mathsf{F}} \prod_{i=0}^{n-1} \sigma(\mathsf{O}(w_{\downarrow 2i+1}))(\Sigma_i) p(s_{i+1} \mid s_i, \Sigma_i)}{\sum_{\rho = s_0 \Sigma_0 \dots s_n \in \mathsf{O}^{-1}(w)} \prod_{i=0}^{n-1} \sigma(\mathsf{O}(w_{\downarrow 2i+1}))(\Sigma_i) p(s_{i+1} \mid s_i, \Sigma_i)} \\
&= \frac{\sum_{\rho = s_0 \Sigma_0 \dots s_n \in \mathsf{O}^{-1}(w) \mid \rho \in \mathsf{F}} \prod_{i=0}^{n-1} p(s_{i+1} \mid s_i, \Sigma_i)}{\sum_{\rho = s_0 \Sigma_0 \dots s_n \in \mathsf{O}^{-1}(w)} \prod_{i=0}^{n-1} p(s_{i+1} \mid s_i, \Sigma_i)}
\end{aligned}$$

which is independent of σ, therefore for any strategy σ' such that at least one path of $\mathbb{M}_{\sigma'}(\mu_0)$ produces w, $\mathsf{Fprop}_{\mathbb{M}_{\sigma'}(\mu_0)}(w) = \mathsf{Fprop}_{\mathbb{M}_\sigma(\mu_0)}(w)$. □

3.2 Solving AA-Diagnosability for CoMCs

While accurate approximate diagnosability is simpler than exact diagnosability for oMC (PTIME vs PSPACE) [6,7], for CoMCs this difference disappears and both problems are EXPTIME-complete. The EXPTIME-completeness of exact diagnosis for CoMC was established in [5]. We will devote this section to the proof of the following theorem:

Theorem 4. *The* AA-*diagnosability problem over CoMCs is* EXPTIME-*complete.*

First, the hardness is obtained directly by applying the proof of Proposition 3 of [5]. This proof relies on a reduction from safety games with imperfect information [9] to establish EXPTIME-hardness of an exact notion of diagnosability. Their proof also applies to AA-diagnosability as, in the system they build, a faulty path is exactly diagnosable iff it is AA-diagnosable.

Proposition 4. *The* AA*-diagnosability problem over CoMCs is* EXPTIME*-hard.*

The most important step to solve AA-diagnosability for CoMC is to develop a good understanding on the strategies optimising AA-disclosure. For starters, with a straightforward adaptation of a proof of [15], we show that one can consider deterministic strategies only.

Lemma 2. *Given $\mathbb{M}$ a CoMC, μ_0 an initial distribution, $\mathsf{S}^\mathsf{F} \subseteq S$ and σ a strategy, there exists a deterministic strategy σ' such that $Disc^{\mathsf{AA}}(\mathbb{M}_\sigma(\mu_0)) = \mathbf{P}_{\mathbb{M}_\sigma(\mu_0)}(\mathsf{F}_\infty)$ implies $Disc^{\mathsf{AA}}(\mathbb{M}_{\sigma'}(\mu_0)) = \mathbf{P}_{\mathbb{M}_{\sigma'}(\mu_0)}(\mathsf{F}_\infty)$.*

Proof. In the proof of Lemma 1 of [15], the authors show that a randomised 'observation based' strategy can be seen as an average over a family of deterministic 'observation based' strategies[3]. A consequence of their equation (2) in our framework is the following: given a strategy σ, for every set of path E, there exists a deterministic strategy σ_{det} such that (a) $\mathsf{Path}(\mathbb{M}_{\sigma_{det}}(\mu_0)) \subseteq \mathsf{Path}(\mathbb{M}_\sigma(\mu_0))$ and (b) $\mathbf{P}_{\mathbb{M}_{\sigma_{det}}(\mu_0)}(E) \geq \mathbf{P}_{\mathbb{M}_\sigma(\mu_0)}(E)$. Using this result with the appropriate set E we will show that if $\mathbb{M}_\sigma(\mu_0)$ is AA-diagnosable then $\mathbb{M}_{\sigma_{det}}(\mu_0)$ is AA-diagnosable.

We define $E_\sigma = \mathcal{V}_{\mathbb{M}_\sigma(\mu_0)} \cup (\mathsf{Path}(\mathbb{M}_\sigma(\mu_0)) \backslash \mathsf{F}_\infty)$ which are the set of infinite σ-compatible paths that are either correct or AA-disclosing. Let σ_{det} be the strategy obtained by applying the result of [15] on the set E_σ. Suppose $\mathbb{M}_\sigma(\mu_0)$ is AA-diagnosable. By definition, this is equivalent to $\mathbf{P}_{\mathbb{M}_\sigma(\mu_0)}(E_\sigma) = 1$. Due to (b), this implies that $\mathbf{P}_{\mathbb{M}_{\sigma_{det}}(\mu_0)}(E_\sigma) = 1$ too. Moreover $\mathcal{V}_{\mathbb{M}_{\sigma_{det}}(\mu_0)} = \mathcal{V}_{\mathbb{M}_\sigma(\mu_0)} \cap \mathsf{Path}(\mathbb{M}_{\sigma_{det}}(\mu_0))$, thanks to Lemma 1 and (a). Thus

$$\begin{aligned} E_\sigma &= \mathcal{V}_{\mathbb{M}_{\sigma_{det}}(\mu_0)} \cup (\mathcal{V}_{\mathbb{M}_\sigma(\mu_0)} \setminus \mathsf{Path}(\mathbb{M}_{\sigma_{det}}(\mu_0)) \cup (\mathsf{Path}(\mathbb{M}_\sigma(\mu_0)) \setminus \mathsf{F}_\infty) \\ &= E_{\sigma_{det}} \cup (\mathcal{V}_{\mathbb{M}_\sigma(\mu_0)} \cup (\mathsf{Path}(\mathbb{M}_\sigma(\mu_0)) \setminus \mathsf{F}_\infty) \setminus \mathsf{Path}(\mathbb{M}_{\sigma_{det}}(\mu_0)) \end{aligned}$$

where $E_{\sigma_{det}} = \mathcal{V}_{\mathbb{M}_{\sigma_{det}}(\mu_0)} \cup (\mathsf{Path}(\mathbb{M}_{\sigma_{det}}(\mu_0)) \setminus \mathsf{F}_\infty)$.

Finally, $\mathbf{P}_{\mathbb{M}_{\sigma_{det}}(\mu_0)}(\mathcal{V}_{\mathbb{M}_\sigma(\mu_0)} \cup (\mathsf{Path}(\mathbb{M}_\sigma(\mu_0)) \setminus \mathsf{F}_\infty) \setminus \mathsf{Path}(\mathbb{M}_{\sigma_{det}}(\mu_0)) = 0$ as no path of this set is σ_{det}-compatible. Therefore $\mathbf{P}_{\mathbb{M}_{\sigma_{det}}(\mu_0)}(E_{\sigma_{det}}) = 1$ which implies that $\mathbb{M}_{\sigma_{det}}(\mu_0)$ is AA-diagnosable. □

We can further restrict the strategies by limiting ourselves to belief-based strategy. This is far from an intuitive result. Indeed, while the AA-diagnosability of an oMC depends heavily on the exact values of the probabilities in the oMC, this result implies that the control only needs to remember the set of states potentially reached with a given observation sequence, not the probabilities with which one is in each state. Remark though that the choice made by the strategy in each belief depends on the probabilities.

[3] In our framework, by definition, every strategy is 'observation based'.

Lemma 3. *detebel Given* $\mathbb{M}$ *a CoMC,* μ_0 *an initial distribution,* $\mathsf{S}^\mathsf{F} \subseteq S$ *and* σ *a deterministic strategy, there exists a deterministic belief based strategy* σ' *such that* $Disc^{\mathsf{AA}}(\mathbb{M}_\sigma(\mu_0)) = \mathbf{P}_{\mathbb{M}_\sigma(\mu_0)}(\mathsf{F}_\infty)$ *implies* $Disc^{\mathsf{AA}}(\mathbb{M}_{\sigma'}(\mu_0)) = \mathbf{P}_{\mathbb{M}_{\sigma'}(\mu_0)}(\mathsf{F}_\infty)$.

Proof. Let $\mathbb{M}$ be a CoMC, μ_0 be an initial distribution and σ be a deterministic strategy such that $\mathbb{M}_\sigma(\mu_0)$ is AA-diagnosable. We define a belief based strategy σ' from σ in the following way. Let $\rho \in \mathsf{fPath}(\mathbb{M}_\sigma(\mu_0))$. We define by E_ρ the set of finite path producing the same belief as ρ, *i.e.* $E_\rho = \{\rho' \in \mathsf{fPath}(\mathbb{M}_\sigma(\mu_0)) \mid B_{\mathbb{M}(\mu_0)}(\mathsf{O}(\rho')) = B_{\mathbb{M}(\mu_0)}(\mathsf{O}(\rho))\}$. We define $\sigma'(B_{\mathbb{M}(\mu_0)}(\mathsf{O}(\rho))) = \bigcup_{\rho' \in E_\rho} \sigma(\mathsf{O}(\rho'))$. In other words, in a given belief, σ' allows anything that σ allowed at least once in this belief. Let us show that $\mathbb{M}_{\sigma'}(\mu_0)$ is AA-diagnosable.

Let two states $q = (s, \Sigma^\bullet, B) \in \mathsf{S}^\mathsf{F}$ and $q' = (s', \Sigma^\bullet, B) \in S \backslash \mathsf{S}^\mathsf{F}$ belonging to a BSCC of $\mathbb{M}_{\sigma'}(\mu_0)$ and reached by two finite paths ρ and ρ' of $\mathsf{fPath}(\mathbb{M}_{\sigma'}(\mu_0))$ with $\mathsf{O}(\rho) = \mathsf{O}(\rho')$. We will show that $d(\mathbb{M}_{\sigma'}(\mathbf{1}_q), \mathbb{M}_{\sigma'}(\mathbf{1}_{q'})) = 1$ using the characterisation given in Proposition 2. More precisely, for any observations sequence $w \in \Sigma^*$, and any pair of distributions on the set of states reached from q and from q' after observing w, we consider the probabilistic language generated by similar distributions in $\mathbb{M}_\sigma$ (*i.e.* distributions giving the same weight to the states of the original CoMC $\mathbb{M}$) and rely on the fact that $\mathbb{M}_\sigma$ is AA-diagnosable to show that the generated languages are different. This implies the distance is 1 thanks to Proposition 2.

Let $w \in \Sigma^*$ such that $\mathbb{P}_{\mathbb{M}_{\sigma'}(\mathbf{1}_q)}(w) > 0$ and $\mathbb{P}_{\mathbb{M}_{\sigma'}(\mathbf{1}_{q'})}(w) > 0$, we denote by B_w, B_q and $B_{q'}$ the beliefs reached after observing w from the beliefs B, $\{q\}$ and $\{q'\}$ respectively, let two distributions μ'_1 and μ'_2 such that $\mathsf{Supp}(\mu'_1) \subseteq B_q$, $\mathsf{Supp}(\mu'_2) \subseteq B_{q'}$. As σ' does not allow events that are never allowed by σ in the same belief, there exists an observation sequence $w_\sigma \in \Sigma^*$ such that $\mathbb{P}_{\mathbb{M}_\sigma(\mu_0)}(w_\sigma) > 0$ and the belief reached in $\mathbb{M}(\mu_0)$ after a path of observation w_σ from the initial distribution is B_w, *i.e.* $B_{\mathbb{M}(\mu_0)}(w_\sigma) = B_w$.

We can thus define initial distributions μ_1 and μ_2 on the set of states reached after observing w_σ in $\mathbb{M}_\sigma$ mimicking the distributions μ'_1 and μ'_2 (*i.e.* for every state $q_0 = (s_0, \Sigma_0, B_w)$ of $\mathbb{M}_{\sigma'}(\mu_0)$, we select some q_1, state of $\mathbb{M}_\sigma(\mu_0)$ associated to a σ-compatible paths ρ that ends in s_0 and such that $\mathsf{O}(\rho) = w_\sigma$, and we set for $i \in \{1,2\}, \mu'_i(q_0) = \mu_1(q_1)$). From Proposition 3 and Proposition 2, there exists a word w_d such that $\mathbb{P}_{\mathbb{M}_\sigma(\mu_1)}(w_d) \neq \mathbb{P}_{\mathbb{M}_\sigma(\mu_2)}(w_d)$. This implies that there exists a word w'_d such that $\mathbb{P}_{\mathbb{M}_{\sigma'}(\mu'_1)}(w'_d) \neq \mathbb{P}_{\mathbb{M}_{\sigma'}(\mu'_2)}(w'_d)$. Indeed, let E be the set of observation sequences of the form $w'a$ where w' is a strict prefix of w_d, $a \in \Sigma$, $\mathbb{P}_{\mathbb{M}_{\sigma'}(\mu'_1)}(w'a) > 0$ and $\mathbb{P}_{\mathbb{M}_\sigma(\mu_1)}(w'a) = 0$. If $\mathbb{P}_{\mathbb{M}_{\sigma'}(\mu'_1)}(E) \neq \mathbb{P}_{\mathbb{M}_{\sigma'}(\mu'_2)}(E)$, this implies our result. Otherwise, by construction of the strategy σ' we have:

$$\begin{aligned}
\mathbb{P}_{\mathbb{M}_{\sigma'}(\mu'_1)}(w_d) &= \mathbb{P}_{\mathbb{M}_\sigma(\mu_1)}(w_d) \times (1 - \mathbb{P}_{\mathbb{M}_{\sigma'}(\mu'_1)}(E)) \\
&\neq \mathbb{P}_{\mathbb{M}_\sigma(\mu_2)}(w_d) \times (1 - \mathbb{P}_{\mathbb{M}_{\sigma'}(\mu'_1)}(E)) \\
&= \mathbb{P}_{\mathbb{M}_\sigma(\mu_2)}(w_d) \times (1 - \mathbb{P}_{\mathbb{M}_{\sigma'}(\mu'_2)}(E)) \\
&= \mathbb{P}_{\mathbb{M}_{\sigma'}(\mu'_2)}(w_d),
\end{aligned}$$

in which case we can choose $w'_d = w_d$. As this holds for any $w \in \Sigma^*$ and pair of distributions μ'_1 and μ'_2, according to Proposition 2 we have $d(\mathbb{M}_{\sigma'}(\mathbf{1}_q), \mathbb{M}_{\sigma'}(\mathbf{1}_{q'})) = 1$. From Theorem 2, we can thus deduce that $\mathbb{M}_{\sigma'}(\mu_0)$ is AA-diagnosable. Therefore belief-based strategies are sufficient to decide AA-diagnosability. □

A naive NEXPTIME algorithm can be obtained from these two lemmas: we guess a deterministic belief-based strategy then verify AA-diagnosability of the exponential oMC generated by the CoMC and the strategy. In the following proposition, we show how to efficiently build a good belief-based strategy, which gives us an EXPTIME algorithm.

Proposition 5. *The* AA-*diagnosability problem over CoMCs is in* EXPTIME.

Proof. Let $\mathbb{M}$ be a CoMC and μ_0 be an initial distribution. This proof is done in two steps.

1. We show that, given two deterministic belief based strategies σ_1 and σ_2 such that σ_1 is less restrictive than σ_2 and a state q belonging to a BSCC of both $\mathbb{M}_{\sigma_1}(\mu_0)$ and $\mathbb{M}_{\sigma_2}(\mu_0)$, then if the paths of $\mathbb{M}_{\sigma_2}(\mu_0)$ that visits q are almost surely AA-disclosing then so are the paths of $\mathbb{M}_{\sigma_1}(\mu_0)$ that visits q. In other words, within a BSCC, the least restrictive a strategy is, the better it is for the purpose of diagnosis.
2. Thanks to the result obtained in the first step, we efficiently build a strategy in the form of a greatest fixed point: we start by the most permissive strategy and iteratively restrict it to prune the BSCC that cause the strategy not to achieve AA-diagnosability.

Let σ and σ' be two deterministic belief-based strategies such that for any belief B of $\mathbb{M}$ $\sigma(B) \subseteq \sigma'(B)$. Let q be a faulty state associated to a belief B and belonging to a BSCC of both $\mathbb{M}_\sigma(\mu_0)$ and $\mathbb{M}_{\sigma'}(\mu_0)$. Assume that there exists a positive measure of paths in $\mathbb{M}_{\sigma'}(\mu_0)$ that visit q and that are not associated to an AA-disclosing observation sequence. Defining $B' = (B \backslash \mathsf{S}^\mathsf{F}) \cup \{q\}$, this is equivalent to saying that the CoMC $\mathbb{M}_{\sigma'}(\mu_1)$, where μ_1 is an initial distribution of support B', is not AA-diagnosable. Therefore we can use the characterisation of Theorem 2. Without loss of generality, as q belongs to a BSCC, we can assume the pair of states given by the characterisation is (q, q') where $q' \notin \mathsf{S}^\mathsf{F}$, is associated to the belief B, belongs to a BSCC of $\mathbb{M}_{\sigma'}(\mu_1)$ and is such that $d(\mathbb{M}_{\sigma'}(\mathbf{1}_q), \mathbb{M}_{\sigma'}(\mathbf{1}_{q'})) < 1$. Let w, π_1 and π_2 be the observation sequence and the two distributions obtained by applying Proposition 2 on the pair of CoMC $(\mathbb{M}_{\sigma'}(\mathbf{1}_q), \mathbb{M}_{\sigma'}(\mathbf{1}_{q'}))$. Let $q'' \notin \mathsf{S}^\mathsf{F}$ be a state belonging to a BSCC of $\mathbb{M}_\sigma(\mu_1)$ reachable from q' by a σ-compatible path with observation sequence ww'. Let π'_1 and π'_2 be the distribution obtained after observing w' starting in π_1 and π_2. As $\forall v \in \Sigma^*, \mathbf{P}_{\mathbb{M}_{\sigma'}(\pi_1)}(v) = \mathbf{P}_{\mathbb{M}_{\sigma'}(\pi_2)}(v)$, we also have $\forall v \in \Sigma^*, \mathbf{P}_{\mathbb{M}_{\sigma'}(\pi'_1)}(v) = \mathbf{P}_{\mathbb{M}_{\sigma'}(\pi'_2)}(v)$. This implies that $\forall v \in \Sigma^*, \mathbf{P}_{\mathbb{M}_\sigma(\pi'_1)}(v) = \mathbf{P}_{\mathbb{M}_\sigma(\pi'_2)}(v)$. Indeed, given $v \in \Sigma^*$, we have

$$\begin{aligned}
\mathbf{P}_{\mathbb{M}_\sigma(\pi_1')}(v) &= \sum_{\rho \in \mathsf{O}^{-1}(v)} \mathbf{P}_{\mathbb{M}_\sigma(\pi_1')}(\rho) \\
&= \sum_{\rho = s_0 \Sigma_0 \dots s_n \in \mathsf{O}^{-1}(v)} \pi_1'(s_0) \prod_{i=0}^{n-1} \sigma(\mathsf{O}(v_{\downarrow 2i+1}))(\Sigma_i) p(s_{i+1} \mid s_i, \Sigma_i) \\
&= \left(\prod_{i=0}^{n-1} \sigma(\mathsf{O}(v_{\downarrow 2i+1}))(\Sigma_i) \right) \sum_{\rho = s_0 \Sigma_0 \dots s_n \in \mathsf{O}^{-1}(v)} \pi_1'(s_0) \prod_{i=0}^{n-1} \frac{T(s_i, s_{i+1})}{\sum_{s'', \mathsf{O}(s'') \in \Sigma_i} T(s_i, s'')} \\
&= \left(\prod_{i=0}^{n-1} \sigma(\mathsf{O}(v_{\downarrow 2i+1}))(\Sigma_i) \right) \sum_{\rho = s_0 \Sigma_0 \dots s_n \in \mathsf{O}^{-1}(v)} \pi_2'(s_0) \prod_{i=0}^{n-1} \frac{T(s_i, s_{i+1})}{\sum_{s'', \mathsf{O}(s'') \in \Sigma_i} T(s_i, s'')} \\
&= \mathbf{P}_{\mathbb{M}_\sigma(\pi_2')}(v).
\end{aligned}$$

As a consequence, $d(\mathbb{M}_\sigma(\mathbf{1}_q), \mathbb{M}_\sigma(\mathbf{1}_{q'})) < 1$. From Theorem 2, this implies that $\mathbb{M}_\sigma(\mu_1)$ is not AA-diagnosable and thus there exists a positive measure of paths in $\mathbb{M}_\sigma(\mu_0)$ that visit q and that are not associated to an AA-disclosing observation sequence. Therefore, having restricted the strategy σ' did not allow to regain AA-diagnosability of the paths visiting q. This means that a strategy achieving AA-diagnosability of the CoMC must ensure that q cannot be reached.

Using this result, we build iteratively the most permissive strategy ensuring AA-diagnosability. We start with the strategy σ_0 allowing everything. Assume we built the strategy σ_k such that any less permissive strategy do not ensure AA-diagnosability. If $\mathbb{M}_{\sigma_k}(\mu_0)$ is not AA-diagnosable, there exists two states s and s' associated to the same belief B that satisfies the characterisation of Theorem 2. W.l.o.g one can assume that both of these states belong to BSCCs of $\mathbb{M}_{\sigma_k}(\mu_0)$. From our preliminary result, we know that any strategy that contains the states s and s' in a BSCC does not ensure AA-diagnosability. As any strategy less permissive than σ_k does not ensure AA-diagnosability, we need to restrict the strategy so that the belief B is not reachable, or that B is not associated to states belonging to a BSCC anymore. The latter is in fact not sufficient as Theorem 2 would still apply on the pair of states (s, s'). Thus we build σ_{k+1} as the most permissive strategy such that $\mathbb{M}_{\sigma_{k+1}}(\mu_0)$ does not contain the belief B, which can easily be done with belief based strategies. This procedure ends when the strategy σ_n that is created either is the most permissive strategy ensuring AA-diagnosability or if one cannot build a strategy removing the problematic belief. This algorithm is in EXPTIME as every step of the procedure can be done in exponential time (verification of AA-diagnosability, identification of the pair of problematic states and creation of the new strategy are all steps that can be done in EXPTIME) and there is at most exponentially many steps as each one of them removes at least one belief from the system, and there are exponentially many beliefs. Therefore, the AA-diagnosability problem can be solved in EXPTIME. □

Remark that the above proof builds the strategy ensuring AA-diagnosability when it exists.

4 Conclusion

This paper considers the accurate approximate notion of disclosure for diagnosability. We establish how to decide AA-diagnosability in CoMC and how to measure the AA-disclosure in oMC. Measuring the AA-disclosure was not developed for CoMC here as the notion is undecidable (straightforward application of the undecidability of the emptiness problem for probabilistic automata).

Opacity is a notion that intuitively appears as some kind of dual to diagnosability. The goal of opacity is to make sure some secret paths of the system are not detected by an observer. Following the idea that some small amount of revealed secret information is not problematic, this line of research favors a quantitative approach to the problem, thus closer to the AA-disclosure problem we studied for oMC. In this endeavour, various measures for the disclosure set have been introduced [1,3,4,22]. Opacity has been studied in an active framework called observable Markov decision processes (oMDP) where the controller is deemed internal to the system and thus makes its choice with more information than just the observation sequence. This framework is thus not equivalent to the CoMC model presented in this paper; the strategy is more powerful. As such, while measuring the disclosure is undecidable (for any disclosure notion) in CoMC, some positive results were established in oMDP [2]. However, as this work only considered the exact notion of disclosure, it would be interesting to see if the approximate approach pushed here could also be applied for oMDP. Moreover, this framework also makes sense for a study of diagnosability as the control defined in oMDP can correspond to designing choices of the system.

A AA-Disclosure Problem for oMC

Theorem 3. *The* AA*-disclosure problem for finite oMC is* PSPACE*-complete.*

We decompose the proof of the theorem in the two following proposition, each establishing one direction.

Proposition 6. *The* AA*-disclosure problem for finite oMC is in* PSPACE.

Proof. To establish this result, we first build an exponential size oMC which contains additional information: the set of states the system could be in after the observation sequence. Then we show that there are two kinds of BSCC in this new oMC: the ones that are reached by paths that almost surely have an AA-disclosing observation sequence, and the ones that are reached by paths that do not correspond to AA-disclosing observation sequences. We can then use the existing results for the AA-diagnosability problem to determine the status of each BSCC. Therefore, computing the AA-disclosure of the oMC is equivalent to computing the probability to reach the "AA-disclosing" BSCC, which can be done in NC in the size of the oMC, thus giving an overall PSPACE algorithm.

Let $\mathcal{M} = (S, p, \mathsf{O})$ be a finite oMC and μ_0 be an initial distribution. We build a new oMC $\mathcal{M}' = (S', p', \mathsf{O}')$ which has the same behaviour as $\mathcal{M}$ but where the states are enriched with an additional information (the set of states the system can be in, given the produced observation sequence):

- $S' = S \times 2^S$;
- For $(s,B),(s',B') \in S'$, $p'((s',B') \mid (s,B)) = p(s' \mid s)$ if $B' = \cup_{q\in B}\mathsf{Supp}(p(q)) \cap \mathsf{O}^{-1}(\mathsf{O}(s'))$ else, $p'((s',B') \mid (s,B)) = 0$;
- For $(s,B) \in S'$, $\mathsf{O}'(s,B) = \mathsf{O}(s)$.

We define the initial distribution μ_0' for $\mathcal{M}'$ by $\mu_0'(s, \mathsf{Supp}(\mu_0) \cap \mathsf{O}^{-1}(\mathsf{O}(s))) = \mu_0(s)$ for all $s \in S$. There is a one-to-one correspondence between the paths of $\mathcal{M}(\mu_0)$ and $\mathcal{M}'(\mu_0')$: every path $\rho = s_0 s_1 \cdots s_n$ of $\mathcal{M}(\mu_0)$ is associated to an unique path $\rho' = (s_0,B_0)(s_1,B_1)\cdots(s_n,B_n)$ with $\mathsf{O}(\rho) = \mathsf{O}(\rho')$, $\mathbf{P}_{\mathcal{M}(\mu_0)}(\rho) = \mathbf{P}_{\mathcal{M}'(\mu_0')}(\rho')$ and B_n contains the set of states of S that can be reached with a path of observation $\mathsf{O}(\rho)$. Due to the latter property, B_n only depends on $\mathsf{O}(\rho)$ and is called the *belief* associated to $\mathsf{O}(\rho)$.

Let $(s,B) \in S'$ such that $s \in \mathsf{S^F}$ and (s,B) belongs to a BSCC of $\mathcal{M}'$. We claim that either for every path ρ ending in (s,B), $\mathbf{P}(\{\rho' \in \mathsf{Path}(\mathcal{M}'(\mu_0')) \mid \rho \preceq \rho' \wedge \mathsf{O}(\rho') \in D^{\mathsf{AA}}\}) = 0$ or for every path ρ ending in (s,B), $\mathbf{P}(\{\rho' \in \mathsf{Path}(\mathcal{M}'(\mu_0')) \mid \rho \preceq \rho' \wedge \mathsf{O}(\rho') \in D^{\mathsf{AA}}\}) = \mathbf{P}(\rho)$. In other words, there are two categories of BSCC composed of faulty states: the good ones, that almost surely accurate approximately disclose the fault, and the bad ones that do not accurate approximately disclose the fault at all. Moreover, a BSCC containing the state (s,B) do not disclose the fault at all iff there exists a state $s' \in B$ such that s' belongs to a BSCC of $\mathcal{M}$, $s' \notin \mathsf{S^F}$ and $d(\mathcal{M}(\mathbf{1}_s), \mathcal{M}(\mathbf{1}_{s'})) < 1$.

Let (s,B) be a state belonging to a BSCC of $\mathcal{M}'$. Assume that for all $s' \in B$ such that s' belongs to a BSCC of $\mathcal{M}$ and $s' \notin \mathsf{S^F}$ we have $d(\mathcal{M}(\mathbf{1}_s), \mathcal{M}(\mathbf{1}_{s'})) = 1$. We denote $B' = (B \backslash \mathsf{S^F}) \cup \{s\}$, and define $\mathcal{M}''$ by removing the path leading to a faulty state (aka, a path either starts faulty or forever remain correct). Then as s belongs to a BSCC of $\mathcal{M}$, we can directly use Theorem 2 to obtain that for any initial distribution μ_1 of support B', we have that $\mathcal{M}''(\mu_1)$ is AA-diagnosable. As the limitation to the states of $B \backslash B'$ and the transformation from $\mathcal{M}$ to $\mathcal{M}''$ can only increase the failure proportion, this ensures that $\mathbf{P}(\{\rho' \in \mathsf{Path}(\mathcal{M}'(\mu_0')) \mid \rho \preceq \rho' \wedge \mathsf{O}(\rho') \in \mathit{Disc}^{\mathsf{AA}}\}) = \mathbf{P}(\rho)$.

Conversely, if there exists a state $s' \in B$ such that s' belongs to a BSCC of B, $s' \notin \mathsf{S^F}$ and $d(\mathcal{M}(\mathbf{1}_s), \mathcal{M}(\mathbf{1}_{s'})) < 1$, then one can rely on the proof of Lemma A of [8] to obtain the result. For the sake of pedagogy, we present the proof here in the simpler case where B does not contain any faulty state beside s. Using Proposition 2 and the correspondence between $\mathcal{M}$ and $\mathcal{M}'$, one deduces that there exists $\rho_{(s,B)} \in \mathsf{fPath}(\mathcal{M}(\mathbf{1}_{(s,B)}))$ and $\alpha > 0$ such that for all $w \in \Sigma^*$ with $\mathsf{O}(\rho) \leq w$

$$\mathbf{P}_{\mathcal{M}'(\mathbf{1}_{(s,B)})}(\{\rho' \in \mathsf{fPath}(\mathcal{M}'(\mathbf{1}_{(s,B)})) \mid \rho_{(s,B)} \preceq \rho' \wedge \mathsf{O}(\rho') = w\}) \quad (1)$$

$$\leq \alpha \mathbf{P}_{\mathcal{M}'(\mathbf{1}_{(s',B)})}(\{\rho' \in \mathsf{fPath}(\mathcal{M}'(\mathbf{1}_{(s',B)})) \mid \mathsf{O}(\rho') = w\}). \quad (2)$$

Therefore, for all $w \in \Sigma^*$ and initial distribution μ_1 of support B we have:

$$\mathsf{Fprop}_{\mathcal{M}'(\mu_1)}(w) \leq \frac{\mathbf{P}_{\mathcal{M}'(\mathbf{1}_{(s,B)})}(w)}{\mathbf{P}_{\mathcal{M}'(\mathbf{1}_{(s,B)})}(w) + \frac{\mu_1(s')}{\mu_1(s)}\mathbf{P}_{\mathcal{M}'(\mathbf{1}_{(s',B)})}(w)} \tag{3}$$

$$\leq \frac{\varepsilon_w + \sum\limits_{\rho | \mathsf{O}(\rho\rho_{(s,B)}) \leq w} \frac{\alpha \mathbf{P}_{\mathcal{M}'(\mathbf{1}_{(s,B)})}(\rho)}{\mathbf{P}_{\mathcal{M}'(\mathbf{1}_{(s,B)})}(\rho_{(s,B)})} \mathbf{P}_{\mathcal{M}'(\mathbf{1}_{(s',B)})}(w^\rho)}{\mathbf{P}_{\mathcal{M}'(\mathbf{1}_{(s,B)})}(w) + \frac{\mu_1(s')}{\mu_1(s)}\mathbf{P}_{\mathcal{M}'(\mathbf{1}_{(s',B)})}(w)} \tag{4}$$

where w^ρ is such that $w = \mathsf{O}(\rho)w^\rho$, the first term $\varepsilon_w = \mathbf{P}_{\mathcal{M}'(\mathbf{1}_{(s,B)})}(\{\rho \in \mathsf{fPath}(\mathcal{M}(\mathbf{1}_{(s,B)}) \mid \nexists \rho_1, \rho_2, \rho = \rho_1\rho_{(s,B)}\rho_2 \wedge \mathsf{O}(\rho) = w\})$ is the probability of the set of paths with observation w that do not contain the infix $\rho_{(s,B)}$ and the second term relies on the bound from Eq. 2 to bound the probability of every other paths. As with probability 1, a path of $\mathcal{M}'(\mathbf{1}_{(s,B)})$ visits (s, B) infinitely often, it will almost surely contain a $\rho_{(s,B)}$ subpath, more precisely: the value $\frac{\varepsilon_w}{\mathbf{P}_{\mathcal{M}'(\mathbf{1}_{(s,B)})}(w)}$ almost surely converges to 0 when $|w|$ diverges to ∞. Let $w \in \Sigma^\omega$, if $\mathsf{Fprop}_{\mathcal{M}'(\mu_1)}(w_{\downarrow n}) \xrightarrow{n\to\infty} 1$ then, for all ρ such that $\mathsf{O}(\rho\rho_{(s,B)}) \leq w$ we have that $\frac{\mathbf{P}_{\mathcal{M}'(\mathbf{1}_{(s',B)})}(w^\rho_{\downarrow n})}{\mathbf{P}_{\mathcal{M}'(\mathbf{1}_{(s,B)})}(w_{\downarrow n})}$ converges to 0, thus, due to Eq. 4, $\varepsilon_{w_{\downarrow n}}$ does not converge to 0, which can only happen with probability 0. Therefore $\mathsf{Fprop}_{\mathcal{M}'(\mu_1)}(w_{\downarrow n})$ almost surely does not converge to 1. This implies that $\mathbf{P}\{\rho' \in \mathsf{Path}(\mathcal{M}'(\mu'_0)) \mid \rho \preceq \rho' \wedge \mathsf{O}(\rho') \in D^{AA}\} = 0$.

This result establishes that the BSCC of $\mathcal{M}'$ are partitioned between the good ones that accurately approximately and almost surely disclose the fault and the bad ones that do not accurately approximately disclose it at all. Moreover, given a state (s_0, B_0) belonging to a BSCC of $\mathcal{M}'$, if there exists a state $s'_0 \in B_0$ such that s'_0 belongs to a BSCC of B, $s'_0 \notin \mathsf{S}^\mathsf{F}$ and $d(\mathcal{M}(\mathbf{1}_{s_0}), \mathcal{M}(\mathbf{1}_{s'_0})) < 1$, then for any state (s_1, B_1) belonging to the same BSCC, one can find a state $s'_1 \in B_1$ satisfying a similar property with respect to s_1. In other words, for every BSCC of $\mathcal{M}'$, we only need to check a single state (s, B) of the BSCC to identify whether the BSCC is disclosing or not. Furthermore, this check can be done by testing the distance 1 between copies of $\mathcal{M}$ starting in s and copies starting in some of the states in B. There is thus at most linearly many tests to do, each of which can be done in polynomial time in the size of $\mathcal{M}$.

Therefore, one can obtain the value of $Disc^{AA}(\mathcal{M}'(\mu'_0))$ by computing the probability to reach the good BSCC, which is known to be possible in PTIME in the size of $\mathcal{M}'$. In fact, as computing this probability amount to solve a linear system of equations, this can even be done in NC [11,21]. The oMC $\mathcal{M}'$ being exponential in the size of $\mathcal{M}$, and as NC blown up to the exponential is equal to PSPACE [10], this yields a PSPACE algorithm. As $Disc^{AA}(\mathcal{M}(\mu_0)) = Disc^{AA}(\mathcal{M}'(\mu'_0))$, this allows us to solve the AA-disclosure problem. □

Proposition 7. *The* AA*-disclosure problem for finite oMC is* PSPACE*-hard.*

Proof. We now establish the hardness by reducing the universality problem for non-deterministic finite automaton (NFA), which is known to be PSPACE-complete [19].

An NFA is a tuple $\mathcal{A} = (Q, \Sigma, T, q_0, F)$ where Q is the set of states, q_0 is the initial state, F is the set of accepting states, Σ is the alphabet and $T \in Q \times \Sigma \times Q$ is the transition function. An NFA is universal if for all $w = a_1 a_2 \dots a_n \in \Sigma^n$, there exists a path $q_0 a_1 q_1 a_2 \dots q_n$ such that $q_n \in F$ and for all $1 \leq i \leq n, (q_{i-1}, a_i, q_i) \in T$.

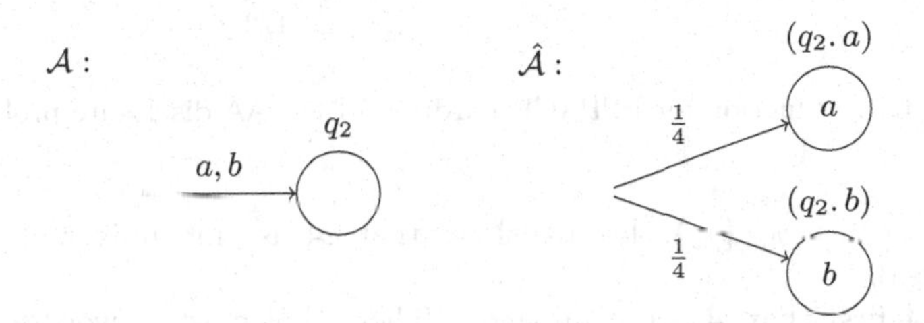

Fig. 3. From NFA $\mathcal{A}$ to incomplete oMC $\hat{\mathcal{A}}$. The label next to the state is its name. We will not always display the state's name so as not to overload the figure.

Let $\mathcal{A} = (Q, \Sigma, T, q_0, F)$ be an NFA. W.l.o.g. we can assume that $F = Q$ and $\Sigma = \{a, b\}$. Our first step is to push the observations onto the states (as shown in Fig. 3). From $\mathcal{A}$ we define the incomplete oMC $\hat{\mathcal{A}} = (S_A, p_A, O_A)$ and the initial distribution μ_0^A such that:

- $S_A = Q \times \Sigma$;
- for $(q, c), (q', d) \in S_A$, if $(q, d, q') \in T$, then $p_A((q', d) \mid (q, c)) = \frac{1}{|S_A|+1}$, else $p_A((q', d) \mid (q, c)) = 0$;
- for $(q, c) \in S_A, O_A(q, c) = c$;
- for $(q', d) \in S_A$, if $(q_0, d, q') \in T$, then $\mu_0^A(q', d) = \frac{1}{|S_A|+1}$, else $\mu_0^A(q', d) = 0$.

This oMC is incomplete as none of the distributions μ_0^A and $p_A(\cdot \mid s)$ (for $s \in S_A$) sum to 1. We now build the oMC $\mathcal{M} = (S, p, O)$ represented in Fig. 4 where

- $S = S_A \cup \{s_\sharp, f_a, f_b, f_\sharp\}$;
- given $s, s' \in S_A$, $p(s' \mid s) = p_A(s' \mid s)$, $p(s_\sharp \mid s) = 1 - \sum_{s' \in S_A} p(s' \mid s)$, for $h \in \{f_a, f_b\}$ and $g \in \{f_a, f_b, f_\sharp\}$, $p(g \mid h) = 1/3$ and $p(f_\sharp \mid f_\sharp) = p(s_\sharp \mid s_\sharp) = 1$;
- for $s \in S_A$, $O(s) = O_A(s)$, $O(s_\sharp) = O(f_\sharp) = \sharp, O(f_a) = a$ and $O(f_b) = b$.

We also define μ_0 as $\mu_0(s) = \mu_0^A(s)$ for $s \in S_A$ and $\mu_0(f_a) = \mu_0(f_b) = \frac{1 - \sum_{s \in S_A} \mu_0(s)}{2}$.

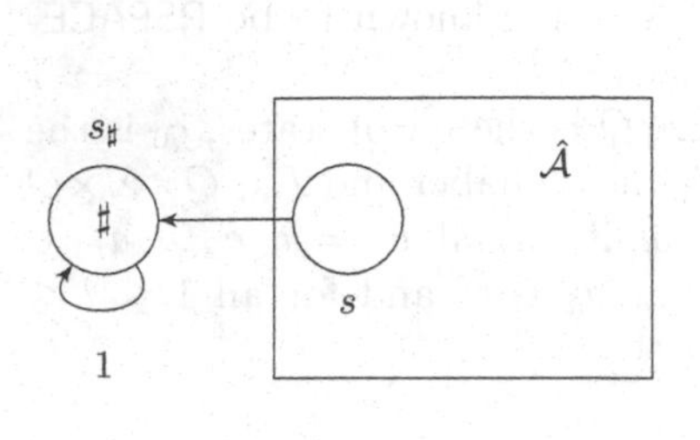

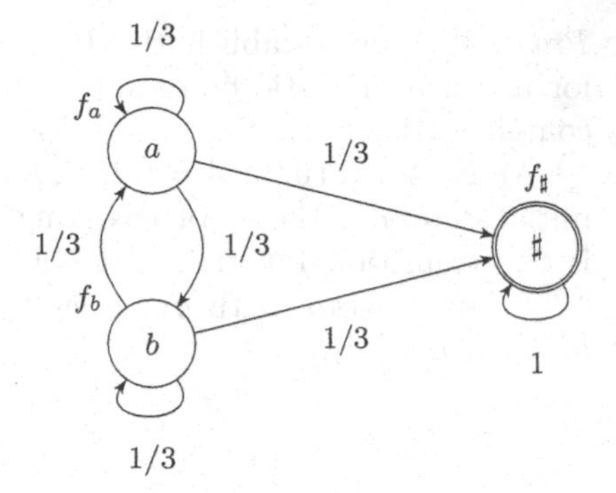

Fig. 4. A reduction for PSPACE-hardness of the AA-disclosure problem.

Choosing $\mathsf{S}^{\mathsf{F}} = \{f_\sharp\}$, let us show that $\mathcal{A}$ is not universal iff $Disc^{\mathsf{AA}}(\mathcal{M}(\mu_0)) > 0$.

Suppose first that $\mathcal{A}$ is not universal. There thus exists a word $w \in \Sigma^*$ such that no path starting in S_A has observation sequence w. As there exists one faulty path ρ (starting in either f_a or f_b) associated to $w\sharp$, we have $\mathsf{Fprop}_{\mathcal{M}(\mu_0)}(w\sharp) = 1$. Therefore $Disc^{\mathsf{AA}}(\mathcal{M}(\mu_0)) \geq \mathbf{P}_{\mathcal{M}(\mu_0)}(\rho) > 0$.

Conversely, assume that $\mathcal{A}$ is universal. Let ρ be a path ending in $f_\sharp$ with observation sequence $\mathsf{O}(\rho) = w\sharp$ for some $w \in \Sigma^*$. As $\mathcal{A}$ is universal, there exists a finite path ρ' in $\hat{\mathcal{A}}$ with observation sequence w. As for every state s of $\hat{\mathcal{A}}$, $p(s_\sharp \mid s) > 0$, ρ' can be extended into a finite path ρ'' ending in $s_\sharp$ with observation $w\sharp$. Thus, $\mathsf{Fprop}_{\mathcal{M}(\mu_0)}(w\sharp) < 1$. Moreover, every path ending with a $\sharp$ remains with probability 1 in either $s_\sharp$ or $f_\sharp$, due to this for every $k \geq 2$, $\mathsf{Fprop}_{\mathcal{M}(\mu_0)}(w\sharp^k) = \mathsf{Fprop}_{\mathcal{M}(\mu_0)}(w\sharp)$. Therefore, $w\sharp^\omega \notin D^{\mathsf{AA}}$. This implies that no infinite path visiting $f_\sharp$ corresponds to an AA-disclosing observation sequence. $f_\sharp$ being the only faulty state, $Disc^{\mathsf{AA}}(\mathcal{M}(\mu_0)) = 0$. □

References

1. Bérard, B., Chatterjee, K., Sznajder, N.: Probabilistic opacity for Markov decision processes. Inf. Process. Lett. **115**(1), 52–59 (2015)
2. Bérard, B., Haddad, S., Lefaucheux, E.: Probabilistic disclosure: maximisation vs. minimisation. In: Proceedings of FSTTCS 2017, volume 93 of LIPIcs, pp. 13:1–13:14. Leibniz-Zentrum für Informatik (2017)
3. Bérard, B., Kouchnarenko, O., Mullins, J., Sassolas, M.: Preserving opacity on interval Markov chains under simulation. In: Proceedings of WODES 2016, pp. 319–324. IEEE (2016)
4. Bérard, B., Mullins, J., Sassolas, M.: Quantifying opacity. Math. Struct. Comput. Sci. **25**(2), 361–403 (2015)

5. Bertrand, N., Fabre, É., Haar, S., Haddad, S., Hélouët, L.: Active diagnosis for probabilistic systems. In: Muscholl, A. (ed.) FoSSaCS 2014. LNCS, vol. 8412, pp. 29–42. Springer, Heidelberg (2014). https://doi.org/10.1007/978-3-642-54830-7_2
6. Bertrand, N., Haddad, S., Lefaucheux, E.: Foundation of diagnosis and predictability in probabilistic systems. In: Proceedings of FSTTCS 2014, volume 29 of LIPIcs, pp. 417–429. Leibniz-Zentrum für Informatik (2014)
7. Bertrand, N., Haddad, S., Lefaucheux, E.: Accurate approximate diagnosability of stochastic systems. In: Dediu, A.-H., Janoušek, J., Martín-Vide, C., Truthe, B. (eds.) LATA 2016. LNCS, vol. 9618, pp. 549–561. Springer, Cham (2016). https://doi.org/10.1007/978-3-319-30000-9_42
8. Bertrand, N., Haddad, S., Lefaucheux, E.: A tale of two diagnoses in probabilistic systems. Inf. Comput. **269**, 104441 (2019)
9. Berwanger, D., Doyen, L.: On the power of imperfect information. In: Proceedings of FSTTCS 2008, volume 2 of LIPIcs, pp. 73–82. Leibniz-Zentrum für Informatik (2008)
10. Borodin, A.: On relating time and space to size and depth. SIAM J. Comput. **6**, 733–744 (1977)
11. Borodin, A., von zur Gathen, J., Hopcroft, J.: Fast parallel matrix and GCD computations. Inf. Control **52**(3), 241–256 (1982)
12. Cabasino, M.P., Giua, A., Lafortune, S., Seatzu, C.: A new approach for diagnosability analysis of petri nets using verifier nets. Trans. Autom. Control **57**(12), 3104–3117 (2012)
13. Cassez, F., Tripakis, S.: Fault diagnosis with static and dynamic observers. Fundamenta Informaticae **88**, 497–540 (2008)
14. Chanthery, E., Pencolé, Y.: Monitoring and active diagnosis for discrete-event systems. IFAC Proc. Vol. **42**(8), 1545–1550 (2009)
15. Chatterjee, K., Doyen, L., Gimbert, H., Henzinger, T.A.: Randomness for free. In: Hliněný, P., Kučera, A. (eds.) MFCS 2010. LNCS, vol. 6281, pp. 246–257. Springer, Heidelberg (2010). https://doi.org/10.1007/978-3-642-15155-2_23
16. Chen, T., Kiefer, S.: On the total variation distance of labelled Markov chains. In: Proceedings of CSL-LICS 2014, pp. 33:1–33:10. ACM (2014)
17. Haar, S., Haddad, S., Melliti, T., Schwoon, S.: Optimal constructions for active diagnosis. J. Comput. Syst. Sci. **83**(1), 101–120 (2017)
18. Jiang, S., Huang, Z., Chandra, V., Kumar, R.: A polynomial algorithm for testing diagnosability of discrete-event systems. Trans. Autom. Control **46**(8), 1318–1321 (2001)
19. Meyer, A.R., Stockmeyer, L.J.: The equivalence problem for regular expressions with squaring requires exponential space. In: SWAT 1972, pp. 125–129. IEEE (1972)
20. Morvan, C., Pinchinat, S.: Diagnosability of pushdown systems. In: Namjoshi, K., Zeller, A., Ziv, A. (eds.) HVC 2009. LNCS, vol. 6405, pp. 21–33. Springer, Heidelberg (2011). https://doi.org/10.1007/978-3-642-19237-1_6
21. Mulmuley, K.: A fast parallel algorithm to compute the rank of a matrix over an arbitrary field. In: STOC 1986, pp. 338–339 (1986)
22. Saboori, A., Hadjicostis, C.N.: Current-state opacity formulations in probabilistic finite automata. Trans. Autom. Control **59**(1), 120–133 (2014)
23. Sampath, M., Lafortune, S., Teneketzis, D.: Active diagnosis of discrete-event systems. Trans. Autom. Control **43**(7), 908–929 (1998)
24. Sampath, M., Sengupta, R., Lafortune, S., Sinnamohideen, K., Teneketzis, D.: Diagnosability of discrete-event systems. Trans. Autom. Control **40**(9), 1555–1575 (1995)

25. Thorsley, D., Teneketzis, D.: Diagnosability of stochastic discrete-event systems. Trans. Autom. Control **50**(4), 476–492 (2005)
26. Thorsley, D., Teneketzis, D.: Active acquisition of information for diagnosis and supervisory control of discrete-event systems. Discret. Event Dyn. Syst. **17**, 531–583 (2007)

Optimizing Reachability Probabilities for a Restricted Class of Stochastic Hybrid Automata via Flowpipe-Construction

Carina Pilch[1(✉)], Stefan Schupp[2], and Anne Remke[1]

[1] Westfälische Wilhelms-Universität Münster, Münster, Germany
{carina.pilch,anne.remke}@uni-muenster.de
[2] Technische Universität Wien, Wien, Austria
stefan.schupp@tuwien.ac.at

Abstract. Stochastic Hybrid automata (SHA) are increasingly used to evaluate the dependability and safety of critical infrastructures. Nondeterminism, which is present in many purely hybrid models, is often only implicitly considered in SHA. This paper instead proposes algorithms for computing optimal reachability probabilities for singular automata with *urgent* transitions and random clocks which follow arbitrary continuous probability distributions. We borrow a well-known approach from hybrid systems reachability analysis, namely flowpipe construction. We extract those valuations of random clocks which ensure reachability of specific goal states from the computed flowpipes and compute reachability probabilities by integrating over these valuations. We compute maximal and minimal probabilities for history-dependent prophetic and non-prophetic schedulers using set-based methods. A case study featuring a series of nondeterministic choices shows the feasibility of the approach.

1 Introduction

The combination of nondeterminism and stochasticity in hybrid models, e.g. in stochastic hybrid automata (SHA), poses a serious challenge to their reachability analysis. Current approaches treat nondeterminism by discretizing the state-space [1,9] or the support of random variables [14], which both lead to overapproximating the reachability probabilities.

In this work, we present an approach to compute optimal (minimum and maximum) reachability probabilities which does not require approximation. We consider singular automata with random clocks as introduced in [25], which extend a sub-class of hybrid automata (HA) by random clocks, that follow absolute continuous probability distributions. They form a subclass of SHA, where the evolution of continuous variables is piece-wise linear and random time delays are included. The resulting model class is highly useful for dependability evaluation and models with similar expressivity have been used to evaluate water sewage plants [11], smart homes [21] and electric vehicles [6,20].

A. Abate and A. Marin (Eds.): QEST 2021, LNCS 12846, pp. 435–456, 2021.
https://doi.org/10.1007/978-3-030-85172-9_23

A novel approach for computing time-bounded reachability for this subclass, i.e., the probability to reach a specified set of goal states within a given time bound, is presented. While the model class maintains both, discrete and continuous nondeterminism, this paper excludes continuous nondeterminism, which occurs, e.g., through time nondeterminism or nondeterministic resets. We resolve discrete nondeterminism by the use of history-dependent non-prophetic and prophetic schedulers. For both scheduler classes, we compute the minimum/maximum reachability probability induced by an optimal scheduler.

To obtain these probabilities, sets of reachable states are computed by flowpipe-construction, where random variables, induced by random clocks, are treated similar to other continuous variables. From the computed sets of reachable states, we extract those sets of values for random variables that lead to predefined goal states. We present dedicated algorithms for computing optimal reachability probabilities for both non-prophetic and prophetic schedulers. Thereby, we tackle the interplay of stochastic and nondeterministic behaviour by combining geometric operations and multi-dimensional integration of the joint probability distribution over all random variables present in the model. The resulting approach is exact up to numerical integration. The feasibility of our approach is shown on a parametrized model of a tank that is filled at a constant rate and drained nondeterministically via one of two different valves with random blocking times. We validate the non-prophetic case with HPNMG [17], which however cannot optimize prophetic schedulers in models with multiple nondeterministic decisions.

Related Work. Reachability for a bounded number of discrete steps is decidable for potentially non-initialized rectangular automata [2,10]. Flowpipe-construction computes reachable state-set using different representations, e.g., polytopes [10].

CEGAR-style abstraction allows the application of model checking methods for probabilistic HA [31]. When decidable subclasses of HA are extended with discrete probability distributions on jumps, reachability is still decidable [30]. For example, [23] extends probabilistic timed automata with continuously distributed resets and applies randomized schedulers to resolve discrete nondeterminism, while discretizing the state-space. Timed automata have also been extended by continuous probability distributions [5], stochastic delays and jumps [4] and analyzed using abstraction in [13], or via transient analysis [3]. For more general classes, incomplete approximative approaches are available [22,26]. For stochastic hybrid systems with a single mode and finite actions, [29] proposes abstractions for uncountable-state discrete-time stochastic processes. [9] present a safe overapproximation for stochastic hybrid systems and [14] discretize the support of random variables and abstracts to Markov decision processes.

History-dependent non-prophetic and prophetic schedulers that resolve discrete nondeterminism, have been applied to stochastic automata [7], where all continuous variables are random clocks. Prophetic and non-prophetic scheduling has been introduced for hybrid Petri nets with general transitions in [24], however, no general implementation exists for the computation of optimal reachability probabilities in the prophetic case. Due to the state-space representation the approach in [24] is restricted to that specific model class. In contrast, this paper provides algorithms to compute optimal reachability probabilities for both, non-prophetic and prophetic schedulers which allows to reduce the computational effort for the prophetic case to the same complexity as for the non-prophetic case.

Outline. Sect. 2 defines singular automata extended by random clocks. The flowpipe-based reachability analysis is explained in Sect. 3 for a fixed scheduler. Section 4 explains how optimal non-prophetic and prophetic schedulers are determined. Section 5 presents a case study. The paper is concluded in Sect. 6.

2 Stochastic Hybrid Automata

Singular automata form a sub-class of hybrid automata [16], in which the derivatives of the continuous variables are constant real values. In the following, we extend singular automata with so-called *random clocks*, for which we further define schedulers to resolve discrete nondeterminism. A singular automaton is a hybrid automaton, where all initial states, invariants and the transition relations are restricted to rectangular sets and where the activities as well as the resets on jumps (see below) are further restricted to a singleton, as presented in [15,16].

Definition 1. *Let $\mathbb{I}$ be the set of all intervals in $\mathbb{R} \cup \{-\infty, \infty\}$ with rational or infinite endpoints and let $d \in \mathbb{N}$. A subset of $\mathbb{R}^d$ is* rectangular *if it is a Cartesian product of d intervals. A rectangular set is a* singleton *if each of its intervals is a singleton, i.e., a set with exactly one element $c \in \mathbb{R}^d$.*

For the definition of singular automata, we refer to [16] and omit labels. Singular automata have been extended by stochastic variables in terms of random clocks [25]. Random clocks r_i evolve like stop-watches with derivative one or zero (i.e., $Act(l)_i \in \{0,1\}, l \in Loc$), are reset only to zero on discrete transitions, and have a random expiration time, which is described by a random variable. Precisely, each random clock is associated with a continuous probability distribution, which describes its expiration time [8]. To indicate expiration, the random clock is reset to zero and its valuation stored in the associated random variable.

Definition 2. *We define an* absolute continuous probability distribution *(CDF) of a real-valued random variable X over a set $D \subseteq \mathbb{R}_{\geq 0}$ as a function $F : D \rightarrow [0,1] \subseteq \mathbb{R}$, where F is absolutely continuous and $F(x) = P(X \leq x)$ equals the probability that X takes on a value less than or equal to x. The corresponding probability density function is denoted f. The set of all CDFs is denoted as $\mathbb{F}$.*

Definition 3. *A singular automaton with random clocks $\mathcal{S} = (Loc, Var', Edg, Act, Inv, Init, \Phi)$ is a singular automaton extended by a set $\mathcal{R} \subseteq Var'$.*

Loc is a finite set of locations *and Var' a finite set of real-valued* variables. *Let $|Var'|$ be denoted as d. A* valuation *v is the image of a function $Var' \rightarrow \mathbb{R}$, which assigns a real-value to each variable, i.e., v is in $\mathbb{R}^d$. The set of all valuations is denoted $\mathbb{V}$ and we refer to a set of valuations by $V \subseteq \mathbb{V}$.*

The set $Edg \subseteq Loc \times (\mathbb{I}^d \times \mathbb{I}^d \times 2^{\{1,\dots,d\}}) \times Loc$ is the finite set of transitions *$(l, (pre, post, jump), l') \in Edg$. Each transition consists of a source location l, a transition relation $(pre, post, jump)$ and a target location l'.*

Act: $Loc \rightarrow \mathbb{R}^d$ *assigns a set of (deterministic)* activities *to each location. We use* $Act(l)_i$ *to refer to the activity for the i-th variable.* $Inv \colon Loc \rightarrow \mathbb{I}^d$ *assigns an* invariant *to each location* $l \in Loc$. *A* state *in a singular automaton with random clocks is a tuple* (l, v) *with location* $l \in Loc$ *and valuation* $v \in \mathbb{V}$. $S = Loc \times \mathbb{V}$ *is the set of all states and* $Init \subseteq Loc \times \mathbb{I}^d$ *is the set of initial states. We use* (l, V) *to refer to a set of states whose elements agree on the location* l *and whose valuations are taken from* $V \subseteq \mathbb{V}$.

The function $\Phi \colon \mathcal{R} \rightarrow \mathbb{F}$ *associates an absolute continuous probability distribution (CDF) to each random clock* $r_i \in \mathcal{R}$. *For each random clock holds:*

1. $\forall (l, v) \in Init : r_i \in \mathcal{R} \Rightarrow v(r_i) = 0$,
2. $\forall l \in Loc \colon r_i \in \mathcal{R} \Rightarrow Act(l)_i \in \{0, 1\}$,
3. $\forall (l, (pre, post, jump), l') \in Edg$ *and* $\forall r_i \in \mathcal{R}$: *if and only if* $i \in jump$, *then* $Act(l)_i = 1, pre = \mathbb{R}^d, post_i = 0$.[1] *Every expiration time of a random clock* r_i *follows* F_i *assigned by* $\Phi(r_i)$.

The state of a singular automaton with random clocks can change in two ways: (i) A *time delay* describes the evolution of time, which only changes the valuations of continuous variables and not the state location; (ii) A *jump* takes a transition to another location and may change both, the location and the variable valuations. The system can only be in states (l, v) with $v \in Inv(l)$.

A transition $(l, (pre, post, jump), l') \in Edg$ can only be taken in a state (l, v) if the valuation v lies in *pre* and further $v \in Inv(l)$ and $v' \in Inv(l')$. When the transition is taken, the value of every variable $x_i \in Var'$ is updated to v' as follows: If $i \notin jump$, $v(x_i)$ is not changed and must lie in $post_i$ and if $i \in jump$, x_i is deterministically set to $post_i$, which is a singleton in this case. Accordingly, this jump leads to a new state $(l', v') \in S$. A transition is called *urgent* if its enabling prevents further passage of time—the current location has to be left immediately. In the remainder of this work, we restrict the model class to urgent transitions to exclude continuous nondeterminism. This is realized via combinations of *pre*-guards and invariant conditions. More formally, for every transition $(l, (pre, post, jump), l') \in Edg$ with $k \notin jump$ $\forall r_k \in \mathcal{R}$ it holds that for $pre = I_0 \times \cdots \times I_{d-1} : \exists i \in \{0, \ldots, d-1\}$: I_i is half-bounded and $\forall j \neq i : I_j = (-\infty, \infty)$. In case of a lower bound guard, i.e., $I_i = [c, \infty), c \in \mathbb{R}$, the invariant equals $Inv(l)_i = [e, c], e \leq c, e \in \mathbb{R} \cup \{-\infty\}$ and analogously for an upper bound guard.

A random clock $\in \mathcal{R}$ expires whenever a transition is taken which resets its value.[2] Furthermore, a random clock may be paused, which changes its activity to zero, while its valuation remains unchanged. Precisely, every expiration delay induces a random variable, which follows the same CDF as the random clock. We define the set $\mathcal{C}$ of random variables induced by a singular automaton with random clocks where $s_{i,j} \in \mathcal{C}$ is the j-th expiration of the random clock r_i. Note

[1] Note that $pre = \mathbb{R}^d$, since the random clocks do not have a pre-guard.

[2] Definition 3 differs from [25], where clocks run backwards. This does not change model semantics, but simplifies the identification of goal states via flowpipe-construction.

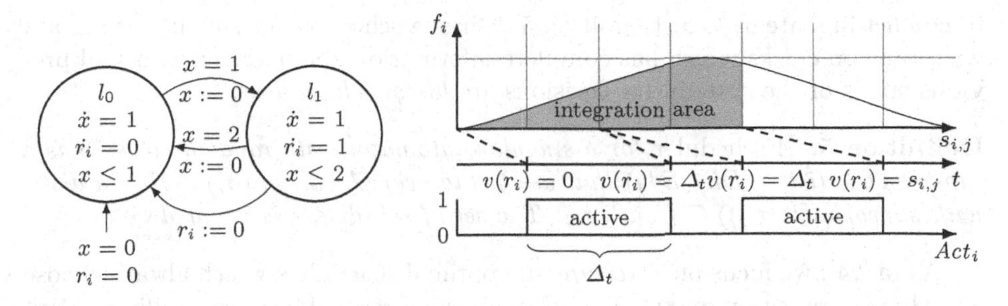

Fig. 1. Singular automaton with random clock r_i and corresponding timeline for activation and expiration of r_i; its j-th expiration is given by a random variable $s_{i,j}$. In location l_1, two outgoing transitions (one deterministic, one stochastic) compete. In the chosen scenario, the random clock expires at $t = 4$, thus, the clock is deactivated once.

that since we exclude Zeno-behavior (infinitely many discrete jumps in finite or zero time) and perform analysis up to a time bound t_{max}, $\mathcal{C}$ is finite. We define a function $\Psi\colon Loc \times \mathcal{C} \to \{0, 1\}$, which assigns an *expiration* status to each $s_{i,j} \in \mathcal{C}$ that holds in a given location $l \in Loc$. In the following, $Var = (Var' \backslash \mathcal{R}) \cup \mathcal{C}$ defines the set of continuous variables present in a singular automaton, where every random clock is replaced by the set of random variables, which it induces. Technically, this corresponds to drawing a sample from F_i when this random clock becomes *active* ($Act(l)_i = 1$) initially and after every expiration.

Figure 1 sketches the j-th expiration of random clock r_i, which is paused at Δ_t without being reset. The depicted expiration delay corresponds to a fresh random variable $s_{i,j}$, which follows density f_i. To resolve the inherent nondeterminism, we consider history-dependent schedulers, which require the history of a state:

Definition 4. *A state $\sigma_i \in S$ is called* reachable *in a singular automaton with random clocks with initial set Init, if there exists a finite sequence $\sigma_0, \ldots, \sigma_i$ of states connected by either time delays or discrete jumps and $\sigma_0 \in Init$. The sequence of alternating time delays and jumps is called* history $h(\sigma_i)$ *of σ_i, where* duration $h_d(\sigma_i)$ *of σ_i denotes the sum of all durations of time delays in $h(\sigma_i)$.*

The set of trajectories is not restricted by alternating delays and jumps, as delays may have duration zero and consecutive time delays can be combined. The set of all histories in a singular automaton with random clocks is denoted *Hist*.

In singular automata with random clocks and only urgent transitions, two or more transitions might be enabled at the same point in time. We call them to be *in conflict*. Conflicts result in discrete nondeterminism and in this work we make use of *schedulers* to resolve it, as proposed in [24] for hybrid Petri nets and in [25] for a similar class of singular automata with random clocks.

We define a *discrete probability distribution* over a set D as a function $\mu : D \to [0, 1] \subseteq \mathbb{R}$ such that $support(\mu) = \{d \in D \mid \mu(d) > 0\}$ is countable and $\sum_{d \in support(\mu)} \mu(d) = 1$. Further, let $Dist(D)$ be the set of discrete probability distributions over D. Let $Confl(\sigma_i) \subseteq Edg$ consist of all transitions that are in conflict in state $\sigma_i \in S$. Definition 5 defines a scheduler for singular automata with random clocks which has complete information about the current and previous states of the system. Its decisions are *history-dependent.*

Definition 5. *A* scheduler *for a singular automaton with random clocks $\mathcal{S}$ is a function $\mathfrak{s}$: Hist $\rightarrow$ Dist(Edg) that assigns to every history $h(\sigma_i) \in$ Hist a distr. with support($\mathfrak{s}(h(\sigma_i))$) $\subseteq$ Confl(σ_i). The set of schedulers is denoted $\mathfrak{S}^n$.*

As in [24], we focus on *deterministic* optimal schedulers which always choose exactly one transition maximizing or minimizing reachability probabilities. Considering random clocks, we further distinguish between *prophetic* and *non-prophetic* schedulers. The schedulers from Definition 5 are non-prophetic, hence they have no information on the future expiration times of the random clocks. In contrast, prophetic schedulers have full information on previous and future expiration times of all random clocks. We denote this set of schedulers as $\mathfrak{S}^p$. For further details on the scheduler classes, we refer to [7, 24].

3 Reachability Analysis

We propose algorithms for computing optimal *time-bounded reachability probabilities* in a restricted class of stochastic hybrid models. This corresponds to maximizing/minimizing the probability of reaching states σ that belong to a specified set of goal states $S^{goal} \subseteq S$ with a duration $h_d(\sigma) \leq t_{max}$. Due to the presence of discrete nondeterminism, every scheduler induces a fully stochastic version of a singular automaton with random clocks.

For a specific class of schedulers $\mathfrak{S} \in \{\mathfrak{S}^n, \mathfrak{S}^p\}$, we obtain a range of probabilities $[p^{\mathfrak{S}}_{min}(S^{goal}, t_{max}), p^{\mathfrak{S}}_{max}(S^{goal}, t_{max})]$, where the bounds are the infimum and supremum, respectively, over the probabilities induced by all schedulers in that class. We refer to these optimal probabilities as minimum and maximum. The set S^{goal} is said to be *reachable* within time bound t_{max} if the minimum probability $p^{\mathfrak{S}}_{min}(S^{goal}, t_{max})$ is larger than zero. History-dependent classic schedulers, as discussed in [7] for stochastic automata, have knowledge on locations, valuations and expiration times and form the most powerful class of schedulers for these models. Since our history-dependent prophetic schedulers have complete information on locations, valuations and expiration times, the induced probability by such an optimal scheduler is optimal over all scheduler classes.

Construction of a Flowpipe. Flowpipe-construction is used to obtain a geometric representation of the set of reachable states without resolving nondeterminism. Our approach first disregards the CDFs of the random clocks and treats each induced expiration delay as continuous variable.

To compute the set of reachable states for a hybrid automaton, successor states resulting from time delays as well as from discrete jumps are computed in a fixpoint iteration. A pseudo-code for this fixpoint computation given in Algorithm 1. Starting from a set of initial states (Lines 1, 2), states reachable by

Algorithm 1. `computeReachability(H)`

1: $R = Init$;
2: $R^{new} = R$;
3: **while** $R^{new} \neq \emptyset$ **do**
4: $S = \text{getStateSet}(R^{new})$;
5: $F = \text{computeTimeDelaySuccessorStates}(S)$;
6: $J = \text{computeJumpSuccessorStates}(F)$;
7: $R^{new} = (R^{new} \setminus S) \cup (J \setminus R)$;
8: $R = R \cup F$;
9: **end while**
10: **return** R;

time delay (Line 5) and by discrete jumps (Line 6) are discovered for each set of states in the working set (Line 4). The computed set of reachable states and the working set are updated accordingly (Lines 7, 8). Algorithm 1 terminates when no new sets of reachable states of H are discovered, i.e., a fixpoint has been reached (Line 3) and R is returned (Line 10). Depending on the subclass of hybrid automata and H itself a fixpoint may not exist, hence in practice further termination criteria, such as a global time bound t_{max} and a maximal number of jump successors computed are used. For singular automata jump-bounded reachability is decidable and returns the exact set of reachable states (c.f. [2]). For models for which Zeno behaviour is excluded, flowpipe-construction terminates with exact result when computing time-bounded reachability. *Convex polytopes* are commonly used as state-set representation for sets of reachable states of singular automata, as they allow to represent those sets exactly. We denote a closed and bounded convex set in the d-dimensional real vector space as *convex polytope*. For further details, refer to [32].

Assume the dynamics $Act(l)_i \in \mathbb{R}$ for a continuous variable x_i in location $l \in Loc$. During a time delay, each variable x_i changes its valuation to $v(x_i)^+ = v(x_i) + Act(l)_i \cdot t, t \geq 0$ (see Fig. 2a). We introduce t as a fresh variable with the constraint $t = 0$ into the set of variable valuations $V \subseteq \mathbb{R}^d$ to obtain a set $V' \in \mathbb{R}^{d+1}$. Time successor states can be computed as the Minkowski sum of the *ray* $Act(l) \cdot t, t \geq 0$ and the set V'. The resulting set $V^+ \in \mathbb{R}^{d+1}$ describes variable valuations reachable by positive time delays depending on t. If an upper bound t_{max} on t is used to compute time-bounded reachability, the ray becomes a line segment $v(x_i) + Act(l)_i \cdot t, t \in [0, t_{max}]$ and the resulting set is bounded (if V was bounded). Elimination of the additional variable t in the description of V^+ yields a projection $V^+ \downarrow_{Var}$ onto the original variables and describes all valuations reachable by positive time elapse starting from V. Finally, the set $(V^+ \downarrow_{Var}) \cap Inv(l)$ yields those states which also admit the invariant condition in location l. Note that enabled urgent transitions bound the duration of a time delay. The computation of jump successors may limit the set of time successor states if an urgent transition is enabled before reaching t.

A transition $(l, (pre, post, jump), l') \in Edg$ with pre being a rectangular set is enabled by a state set (l, V^{pre}) if $V^{pre} \subseteq pre$ holds. The deterministic reset of

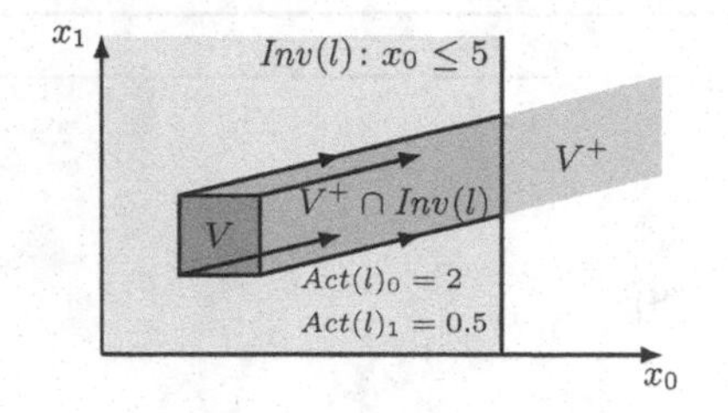

(a) Projected illustration of computing successor states of V for a time delay.

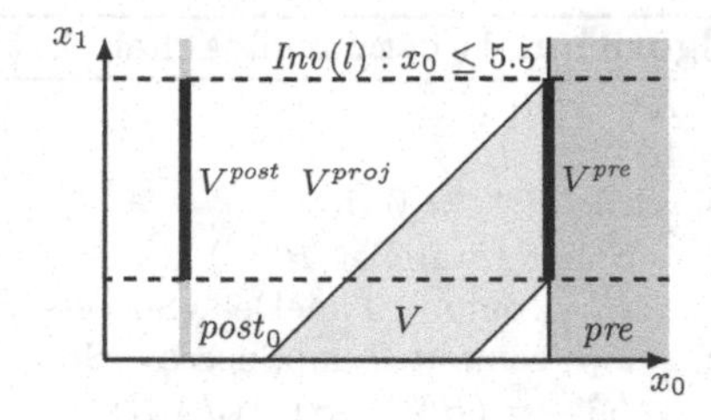

(b) Computation of jump successor states of V with $0 \in jump$ and $post_0$ resets x_0.

Fig. 2. Illustration of time- and jump-successor computation.

a variable x_i to a value in $post_i$ as described in Sect. 2 is performed by projection of V^{pre} on Var/x_i (e.g., via Fourier-Motzkin variable elimination) to obtain $V^{proj} = V^{pre} \downarrow_{Var/x_i}$, which effectively removes all constraints on x_i and returns a convex polytope unbounded in x_i. The result V^{proj} of the projection is intersected with the hyperplane representing the equation $x_i = post_i$. An intersection of the resulting set $V^{post} = V^{proj} \cap post$ with $Inv(l')$ ensures the invariant condition in the target location is satisfied. The whole process is illustrated in Fig. 2b. For urgent transitions it suffices to compute jump successors up to the first time point where an urgent transition is enabled.

The alternating computation of all time delays and all valid jump successor states for a hybrid automaton $\mathcal{H}$ eventually yields a set of state sets (*segments*) whose union, the so-called *flowpipe*, represents the set of reachable states of $\mathcal{H}$. The computation traverses the *reachability tree* in which nodes represent time delays and the parent-child relation represents a jump. Following [27], we define the reachability tree, which is finite for time-bounded reachability in singular automata without Zeno-behavior:

Definition 6. *For a hybrid automaton* $(Loc, Var, Edg, Act, Inv, Init)$ *with dimension* $d = |Var|$ *a (finite)* reachability tree *is defined as tree* $(N, E, Sfunc)$ *with the following components: a finite set* N *of* nodes *and a* root node $n_{root} \in N$*; a set* $E \subseteq N \times N$ *of* edges*; a function* Sfunc : $N \rightarrow (Loc \times 2^{\mathbb{R}^d})$ *that assigns a flowpipe segment* (l, V) *to each node as data.*

Computation of Probabilities. We compute time-bounded reachability probabilities by extracting valuations of all random variables present in the model from the reachable goal states. Then, we integrate the joint probability density of the random variables over the sets of extracted valuations. We detail the main steps of this computation in the following (see also Fig. 3):

Reduction. In the first step, we reduce the flowpipe segments to states reachable in S^{goal} before t_{max}. To bound global time, a fresh clock t_G is introduced which is never reset. Adding an invariant $t_G \leq t_{max}$ to each location ensures that all segments of the computed flowpipe comply with the global time bound.

In our approach, a goal state (l^{goal}, v^{goal}) in S^{goal} is a tuple of a goal location l^{goal} and a goal valuation v^{goal}. We extend the notation to sets of goal states (l^{goal}, V^{goal}). A state set (l, V) contains goal states from the set (l^{goal}, V^{goal}) if $l = l^{goal}$ and $V \cap V^{goal} \neq \emptyset$. We represent V^{goal} as a convex polytope.

Projection. For the computation of time-bounded reachability probabilities, we consider only the random variables and their corresponding probability distributions. Hence, we exclude all other variables by projecting the non-empty state set $(l, V \cap V^{goal})$ onto the n-dimensional sub-space of the random variables ($n = |\mathcal{C}|$) to obtain $(l, V') = (l, (V \cap V^{goal}) \downarrow_{\mathcal{C}})$.

Extension. Recall that $\Psi(l, s_{i,j})$ denotes the expiration status of each random variable $s_{i,j} \in \mathcal{C}$ in location $l \in Edg$. If expired, the set of valuations stored for $s_{i,j}$ describes all possible valuations which allow reaching a goal state. These are required to compute the overall reachability probability. If a random variable is not expired when reaching a goal state, its expiration time lies in the future and is not relevant for reaching a goal state. After a goal state has been reached where $v(s_{i,j}) = k$, the random variable can expire at any point in time and thus any value between k and t_{max} assigned to the random variable $s_{i,j}$ leads to the goal state[3]. To include all values in $[k, t_{max}]$, current valuations of $s_{i,j}$ are extended to the time bound t_{max} to reflect this case. Technically, this corresponds to computing the convex closure (i.e., convex hull) of the union of V' and the set V'' in which all valuations of $s_{i,j}$ are set to t_{max}. We use $V_{\mathcal{C}}^{goal}$ to refer to the set $cHull(V' \cup V'')$ computed in this step and use $S_{\mathcal{C}}^{goal} = \{(l, V_{\mathcal{C}}^{goal})\}$ for the corresponding state set. The function *pre-process*: $Loc \times 2^{\mathbb{R}^d} \to Loc \times 2^{\mathbb{R}^n}$ then maps state sets to their reduced, projected and extended equivalents.

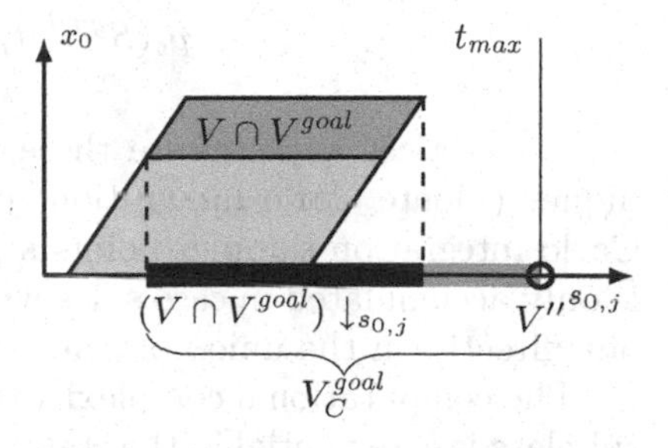

Fig. 3. Projection of a set of reachable valuations (light gray) $V \cap V^{goal}$ satisfying a goal condition (gray) on the dimension of $s_{0,j}$. The resulting one-dimensional set $V' = (V \cap V^{goal}) \downarrow_{s_{0,j}}$ (thick black) is extended to t_{max} (thick gray) by convex union with V''.

Scheduling. The previous steps yield a set of n-dim. state sets $S_{\mathcal{C}}^{goal}$ that indicate which values assigned to the n random variables lead to goal states within the allowed time bound. So far, we consider this set of state sets independently of the chosen scheduler. Given a scheduler $\mathfrak{s} \in \mathfrak{S}$, only a subset $S^{\mathfrak{s}} \subseteq S_{\mathcal{C}}^{goal}$ of these assignments results from the decisions of $\mathfrak{s}$. This is described in detail in Sect. 4.

Integration. A fixed scheduler $\mathfrak{s} \in \mathfrak{S}$ results in a subset $S^{\mathfrak{s}}$ and each corresponding valuation set $V^{\mathfrak{s}}$ is represented as a convex polytope $\mathcal{P}^{\mathfrak{s}}$ over the space of the

[3] For a fixpoint in time (at which the goal state is reached) $v(s_{i,j}) = k$ is a singleton since the random variable starts at $s_{i,j} = 0$, is never reset, and follow clock-dynamics.

random variables. We compute the probability $p_{\mathfrak{s}}(S^{goal}, t_{max})$ by integrating the joint probability distribution of all random variables. For a specific scheduler $\mathfrak{s}$, the region of integration is given by the union $U^{\mathfrak{s}} = \bigcup_{(l,V^{\mathfrak{s}})\in S^{\mathfrak{s}}} V^{\mathfrak{s}}$ over all $V^{\mathfrak{s}}$, i.e., the union[4] of polytopes $\mathcal{P}^{\mathfrak{s}}$.

Then, $p_{\mathfrak{s}}(S^{goal}, t_{max})$ equals the probability that the value of every random variable s_i for $0 \leq i \leq n-1$ lies within the union $U^{\mathfrak{s}}$. Let $\mathbf{s} = (s_0, \ldots, s_{n-1})$ denote an n-dimensional point (i.e., one specific assignment of values to the random variables). Let $G(\mathbf{s}) = \prod_{i=0}^{n-1} f_i(s_i)$ be the joint probability density function which, due to the independence of the random variables, equals the product over the n probability density functions f_i. The resulting probability is:

$$p_{\mathfrak{s}}(S^{goal}, t_{max}) = \int_{U^{\mathfrak{s}}} G(\mathbf{s})\, d\mathbf{s}. \tag{1}$$

In practice, we compute these probabilities by numerical integration techniques (Monte Carlo integration) in the same way as described in [18]. Monte Carlo integration samples points $\mathbf{s}$ in the state-space and the joint density $G(\mathbf{s})$ is only accumulated in case $\mathbf{s}$ lies within $U^{\mathfrak{s}}$. Thus, the integration can be carried out directly on the union of convex polytopes [19].

The computational complexity of the flowpipe-construction for this restricted subclass is exponential in the state-space dimension d. The exponential complexity results from required polytope representation conversions. The operations *projection* and *extension* are also exponential in d (due to representation conversion). Integration is polynomial in number of random variables n, the number of discrete jumps and in the complexity of the polytope representation.

4 Optimal Schedulers

The probability that we obtain for time-bounded reachability depends on the specific scheduler. We want to find the optimal probabilities $p^{\mathfrak{S}}_{min}(S^{goal}, t_{max})$ and $p^{\mathfrak{S}}_{max}(S^{goal}, t_{max})$ over all schedulers of a given class $\mathfrak{S} \in \{\mathfrak{S}^n, \mathfrak{S}^p\}$ for a certain property described by goal states. For a fixed goal state, an optimal scheduler $\mathfrak{s}$ either maximizes or minimizes the reachability probability. The decisions taken by a certain scheduler relate to certain branches in the computed reachability tree (c.f. Definition 6) and thus every decision resolves a nondeterministic conflict between different branches. Accordingly, only those subsets of reachable states which originate from the scheduler decision in the chosen branch are reachable for this scheduler. Hence, every scheduler decision reduces the set of reachable states to the subset $S^{\mathfrak{s}}$ over which is integrated.

The number of nodes in the reachability tree is linear in the total number of taken jumps. For every node the operations *reduction*, *projection*, and *extension* are executed (at most) once. Next, we introduce the computation of time-bounded reachability probabilities for non-prophetic and prophetic schedulers. A proof of correctness of the presented algorithms can be found in Appendix A.

[4] Note that this is not the convex closure of the union.

Algorithm 2. `computeNonProphetic(`$\mathcal{S}, S^{goal}, t_{max}$`)`

1: *root* = computeReachability($\mathcal{S}, t_{max}$);
2: *polySched* = collPolys(*root*, $S^{goal}, t_{max}, 0, \emptyset$);
3: *optimalProb* = 0;
4: **for** each *polytopes* $\in$ *polySched* **do**
5: *currentProb* = computeProbOverUnion(*polytopes*, $\mathcal{S}$.*distributions*);
6: *optimalProb* = max(*currentProb*, *optimalProb*);
7: **end for**
8: **return** *optimalProb*;

4.1 Optimal Non-prophetic Scheduler

Each history-dependent non-prophetic scheduler $\mathfrak{s} \in \mathfrak{S}^n$ has knowledge about the history and the current state of a model. However, it does not know the actual variable assignment $\mathbf{s}$ of the random variables, it only knows the set S_C^{goal} and thus cannot make decisions based on $\mathbf{s}$. For every conflict, which leads to branching the reachability tree, scheduler $\mathfrak{s}$ can choose only a specific sub-tree. The resulting set of paths (from the root to leaf-nodes) through the selected sub-trees yields a set of nodes $N^{\mathfrak{s}} \subseteq N$. The set $S^{\mathfrak{s}}$ is given as $S^{\mathfrak{s}} = \{\textit{pre-process}(\textit{Sfunc}(n)) \mid n \in N^{\mathfrak{s}}\}$.

The maximum non-prophetic scheduler $\mathfrak{s}_{max} \in \mathfrak{S}^n$ resolves nondeterminism which results in the set $S^{\mathfrak{s}}_{max}$, for which $p^{\mathfrak{S}^n}_{max}(S^{goal}, t_{max})$ is maximized, i.e.,

$$p^{\mathfrak{S}^n}_{max}(S^{goal}, t_{max}) = \max\left(p_{\mathfrak{s}}(S^{goal}, t_{max}) \mid \mathfrak{s} \in \mathfrak{S}^n\right). \quad (2)$$

The minimum case for $p^{\mathfrak{S}^n}_{min}(S^{goal}, t_{max})$ is defined analogously. When computing optimal reachability probabilities the result of every non-prophetic scheduler is compared and the optimal one is determined by iterating over the reachability tree and collecting the corresponding state sets for every scheduler decision. Recall from Eq. 1, that we need to integrate over the union of all state sets $S^{\mathfrak{s}}$ for a specific scheduler, since its decisions can lead to multiple state sets.

Algorithm 2 presents pseudo-code for the computation of the maximum probability obtained by the optimal non-prophetic scheduler. It expects a singular automaton with random clocks $\mathcal{S}$, the desired set of goal states S^{goal} and the time bound t_{max} as input. Line 1 calls the flowpipe-construction, which returns the *root* of the computed reachability tree, where every node holds the corresponding flowpipe segment. The function `collPolys(` *root*, $S^{goal}, t_{max}, 0, \emptyset$`)` called in Line 2 collects the polytopes, separated by schedulers, such that each entry in *polySched* represents one scheduler and contains the set of polytopes which represent those goal states that are reachable by this scheduler. Lines 4–7 loop over all entries of *polySched* and compute the probability *currentProb* to reach a goal state for each scheduler in Line 5. This is realized by integration over the union of the polytopes (see Eq. 1). If the probability is larger than the maximum probability *optimalProb* that was computed so far, the latter is updated (Line 6). Finally, the optimal probability is returned in Line 8.

Function `collPolys(`*node*, S^{goal}, t_{max}, *index*, *polySched*`)` (see Algorithm 3) is called for the root nodes in the reachability tree *node*. Given the set of goal states

Algorithm 3. `collPolys`($node, S^{goal}, t_{max}, index, polySched$)

1: **if** ($Sfunc(node) \cap S^{goal} \neq \emptyset$) **then**
2:     $polytope$ = preProcess($Sfunc(node)$);//reduce, project, extend, see Sec. 3.2
3:     $polySched[index]$.insert($polytope$);
4: **end if**
5: $conflChildren$ = getConflictingChildren($node$);
6: $nonConflChildren = node$.children $\setminus$ $conflChildren$;
7: **for** (each $child \in nonConflChildren$) **do**
8:     $polySched$ = collPolys($child, S^{goal}, t_{max}, index, polySched$);
9: **end for**
10: $numSched = |polySched|$;
11: **for** (i=1, ..., $|conflChildren| - 1$) **do**
12:     $newIdx = numSched + i - 1$;
13:     $polySched[newIdx] = (polySched[index])$;
14:     $polySched$ = collPolys($conflChildren[i], S^{goal}, t_{max}, newIdx, polySched$);
15: **end for**
16: **return** $polySched$;

S^{goal}, a time bound t_{max} , the algorithm stores polytopes which represent the integration domain for each scheduler identified by *index* in the array *polySched*. Initially, the function is called with $index = 0$ and an empty array. At first, Line 1 checks if the state set referenced by the current node *node* contains goal states. If this is the case, in Line 2 the state set is reduced to the reachable goal states, projected and extended to t_{max} (see also Sect. 3). The result is inserted into the array of polytopes in *polySched* at index *index* (Line 3) to indicate that the segment is reachable by the scheduler identified by the current *index*. In Lines 5, 6 the set of child nodes is separated into conflicting and non-conflicting nodes. For the recursive call of `collPolys` on all non-conflicting child-nodes (Lines 7–9), the given value of *index* is passed, as no separation of schedulers is required if no conflict occurs. If there are child nodes in conflict Lines 11–15 loop over the respective nodes and the current polytope vector in *polySched* at index *index* is copied. Thus, for each decision over the conflicting children, a new scheduler is instantiated based on the previously taken decisions. For each conflicting child node, the recursive function is then called (Line 14) with according index of the new scheduler. Reachability probabilities for the optimal minimum scheduler can be computed similarly.

4.2 Optimal Prophetic Scheduler

Prophetic schedulers have knowledge on all future expiration times of the random variables. Hence, a prophetic scheduler can take decisions based on all random variables present in the model. Precisely, it can take different decisions for different assignments **s** and can optimize decisions for any given **s**. This results in infinitely many prophetic schedulers, if at least one random variable is present in the model. In contrast, for a fixed time bound and excluding Zeno-behaviour the number of non-prophetic schedulers is finite, due to the finite reachability tree.

Algorithm 4. `computePropheticMinimum`$(\mathcal{S}, S^{goal}, t_{max})$

1: $root$ = computeReachability$(\mathcal{S}, t_{max})$;
2: $polySched$ = collPolys$(root, S^{goal}, t_{max}, 0, \emptyset)$;
3: **for** each $polytopes \in polySched$ **do**
4: $union$ = computeUnion$(polytopes)$;
5: $unitedPolys$.insert$(union)$;
6: **end for**
7: $optimalProb$ = computeProbOverIntersect$(unitedPolys, \mathcal{S}.distributions)$;
8: **return** $optimalProb$;

Minimum Probabilities. An optimal prophetic scheduler that *minimizes* the probability to reach a goal state aims to *evade* those paths in the reachability tree leading to goal states for the given **s**. Recall from Eq. 1 that the probability obtained by a specific scheduler $\mathfrak{s}$ is given by integration over the union of all state sets $S^{\mathfrak{s}}$ reachable due to decisions of $\mathfrak{s}$. Only those **s**, for which the scheduler cannot take any decision to avoid reaching a goal state, have to be included in the region of integration when computing the minimum probability. For any other **s**, there always exists a decision leading to a path which evades goal states.

The set of states that is reachable by all prophetic schedulers and hence, cannot be avoided, is given by the intersection $\bigcap_{\mathfrak{s}\in\mathfrak{S}^p} S^{\mathfrak{s}}$. If this intersection is empty, for every **s** decisions are possible which avoid goal states. Accordingly, the *minimum* probability results from integration over the above intersection.

Algorithm 4 shows pseudo-code for computing the minimum probability obtained by the optimal prophetic scheduler. Similarly to Algorithm 2, Lines 1–2 collect polytopes for each scheduler. Lines 3–6 loop over each entry in *polySched*, for which the union of the corresponding polytope set is computed and stored (Lines 4, 5). The minimum probability is computed over the intersection of the previously united polytopes in Line 7 and finally returned.

Maximum Probabilities. In contrast, the *maximum* probability is computed by integration over the union of all polytopes representing $S^{\mathfrak{s}}$ for all $\mathfrak{s} \in \mathfrak{S}^p$: For any **s** that lies within the union it is possible to take decisions leading to a goal state. The region of integration is thus given as the set of polytopes representing all valuations in $\bigcup_{\mathfrak{s}\in\mathfrak{S}^p} S^{\mathfrak{s}}$, which is equal to $S^{goal}_{\mathcal{C}}$. Recall from Sect. 3, that for each scheduler $\mathfrak{s}$ these valuations are given by $U^{\mathfrak{s}}$, which is again represented by a union of polytopes. Hence, we integrate over the set

$$\bigcup_{\mathfrak{s}\in\mathfrak{S}^p} U^{\mathfrak{s}} = \bigcup_{\mathfrak{s}\in\mathfrak{S}^p} \bigcup_{(l,V^{\mathfrak{s}})\in S^{\mathfrak{s}}} V^{\mathfrak{s}} = \bigcup_{(l,V)\in S^{goal}_{\mathcal{C}}} V. \tag{3}$$

To compute the maximum probability of an optimal prophetic scheduler, we thus do not need to separate sets of states by schedulers, but can process flowpipe segments which intersect S^{goal} (see Algorithm 5). Line 1 calls the flowpipe-construction and Line 2 collects all flowpipe segments stored in the reachability tree. Lines 3–8 loop over all of these flowpipe segments. Line 4 checks for the

Algorithm 5. `computePropheticMaximum`$(\mathcal{S}, S^{goal}, t_{max},)$

1: $root$ = computeReachability($\mathcal{S}, t_{max}$);
2: $flowpipe$ = getAllSetsFromTree($root$);
3: **for** each $segment \in flowpipe$ **do**
4: **if** ($segment \cap S^{goal} \neq \emptyset$) **then**
5: $polytope$ = preProcess($segment$); // reduce, project, extend, see Sec. 3.2
6: $polytopes$.insert($polytope$);
7: **end if**
8: **end for**
9: $optimalProb$ = computeProbOverUnion($polytopes, \mathcal{S}.distributions$);
10: **return** $optimalProb$;

Table 1. Maximum reachability probabilities for non-prophetic and prophetic schedulers, goal states $S^{goal} = \{(l, v) \in S | v(x_0) \geq 18\,l\}$, error estimates ($\pm$) and computation times for different time bounds t_{max}, from the flowpipe approach.

		t_{max}	7 h	8 h	9 h	10 h	11 h
		n	2	4	5	6	8
		#locs SA	22	34	53	83	119
Non-proph.	Flowpipe	$p^{\mathfrak{S}^n}_{max}(S^{goal}, t_{max})$	0	0.534 404	0.626 397	0.644 603	0.729 008
		$\pm$	0	$4.306 \cdot 10^{-4}$	$5.887 \cdot 10^{-4}$	$7.883 \cdot 10^{-4}$	$5.343 \cdot 10^{-3}$
		computation time	0.20 s	1.88 s	10.39 s	54.71 s	2296.73 s
	HPnG [24]	$p^{\mathfrak{S}^n}_{max}(S^{goal}, t_{max})$	0	0.534 718	0.626 442	0.644 498	0.737 537
		$\pm$	0	$3.090 \cdot 10^{-5}$	$4.382 \cdot 10^{-5}$	$3.291 \cdot 10^{-5}$	$1.417 \cdot 10^{-5}$
		Computation time	0.01 s	4.47 s	12.88 s	20.49 s	51.98 s
Proph.	Flowpipe	$p^{\mathfrak{S}^p}_{max}(S^{goal}, t_{max})$	0	0.604 091	0.648 847	0.651 300	0.755 494
		$\pm$	0	$5.667 \cdot 10^{-4}$	$6.148 \cdot 10^{-4}$	$1.095 \cdot 10^{-3}$	$3.050 \cdot 10^{-3}$
		Computation time	0.19 s	1.67 s	9.19 s	55.04 s	2289.94 s

current segment if it contains goal states. In this case, the segment is reduced to reachable goal valuations, projected and extended (Line 5). The resulting polytope is added to the list of *polytopes* in Line 6. The maximum probability *optimalProb* is computed by integration over the union of all *polytopes* in Line 9.

5 Case Study

We developed a prototype implementation for the maximum case of the presented approach, which uses the libraries HyPro [28] and GNU Scientific Library (GSL) [12]. The current implementation relies on the transformation presented in [25] to compute the set of induced random variables up to time t_{max}.

Figure 4 shows the model of a tank which is constantly filled and can be drained by one of two valves. The fluid in the tank initially equals 4 l (liters) and its capacity is 20 l. The tank is filled with a rate of 4 l h^{-1} (liters per hour). The first valve drains 6 l h^{-1} and the second one drains 4 l h^{-1}. The controller chooses nondeterministically to activate exactly one of the valves, as soon as the fluid level reaches 16 l. If activated, the first valve remains active for 2 h and then is blocked for a random period of time, which is uniformly distributed. If the second valve is activated it stays active for 1 h and is also blocked for a uniformly distributed period of time. A blocked valve cannot be activated.

The above has been modeled as a singular automaton (see Appendix B). All random variables follow a continuous uniform distribution on the interval [0 h, 6 h]. We define the set of goal states $S^{goal} = \{(l, v) \in S | v(x_0) \geq 18\,\text{l}\}$, i.e., a goal state is reached when the fluid level of the tank exceeds 18 l.

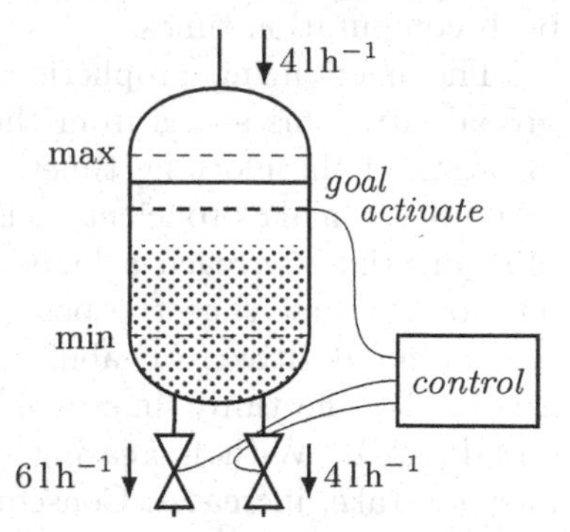

Fig. 4. A tank with constant inflow and two controlled draining valves. After usage each valve is blocked for a random time.

To validate our non-prophetic method, we analyzed the tank system modelled as a hybrid Petri net with general transitions (HPnG), using the non-prophetic approach from [24], implemented in the tool HPNMG [17]. We transformed the HPnG model into a singular automaton model via [25] to ensure it matches the original model. We were not able to validate prophetic results, as the prophetic approach from [24] only works for one nondeterministic decision; the tool ProHVer [14] was only able to compute prophetic overapproximations for the two smallest model instances (c.f. Appendix C). All experiments were performed on a machine with an Intel Core I7 CPU (4×1.8 GHz) and 16 GiB memory.

Resulting probabilities for both schedulers classes, error bounds and computation times are summarized in Table 1. Choosing a time bound t_{max} results in a model with n random variables and *#locs SA* discrete states (i.e., locations) for the singular automaton, as indicated in Table 1. The number of random variables increases with t_{max}, since over time further random clocks are activated modeling the blocking time of a valve. Note that the number of discrete states in both the HPnG and the automaton grows due to our approach, which requires a fresh random clock for every instance of the blocking delay of the valves.

Our evaluation shows that the non-prophetic results computed by both approaches match for $t_{max} \leq 10$ h, when taking into account the statistical errors from Monte Carlo integration. For $t_{max} = 11$ h the observed difference between both approaches is larger than the estimated errors provided. We stress that the error is a statistical estimate and not a strict error bound. Hence, this may occur. However, note that up to integration our computations are exact and the error only results from the last step of computing probabilities. Additional samples for numerical integration can be used to refine the result.

Computation times for the flowpipe approach can only compete with the HPnG-based approach for $t_{max} \leq 9$ h, which was to be expected, since the former is much more general and in particular capable of prophetic scheduling. Both approaches share the computational effort for projecting state sets onto the domain of the random variables and integrating over reachable sets of states. Additionally, the flowpipe approach needs to extend the clock domain to time t_{max}. The latter is expensive due to required representation conversions, especially for higher dimensions, which results in an increasing difference between both computation times.

The maximizing prophetic scheduler always reaches a higher reachability probability. This stems from the knowledge of the prophetic scheduler on the duration of the blocking times of the valves in Fig. 4 and allows the prophetic scheduler to make proficient decisions. Hence, it chooses the valve with the longer blocking time to reach a high level of fluid. At time $t = 8\,\mathrm{h}$ either scheduler has for the first time the possibility of exceeding the threshold of $18\,\mathrm{l}$, which results in a positive reachability probability. At this time a scheduler can choose a valve for the third time, which leads to the possibility that both valves are still blocked. With increasing t_{max}, the number of decisions which a scheduler can take, increases. Consequently, a maximizing scheduler (both prophetic and non-prophetic) has more opportunity to influence the reachability probability, as reflected in the increasing reachability probabilities for both scheduler classes. The computation times for the prophetic approach are in general slightly smaller than for the non-prophetic approach, which is to be expected as the computation of the maximizing reachability probabilities in the prophetic case is simplified to an integration over the union of reachable state sets. The code (http://go.wwu.de/bes9y) to replicate the results, including the models, is submitted as an artifact.

6 Conclusion

We present a novel flowpipe construction-based approach to analyze singular automata with random clocks, excluding continuous nondeterminism. We introduce algorithms to compute reachability probabilities for optimal history-dependent (non-)prophetic schedulers, both with similar computational complexity.

Our approach overcomes the requirement for state-space discretization, building on a combination of geometric operations on multidimensional polytopes and integration of the joint probability distribution. This allows us to optimize the probability of reaching a set of goal state within a given time bound. The computed results are exact up to numerical errors from integration. We have shown the feasibility of the presented method on a small case study and validated the non-prophetic case with the analytical approach from [24].

This Petri net-based approach (which exploits restrictions of HPnGs), does not scale for multiple decisions in the prophetic case and hence cannot be used for validation. Instead, the newly presented approach is equally efficient for both, non-prophetic and prophetic scheduling. We are able to compute results efficiently for 8 random variables, resulting in a state-space with 12 dimensions.

Furthermore, we expect the presented approach to be extensible to more powerful model classes, including e.g., time nondeterminism and rectangular flows, which will be investigated in future work.

We further plan to investigate the influence of (possibly symbolic) state-space representations (and conversions between several representations) on the performance of the flowpipe construction to improve the efficiency of our algorithms.

Appendix A Proof of Correctness

Lemma 1. *Given an integration oracle, Algorithms 2 and 3 compute the maximum non-prophetic reachability probability.*

$$p_{max}^{\mathfrak{S}^n}(S^{goal}, t_{max}) = \max\left(p_{\mathfrak{s}}(S^{goal}, t_{max}) \mid \mathfrak{s} \in \mathfrak{S}^n\right). \tag{4}$$

Proof (Proof of termination). Algorithm 2 loops over all non-prophetic schedulers in $\mathfrak{S}^n$. The number of schedulers $|\mathfrak{S}^n|$ is finite as the number of discrete nondeterministic choices is finite in finite time t_{max}. Hence, Algorithm 2 terminates. Algorithm 2 calls Algorithm 3 (Line 2) which recursively traverses the finite (c.f. Sect. 3) reachability tree, and hence also terminates.

Proof (Proof of optimality).
Algorithm 3 collects the complete state-space recursively for all schedulers. This follows directly from the correctness of bounded reachability analysis for singular automata (c.f. [2]). As we treat random variables as stopwatch variables, the resulting model is singular.

For each possible scheduler $\mathfrak{s} \in \mathfrak{S}^n$(represented by the variable *index*), Algorithm 3 traverses the reachability tree (N, E, Sfunc). The traversal follows the set of nodes $N^{\mathfrak{s}} \subseteq N$ which is induced by that scheduler $\mathfrak{s}$, where $S^{\mathfrak{s}} = \{\textit{pre-process}(\mathit{Sfunc}(n)) \mid n \in N^{\mathfrak{s}}\}$ are in S^{goal} (c.f. Sect. 4.1).

The algorithm recursively collects the sets of valuations $V_{\mathfrak{s}}$ for each random clock in $N^{\mathfrak{s}}$ (c.f. Lines 8, 14). The mapping *polySched* stores an assignment of polytopes (i.e., the set of valuations for all random clocks) for each scheduler (Lines 1–4).

Algorithm 2 iterates over said mapping (Lines 4–7) and computes for each scheduler $\mathfrak{s} \in \mathfrak{S}^n$ the resulting reachability probability $p_{\mathfrak{s}}(S^{goal}, t_{max})$ (Line 6).

Note that a non-prophetic scheduler does not have knowledge on future expiration times of random clocks. Hence, every choice induces a branching in the reachability tree and the probability can only be optimised by separately comparing the probability of the reachable state-space of the respective branch in the reachability tree. As Line 6 compares the probability induced by all possible non-prophetic schedulers over the complete state-space, optimality follows directly.

The correctness of Lemma 1 follows from termination and optimality. The minimum case is analogous.

Lemma 2. *Given an integration oracle, Algorithm 4 computes the minimum prophetic reachability probability:*

$$p_{min}^{\mathfrak{S}^p}(S^{goal}, t_{max}) = \int_{\bigcap_{\mathfrak{s} \in \mathfrak{S}^p} U^{\mathfrak{s}}} G(\mathbf{s})\ ds. \tag{5}$$

Lemma 3. *Given an integration oracle, Algorithm 5 computes the maximum prophetic reachability probability:*

$$p_{max}^{\mathfrak{S}^p}(S^{goal}, t_{max}) = \int_{\bigcup_{\mathfrak{s} \in \mathfrak{S}^p} U^{\mathfrak{s}}} G(\mathbf{s}) \, ds, \tag{6}$$

where

$$\bigcup_{\mathfrak{s} \in \mathfrak{S}^p} U^{\mathfrak{s}} = \bigcup_{\mathfrak{s} \in \mathfrak{S}^p} \bigcup_{(l, V^{\mathfrak{s}}) \in S^{\mathfrak{s}}} V^{\mathfrak{s}} = \bigcup_{(l,V) \in S_C^{goal}} V. \tag{7}$$

Proof (Proof of termination). The set of prophetic schedulers $\mathfrak{S}^p$ is finite. Algorithm 4 calls Algorithm 3, which terminates (c.f. the non-prophetic case) and computes a finite union $U^{\mathfrak{s}}$ of polytopes for each scheduler $\mathfrak{s} \in \mathfrak{S}^p$. Hence, it also terminates. For Algorithm 5 termination follows directly from the finiteness of the traversed reachability tree, as no iteration over schedulers is needed.

In contrast to the non-prophetic case, the prophetic scheduler can base its decision on the future expiration times of the random clock (c.f. Sect. 4.2) and is able to find the optimal decision for every expiration time.

Proof (Proof of minimality).

Algorithm 4 calls Algorithm 3 (Line 2) which computes the complete state-space (again c.f. [2]) and yields the mapping from schedulers to resulting valuations for random clocks.

For each scheduler $\mathfrak{s} \in \mathfrak{S}^p$, it unites all clock valuations that allow reaching the goal states for this scheduler (Lines 3–6), thus computing $U^{\mathfrak{s}}$. Line 7 computes the minimum reachability probability $p_{min}^{\mathfrak{S}^p}(S^{goal}, t_{max})$ that yields from integration over the intersection $\bigcap_{\mathfrak{s} \in \mathfrak{S}^p} U^{\mathfrak{s}}$ over the (united) clock valuations for all schedulers in $\mathfrak{S}^p$.

For the clock valuations in the intersection $\bigcap_{\mathfrak{s} \in \mathfrak{S}^p} U^{\mathfrak{s}}$ all schedulers lead to goal states. Thus integrating over this intersection yields the minimum probability, since (due to completeness) no scheduler exists that avoids goal states for those clock valuations. This proves minimality.

Proof (Proof of maximality).

Algorithm 5 computes the reachable state-space as a flowpipe, without distinguishing between schedulers $\mathfrak{s} \in \mathfrak{S}^p$. In this special case, the computation does not need to distinguish between schedulers, as the union abstracts from individual schedulers, anyhow (c.f. Sect. 4.2). Again, completeness follows from the correctness of bounded reachability analysis for singular automata (c.f. [2]). Hence, all schedulers in $\mathfrak{S}^p$ that lead to reachable goal states in S^{goal} are considered within the computed state-space.

The algorithm then takes the union over all possible clock valuations S_C^{goal} leading to reachable goal states (c.f. Sect. 4.2) and it is impossible that another scheduler exists that leads to goal states and is not considered in the union. Thus, integrating over this union results in the maximal probability.

The correctness of Lemma 2 follows from termination and minimality and the correctness of Lemma 3 follows from termination and maximality.

Appendix B Singular Automaton for Case study

See Fig. 5.

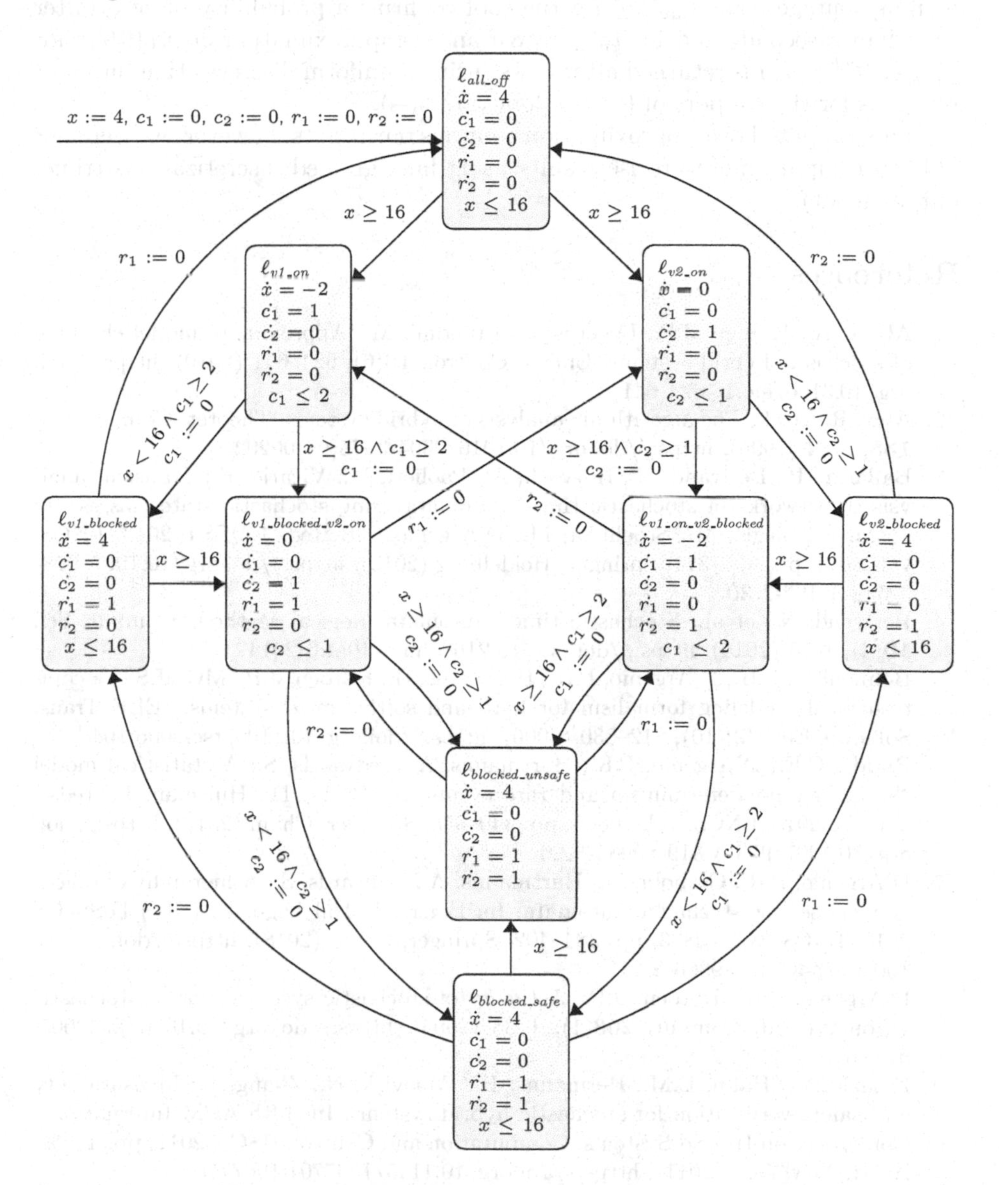

Fig. 5. Singular automaton with random clocks for tank model. x models the fluid level of the tank. c_1 and c_2 model the active time for the first and second valve. r_1 and r_2 are the random clocks, modeling the random blocking time for the valves.

Appendix C Validation of Prophetic Probabilities

Since HPNMG is not able to validate the prophetic case, we tried the tool PROHVER [14] for comparison and were able to compute prophetic overapproximations of reachability probabilities for the two smallest model instances.

For larger time bounds we were not able to obtain results due to time- and memory outages. For $t_{max} \leq 7\,\text{h}$ the tool confirms a probability of zero (after less than a second) and for $t_{max} = 8\,\text{h}$ an overapproximation of 0.719 998 for $p^{\mathfrak{S}^p}_{max}(S^{goal}, t_{max})$ is returned after 648 s (using a uniform discretization into four intervals for the support of the random variables).

The computed overapproximation confirms our results, however, we were not able to compute more precise results using finer-grained discretizations (timeout/memout).

References

1. Abate, A., Katoen, J.P., Lygeros, J., Prandini, M.: Approximate model checking of stochastic hybrid systems. Eur. J. Control. **16**(6), 624–641 (2010). https://doi.org/10.3166/ejc.16.624-641
2. Alur, R., et al.: The algorithmic analysis of hybrid systems. Theoret. Comput. Sci. **138**, 3–34 (1995). https://doi.org/10.1016/0304-3975(94)00202-t
3. Ballarini, P., Bertrand, N., Horváth, A., Paolieri, M., Vicario, E.: Transient analysis of networks of stochastic timed automata using stochastic state classes. In: Joshi, K., Siegle, M., Stoelinga, M., D'Argenio, P.R. (eds.) QEST 2013. LNCS, vol. 8054, pp. 355–371. Springer, Heidelberg (2013). https://doi.org/10.1007/978-3-642-40196-1_30
4. Bertrand, N., et al.: Stochastic timed automata. Logical Methods Comput. Sci. **10**(4), 1–73 (2014). https://doi.org/10.2168/lmcs-10(4:6)2014
5. Bohnenkamp, H., D'Argenio, P.R., Hermanns, H., Katoen, J.P.: MODEST: a compositional modeling formalism for hard and softly timed systems. IEEE Trans. Software Eng. **32**(10), 812–830 (2006). https://doi.org/10.1109/tse.2006.104
6. Budde, C.E., D'Argenio, P.R., Hartmanns, A., Sedwards, S.: A statistical model checker for nondeterminism and rare events. In: Beyer, D., Huisman, M. (eds.) TACAS 2018. LNCS, vol. 10806, pp. 340–358. Springer, Cham (2018). https://doi.org/10.1007/978-3-319-89963-3_20
7. D'Argenio, P.R., Gerhold, M., Hartmanns, A., Sedwards, S.: A hierarchy of scheduler classes for stochastic automata. In: Baier, C., Dal Lago, U. (eds.) FoSSaCS 2018. LNCS, vol. 10803, pp. 384–402. Springer, Cham (2018). https://doi.org/10.1007/978-3-319-89366-2_21
8. D'Argenio, P.R., Katoen, J.P.: A theory of stochastic systems part I: stochastic automata. Inf. Comput. **203**(1), 1–38 (2005). https://doi.org/10.1016/j.ic.2005.07.001
9. Fränzle, M., Hahn, E.M., Hermanns, H., Wolovick, N., Zhang, L.: Measurability and safety verification for stochastic hybrid systems. In: 14th ACM International Conference on Hybrid Systems: Computation and Control, HSCC 2011, pp. 43–52. ACM, New York (2011). https://doi.org/10.1145/1967701.1967710
10. Frehse, G.: PHAVer: algorithmic verification of hybrid systems past HyTech. In: Morari, M., Thiele, L. (eds.) HSCC 2005. LNCS, vol. 3414, pp. 258–273. Springer, Heidelberg (2005). https://doi.org/10.1007/978-3-540-31954-2_17

11. Ghasemieh, H., Remke, A., Haverkort, B.R.: Analysis of a sewage treatment facility using hybrid petri nets. In: 7th EAI International Conference on Performance Evaluation Methodologies and Tools, VALUETOOLS 2013, pp. 165–174. ICST (2013)
12. Gough, B.: GNU Scientific Library Reference Manual. Network Theory Ltd. (2009)
13. Hahn, E.M., Hartmanns, A., Hermanns, H.: Reachability and reward checking for stochastic timed automata. Electron. Commun. EASST **70** (2014). https://doi.org/10.14279/tuj.eceasst.70.968
14. Hahn, E.M., Hartmanns, A., Hermanns, H., Katoen, J.P.: A compositional modelling and analysis framework for stochastic hybrid systems. Formal Methods Syst. Des. **43**(2), 191–232 (2013). https://doi.org/10.1007/s10703-012-0167-z
15. Henzinger, T.A.: The theory of hybrid automata. In: Inan, M.K., Kurshan, R.P. (eds.) Verification of Digital and Hybrid systems, NATO ASI Series, vol. 170, pp. 265–292. Springer, Heidelberg (2000). https://doi.org/10.1007/978-3-642-59615-5_13
16. Henzinger, T.A., Kopke, P.W., Puri, A., Varaiya, P.: What's decidable about hybrid automata? J. Comput. Syst. Sci. **57**(1), 94–124 (1998). https://doi.org/10/cpnnbv
17. Hüls, J., Niehaus, H., Remke, A.: HPNMG: a C++ tool for model checking hybrid petri nets with general transitions. In: 12th International NASA Formal Methods Symposium, NFM 2020. LNCS, vol. 12229, pp. 369–378. Springer, Cham (2020). https://doi.org/10.1007/978-3-030-55754-6_22
18. Hüls, J., Pilch, C., Schinke, P., Delicaris, J., Remke, A.: State-space construction of hybrid petri nets with multiple stochastic firings. In: Parker, D., Wolf, V. (eds.) QEST 2019. LNCS, vol. 11785, pp. 182–199. Springer, Cham (2019). https://doi.org/10.1007/978-3-030-30281-8_11
19. Hüls, J., Pilch, C., Schinke, P., Niehaus, H., Delicaris, J., Remke, A.: State-space Construction of Hybrid Petri Nets with Multiple Stochastic Firings. arXiv.org (2020)
20. Hüls, J., Remke, A.: Coordinated charging strategies for plug-in electric vehicles to ensure a robust charging process. In: 10th EAI International Conference on Performance Evaluation Methodologies and Tools, VALUETOOLS 2016. ICST (2016)
21. Hüls, J., Remke, A.: Energy storage in smart homes: grid-convenience versus self-use and survivability. In: 24th IEEE International Symposium on Modeling, Analysis and Simulation of Computer and Telecommunication Systems, pp. 385–390. IEEE (2016)
22. Koutsoukos, X.D., Riley, D.: Computational methods for verification of stochastic hybrid systems. IEEE Trans. Syst. Man Cybern. Part A Syst. Hum. **38**(2), 385–396 (2008). https://doi.org/10.1109/tsmca.2007.914777
23. Kwiatkowska, M., Norman, G., Segala, R., Sproston, J.: Verifying quantitative properties of continuous probabilistic timed automata. In: Palamidessi, C. (ed.) CONCUR 2000. LNCS, vol. 1877, pp. 123–137. Springer, Heidelberg (2000). https://doi.org/10.1007/3-540-44618-4_11
24. Pilch, C., Hartmanns, A., Remke, A.: Classic and non-prophetic model checking for hybrid petri nets with stochastic firings. In: 23rd ACM International Conference on Hybrid Systems: Computation and Control, HSCC 2020. pp. 1–11. ACM, New York (2020). https://doi.org/10.1145/3365365.3382198

25. Pilch, C., Krause, M., Remke, A., Ábrahám, E.: A transformation of hybrid petri nets with stochastic firings into a subclass of stochastic hybrid automata. In: Lee, R., Jha, S., Mavridou, A., Giannakopoulou, D. (eds.) NFM 2020. LNCS, vol. 12229, pp. 381–400. Springer, Cham (2020). https://doi.org/10.1007/978-3-030-55754-6_23
26. Prandini, M., Hu, J.: A stochastic approximation method for reachability computations. In: Blom, H.A.P., Lygeros,, J. (eds.) Stochastic Hybrid Systems: Theory and Safety Critical Applications, LNCIS, vol. 337, pp. 107–139. Springer, Heidelberg (2006). https://doi.org/10/fbxq4h
27. Schupp, S.: State Set Representations and Their Usage in the Reachability Analysis of Hybrid Systems. Dissertation, RWTH Aachen University, Aachen (2019). http://publications.rwth-aachen.de/record/767529
28. Schupp, S., Ábrahám, E., Makhlouf, I.B., Kowalewski, S.: HyPro: A C++ library of state set representations for hybrid systems reachability analysis. In: Barrett, C., Davies, M., Kahsai, T. (eds.) NFM 2017. LNCS, vol. 10227, pp. 288–294. Springer, Cham (2017). https://doi.org/10.1007/978-3-319-57288-8_20
29. Soudjani, S.E.Z., Gevaerts, C., Abate, A.: FAUST2: formal abstractions of uncountable-state stochastic processes. In: 21st International Conference on Tools and Algorithms for the Construction and Analysis of Systems, TACAS 2015. LNCS, vol. 9035, pp. 272–286. Springer, Heidelberg (2015). https://doi.org/10.1007/978-3-662-46681-0_23
30. Sproston, J.: Decidable model checking of probabilistic hybrid automata. In: Joseph, M. (ed.) FTRTFT 2000. LNCS, vol. 1926, pp. 31–45. Springer, Heidelberg (2000). https://doi.org/10.1007/3-540-45352-0_5
31. Zhang, L., She, Z., Ratschan, S., Hermanns, H., Hahn, E.M.: Safety verification for probabilistic hybrid systems. Eur. J. Control. **18**(6), 572–587 (2012). https://doi.org/10.3166/ejc.18.572-587
32. Ziegler, G.: Lectures on Polytopes. Graduate Texts in Mathematics, vol. 152. Springer, New York (1995). https://doi.org/10.1007/978-1-4613-8431-1

Attack Trees vs. Fault Trees: Two Sides of the Same Coin from Different Currencies

Carlos E. Budde[1(✉)], Christina Kolb[1], and Mariëlle Stoelinga[1,2]

[1] Formal Methods and Tools, University of Twente, Enschede, The Netherlands
{c.e.budde,c.kolb,m.i.a.stoelinga}@utwente.nl
[2] Department of Software Science, Radboud University, Nijmegen, The Netherlands

Abstract. This work compares formal approaches to define and operate with attack trees and fault trees. We start by investigating similarities between the syntactic structure, semantics, and qualitative analysis, of static attack trees and fault trees. Then we point out differences of the analysis methods and metrics between the two formalisms, providing a deeper insight for their dynamic variants. Finally, we overview several extensions and categorise them using the new concept of dimension, which allows us to compare these extensions and point out research gaps.

1 Introduction

Attack trees (ATs) and fault trees (FTs) are popular formalisms that support the identification, documentation, and analysis of security (resp. safety) risks. They are part of system-engineering frameworks such as SysML-Sec [28], and count with commercial tools such as Isograph's AttackTree and FaultTree+ [14,15].

The popularity of these formalisms in industry is due to their capacity to represent complex processes succinctly and to the desired level of detail. In AT (resp. FT) analysis, a hierarchical diagram is designed to systematically map security (resp. safety) hazards. The resulting model gives insight into the vulnerabilities of the system, which can then be countered cost-efficiently [18,34].

The Origins. This analogous procedural approach is no coincidence: ATs were inspired on FTs. The latter were introduced in 1961 at Bell Labs to study ballistic missiles [30,33]. In 1990 FT analysis was "about 39 years old, and has become a well-recognized tool worldwide" [9]. In contrast, Weiss introduced threat logic trees—the origin of ATs—in 1991, and its "similarity... to fault trees suggests that graph-based security modelling has its roots in safety modelling" [21].

Ever since, these formalisms increased their modelling and analysis power to best satisfy the needs of the safety or security application domain. This has separated the syntax and semantics of new FT- and AT-based formalisms, in spite of

This work was partially funded by NWO project 15474 (*SEQUOIA*) and ERC Consolidator Grant 864075 (*CAESAR*).

A. Abate and A. Marin (Eds.): QEST 2021, LNCS 12846, pp. 457–467, 2021.
https://doi.org/10.1007/978-3-030-85172-9_24

their sharing the modelling principle of top-down hierarchical decomposition. In this work we study this disjoint evolution from the perspective of formal methods. We first show in Sect. 2 that the syntax of their so-called static versions, as wells as their corresponding semantics and qualitative analysis, are mathematically equivalent. The only distinction between static FTs and ATs as a formalism is their domain ontology: safety vs. security. This is the root of their subsequent differentiation, which we study in Sect. 3. To compare their extensions in a systematic manner we introduce the notion of dimension, which allows us to contrast parallel (even symmetrical) evolutions. The work concludes in Sect. 4.

Formalism. A *formalism* is defined by **1.** an (unambiguous) *syntax* to represent its elements, called *models*; **2.** a *semantics* that maps each model to a (unique) mathematical object; **3.** a *domain ontology* in which the models are interpreted. The three parts of this preliminary definition are formalised in the sequel.

Related Work. Surveys on FTs are [18] and [30]. The latter covers modelling and analysis tools. The former pinpoints limitations of FTs to assess the reliability of static systems (only), and mentions extensions that overcome them, e.g. dynamic FTs [8], state-event FTs [19], and Stochastic Hybrid FT Automata [6]. Standard FT analysis and its extensions are also extensively discussed in [30], including technical details of different FT models and analyses.

Surveys on attack trees, [21] and [34], present the state of the art in graphical-hierarchical attack/defense modelling. The latter is a modern and comprehensive summary on the use of formal methods to enhance security evaluation. It references software tools, and discusses steps for industrial technology transfer.

2 Similarities Between Fault and Attack Trees

ATs and FTs follow the same modelling principle: an expert panel identifies a main event of interest—one *top element*—and refine it logically to the level of well-understood basic components or actions—the set of *basic elements*—[21,30]. This analogous model-building process results in identical syntactic structures.

2.1 Syntactic Structure: Static FTs and ATs

The vanilla version of FTs and ATs, so-called *static* fault or attack trees, have the same syntax. We unify them under the concept of *logical-decomposition tree*.

Definition 1 (LDT). *A* logical-decomposition tree *(*LDT *for short) is a tuple* $T = \langle N, t, ch \rangle$ *where:* (i) N *is a finite set of* nodes*;* (ii) $t \colon N \to \{AND, OR, LEAF\}$ *gives the* type *of each node;* (iii) $ch \colon N \to 2^N$ *gives the (possibly empty) set of* children *of a node. Moreover,* T *satisfies the following constraints:* (a) $\langle N, E \rangle$ *is a connected directed acyclic graph (*DAG*), where* $E = \{(v, u) \in N^2 \mid u \in ch(v)\}$*;* (b) T *has a unique root, denoted* R_T*:* $\exists! R_T \in N. \forall v \in N.\ R_T \notin ch(v)$*;* (c) $LEAF$ *nodes* $N_L \subseteq N$ *are the leaves of* T*:* $\forall v \in N.\ t(v) = LEAF \Leftrightarrow v \in N_L \Leftrightarrow ch(v) = \varnothing$*.*

So, syntactically, static FTs and ATs are LDTs (VOT gates in static FTs are syntactic sugar of AND and OR gates). LDTs can be proper trees or DAGs: the difference is that in proper trees, each child node has exactly one parent. Node $v \in N$ is *the parent* of $u \in N$ iff u is a child of v, i.e. T has the edge $v \to u$. By definition, *LEAF* nodes are not parents: parent nodes are called *gates*.

Logical Gates. An LDT represents the logical decomposition of events via disjunction and conjunction, which can be interpreted as a Boolean function. Consider e.g. a wooden gate that can break due to rotten wooden planks OR rusty hinges (or both); and the hinges rust if they are made of iron AND the environment is humid AND sufficient time elapses. This decomposition is safety-oriented. From an analogous security perspective, the wooden gate is breached by dislodging the hinges OR cracking the wooden planks, and hinges can be dislodged if the alloy is fragile AND the attacker has a crowbar AND enough strength. Both cases yield the LDT $\langle\{a, b, c, d, g_1, g_2\}, t, ch\rangle$ with leaves $N_L = \{a, b, c, d\}$ and gates $t(g_1) = AND$, $t(g_2) = OR$, s.t. $ch(g_1) = \{a, g_2\}$, $ch(g_2) = \{b, c, d\}$. This represents the Boolean function $\lambda\, abcd\, .\, a \vee (b \wedge c \wedge d)$ whose atoms take safety or security meaning: this is denoted OR(a, AND(b, c, d)) and visualised as Fig. 1a.

Visualisation. FTs and ATs are graphical formalisms, drawn as in Fig. 1 with the root on top, and every child connected to an (upper) parent by a line. Leaves are circles, and logical gates resemble their electronic-circuit counterparts.

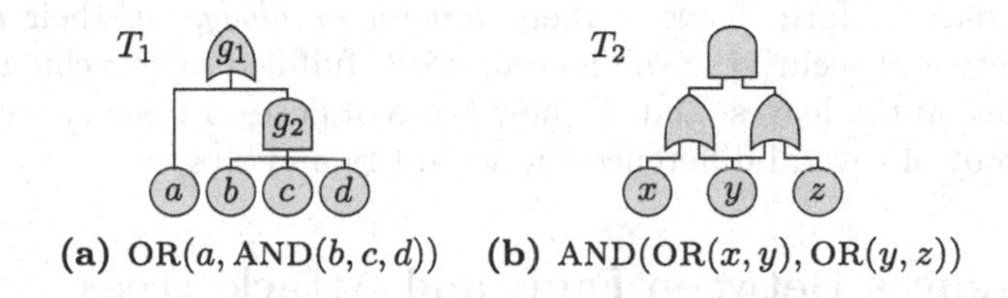

Fig. 1. T_1 has tree structure; the OR gate g_1 is the root; g_2 is an AND gate. T_2 has DAG structure: y has two parents.

2.2 Semantics and Analysis

Once an LDT model T has been created, it is studied to find safety/security vulnerabilities of the system. For this, T is bestowed with formal semantics.

Structure Function. These semantics can be given via a function $f_T : 2^{N_L} \to \mathbb{B}$ that indicates whether a set of basic elements trigger the top element of T. That is, $f_T(A) = \top$ iff the Boolean function represented by T is satisfied by the mapping $(A \mapsto \top) \cup ((N_L \setminus A) \mapsto \bot)$, where $\setminus$ denotes set difference. For instance for T_1 in Fig. 1a, to evaluate $f_{T_1}(\{a, c\})$ one maps a and c to $\top$, b and d to $\bot$, and evaluates the Boolean function represented by T_1—$\lambda\, abcd\, .\, a \vee (b \wedge c \wedge d)$—which returns $\top$. This so-called *structure function* f_T is given by the syntactic structure of T, and hence it is analogous for static FTs and ATs [17,30].

Evidence. The set $A \subseteq N_L$ on which f_T is evaluated is called *evidence*: for ATs it represents the steps carried out by an attacker; for FTs it represents elements that have failed. If $f_T(A) = \top$ then A is called *valid evidence*; else it is *invalid.* Valid evidence A is called *minimal* if no proper subset of A is valid. For instance in Fig. 1, $\{a\}$ and $\{x, z\}$ are minimal evidence of T_1 and T_2 resp., $\{a, b\}$ is also valid (but not minimal) evidence, and $\{x\}$ is invalid. In FT analysis, minimal evidence is also called "minimal cut set" or "prime implicant."

Formal Semantics. Static FTs and ATs are *coherent*: adding basic elements to evidence preserves its validity [3]. That is, if $f_T(A) = \top$, then $f_T(A \cup \{a\}) = \top$ for every $a \in N_L$. Thus, all valid evidence of T—i.e. that can trigger its top element—is characterised by the collection of minimal evidence. This gives rise to the formal semantics of T: $[\![T]\!] = \{A \subseteq N_L \mid f_T(A) = \top \wedge A \text{ is minimal}\}$.[1]

Qualitative Analysis. Since $[\![T]\!]$ subsumes all ways to trigger the top element of T, its computation provides key information on the vulnerability of the system. For FTs, any $A \in [\![T]\!]$ with few elements pinpoints a safety hazard—where a few basic failures can trigger a system-level failure—and likewise for ATs. The amount of subsets in $[\![T]\!]$ can be exponential on the number of nodes in T [24]. Since this hinders computations, and given the interest in small sets from $[\![T]\!]$, FT analysis sometimes bounds the size of the minimal evidence to compute [32].

Thus, static FTs and ATs are mathematically equivalent: their syntax is given by an LDT, T, and their semantics by the set of minimal evidence, $[\![T]\!]$. What sets them apart as formalisms is their *domain ontology*, i.e. their application domain: safety and security have different goals, fulfilled by enriching LDTs with (a) attributes on the leaves, and (b) new types of gates. In Sect. 3 we show how this is the root of several differences between FTs and ATs.

3 Differences Between Fault and Attack Trees

The aforementioned similarities apply only to *static* FTs and ATs. Later extensions to these formalisms, e.g. to include notions of complement and dynamic behaviour, have caused them to grow in different directions. We discuss this in Sect. 3.2 but first we show, in Sect. 3.1, that the different goals of the safety/security domains break the analogies even for static FTs and ATs.

3.1 Analyses that Differ for Static FTs and ATs

Quantitative Analysis. Beyond the constitution of each set $A \in [\![T]\!]$, it is useful to quantify their relevance. For example, if every basic element a_i has a probability $p_i \in [0, 1]$ of occurrence, one can compute the total probability of some evidence A [26]. Similarly, values $\lambda_i \in \mathbb{R}_{>0}$ can describe *the rate* at which these basic probabilities increase with time. Then one can measure the system unreliability, i.e. the probability of triggering the top element at various mission

[1] There are other equivalent ways to define static AT/FT semantics, e.g. bundles [25].

times. Rates also enable time-only measurements, such as the mean time it takes for some evidence $A \in [\![T]\!]$ to be observed [30].

All these quantitative queries, that deal with the probability and frequency of occurrence of events, are characteristic of FT analysis [9,24,30,32]. The reason is that it is feasible and useful to estimate e.g. the mean time to failure (MTTF) of machine components: this allows engineers to compute safe, cost-optimal policies for inspection, maintenance, and replacement of company assets [29].

In contrast, the probability of basic attacks in ATs are very hard to know [11]. Unknown vulnerabilities may increase it, also its frequency, and the conditional probability tables are usually not-knowable. Therefore, it is typical to query the max (rather than total) attack probability [34]. This is also cost-driven: rather than try everything, an attacker may only choose the most promising attack.

The time for an attack is also described differently than for a failure: whereas failures typically have MTTF or rate values, basic attacks steps can be given [min, max] intervals, and further differentiate activation from execution time [23].

Finally, quantitative analyses in ATs can be richer than in FTs, exploring attribute domains beyond time and probability. These include the cost to carry out certain attacks, the skill or psychological profile required to do it, the max damaged caused, and Pareto analyses thereof [2,11,23].

Propagation of Values Through Logical Gates. When the AT or FT is a proper tree, quantitative queries can be computed bottom-up directly on its syntactic structure. For this, the values of the basic elements are propagated upwards in the tree [25]. For instance if basic elements a and b cost resp. €3 and €7, then the cost of OR(a, b) is the min, €3, and the cost of AND(a, b) is the sum, €10.

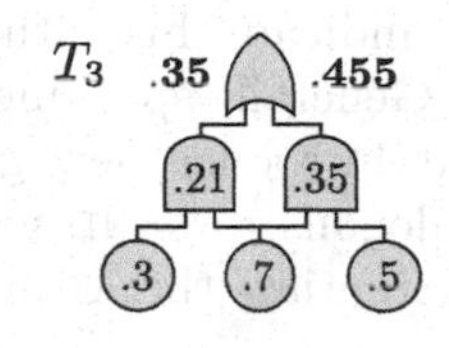

Fig. 2. Probability computation: AT (left, red) vs. FT (right, blue) (Color figure online)

However, here too we find a remarkable difference between static FTs and ATs, that is overlooked by many reviews in spite of its apparent impact in quantitative analyses. In FTs, the "probability of failure" asks for *total probability*, so the (probability) value of an OR gate is the sum of the values of its children, minus the value of their intersection. Instead and as indicated above, attacks are characterised by their *max probability*, so the value of an OR gate in an AT is the max value among its children. This is illustrated in Fig. 2: the values in the basic elements are given; the probability of an AND gate is the product of its children; but if T_3 is an AT, its (max) attack probability is .35; and if it is an FT, its (total) failure probability is .455.

This different propagation of values also affects conjunctive gates (AND), most notably with time attributes. The basic elements in static FTs refer to failures in components, which are under simultaneous use and therefore whose degradation is concurrent. Thus the MTTF of an AND gate is the max across the MTTF of its children. ATs, in contrast, have more ways to describe a conjunction of activities.

Table 1. Overview of extensions to the FT and AT formalisms

Form.	Extensions			Main features
FTs	**DFT**	[8]	Dynamic FTs	FTs + PAND + SPARE + FDEP
	RFT	[4,7]	Repairable FTs	FTs + repair boxes
	E-DFT	[10]	Extended DFTs	DFTs + gen. SPAREs + triggers
	SE-FT	[27]	State/Event FTs	FTs + Petri nets in leaves
ATs	**SAND-AT**	[16]	SAND attack trees	ATs with sequential AND
	ADTree	[20]	Attack–defense trees	ATs + defenses
FTs & ATs	**BDMP**	[5]	Boolean Markov proc.	FTs + ATs + triggers + repairs
	CFT	[31]	Component FTs	ATs + FTs with modular structure
	AFT	[22]	Attack-Fault Trees	SAND-ATs + DFTs
	FT-AT	[12]	FTs integrated to ATs	FTs whose BEs are refined as ATs

In particular they could require time-exclusion: consider one attacker who must perform multiple actions, e.g. deactivating an alarm and silencing the dog. Here, the time to execute all attacks is not the max, but the sum of the times [1]. ATs can indicate this with a new gate: *sequential-AND* (SAND).

Generally speaking, static ATs and FTs have begotten independent extensions that introduce new gates. For instance, SEQ enforcers in FTs can be seen as analogous to SAND gates in ATs. In general, however, these extensions further differentiate the AT and FT formalisms, as we discuss in Sect. 3.2.

3.2 Extensions of the Formalisms

So far we considered (only) static FTs and ATs, pointing out their similarities and differences. In this section we revise several extensions that grow beyond them. We first overview some prominent formalisms in Table 1, and then refine the comparison by defining and making use of the concept of dimension.

Safety Extensions of FTs. The first formalisms in Table 1 are important extensions of FTs: *DFTs* are static FTs plus PAND gates (that fail if all children fail in left-to-right order), SPARE gates (for spare parts), and FDEPs (that model common-cause failures); *RFTs* are static FTs with repair boxes, that can repair failed BEs; and *E-DFTs* are generalised DFTs with triggers, which allow arbitrary subtrees as spares, and whose FDEPs can trigger gates as well as BEs.

FTs + Security. While DFTs, RFTs, and E-DFTs, improve safety modelling in fault trees, other extensions can cover security aspects. For instance, *SE-FTs* have Petri nets as basic elements. These are more versatile than the usual two-state BEs, allowing state changes that can model safety *and security* hazards.

Security Extension of ATs. There also exist extensions to improve security modelling of attack trees: *SAND-ATs* add dynamic behaviour to ATs, by forcing attacks to occur in a specific order via SAND gates [2]; and *ADTrees* model protections against attacks via special defense nodes.

Combinations of FTs and ATs. All those formalisms extend either FTs or ATs, but there also exist formalisms that combine them. *BDMPs* can have ATs as subtrees of FTs and vice versa, and allow propagations of failures/attacks (and repairs) via triggers among gates. *CFTs* add modular FTs to ATs, to foster large-system analysis via decoupled studies of smaller components. In contrast, *AFTs* trade scalability for versatility, by merging DFTs (with all its dynamic gates) with ATs plus SAND gates. Finally, *FT-ATs* refine the BEs in fault trees via attack trees, modelling attackers that try to force a system failure.

Note that, interestingly and to the best of our knowledge, no formalism that combines FTs with ATs includes defenses. More importantly, we find independent extensions that overlap in some modelling goals, e.g. RFTs and the repairs of BDMPs. We compare these (partial) overlaps via dimensions.

Dimensions. An extension augments the modelling power of FTs/ATs. Some extensions reach to each other, e.g. AFTs and BDMPs are in both domain ontologies (safety and security). But other extensions are parallel: compare DFTs to SAND-ATs, both of which make the order of events relevant but without crossing the safety/security line. We thus identify different ways to classify the space of formalisms, where a *dimension* d defines (not necessarily exclusive) classes that are orthogonal to those defined by another dimension d'. For example, the domain ontology can be seen as a *domain* dimension: it defines the classes `safety` and `security` s.t. the formalisms {FT, DFT} are in `safety`, {AT,SAND-AT} are in `security`, and {AFT, BDMP} are in both. Further dimensions to classify formalisms include *dynamics*—the order of events matters or not—and *complement*—there is a single type of event (e.g. attacks), or complementary types (also defenses). Figure 3 shows a scheme of this 3D classification.

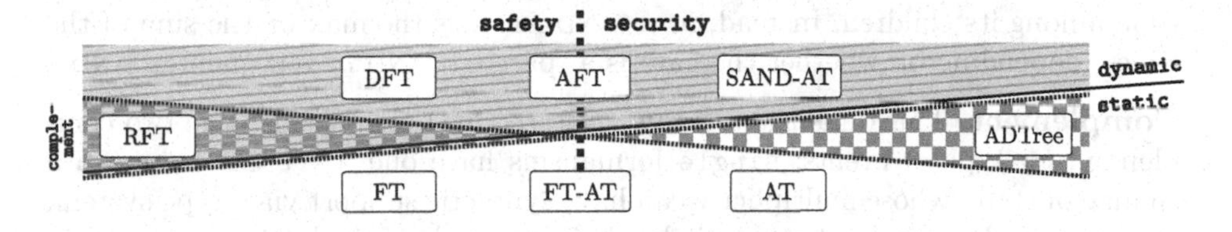

Fig. 3. Dimensional split of formalisms: *domain*, *dynamics*, and *complement*.

Such concept of dimension resembles that of an *ontology* in information science [13]. For us, different dimensions yield orthogonal classifications of the same set of individuals. These individuals are the formalisms within scope: we use Table 1 as the scope, but Definition 2 generalises to any FT/AT extension.

[2] SANDs in ATs *force a sequence of events*, similarly to SEQ enforcers in certain flavours of FTs; this differs from PAND gates in DFTs, which *observe the order of events*.

Table 2. Dimensional split of formalisms in Table 1

Dimension		FT	DFT	RFT	E-DFT	SE-FT	BDMP	CFT	AFT	FT-AT	SAND-AT	ADTree	AT
dom.	safety	✓	✓	✓	✓	✓	✓	✓	✓	✓			
	security					✓	✓	✓	✓	✓	✓	✓	✓
dyn.	static	✓						✓		✓		✓	✓
	dynamic		✓	✓	✓	✓	✓		✓		✓		
cmp.	single	✓	✓		✓	✓		✓	✓	✓	✓		✓
	dual			✓			✓					✓	

Definition 2 (Dimension). *A* dimension *is an ontology with two or more non-empty classes, whose individuals are the formalisms from Table 1. A* dimensional base $\mathbb{D} = \{d_i\}_{i=1}^n$ *is a finite set of orthogonal* dimensions.

So far we have compared formalisms exclusively from the perspective of the *domain* dimension: we now turn our attention to *dynamics* and *complement*.

Note, however, that these three dimensions—that Table 2 defines in our full scope—are not exhaustive. We identify at least an extra *complexity* dimension, sensitive to the number of states of the basic elements. In terms of *complexity*, FTs and ATs are `simple` (binary states), while SE-FTs and Fault Maintenance Trees [29] are `complex` (its basic elements are resp. Petri nets and Erlang chains).

Dynamics. This dimension classifies formalisms based on whether their semantics caters for order. The broadest possible classes are `static` and `dynamic`. The success of the top element in a `static` formalism does not depend on the order in which the basic elements occur. This includes FTs, ATs, CFTs, AT-FTs, and ADTrees. Other formalisms in Table 1 are `dynamic`: they either enforce an order, e.g. SAND gates and SEQ enforcers; or the propagation of success in some gates depends on it, e.g. PAND gates in DFTs. Besides a richer semantics (that affects qualitative analyses), `dynamic` formalisms have more complex quantitative analyses. In a static AT, the attack time of a conjunctive gate is the max time among its children. Instead, in a SAND-AT, it is the max *or* the sum of the times, depending on whether the gate is a "parallel" AND *or* a sequential-AND.

Complement. This dimension has two classes: `dual` formalisms have two complementary type of events; `single` formalisms have one. By *event* we mean a change of state, whose multiplicity can have syntactic support via a type system, or it can reside entirely at semantic level. An example of the latter are repairs in RFTs: their (single-typed) basic elements can transit in both directions between their active and failed states. An example of dual events via types are attack- vs. defense-nodes in ADTrees: given an attack, if the counter-defense occurs, then the state of the corresponding gate changes first to "attacked" and then to "not attacked." This differs from the absence of an attack for quantitative queries, e.g. to compute attack cost. In Table 1, the only formalisms in the `dual` class of this *complement* dimension are ADTree, RFT, and BDMP. All the rest are `single`: only one change of state can happen, namely a failure (resp. an attack) that involves a transition from an active to a failed (resp. attacked) state.

Finally, we note that comparing the classification of different dimensions helps to spot research gaps. For instance, from the five formalisms in both classes of the *domain* dimension, only BDMPs are `dual` as per *complement*. Since that comes from repairs of failed basic elements, we know that no formalism in Table 1 that combines `safety` and `security` includes defenses, as pointed out earlier.

4 Conclusions and Future Work

We have compared FTs against ATs, showing how they model system vulnerabilities in the same mathematical *static way*. However, their different domain ontologies—safety for FTs, security for ATs—gives rise to different quantitative analyses. This shows in the algebra used to propagate values through gates, e.g. to compute the probability of a failure vs. that of an attack. Moreover, new gates have been added to FTs and ATs, extending these formalisms in directions that sometimes cross each other. We introduced the concept of dimension to classify these extensions, thus generalising the safety/security dichotomy.

These studies can be deepened by finding new dimensions to compare formalisms. Our dimensional split offers a high-level view that helps to spot research gaps. In particular, we found no formalism that merges ATs and FTs, that also includes defenses against attacks. Neither have we found formalisms with clearly-differentiated AT/FT submodules, such as FT-ATs, that also offer dynamic gates and repairs, such as BDMPs. The industrial relevance of model visualisation, plus the need for versatile modelling, makes this gap a promising line of research.

References

1. Arnold, F., Hermanns, H., Pulungan, R., Stoelinga, M.: Time-dependent analysis of attacks. In: Abadi, M., Kremer, S. (eds.) POST 2014. LNCS, vol. 8414, pp. 285–305. Springer, Heidelberg (2014). https://doi.org/10.1007/978-3-642-54792-8_16
2. Aslanyan, Z., Nielson, F.: Pareto efficient solutions of attack-defence trees. In: Focardi, R., Myers, A. (eds.) POST 2015. LNCS, vol. 9036, pp. 95–114. Springer, Heidelberg (2015). https://doi.org/10.1007/978-3-662-46666-7_6
3. Barlow, R.E., Proschan, F.: Statistical Theory of Reliability and Life Testing: Probability Models. International Series in Decision Processes. Holt, Rinehart and Winston, New York (1975)
4. Bobbio, A., Codetta-Raiteri, D.: Parametric fault trees with dynamic gates and repair boxes. In: RAMS, pp. 459–465. IEEE (2004). https://doi.org/10.1109/RAMS.2004.1285491
5. Bouissou, M.: BDMP (Boolean logic Driven Markov Processes) as an alternative to Event Trees. In: ESREL 2008 (2008)
6. Chiacchio, F., D'Urso, D., Compagno, L., Pennisi, M., Pappalardo, F., Manno, G.: SHyFTA, a stochastic hybrid fault tree automaton for the modelling and simulation of dynamic reliability problems. Expert Syst. Appl. **47**, 42–57 (2016). https://doi.org/10.1016/j.eswa.2015.10.046

7. Codetta-Raiteri, D., Iacono, M., Franceschinis, G., Vittorini, V.: Repairable fault tree for the automatic evaluation of repair policies. In: DSN, pp. 659–668. IEEE Computer Society (2004). https://doi.org/10.1109/DSN.2004.1311936
8. Dugan, J., Bavuso, S., Boyd, M.: Fault trees and sequence dependencies. In: ARMS, pp. 286–293. IEEE (1990). https://doi.org/10.1109/ARMS.1990.67971
9. Ericson, C.A.: Fault tree analysis - A history. In: 17th International System Safety Conference, pp. 1–9 (1999)
10. Arnold, F., Belinfante, A., Van der Berg, F., Guck, D., Stoelinga, M.: DFTCalc: a tool for efficient fault tree analysis. In: Bitsch, F., Guiochet, J., Kaâniche, M. (eds.) SAFECOMP 2013. LNCS, vol. 8153, pp. 293–301. Springer, Heidelberg (2013). https://doi.org/10.1007/978-3-642-40793-2_27
11. Fila, B., Wideł, W.: Attack–defense trees for abusing optical power meters: a case study and the OSEAD tool experience report. In: Albanese, M., Horne, R., Probst, C.W. (eds.) GraMSec 2019. LNCS, vol. 11720, pp. 95–125. Springer, Cham (2019). https://doi.org/10.1007/978-3-030-36537-0_6
12. Fovino, I.N., Masera, M., De Cian, A.: Integrating cyber attacks within fault trees. Reliab. Eng. Syst. Saf. **94**(9), 1394–1402 (2009). https://doi.org/10.1016/j.ress.2009.02.020
13. Guarino, N.: Formal ontology, conceptual analysis and knowledge representation. Int. J. Hum.-Comput. Stud. **43**(5), 625–640 (1995). https://doi.org/10.1006/ijhc.1995.1066
14. Isograph: AttackTree. https://www.isograph.com/software/attacktree/
15. Isograph: FaultTree+. https://www.isograph.com/software/reliability-workbench/fault-tree-analysis-software/fault-tree-analysis/
16. Jhawar, R., Kordy, B., Mauw, S., Radomirović, S., Trujillo-Rasua, R.: Attack trees with sequential conjunction. In: Federrath, H., Gollmann, D. (eds.) SEC 2015. IAICT, vol. 455, pp. 339–353. Springer, Cham (2015). https://doi.org/10.1007/978-3-319-18467-8_23
17. Jürgenson, A., Willemson, J.: Computing exact outcomes of multi-parameter attack trees. In: Meersman, R., Tari, Z. (eds.) OTM 2008. LNCS, vol. 5332, pp. 1036–1051. Springer, Heidelberg (2008). https://doi.org/10.1007/978-3-540-88873-4_8
18. Kabir, S.: An overview of fault tree analysis and its application in model based dependability analysis. Expert Syst. Appl. **77**, 114–135 (2017). https://doi.org/10.1016/j.eswa.2017.01.058
19. Kaiser, B., Gramlich, C., Förster, M.: State/event fault trees–a safety analysis model for software-controlled systems. Reliab. Eng. Syst. Saf. **92**(11), 1521–1537 (2007). https://doi.org/10.1016/j.ress.2006.10.010
20. Kordy, B., Mauw, S., Radomirović, S., Schweitzer, P.: Foundations of attack–defense trees. In: Degano, P., Etalle, S., Guttman, J. (eds.) FAST 2010. LNCS, vol. 6561, pp. 80–95. Springer, Heidelberg (2011). https://doi.org/10.1007/978-3-642-19751-2_6
21. Kordy, B., Piètre-Cambacédès, L., Schweitzer, P.: DAG-based attack and defense modeling: don't miss the forest for the attack trees. Comput. Sci. Rev. **13–14**, 1–38 (2014). https://doi.org/10.1016/j.cosrev.2014.07.001
22. Kumar, R., Stoelinga, M.: Quantitative security and safety analysis with attack-fault trees. In: 18th International Symposium on HASE, pp. 25–32 (2017)
23. Kumar, R., Ruijters, E., Stoelinga, M.: Quantitative attack tree analysis via priced timed automata. In: Sankaranarayanan, S., Vicario, E. (eds.) FORMATS 2015. LNCS, vol. 9268, pp. 156–171. Springer, Cham (2015). https://doi.org/10.1007/978-3-319-22975-1_11

24. Lee, W., Grosh, D., Tillman, F., Lie, C.: Fault tree analysis, methods, and applications – a review. IEEE Trans. Reliab. **R-34**(3), 194–203 (1985). https://doi.org/10.1109/TR.1985.5222114
25. Mauw, S., Oostdijk, M.: Foundations of attack trees. In: Won, D.H., Kim, S. (eds.) ICISC 2005. LNCS, vol. 3935, pp. 186–198. Springer, Heidelberg (2006). https://doi.org/10.1007/11734727_17
26. Rauzy, A.: New algorithms for fault trees analysis. Reliab. Eng. Syst. Saf. **40**(3), 203–211 (1993). https://doi.org/10.1016/0951-8320(93)90060-C
27. Roth, M., Liggesmeyer, P.: Modeling and analysis of safety-critical cyber physical systems using state/event fault trees. In: SAFECOMP (2013)
28. Roudier, Y., Apvrille, L.: SysML-Sec: a model driven approach for designing safe and secure systems. In: MODELSWARD, pp. 655–664. IEEE (2015)
29. Ruijters, E., Guck, D., Drolenga, P., Peters, M., Stoelinga, M.: Maintenance analysis and optimization via statistical model checking. In: Agha, G., Van Houdt, B. (eds.) QEST 2016. LNCS, vol. 9826, pp. 331–347. Springer, Cham (2016). https://doi.org/10.1007/978-3-319-43425-4_22
30. Ruijters, E., Stoelinga, M.: Fault tree analysis: a survey of the state-of-the-art in modeling, analysis and tools. Comput. Sci. Rev. **15–16**, 29–62 (2015). https://doi.org/10.1016/j.cosrev.2015.03.001
31. Steiner, M., Liggesmeyer, P.: Combination of safety and security analysis - finding security problems that threaten the safety of a system (2016)
32. Vesely, W., Stamatelatos, M., Dugan, J., Fragola, J., Minarick, J., Railsback, J.: Fault tree handbook with aerospace applications. NASA Office of Safety and Mission Assurance, version 1.1 (2002)
33. Watson, H.: Launch control safety study. Techical report Section VII, Vol. 1, Bell Labs (1961)
34. Wideł, W., Audinot, M., Fila, B., Pinchinat, S.: Beyond 2014: formal methods for attack tree-based security modeling. ACM Comput. Surv. **52**(4) (2019). https://doi.org/10.1145/3331524

Correction to: DSMC Evaluation Stages: Fostering Robust and Safe Behavior in Deep Reinforcement Learning

Timo P. Gros, Daniel Höller, Jörg Hoffmann, Michaela Klauck, Hendrik Meerkamp, and Verena Wolf

Correction to:
Chapter "DSMC Evaluation Stages: Fostering Robust and Safe Behavior in Deep Reinforcement Learning" in: A. Abate and A. Marin (Eds.): *Quantitative Evaluation of Systems*, LNCS 12846, https://doi.org/10.1007/978-3-030-85172-9_11

In an older version of this paper, there was a mistake in line 12 of the algorithm on page 206. This was corrected.

The updated version of this chapter can be found at
https://doi.org/10.1007/978-3-030-85172-9_11

A. Abate and A. Marin (Eds.): QEST 2021, LNCS 12846, p. C1, 2021.
https://doi.org/10.1007/978-3-030-85172-9_25

Author Index

Printed in the United States
by Baker & Taylor Publisher Services